Proof Theory and Automated Deduction

APPLIED LOGIC SERIES

VOLUME 6

Managing Editor

Dov M. Gabbay, *Department of Computing, Imperial College, London, U.K.*

Co-Editor

Jon Barwise, *Department of Philosophy, Indiana University, Bloomington, IN, U.S.A.*

Editorial Assistant

Jane Spurr, *Department of Computing, Imperial College, London, U.K.*

SCOPE OF THE SERIES
Logic is applied in an increasingly wide variety of disciplines, from the traditional subjects of philosophy and mathematics to the more recent disciplines of cognitive science, computer science, artificial intelligence, and linguistics, leading to new vigor in this ancient subject. Kluwer, through its Applied Logic Series, seeks to provide a home for outstanding books and research monographs in applied logic, and in doing so demonstrates the underlying unity and applicability of logic.

The titles published in this series are listed at the end of this volume.

Proof Theory and Automated Deduction

by

JEAN GOUBAULT-LARRECQ

G.I.E. Dyade / Inria,
Rocquencourt, France

and

IAN MACKIE

École Polytechnique,
Palaiseau, France

KLUWER ACADEMIC PUBLISHERS
DORDRECHT / BOSTON / LONDON

A C.I.P. Catalogue record for this book is available from the Library of Congress

ISBN 1-4020-0368-4
Transferred to Digital Print 2001

Published by Kluwer Academic Publishers,
P.O. Box 17, 3300 AA Dordrecht, The Netherlands.

Sold and distributed in the U.S.A. and Canada
by Kluwer Academic Publishers,
101 Philip Drive, Norwell, MA 02061, U.S.A.

In all other countries, sold and distributed
by Kluwer Academic Publishers,
P.O. Box 322, 3300 AH Dordrecht, The Netherlands.

Logo design by L. Rivlin

Printed on acid-free paper

EDITORIAL PREFACE

The last twenty years have witnessed an accelerated development of pure and applied logic, particularly in response to the urgent needs of computer science. Many traditional logicians have developed interest in applications and in parallel a new generation of researchers in logic has arisen from the computer science community.

A new attitude to applied logic has evolved, where researchers tailor a logic for their own use in the same way they define a computer language, and where automated deduction for the logic and its fragments is as important as the logic itself.

In such a climate there is a need to emphasise algorithmic logic methodologies alongside any individual logics. Thus the tableaux method or the resolution method are as central to todays discipline of logic as classical logic or intuitionistic logic are.

From this point of view, J. Goubault and I. Mackie's book on Proof Theory and Automated Deduction is most welcome. It covers major algorithmic methodologies as well as a variety of logical systems. It gives a wide overview for the applied consumer of logic while at the same time remains relatively elementary for the beginning student.

A decade ago I put forward my view that a logical system should be presented as a point in a grid. One coordinate is its philosphy, motivation, its accepted theorems and its required non-theorems. The other coordinate is the algorithmic methodology and execution chosen for its effective presentation. Together these two aspects constitute a 'logic'.

This book is a good start for teaching this viewpoint.

I am very happy to welcome it.

Dov M. Gabbay
London, 1996

CONTENTS

List of Figures

PREFACE

This book evolved from a set of support lecture notes for the course "Théorie de la preuve et démonstration automatique", given at the DEA I.M.A. at École Polytechnique by the two authors. The purpose of this book is to give the student basic notions in logic, with a particular stress on proof theory, as opposed to, say, model theory or set theory; and to show how they are applied in computer science, and mostly the particular field of automated deduction, i.e. the automated search for proofs of mathematical propositions.

As for the material of the course, we have chosen to give an in-depth analysis of the basic notions, instead of giving a mere sufficient analysis of basic and less basic notions. We often derive the same theorem by different methods, showing how different mathematical tools can be used to get at the very nature of the objects at hand, and how these tools relate to each other. Instead of presenting a linear collection of results, we have tried to show that all results and methods are tightly interwoven. We believe that understanding how to travel along this web of relations between concepts is more important than just learning the basic theorems and techniques by rote.

Because we insisted on presenting the basic notions thoroughly, we had to sacrifice more complicated notions. Because our main aim is to show what we can do in and around first-order logic, there is almost no mention of more complicated frameworks, like higher-order logics or multiple-valued or quantum logics. We have nonetheless sprinkled the various chapters with various mentions of other logical frameworks, proof methods, or simply observations of interest to the curious reader. Some are presented in exercises. Some others are discussed in a special section on digressions at the end of certain chapters. Finally, some other ideas are discussed in the main text itself, in paragraphs written in small type like the following:

> Such paragraphs contain material relevant to the currently discussed topic, but are often a bit technical for a first reading.

Exercises are roughly categorized in:

▶ **EXERCISE 0.1**
easy exercises, which are direct applications of the results in the main text,

 ▶ **EXERCISE 0.2**
and harder exercises, whose solution demands more imagination from the reader.

The reader will find the solutions to all of the exercises in Appendix A.

We also use the danger sign (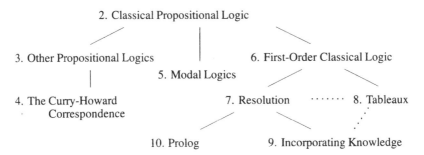) to signal important points. Logic is indeed quite a subtle matter, and it is easy to acquire wrong intuitions on the subject.

As with most scientific books, there are several ways in which this book can be read. If you are already versed in logic or automated theorem proving, and you want to complete your knowledge on the subject, you might want to read it from beginning to end, possibly skipping some sections you think you know enough about, or which you don't understand at first. But, in general, you will probably be more interested in one or two topics only, and it is better to directly go read the relevant sections. Most sections refer to concepts introduced earlier in the book, and you should then consult the necessary definitions and theorems on demand. The rough graph of dependencies between chapters is the following:

1. Introduction

2. Classical Propositional Logic

3. Other Propositional Logics   6. First-Order Classical Logic

5. Modal Logics

4. The Curry-Howard   7. Resolution   · · · · · · ·   8. Tableaux
   Correspondence

10. Prolog   9. Incorporating Knowledge

where dotted lines represent looser connections than solid lines.
Happy reading!

Jean Goubault-Larrecq
G.I.E. Dyade, Rocquencourt

Ian Mackie
Ecole Polytechnique, Palaiseau

ACKNOWLEDGEMENTS

Thanks to Eric Goubault, who provided some valuable information on CCS and Hennessy-Milner logic; to Thomas Tamisier, for explaining the ins and outs of model-checking of temporal formulas with BDDs; to Mourad Debabi, who corrected some errors in early drafts of the document; to Maribel Fernández for proof reading and correcting several errors; to Henri Fraïssé, who helped on several matters related to rewriting; to Jon Freeman for sending us a draft copy of his Ph.D. dissertation, where in-depth information on the Davis-Putnam method can be found; to Stavros Tripakis, Cédric Fournet, Pierre Weis, Chris Hankin and Simon Gay, who found a few errors in the first versions of this document; and to Jean-Paul Billon, with whom the first author had many enthusiastic discussions.

Paul Taylor's prooftree macros were used throughout the text.

Finally, we would like to express our thanks to Jane Spurr for all her help in getting this text into camera-ready form.

CHAPTER 1

INTRODUCTION

We all have our own intuitions about what is *true* and what is *false*. For example it is the case that $1 + 2 = 3$ is *true* while $2 + 2 = 3$ is *false*, and there are formal ways to show that this is the case. There are other entities that we can deem to be true or false, for example "someone likes chocolate" might be true—it depends on whether we can find evidence of the existence of such a person.

These kinds of expressions are called *atomic sentences* and we will always assume that we know the truth value of these. From atomic sentences we can build more complicated ones; for example "someone likes chocolate *and* $1 + 2 = 3$". We reason about the truth of such sentences by looking at the truth value of the components and reasoning about the way they are connected together. In this example we used the connective "and"; other ones that we will use are "or", "not" and "implies" (if ... then ...). All of these connectives follow our intuitions: "A and B" is true just when both A and B are true; "A or B" is true if either of A or B is true; "not A" is true if A is false; "A implies B" is true if when A is true then B is also true. We shall see later in fact that "implies" can be rather subtle.

Rules that allow us to deduce more facts from previously known facts are called *deduction rules*—if the premises of a rule are true, then the conclusion of the rule is also true (provided of course that the rule is *valid*). For example, if A and B are true, we can deduce that A is true, since if both A and B are true, then it *must* be the case that A is true—it is one of the assumptions! Notice that we are not at all interested in what A and B are—they could be anything at all—we are only interested in the *form* of the sentences that we write.

In addition to the above connectives, there are *quantifiers* that allow us to specify the domain of discourse. These are "for all" which specifies a *universal* quantification and "there exists" which specifies an *existential* quantification. Think of "for all" as really meaning *all*, and "there exists" as meaning *at least one*. For example, A implies A is always true no matter what A is, so we could deduce that "for all A, A implies A". We could also say that "there exists A, A implies A", since if it is the case that the sentence is true for all A, then it is certainly true for one such A.

One should take great care with quantifiers. Consider the sentence "if n is a number, then there is another number m such that $n^2 = m$". Intuitively, what we mean here is that *for all* numbers n, *there exists* a number m such that $n^2 = m$; and we certainly don't mean that *there exists* a number n such that *for all* m, $n^2 = m$.

It is a fruitful exercise for the reader to try to write down some sentences, in

1

natural language, that are true using these connectives.

During *logical* reasoning there is an element of *structural* reasoning that allows us to manipulate a sentence. For example we would expect the truth value of "A and B" to be the same as "B and A". However, what about this sentence: I watch the TV and I break the TV. I would certainly have difficulty in watching the TV after I broke it! Hence there are circumstances in which we would prefer not to have the commutativity rule.

"Roses are red and violets are blue therefore we can deduce that I like logic". In this sentence, the premises are true, and the conclusion is true, so is it a valid deduction? The answer depends on whether we are required to use the assumptions (the information) during the deduction or not. Different kinds of logic have been introduced to express these kinds of entities. It is not the formal power of different logics that is at issue, but rather how useful they are at representing different real life situations.

Systems of formal logic have been developed so that we can reason about sentences built from these kinds of connectives in a precise way. There are a collection of formal systems that capture different concepts. Here are just a few of the well known ones:

- Classical logic

- Intuitionistic logic

- Temporal logic

- Modal logic

- Linear logic

Think of these as a set of tools that can be applied to different situations. The right one to use depends on which concept we are interested in modeling. Classical logic can be regarded as *every day* logic, and fits with most of our daily reasoning; it is based wholly on the notion of truth. Intuitionistic logic on the other hand is more about *proof* than truth—a sentence is true when we can provide a constructive proof of the statement. The example that one finds to distinguish intuitionistic logic from classical logic is the sentence 'A or not A'. Classically, this is always true, since whatever A is (true or false), the resulting sentence is true. In intuitionistic logic we are unable to *prove* this sentence, since in general it is not the case that we can find a proof of either 'A' or 'not A'. Temporal logic captures the notion of time in a proof and allows us to express ideas such that sentences become true at a certain point in time. More generally, modal logics explore alternative modes of truth, where truth may depend on the state we are in, whether it be time, a set of beliefs or the current state of a machine. Linear logic is a refined logic that captures the notion of *resources* in a proof. A proof of a sentence in linear logic requires that we have correctly used all the assumptions. All of these logics play a major rôle in computer science, for example: intuitionistic logic connects strongly with type systems for functional programming; temporal logic is used for concurrent system specification and verification; modal logic for artificial intelligence; and linear logic is currently being used to study the dynamics of the evaluation of programs.

Here we are not so concerned with modeling real life situations in a logical system, but more on understanding the logical systems themselves. For example, we might have a rule that says: under the assumption that the atomic sentences A and B *are* true, we can deduce that the sentence "A and B" *is* true; why A and B are true is not part of our reasoning. To express sentences like this we use a particular *syntax* for the logical connectives. In general we will write $\wedge$ for "and", $\vee$ for "or", $\neg$ for "not", $\Rightarrow$ for "implies", $\forall$ for "for all", and $\exists$ for "there exists". We will also introduce some additional connectives upon demand.

The kind of rules that we desire to reason about such sentences are going to tell us how we can introduce these connectives, and also how to reason about the components of the sentences by eliminating the connectives.

Mathematics gains a lot from a formal syntax. We have seen some examples above that were written in English, but we could do much better in a language that is more suited to express our ideas. Roughly speaking, we want to get rid of words that are not important; just capturing the parts of the sentence that we are interested in. For example, the following equation from set theory, expressing distributivity of intersection over union:

$$X \cap (Y \cup Z) = (X \cap Y) \cup (X \cap Z)$$

could be written in English as:

> Given three sets X, Y and Z, then the set obtained by taking the union of the sets Y and Z and then the intersection of this set with the set X is equal to the set obtained by first taking the intersection of the sets X, Y and X, Z and then taking the union of these two sets.

The gains are self evident: mathematical syntax is briefer, more concise and more precise (at least when we have understood what the mathematical symbols mean!). The same phenomenon arises in logic too; it is a mathematical discipline. Generally, one can write a logical sentence representing a natural language one by arranging it in a more structured way.

Here is an example of a logical sentence:

$$\forall n, \text{divisible-by-2}(n) \Rightarrow \text{even}(n)$$

which represents the natural language sentence:

> All numbers that are divisible by 2 are even numbers.

Suppose further that we know that divisible-by-2(10) is true, then we can deduce that even(10). This kind of reasoning is fundamental to logical reasoning. More generally this rule follows the form 'given both A and $A \Rightarrow B$ then we can deduce B', and is called *modus ponens*. Rules of this form will generally be written in the form:

$$\frac{A \qquad A \Rightarrow B}{B}$$

where we write the assumptions above the line (the things that we know are true) and the conclusion below the line (the things that we can deduce to be true).

There are other kinds of rules like this that allow us to reason. You will find many of these throughout this book. Here is one more to give the flavour of the kinds of rules that we will use:

$$\frac{\neg B \qquad A \Rightarrow B}{\neg A}$$

This rule says that if we know that if A is true then B is also true, and we also know that B is *not* true, then it must be the case that A is also not true. This rule is called *modus tollens*. To understand this rule a little better, we give an example. Suppose we have the following facts: "If it is raining then I am inside", and "I am outside". Then the above rule states that we can deduce that it is not raining.

The purpose of a formal system is to allow us to obtain *proofs* of sentences. There are many different proof systems that have been developed to express these kinds of rules and to reason about them in different ways. We will study many of these including Tableaux, Resolution and Natural Deduction. Common to all these methods is the production of a proof of a particular sentence in some logic. The different systems allow us to manipulate the sentence in different ways to construct a proof. It is worth mentioning that proofs themselves are worthy of study. The formal study of proofs is called *proof theory*, which is really the heart of logic, or *the Logic of Logic*. It is essentially a *syntactical* discipline in that we are interested in the structure and properties of manipulation at a syntactical level.

Reasoning in a formal system of course must be founded upon a set of rules that are correct; that is each rule of the formal system must preserve the mathematical truth value of the sentence. For this we need a notion of *semantics*—a mathematical object that we use to interpret the logical system. The property that all the rules are correct (or *valid*) with respect to some semantics is called *soundness* and is one of the most important properties of a proof system.

There are many different kinds of semantics that we can use to model proof systems, and we shall see several of these in this book. Different logics lend themselves towards different kinds of semantics. For example classical logic and truth tables, intuitionistic logic and Heyting (functional) semantics, etc.

There is another property that we may require of our logical system, and that is *completeness*; that is to say we can prove *all* the sentences that are true in the system. Having a logical system that is sound and complete with respect to some mathematical model gives us total freedom to use the proof system in any way that we like—it is not possible to apply a rule to change the meaning of the sentence, and (if we can find it!) there is always a proof of a true sentence.

It is therefore not so hard to believe that the process of producing proofs in these frameworks is a mechanical process that can be automated. We can represent logical sentences as data types in a programming language, and express the rules of deduction as functions over these data types. This is the topic of this book—*automated deduction*. Of course, some proof methods are going to be more appropriate than others, for example, resolution is one of the most successful implementations, and is the basis of the programming language Prolog. We will discuss many proof techniques and mention implementation difficulties.

Logic plays a major rôle in computer science. We can use it for specifications of programs (pre-conditions, post-conditions, loop invariants, etc.) as a way of reasoning about computation and proving certain properties of programs with respect to the specification. Here we are interested in using a much stronger connection between logic and computation where the logic is utilised much more directly. There are two paradigms interfacing logic and computation:

1. Logic Programming

   - Programs *are* theories (over some base logic, for example *Horn clauses*).
   - Computation *is* proof search.

   A computer is given a program (information in the form of logical sentences) and a query (a logical sentence) on this information. Computation is then the process of finding a proof of the query; in other words: is there sufficient information in the program to determine this query. The most significant language to follow this paradigm is Prolog.

2. Typed Functional Programming

   - Types of programs *are* logical formulas.
   - Programs *are* proofs.
   - Computation *is* the normalisation (or cut-elimination) process of proofs.

   The idea is that a computer is given a program in terms of a proof — a proof object corresponding to a sentence — (following the intuition that "proving is like programming") and we require as output the *normal form* of that same proof. Each of the modules of a program are composed together and the computation proceeds (cf. eliminating lemmas in a proof). The modules of the program are plugged together in such a way that the *types* fit. This process can be checked automatically by the process of *type reconstruction*. There are many programming languages that follow this paradigm, for example Standard ML, CAML, etc.

Here we primarily follow the first paradigm, hinting a little at the second to allow a more general picture of the subject to be obtained.

Logic is a mathematical discipline, and the only way to know what it is and how to do it is to play with it. There are many examples and exercises in this text that we encourage the reader to try, and there are many texts that are full of exercises that will do nothing more than reinforce the reader's understanding. Some recommended texts are: (Lemmon, 1965) and (Hodges, 1976).

1   OVERVIEW

The chapters of this book are structured in the following way.

Chapter 2: Classical Propositional Logic

In this chapter we formally present the syntax and semantics for propositional classical logic and establish soundness and completeness results. A variety of deduction systems (for example natural deduction, sequent calculus, Hilbert style, etc.) for this logic are exposed and compared.

The process of automated proof search for classical logic is studied and again a sequence of techniques will be presented: resolution, tableaux, the Davis-Putnam method and the use of Binary Decision Diagrams.

The basic concepts, results and methods presented in this chapter are the fundamental tools that we shall keep on using for most of the book.

Chapter 3: Other Propositional Logics

Classical logic provides a basis for *everyday logic*. In this chapter we will look at some other logics that have been proposed that capture different concepts. We begin with intuitionistic logic which is a logic based on *proof* rather than truth. We will try to mirror some of the results on the previous chapter to give a deeper understanding of the material. Linear logic has recently been proposed to be the "logic of computation" which justifies its inclusion in this book. We show some basic properties of this system, and show how it is related to classical and intuitionistic logic by providing translations of these logics into linear logic and examining the properties of the resulting proofs.

Chapter 4: The Curry-Howard Correspondence

The Curry-Howard Correspondence (or *proofs-as-programs* paradigm) exhibits a remarkable connection between logic and (typed functional) programs. In this chapter we study the full correspondence between intuitionistic logic and the simply typed λ-calculus. We also mention other instances of this isomorphism and some applications of this theory.

Chapter 5: Modal and Temporal Logics

Modal and temporal logics are the study of *modes* of truth. We will study various modal and temporal logics: syntax, semantics and proof systems. There are a number of modal logics that play a major rôle in computer science that we study. Some of them, like S4, come from philosophical logic and have found recent application in the modelisation of non-monotonic, or defeasible, or default reasoning. Some others are the fundamental tools to specify and verify the essential features of sequential or parallel hardware and software, like Hennessy-Milner logic, CTL (computation tree logic) and dynamic logic. Regarding the latter, it will be more interesting not to find proofs, but rather to check whether a given implementation—a model—really obeys the specification: this is called *model-checking*.

Chapter 6: First-Order Classical Logic

In this chapter, we introduce a definitely more expressive logic than the propositional logics we saw in the previous chapters. This is done by refining the language of logic, and replacing propositional variables by more elaborate formulas describing properties of values in a domain of interest. The result is called *first-order logic*, or *predicate calculus*. Most propositional logics can be extended to first-order logics (and even to higher-order logics). We present here *classical* first-order logic.

Chapter 7: Resolution

The core of the most successful implementation of logic as a programming language, and of many automated proof systems for first-order classical logic, is the proof procedure of Resolution. In this chapter we will study resolution and techniques associated with it. There, we introduce the process of *unification*, an integral part of resolution but also of all other automated proof methods for first-order logic. We shall then introduce a number of different improvements on the basic resolution principle, including deletion strategies, set-of-support and semantic resolution, linear resolution and the use of orderings, all aiming at cutting down the size of the search space.

Chapter 8: Tableaux, Connections and Matings

First-order tableaux methods are in a sense dual to resolution, and have advantages and drawbacks of their own. Whereas our presentation of resolution, following traditional presentations, rests mainly on semantical considerations, tableaux are best explained from proof systems. Whereas resolution finds proofs using the cut rule only, tableaux find proofs using all rules but the cut rule.

However, this opposition will become more of an illusion as we progress towards alternative implementations of tableaux, including the connection method and the method of matings, to culminate in model-elimination, a special case of resolution in fact.

Chapter 9: Incorporating Knowledge

If we are interested in finding proofs automatically, some knowledge about the domains of values we are dealing with is necessary in the prover itself.

One of the most important relations a prover has to know about is that of equality; *rewriting* is the name of a family of techniques designed to handle equality, and (ordered) *paramodulation* is the result of integrating it in a resolution-based theorem prover. We shall devote a great part of this chapter to equality and rewriting in particular.

Other kinds of theories that a prover can handle directly include equational theories, sort theories, theories of polynomials and various others, which we shall then touch upon briefly.

Chapter 10: Logic Programming Languages

The final chapter studies some applications of material of this book in the form of programming languages based on the theory exposed. In particular, we will look at the language Prolog which can be seen as an implementation of theorem proving in classical logic. We will then look at several extensions to logic programming, for example constraints and parallelism.

EXERCISES

There are no formal exercises for this introduction, however the reader might like to try (for fun!) the following.

▶ **EXERCISE 1.1**

Try to convert some of the following natural language sentences into logical sentences. We also suggest that the reader might like to find additional examples and exercises in texts that cover this topic.

1. If you go to Asterix Park then you might get wet.

2. If you have my login and password then you are able to delete all my files.

3. Either it is raining and I stay home to watch TV or I go to École Polytechnique to work.

4. You can pass this course only if you attend the lectures.

5. 0 is a number and if n is a number then $n + 1$ is also a number.

6. Write the following menu as a single logical formula:

entrée
Tarte aux oignons
or
Salade campagnarde
plat
Saumon grillé
or
Escalope de veau
with frites *or* pâtes à volonté
dessert
Mousse au chocolat
or
Glace (sorbet poire *or* citron selon la saison)
with sauce chocolat noir

Are any of the connectives that you have used being used for slightly different purposes?

▶ **EXERCISE 1.2**

Are the following logical sentences equivalent? Justify each one using informal reasoning. You might like to formally verify each one after reading the following chapter.

1. $A \wedge (B \wedge C)$ and $(A \wedge B) \wedge C$.

    I.e. is $\wedge$ associative? What about the other binary connectives $\vee$ and $\Rightarrow$. Are the connectives commutative?

2. $(A \vee B) \Rightarrow C$ and $A \Rightarrow (B \Rightarrow C)$.

3. $\neg(A \vee B)$ and $A \wedge B$.

4. $A \Rightarrow B$ and $\neg B \Rightarrow \neg A$.

CHAPTER 2

CLASSICAL PROPOSITIONAL LOGIC

As we have said in the previous chapter, a logic is a syntax (a way of building the set of formulas of interest), a semantics (a description of what these formulas mean), and a deduction system (allowing us to compute the meaning of formulas by deriving proofs). We hereby expose the corresponding syntax, semantics and some deduction systems for classical propositional logic, one of the simplest (and also least expressive) logics. This study then leads to automated proof methods for classical propositional logic, which we then describe: the tableaux method, propositional resolution, the Davis-Putnam method, and binary decision diagrams (BDDs). We end the chapter with a few digressions, which the reader might want to skip on first reading.

1 SYNTAX

The syntax of propositional logic is based on *proposition variables* that we represent by capital letters taken from the beginning of the alphabet, A, B, C, etc., possibly subscripted or primed. These proposition letters are meant to represent basic properties, which are either true or false. These variables are then combined by using *logical connectives* to yield *formulas* or *propositions*, which we denote by capital Greek letters like Φ, Ψ. The connectives are:

- *conjunction*, also written $\wedge$ (pronounced "and"). The conjunction of two formulas Φ and Φ' is written $\Phi \wedge \Phi'$, and we intend this to be true whenever both Φ and Φ' are true.

- *disjunction* $\vee$ (pronounced "or"). We write $\Phi \vee \Phi'$ for the disjunction of two formulas Φ and Φ'. It is meant to be true whenever at least one of Φ and Φ' is true, including the case where both Φ and Φ' are true.

- *negation* $\neg$ (pronounced "not"). We write $\neg\Phi$ for the negation of Φ, which is meant to be true whenever Φ is not.

- *implication* $\Rightarrow$ (pronounced "implies"). $\Phi \Rightarrow \Phi'$ is the logical implication of Φ' by Φ. It is meant to be true if and only if Φ' is true whenever Φ is true. When Φ is false, this statement is vacuously true, so $\Phi \Rightarrow \Phi'$ is true if and only if Φ is false, or Φ' is true.

Although $\Phi \Rightarrow \Phi'$ reads "Φ implies Φ'", it is bad practice to use "Φ implies Φ'" and $\Phi \Rightarrow \Phi'$ as interchangeable phrases. Indeed, the former is a statement saying that whenever Φ is true, Φ' is. The second is a formula, i.e. an object that we can manipulate, and whose value will be the answer to the question "does Φ implies Φ'?" The deduction theorem will make the link between both by saying that "Φ implies Φ'" and "$\Phi \Rightarrow \Phi'$ *is valid*" can be used interchangeably.

More formally:

Definition 2.1 *Let $\mathcal{X}$ be an infinite set of so-called* proposition variables. *The set $\mathcal{F}$ of all* propositional formulas *is the least set such that all variables are formulas, and if Φ and Φ' are formulas, then $\Phi \wedge \Phi'$, $\Phi \vee \Phi'$, $\Phi \Rightarrow \Phi'$ and $\neg \Phi$ are formulas.*

where we still need to define what $\Phi \wedge \Phi'$, $\Phi \vee \Phi'$, $\Phi \Rightarrow \Phi'$ and $\neg \Phi$ really are. We choose the following:

Definition 2.2 *Let $\wedge$, $\vee$, $\neg$, $\Rightarrow$ be four distinct objects. Then $\Phi \wedge \Phi'$ is the triple $(\wedge, \Phi, \Phi')$; $\Phi \vee \Phi'$ is the triple $(\vee, \Phi, \Phi')$; $\neg \Phi$ is the couple $(\neg, \Phi)$; $\Phi \Rightarrow \Phi'$ is the triple $(\Rightarrow, \Phi, \Phi')$.*

This defines formulas as tree-like objects, where each node is either a variable, or a tuple. If it is a tuple, we say that its first component is its *label*, and its other components are its *sons*. Moreover, we say that the set of *subformulas* of a formula is the smallest set such that the formula itself is in it, and it contains all subformulas of the sons of the formula. (If we identify formulas, i.e. trees, with their root nodes, then subformulas are just the nodes accessible from the root.) A subformula of Φ is said to be *proper* if it is not Φ.

From a computer science perspective, formulas are in fact not trees, but directed acyclic graphs, i.e. trees "with identical subtrees coalesced in a single node". Take for example $(A \vee B) \Rightarrow (A \vee B)$, where A and B are variables. This is by definition the triple $(\Rightarrow, \Phi, \Phi)$ where Φ is the node representing $A \vee B$. Then, the two occurrences of Φ are trivially equal, so they actually form a single node. In graphical terms, this yields:

We don't need to represent formulas in the memory of a computer as directed acyclic graphs, and trees are perfectly all right. Indeed, equality is then represented by a recursive function that traverses both graphs in parallel, instead of being represented by equality of addresses.

The kind of objects as above are actually called trees by some logicians. Mathematically speaking, these trees, or directed acyclic graphs, are just instances of elements of the free algebra generated by the symbols $\wedge$, $\vee$, $\neg$, $\Rightarrow$ over the variables in $\mathcal{X}$. We shall define free algebras in Chapter 6, Section 1.1.

Some conventions are needed to print and read formulas faithfully as text. Indeed, it is not clear how we should interpret, say, $\Phi \wedge \Phi' \vee \Phi''$: is it the tree labelled $\vee$, having sons $\Phi \wedge \Phi'$ and Φ'', or the tree labelled $\wedge$ with Φ and $\Phi' \vee \Phi''$ as sons? Clearly, this is a just a parsing problem. We resolve it by judiciously mixing precedences and uses of parentheses. We assume that:

- if Φ is a printed representation of a formula, then (Φ) is a printed representation of the same formula. (This already solves the problem of parsing formulas in a unique way, provided that we agree on using parentheses heavily.)

- $\neg$ binds tighter than $\wedge$, which binds tighter than $\vee$, which binds tighter than $\Rightarrow$. Hence $\Phi \wedge \Phi' \vee \Phi''$ is shorthand for $(\Phi \wedge \Phi') \vee \Phi''$, $\neg\Phi \vee \Phi'$ stands for $(\neg\Phi) \vee \Phi'$, and $\Phi \wedge \Phi' \Rightarrow \Phi'' \vee \Phi'''$ stands for $(\Phi \wedge \Phi') \Rightarrow (\Phi' \vee \Phi''')$, in particular.

- $\wedge$ and $\vee$ are left-associative, i.e. $\Phi \wedge \Phi' \wedge \Phi''$ stands for $(\Phi \wedge \Phi') \wedge \Phi''$ and $\Phi \vee \Phi' \vee \Phi''$ stands for $(\Phi \vee \Phi') \vee \Phi''$.

- $\Rightarrow$ is right-associative, i.e. $\Phi \Rightarrow \Phi' \Rightarrow \Phi''$ stands for $\Phi \Rightarrow (\Phi' \Rightarrow \Phi'')$.

From this set of connectives, we may define *abbreviations* for other interesting operations. For example, *logical equivalence* $\Leftrightarrow$ can be defined by: $\Phi \Leftrightarrow \Phi'$ is shorthand for $(\Phi \Rightarrow \Phi') \wedge (\Phi' \Rightarrow \Phi)$. Indeed, we want $\Phi \Leftrightarrow \Phi'$ to be true if and only if, whenever Φ is true, so is Φ', and whenever Φ' is true, so is Φ; this is exactly the intended meaning of $(\Phi \Rightarrow \Phi') \wedge (\Phi' \Rightarrow \Phi)$.

Other less used, but nonetheless interesting connectives are Sheffer's stroke $|$, such that $\Phi \mid \Phi'$ is true whenever not both Φ, Φ' are true (this is well-known by electrical engineers, who call it the NAND function, since it has the same semantics as $\neg(\Phi \wedge \Phi')$); or dually, Peirce's dagger $\downarrow$, also known as the NOR function ($\Phi \downarrow \Phi'$ is an abbreviation for $\neg(\Phi \vee \Phi')$). These two are most interesting in electronics where it is important to have as few different types of gates on a silicon die; indeed, one of them alone is enough to define all the others in classical logic. (See Exercises 2.4, 2.5, and 2.6.)

On the other hand, we could have defined fewer connectives and let the remaining ones be abbreviations for the previous ones. For example, we might have defined only $\neg$ and $\vee$, and let $\Phi \wedge \Phi'$ be $\neg((\neg\Phi) \vee (\neg\Phi'))$, and $\Phi \Rightarrow \Phi'$ be $\neg\Phi \vee \Phi'$. This would not have changed anything for classical logic, which we develop in the sequel. But it would have changed a lot if we were to consider intuitionistic logic, for instance. (See Chapter 3.)

Definition 2.3 *The set of* free variables $\mathrm{fv}(\Phi)$ *of a formula* Φ *is defined by structural induction by:*

- $\mathrm{fv}(A) = \{A\}$ *for any propositional variable* A,

- $\mathrm{fv}(\Phi \wedge \Phi') = \mathrm{fv}(\Phi \vee \Phi') = \mathrm{fv}(\Phi \Rightarrow \Phi') = \mathrm{fv}(\Phi) \cup \mathrm{fv}(\Phi')$,

- $\mathrm{fv}(\neg\Phi) = \mathrm{fv}(\Phi)$.

In short, a variable is free in Φ if and only if it occurs at some leaf of Φ.

> This is the first definition where use a principle of so-called structural induction. This is a general induction principle that holds not only of finite tree-like objects, but of any set S defined as the *least* one containing some basis set S_0, which is stable by a set F of operations. (i.e. such that if $x_1, x_2, \ldots, x_n$ are in S, and f is an n-ary operation in F, then $f(x_1, x_2, \ldots, x_n)$ is in S.)
>
> It is a straightforward exercise to prove the principle of *proof by structural induction*: if some property P holds of all elements of S_0, and if whenever P holds of $x_1, \ldots, x_n$, P holds of $f(x_1, \ldots, x_n)$, for all f in F, then P holds of all elements of S. (To prove it, consider the set of all x such that P holds of x: it must contain S, as S is the least set verifying these properties.)
>
> We used here the principle of *definition by structural induction*: any function g defined by its values on elements of S_0, and by the value it takes on $f(x_1, \ldots, x_n)$ as a function of the value it takes on $x_1, \ldots, x_n$, for every f in F, is defined on all of S. (To prove it, use the principle of proof by structural induction on the property $P(x)$ defined as "g is well-defined at x".)
>
> This relates to the principles of proof and definition by induction in the integers by considering as set S_0 of base values the set $\{0\}$, and as set F of operations the set $\{s\}$, where s is the successor function mapping x to $x + 1$.

Another useful notion is the following:

Definition 2.4 *A* substitution σ *is a function from propositional variables to formulas, such that* $\sigma(A) = A$ *for all but finitely many variables* A.

The domain *of* σ, $\mathrm{dom}\ \sigma$, *is the (finite) set of variables* A *such that* $\sigma(A) \neq A$. *We write* $[\Phi_1/A_1, \ldots, \Phi_n/A_n]$ *the substitution of domain* $\{A_1, \ldots, A_n\}$ *that maps each variable* A_i *to* Φ_i, $1 \leq i \leq n$, *where the* A_is *are assumed pairwise distinct. In particular,* $[]$ *is the* empty substitution.

Any substitution σ *extends uniquely to a function from formulas to formulas, by:*

- *any variable* A *maps to* $\sigma(A)$,

- *if* Φ *and* Ψ *map respectively to* Φ' *and* Ψ', *then* $\Phi \wedge \Psi$ *maps to* $\Phi' \wedge \Psi'$, $\Phi \vee \Psi$ *maps to* $\Phi' \vee \Psi'$, $\neg\Phi$ *maps to* $\neg\Phi'$, *and* $\Phi \Rightarrow \Psi$ *maps to* $\Phi' \Rightarrow \Psi'$.

We write $\Phi\sigma$ *for the result of applying this function to the formula* Φ, *and we call it the* application *of* σ *to* Φ.

In short, σ can be understood as a finite set of replacement operations Φ_i/A_i, substituting Φ_i for A_i. $\Phi\sigma$ is the formula Φ where all the variables A_i are replaced by Φ_i. This replacement must occur *in parallel*: for example, applying $[B/A, C/B]$ on A yields B, not C, as would occur if we had applied B/A, then C/B.

Then:

Theorem 2.5 *Let σ be $[\Phi_1/A_1, \ldots, \Phi_n/A_n]$, and σ' be $[\Phi'_1/A'_1, \ldots, \Phi'_{n'}/A'_{n'}]$.*
Then there is a unique substitution σ'' such that $\Phi\sigma'' = (\Phi\sigma)\sigma'$ for every formula Φ.

Proof: By definition, if σ'' exists, it must map every variable A to $(A\sigma)\sigma'$, i.e. to $\sigma(A)\sigma'$. So it is unique. Now, let σ'' be this very substitution: the proof that $\Phi\sigma''$ is equal to $(\Phi\sigma)\sigma'$ is an easy structural induction on Φ. □

This σ'' is written $\sigma\sigma'$, or $\sigma' \circ \sigma$. This theorem entitles us to the following definition:

Definition 2.6 Substitution composition *is the operation $\circ$ defined by $\Phi(\sigma' \circ \sigma) = (\Phi\sigma)\sigma'$ for every Φ. We also write $\sigma\sigma'$ for $\sigma' \circ \sigma$, so that $\Phi(\sigma\sigma') = (\Phi\sigma)\sigma'$.*

Substitution composition is *not* function composition! Indeed, the image of a variable A is $\sigma(A)\sigma'$, not $\sigma'(\sigma(A))$. In fact, the latter would be meaningless if $\sigma(A)$ were not a variable.

Theorem 2.7 *Substitution composition is associative, and has the empty substitution as unit element.*

Proof: Note that the empty substitution is actually the identity function on variables, so that by structural induction on Φ, $\Phi[] = \Phi$ for every formula Φ.

In particular, $\sigma[]$ is the only substitution such that $\Phi(\sigma[]) = (\Phi\sigma)[] = \Phi\sigma$ for all Φ, so $\sigma[]$ must be σ. Similarly, $[]\sigma$ is the only substitution such that $\Phi([]\sigma) = (\Phi[])\sigma = \Phi\sigma$ for all Φ, so $[]\sigma$ must be σ. This proves that $[]$ is a unit element for substitution composition.

Finally, if σ, σ' and σ'' are three substitutions, $(\sigma\sigma')\sigma''$ is the only substitution such that $\Phi((\sigma\sigma')\sigma'') = (\Phi(\sigma\sigma'))\sigma'' = ((\Phi\sigma)\sigma')\sigma''$ for all Φ; but $\sigma(\sigma'\sigma'')$ is the only substitution such that $\Phi(\sigma(\sigma'\sigma'')) = (\Phi\sigma)(\sigma'\sigma'') = ((\Phi\sigma)\sigma')\sigma''$ for Φ. By uniqueness, we must have $(\sigma\sigma')\sigma'' = \sigma(\sigma'\sigma'')$. As σ, σ' and σ'' are arbitrary, substitution composition is associative. □

2 SEMANTICS

Without saying so explicitly, we have already introduced the semantics of formulas. For example, we have said that we intended $\Phi \wedge \Phi'$ to be true whenever both Φ and Φ' were true. Inherently, we have tried to convey some *meaning* to the formulas, and this is the role of semantics. It is mainly here that *classical* logic differs from all others.

In classical propositional logic, every formula is intended to be either true or false. Formally, this means that the *set of truth-values* is the set $\mathbb{B} = \{\top, \bot\}$ of Booleans, where $\top \neq \bot$. The meaning of the connectives can then be defined as functions from Boolean values to Boolean values. These functions are usually represented by *truth-tables*. Here are the truth-tables for all the basic connectives:

$\wedge$	$\bot$	$\top$
$\bot$	$\bot$	$\bot$
$\top$	$\bot$	$\top$

$\vee$	$\bot$	$\top$
$\bot$	$\bot$	$\top$
$\top$	$\top$	$\top$

$\neg$	
$\bot$	$\top$
$\top$	$\bot$

$\Rightarrow$	$\bot$	$\top$
$\bot$	$\top$	$\top$
$\top$	$\bot$	$\top$

For example, if Φ is assumed true ($\top$), and Φ' is assumed false ($\bot$), then $\Phi \Rightarrow \Phi'$ must be false. (The first argument denotes the column, and the second argument the row in the truth-table.) This defines binary functions $\overline{\wedge}$, $\overline{\vee}$ and $\overline{\Rightarrow}$ from $\mathbb{B} \times \mathbb{B}$ to $\mathbb{B}$, and one unary function $\overline{\neg}$ from $\mathbb{B}$ to $\mathbb{B}$.

The meaning of a formula Φ depends on the truth-values that we assume for its free variables, so we define formally:

Definition 2.8 *An* assignment *or* interpretation ρ *is any function from the set $\mathcal{X}$ of propositional variables to the set $\mathbb{B}$ of Boolean values.*

The semantics $[\![\Phi]\!]\rho$ *of a formula Φ under the assignment ρ is defined by structural induction as follows:*

- $[\![A]\!]\rho = \rho(A)$ *if A is a propositional variable,*

- $[\![\Phi \wedge \Phi']\!]\rho = [\![\Phi]\!]\rho \,\overline{\wedge}\, [\![\Phi']\!]\rho,$

- $[\![\Phi \vee \Phi']\!]\rho = [\![\Phi]\!]\rho \,\overline{\vee}\, [\![\Phi']\!]\rho,$

- $[\![\Phi \Rightarrow \Phi']\!]\rho = [\![\Phi]\!]\rho \,\overline{\Rightarrow}\, [\![\Phi']\!]\rho,$

- $[\![\neg\Phi]\!]\rho = \overline{\neg}[\![\Phi]\!]\rho.$

We shall then say that a formula Φ is true *under the assignment ρ* whenever $[\![\Phi]\!]\rho = \top$, and is false *under ρ* whenever $[\![\Phi]\!]\rho = \bot$.

Notice that we stress the fact that a formula can only be true or false under a given assignment. This is important, as many wrong proof arguments stem from a confusion as to what "true" and "false" mean. It is all too easy to confuse the fact of being true for some assignment with the fact of being true for all assignments. This is why we use new terms to denote the latter:

Definition 2.9 *Let Φ be a propositional formula, and ρ be an assignment.*

We say that ρ is a model *of Φ, or that ρ satisfies Φ, and we write $\rho \models \Phi$, if and only if $[\![\Phi]\!]\rho = \top$.*

We say that a set Γ of formulas entails *Φ, and we write $\Gamma \models \Phi$, if and only if all assignments that satisfy all the formulas in Γ at once (the models of Γ) are also models of Φ, i.e. whenever $\rho \models \Psi$ for all $\Psi \in \Gamma$, then $\rho \models \Phi$.*

Φ *is* valid *if and only if it is true under* all *assignments* ($[\![\Phi]\!]\rho = \top$ *for every* ρ, *written* $\models \Phi$), *and is* invalid *otherwise. A valid propositional formula is also called a* tautology.

Φ *is* satisfiable *if and only if it is true under* some *assignment* ($[\![\Phi]\!]\rho = \top$ *for some* ρ, *i.e., it has a model), and is* unsatisfiable *otherwise. An unsatisfiable propositional formula is also called an* antilogy.

All valid formulas are satisfiable, and all unsatisfiable formulas are invalid. This divides up the space of formulas in three categories: the valid formulas (always true), the unsatisfiable formulas (always false), and the formulas that are both invalid and satisfiable (sometimes true, sometimes false). This yields the following diagram:

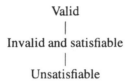

Valid
|
Invalid and satisfiable
|
Unsatisfiable

Second, validity and unsatisfiability correspond through negation: Φ is valid if and only if $\neg\Phi$ is unsatisfiable, Φ is unsatisfiable if and only if $\neg\Phi$ is valid. This means that negation $\neg$ maps the diagram above to the same, turned upside-down. Observe how the "invalid and satisfiable" point remains at the same place. This is why it is dangerous to use "true" in place of "valid" and "false" in place of "unsatisfiable": although there are only two truth-values, there are at least three distinct validity states for a formula. (See Exercise 2.3.)

The following expresses that, in a sense, assignments at the semantic level behave just as substitutions at the syntactic level:

Theorem 2.10 *For every substitution* σ *and every assignment* ρ, *let* $\sigma\rho$ *be the assignment mapping each variable* A *to* $[\![\sigma(A)]\!]\rho$.

Then, for every formula Φ, *for every substitution* σ *and every assignment* ρ, $[\![\Phi\sigma]\!]\rho = [\![\Phi]\!](\sigma\rho)$.

Proof: By structural induction on Φ. If Φ is a variable A, this follows by definition of $\sigma\rho$. If Φ is $\Phi' \wedge \Phi''$, and $[\![\Phi'\sigma]\!]\rho = [\![\Phi']\!](\sigma\rho)$, $[\![\Phi''\sigma]\!]\rho = [\![\Phi'']\!](\sigma\rho)$, then $[\![\Phi\sigma]\!]\rho = [\![\Phi'\sigma]\!]\rho \overline{\wedge} [\![\Phi''\sigma]\!]\rho = [\![\Phi']\!](\sigma\rho) \overline{\wedge} [\![\Phi'']\!](\sigma\rho) = [\![\Phi]\!](\sigma\rho)$. The other cases are similar. $\qquad\square$

Corollary 2.11 *Let* Φ *be a propositional formula.*

If Φ *is valid (resp. unsatisfiable), then* $\Phi\sigma$ *is valid (resp. unsatisfiable) for every substitution* σ.

This corollary allows us to, say, prove that $\Phi \Rightarrow \Phi$ is valid for every formula Φ by just proving that $A \Rightarrow A$, where A is a variable, is valid. All this is rather obvious, but it is better to state it explicitly.

▶ **EXERCISE 2.1**

Show that $\Phi \Rightarrow \Phi'$ is true whenever $\neg\Phi \vee \Phi'$ is, i.e. that $[\![\Phi \Rightarrow \Phi']\!]\rho = [\![\neg\Phi \vee \Phi']\!]\rho$ for every assignment ρ.

▶ **EXERCISE 2.2**

Show that $\Phi \Leftrightarrow \Phi'$ is true whenever Φ and Φ' have the same truth-values. (Recall that $\Phi \Leftrightarrow \Phi'$ is an abbreviation for $(\Phi \Rightarrow \Phi') \wedge (\Phi' \Rightarrow \Phi)$, and draw its truth-table.)

▶ **EXERCISE 2.3**

Which of the following formulas are valid? Which are unsatisfiable? And which are both invalid and satisfiable?

$$
\begin{array}{ll}
(i) & A \Rightarrow B \Rightarrow A \\
(ii) & (A \Rightarrow B \Rightarrow C) \Rightarrow (A \Rightarrow B) \Rightarrow (A \Rightarrow C) \\
(iii) & ((A \Rightarrow B) \Rightarrow A) \Rightarrow A \\
(iv) & ((A \vee B) \wedge \neg C) \Rightarrow B \\
(v) & \neg A \wedge ((A \Rightarrow B) \Rightarrow A) \\
(vi) & ((A \wedge \neg B) \Rightarrow C) \Rightarrow B \\
(vii) & A \Rightarrow (\neg A \vee B) \\
(viii) & (A \Rightarrow B) \vee (B \Rightarrow A) \\
(ix) & ((A \Rightarrow B) \wedge (C \Rightarrow D)) \Rightarrow ((A \wedge C) \Rightarrow (B \vee D)) \\
(x) & ((A \Rightarrow B) \vee (C \Rightarrow D)) \Rightarrow ((A \wedge C) \Rightarrow (B \vee D))
\end{array}
$$

▶ **EXERCISE 2.4**

Let $\Phi \mid \Phi'$ be an abbreviation for $\neg(\Phi \wedge \Phi')$. Draw its truth-table.

Now, let $\mathcal{F}'$ be the set of formulas built on variables in $\mathcal{X}$ with the only binary operator $\mid$ (Sheffer's stroke). Show that we can then define $\Phi \wedge \Phi'$, $\Phi \vee \Phi'$, $\neg\Phi$ and $\Phi \Rightarrow \Phi'$ as abbreviations of formulas in $\mathcal{F}'$.

The purpose of this exercise is to show that we are not tied to a unique set of basic connectives to define classical propositional logic, since we might have chosen Sheffer's stroke instead of the more conventional connectives we used.

▶ **EXERCISE 2.5**

Let $\Phi \downarrow \Phi'$ be an abbreviation for $\neg(\Phi \vee \Phi')$. Draw its truth-table, and define $\wedge$, $\vee$, $\neg$ and $\Rightarrow$ in terms of $\downarrow$ (Peirce's dagger) only, as in Exercise 2.4.

▶ **EXERCISE 2.6**

Let if Φ then Φ' else Φ'' be an abbreviation for $(\Phi \Rightarrow \Phi') \wedge (\neg\Phi \Rightarrow \Phi'')$, **T** be one for $A \vee \neg A$, **F** one for $A \wedge \neg A$, where A is a given variable. Show that the semantics of **T** and **F** do not depend on the choice of A, and that all connectives can be defined in terms of the if/then/else connective, and **T** and **F** only.

3 DEDUCTION SYSTEMS

Now, having a semantics for classical propositional logic is one thing, but it is hard to use it to decide whether a given formula Φ is valid or not, satisfiable or not. Although it is possible to check all relevant assignments, this may require a tremendous amount of time. The number of different assignments to test is indeed 2^n, where n is the number of free variables of Φ: this is finite, but becomes rapidly too large to be handled in practice.

We therefore need other ways of expressing valid (resp. unsatisfiable) formulas. One of the most interesting ones is to examine well-formed proofs, and to look at their conclusions, also called *theorems*. By doing this, we hope (assuming for now that the theorems are exactly the tautologies) that all, or at least some theorems have short proofs, or at least proofs that are shorter than a truth-table. This has two distinct advantages: first, a short proof is better than a long list of truth-values for explanatory purposes; second, if a theorem has a short proof, then it will be easier to find it, whether by hand or by machine, than to check all possible assignments.

Mirroring the fact that propositions are written in the language of propositional logic, proofs are written in a language called a *deduction system*. There are many of them, which are all equally expressive, and we shall define a few of them in this section.

One question that immediately pops up is: is there any relation between tautologies and propositional theorems? It will turn out that in any of the deduction systems that we present, the theorems are precisely the tautologies, which is what we intended to get in the first place. We shall prove this for every deduction system we present. This will be decomposed into a proof of *soundness* (all theorems are valid), and one of *completeness* (all tautologies can be proved).

> Although this is a desirable fact, be aware that some logics do not enjoy this property. In particular, higher-order logic with its standard semantics has no sound and complete effective deduction system, i.e. any deduction system for higher-order logic must fail to prove some valid formulas (incompleteness), or fail to prove only valid formulas (unsoundness), or the deduction rules might not be recognisable by computer (effectiveness). This is a consequence of Gödel's incompleteness theorem (Gödel, 1931), which we shall mention as Theorem 6.36, page 221.

The word "completeness" is one of the most overloaded words in mathematics. For example, a uniform space (in particular a metric space) is complete whenever all Cauchy sequences have a limit, a lattice is complete whenever it has least upper bounds and greatest lower bounds of every finite or infinite family of elements, and so on. In logic alone, "completeness" also has several different meanings, that the reader should not confuse. For now, we are just interested in one of these meanings: a *proof system* is complete with respect to a *semantics* if and only if every valid proposition (as defined by the semantics) has a proof (in the system).

3.1   Hilbert-style Systems

David Hilbert was one of the first mathematicians to become interested in the mechanisation of proofs and of theorem discovery. He tried to provide a rigorous format for proofs by following the way most mathematicians present their hand-made proofs. Although these proofs are usually written on two-dimensional sheets of paper, they really are sequences of statements, either axioms or consequences of previous statements, inferred by well-defined deduction rules.

Definition 2.12 *A* Hilbert-style deduction system *is a couple* $(\mathcal{A}, \mathcal{R})$, *where* $\mathcal{A}$ *is a set of formulas called the* axioms, *and* $\mathcal{R}$ *is a set of* deduction rules *or* rules of inference, *i.e. relations between sets of formulas (the set of premises) and formulas (the conclusion).*

A derivation *from a set* Γ *of hypotheses is any non-empty sequence* $\Phi_1, \Phi_2, \ldots,$ $\Phi_n, n > 0$, *of formulas such that for every* i *between* 1 *and* n, Φ_i *is either an axiom* $(\Phi_i \in \mathcal{A})$, *an element of* Γ, *or is deduced from previous formulas in the sequence (i.e.,* $\Delta \, r \, \Phi_i$, *for some* $r \in \mathcal{R}$ *and some* $\Delta \subseteq \{\Phi_1, \ldots, \Phi_{i-1}\}$*).*

A proof π *of a formula* Φ *from* Γ *is a derivation from* Γ *ending in* Φ. *If there is a proof of* Φ *from* Γ, *then we say that* Γ proves Φ, *or that* Φ *is* provable *from* Γ, *and we write* $\Gamma \vdash^{(\mathcal{A}, \mathcal{R})} \Phi$.

A theorem *is a formula that is provable from the empty set of hypotheses.*

We shall also use the following notation: if Γ and Γ' are two sets of formulas, the notation Γ, Γ' denotes the union of Γ and Γ'; if Φ is a formula, Γ, Φ and Φ, Γ denote $\Gamma \cup \{\Phi\}$; the empty set is denoted by a blank space. This is customary notation in logic.

An example of a Hilbert-style system for propositional logic is Andrews' system $\mathcal{P}$ (Andrews, 1986), defined as follows:

- Axioms:

    (1) $\Phi \vee \Phi \Rightarrow \Phi$ for every formula Φ,

    (2) $\Phi \Rightarrow \Phi' \vee \Phi$ for all formulas Φ, Φ',

    (3) $(\Phi \Rightarrow \Phi') \Rightarrow (\Phi'' \vee \Phi \Rightarrow \Phi' \vee \Phi'')$ for all Φ, Φ', Φ'',

- Rules:

    (MP) from Φ and $\Phi \Rightarrow \Phi'$, deduce Φ', for all Φ, Φ'. (Formally, $\mathcal{R} = \{(MP)\}$, where (MP) maps all pairs $\{\Phi, \Phi \Rightarrow \Phi'\}$ to Φ'.)

Additional side-conditions in system $\mathcal{P}$ are that only $\neg$ and $\vee$ are primitive connectives in this system, so that $\Phi \Rightarrow \Phi'$ is really an abbreviation for $\neg\Phi \vee \Phi'$, $\Phi \wedge \Phi'$ is an abbreviation for $\neg(\neg\Phi \vee \neg\Phi')$.

In particular, this also means that (1), (2) and (3) are abbreviations for $\neg(\Phi \vee \Phi) \vee \Phi$, $\neg\Phi \vee (\Phi' \vee \Phi)$ and $\neg(\neg\Phi \vee \Phi') \vee (\neg(\Phi'' \vee \Phi) \vee (\Phi' \vee \Phi''))$, respectively.

Rule (MP), also called the *modus ponens* rule, is the only rule of inference. As promised, the system is sound:

Theorem 2.13 (Soundness) *Let Γ be a set of formulas, and Φ be a formula. If $\Gamma \vdash^{\mathcal{P}} \Phi$, then $\Gamma \models \Phi$.*

Proof: By induction on the length of a proof of Φ from Γ.

If this length is 0, then Φ is either an element of Γ or an axiom. If it is an element of Γ, obviously $\Gamma \models \Phi$. If it is an axiom, we actually prove that $\models \Phi$ (Φ is valid), hence $\Gamma \models \Phi$ by inspection of the truth-tables.

If the length is n, and the theorem holds for all shorter proofs, then Φ was deduced by (MP) from two formulas Φ' and $\Phi' \Rightarrow \Phi$ that were proved in less than n steps from Γ. By induction hypothesis, $\Gamma \models \Phi'$ and $\Gamma \models \Phi' \Rightarrow \Phi$. By inspection of the truth-table of $\Rightarrow$, $\Phi' \Rightarrow \Phi$ can be true under an interpretation that makes Φ' true only when Φ is true, so that $\Gamma \models \Phi$. $\square$

The fact that the system is sound entails that it is *consistent*:

Theorem 2.14 (Consistency) *System $\mathcal{P}$ is absolutely consistent, i.e. there is a formula Φ which cannot be proved by $\mathcal{P}$.*

System $\mathcal{P}$ is consistent with respect to negation, i.e. there is no formula Φ for which both Φ and $\neg\Phi$ are provable.

Proof: We prove that $\mathcal{P}$ is consistent with respect to negation, then as the set of formulas is non-empty, we trivially deduce its absolute consistency.

Assume on the contrary that there is a formula such that both Φ and $\neg\Phi$ are provable. By the Soundness Theorem, both Φ and $\neg\Phi$ are valid. We deduce $\top = \bot$, a contradiction. $\square$

The notion of absolute consistency may seem strange. It is the simplest notion of consistency that we can impose on a system without referring to logics. It enforces the fact that there are at least two non-equivalent formulas, or else the system would be rather trivial.

Notice also how we used models to give a short proof of consistency, a syntactically unpalatable problem: trying to prove the same thing using only the rules of deduction, and induction on the lengths of proofs can be done, but would hardly yield a short proof of the fact.

The fact that the system is complete is less obvious. In fact, proving a deduction system complete with respect to a semantics is usually rather difficult. The case of propositional logic is the simplest, but $\mathcal{P}$ is very unnatural to work with, and so we won't prove completeness here. We refer the interested reader to (Andrews, 1986, Pages 10–22).

Theorem 2.15 (Completeness) *Let Φ be a formula, and Γ be a set of formulas. If $\Gamma \models \Phi$, then $\Gamma \vdash^{\mathcal{P}} \Phi$.*

Notice in particular that this holds even in the case where Γ is an infinite set of formulas.

From the Completeness Theorem, we draw a few important conclusions. Actually, these conclusions are usually needed to prove the Completeness Theorem first, but they can also be deduced from it.

The first important consequence of the Completeness Theorem is the fact that $\Rightarrow$ is, as we expect, strongly connected to the notion of deducibility:

Theorem 2.16 (Deduction Theorem) *Let Γ be a set of formulas, Φ and Φ' be formulas. Then $\Gamma, \Phi \vdash^{\mathcal{P}} \Phi'$ if and only if $\Gamma \vdash^{\mathcal{P}} \Phi \Rightarrow \Phi'$.*

Proof: (This can be proved purely syntactically, i.e. by transforming proofs by induction on their lengths, see (Andrews, 1986).)

By soundness and completeness, $\Gamma, \Phi \vdash^{\mathcal{P}} \Phi'$ if and only if $\Gamma, \Phi \models^{\mathcal{P}} \Phi'$. On the other hand, $\Gamma \vdash^{\mathcal{P}} \Phi \Rightarrow \Phi'$ if and only if $\Gamma \models \Phi \Rightarrow \Phi'$. It therefore suffices to show that $\Gamma, \Phi \models \Phi'$ if and only if $\Gamma \models \Phi \Rightarrow \Phi'$.

If $\Gamma, \Phi \models \Phi'$, let ρ be an arbitrary assignment such that $\rho \models \Psi$ for all $\Psi \in \Gamma$: if $\rho \models \Phi$, then $\rho \models \Phi'$; by inspection of the truth-table for $\Rightarrow$, it follows that $\rho \models \Phi \Rightarrow \Phi'$. Since ρ is arbitrary, $\Gamma \models \Phi \Rightarrow \Phi'$.

Conversely, if $\Gamma \models \Phi \Rightarrow \Phi'$, take ρ such that $\rho \models \Phi$ for all $\Phi \in \Gamma$, and $\rho \models \Phi$; then, by inspection of the truth-table for $\Rightarrow$, we must have $\rho \models \Phi'$. Since ρ is arbitrary, $\Gamma, \Phi \models \Phi'$.   □

Another important consequence of the Completeness Theorem is the following:

Theorem 2.17 (Compactness) *If $\Gamma \models \Phi$, then there is a finite subset Δ of Γ such that $\Delta \models \Phi$.*

Proof: (This is actually needed to prove the Completeness Theorem in the general case, see (Andrews, 1986, Page 38). We shall prove it without using the Completeness Theorem in Section 3.2.)

If $\Gamma \models \Phi$, then $\Gamma \vdash^{\mathcal{P}} \Phi$ by completeness, hence there is a finite subset Δ of Γ such that $\Delta \vdash^{\mathcal{P}} \Phi$; indeed, any proof is finite, so that proving Φ can use but a finite number of hypotheses in Γ. By soundness, $\Delta \models \Phi$.   □

Notice that compactness is not a property of $\mathcal{P}$, but of propositional logic independently of any system of inference it may have.

Another consequence is decidability of propositional logic:

Theorem 2.18 (Decidability) *There is an algorithm which, given a finite set Γ of formulas and a formula Φ, decides whether $\Gamma \vdash^{\mathcal{P}} \Phi$.*

Proof: To decide $\Gamma \vdash^{\mathcal{P}} \Phi$ means to decide $\Gamma \models \Phi$, by soundness and completeness of the deduction system. Moreover, we can assume Γ empty without loss of generality (otherwise use the Deduction Theorem as in the "Γ finite" case of the proof of the Completeness Theorem). Now, $[\![\Phi]\!]\rho$ only depends on the values of

ρ on the free variables of Φ, by an easy structural induction on Φ. So, enumerate the restrictions of assignments to $\mathrm{fv}(\Phi)$ (there are finitely many), and evaluate Φ on each. This takes finite time.   □

 Although the axioms of $\mathcal{P}$ are effectively recognisable (we can decide whether a given formula is an axiom of $\mathcal{P}$ or not by machine), this does not entail decidability of deducibility in $\mathcal{P}$. A procedure for deciding whether a formula Φ is provable in $\mathcal{P}$ might enumerate all proofs and check whether they indeed prove Φ, but System $\mathcal{P}$ does not offer an obvious bound to the number of proofs we have to test, so this procedure may not terminate. On the other hand, enumerating all partial assignments *does* terminate.

Of course, enumerating all assignments of the free variables may take quite a long time. If there are n propositional variables in Φ, there are 2^n different assignments, when restricted to the free variables. We shall see more efficient methods in Section 4.

> It is a curious point that the proof of decidability comes not from the syntactic, almost operational view of the logic, i.e. the deduction system, but from the abstract, semantic view. This is mainly because Hilbert-style systems do not lend themselves well to any meta-theoretic study. Sequent systems are better-behaved in this respect. (See Section 3.3.)

▶ **EXERCISE 2.7**
Let Φ be a formula and σ be a substitution. Show that, if $\vdash^{\mathcal{P}} \Phi$, then $\vdash^{\mathcal{P}} \Phi\sigma$.

▶ **EXERCISE 2.8**
Prove that if $\Gamma \vdash^{\mathcal{P}} \Phi \Rightarrow \Phi'$ and $\Gamma \vdash^{\mathcal{P}} \Phi'' \vee \Phi$, then $\Gamma \vdash^{\mathcal{P}} \Phi' \vee \Phi''$. (If possible, without any recourse to semantics.)

 ▶ **EXERCISE 2.9**
Prove $\Phi \vee \neg\Phi$ and $\Phi \Rightarrow \neg\neg\Phi$ in $\mathcal{P}$.

 ▶ **EXERCISE 2.10**
Prove $\neg\neg\Phi \Rightarrow \Phi$ in $\mathcal{P}$. (You may use Exercise 2.9.)

▶ **EXERCISE 2.11**
Let $\mathcal{SKC}$ be the deduction system whose axioms are:

(K) $\Phi \Rightarrow \Phi' \Rightarrow \Phi$,

(S) $(\Phi \Rightarrow \Phi' \Rightarrow \Phi'') \Rightarrow (\Phi \Rightarrow \Phi') \Rightarrow (\Phi \Rightarrow \Phi'')$,

(C) $\neg\neg\Phi \Rightarrow \Phi$

and whose sole inference rule is (MP). We assume that $\Rightarrow$ and $\mathbf{F}$ are the only primitive propositional connectives, and that $\neg$ is defined as $\neg\Phi = \Phi \Rightarrow \mathbf{F}$, $\vee$ and $\wedge$ are defined by $\Phi \vee \Phi' = \neg\Phi \Rightarrow \Phi'$, $\Phi \wedge \Phi' = \neg(\Phi \Rightarrow \neg\Phi')$.

Prove the Deduction Theorem for $\mathcal{SKC}$.

▸ **EXERCISE 2.12**

Let $\mathcal{SKC}$ be the system of Exercise 2.11. Show that $\Phi \Rightarrow \neg\neg\Phi$ is a theorem in $\mathcal{SKC}$. (Hint: use (C) and the Deduction Theorem for $\mathcal{SKC}$.)

▸ **EXERCISE 2.13**

Let $\mathcal{SKC}$ be the system of Exercise 2.11. Show that we can deduce $\Phi \Rightarrow \Phi''$ from $\Phi \Rightarrow \Phi'$ and $\Phi' \Rightarrow \Phi''$. (Use the Deduction Theorem for $\mathcal{SKC}$.)

▸ **EXERCISE 2.14**

Let $\mathcal{SKC}$ be the system of Exercise 2.11. Show that $\neg\Phi \vee \Phi'$ and $\Phi \Rightarrow \Phi'$ are logically equivalent in $\mathcal{SKC}$. (I.e., we can deduce each one from the other.)

Conclude that if Ψ and Ψ' are two formulas that differ only by replacing subformulas of the form $\neg\Phi \vee \Phi'$ by $\Phi \Rightarrow \Phi'$ (or conversely), then Φ and Φ' are logically equivalent in $\mathcal{SKC}$. (You may use Exercises 2.11 and 2.12.)

▸ **EXERCISE 2.15**

Let $\mathcal{SKC}$ be the system of Exercise 2.11. Show that it is sound and complete. (Hint: prove the axioms of $\mathcal{P}$ in $\mathcal{SKC}$ for the completeness part. You may use the results of Exercises 2.11, 2.12, 2.13 and 2.14.)

3.2 Natural Deduction

Hilbert-style systems are not easy to work with, either to prove propositions by hand or by computer, or to prove meta-theorems on propositional logic itself.

The main defect of Hilbert-style systems is the fact that we cannot write down auxiliary assumptions. That is, to prove $\Phi \Rightarrow \Phi'$, there is no mechanism that allows us to posit Φ temporarily, to allow us to prove Φ'. (The Deduction Theorem however tells us that it would be sound to do this.)

Natural deduction systems remedy this situation, and can be presented in several different formats, all equivalent. These systems were first invented by Gerhard Gentzen in the 1930s (Szabo, 1969). Gentzen was then unable to derive the metamathematical results that interested him (consistency, mainly) by purely syntactical manipulations of natural deductions, and proceeded to invent sequent systems to this purpose (Section 3.3). Dag Prawitz (1965) then showed in the sixties that all these results could actually have been derived with natural deduction systems.

A first way of presenting natural deduction systems is to say that proofs are two-dimensional diagrams of formulas connected by deduction rules. The deduction

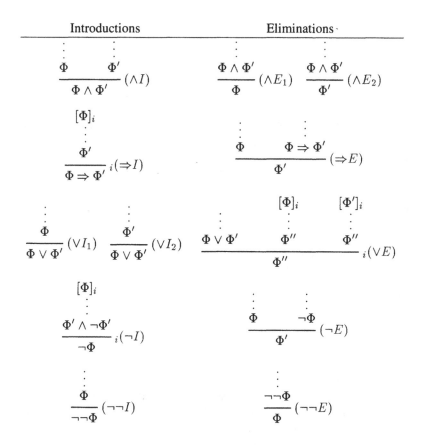

Figure 2.1. Natural deduction

rules for classical propositional logic are classified into *introduction rules* and *elim-*

ination rules (See Figure 2.1). Three vertical dots ($\vdots$) abbreviate a whole derivation tree. The rules combine derivation trees to produce new derivation trees.

For example, if (π) is a derivation that proves Φ, and (π') is a derivation that proves Φ', then the following derivation proves $\Phi \wedge \Phi'$:

$$\begin{array}{cc} (\pi) & (\pi') \\ \vdots & \vdots \\ \Phi & \Phi' \end{array}$$
$$\overline{\qquad \Phi \wedge \Phi' \qquad} \; (\wedge I)$$

$(\wedge I)$ is the rule of introduction of the conjunction connective. From two proofs of two formulas, it builds a proof of a conjunctive formula. Symmetrically, $(\wedge E_1)$ and $(\wedge E_2)$ take a derivation of a conjunctive formula, and return a subformula of it under the $\wedge$ connective: these rules therefore eliminate one $\wedge$ sign from the formula.

The rule $(\neg E)$ has as a conclusion an arbitrary formula Φ': indeed, from Φ and $\neg\Phi$, a contradiction, we may deduce whatever we wish.

The other rules are then mostly self-explanatory, except for those using the strange $[\;]_i$ notation, that is, $(\Rightarrow I)$, $(\neg I)$ and $(\vee E)$. The bracket represents an assumption that has been *discharged*. Look at the $(\Rightarrow I)$ rule. It represents the fact that if (π) is a proof of Φ' using Φ as an assumption, i.e. :

$$\begin{array}{c} \Phi \\ \vdots \\ (\pi) \\ \vdots \\ \Phi' \end{array}$$

then we can deduce from this a proof of $\Phi \Rightarrow \Phi'$ without using Φ as auxiliary assumption, so we can cross out Φ—discharge it— and add a new line below the proof. The notation for discharging Φ is $[\Phi]_i$, where i is a unique tag (say, an integer) that we use to associate the inference of $\Phi \Rightarrow \Phi'$ to the occurrences of Φ that we discharge. Discharging Φ in the proof above then yields:

$$\begin{array}{c} [\Phi]_1 \\ \vdots \\ (\pi) \\ \vdots \\ \Phi' \end{array}$$
$$\overline{\qquad \Phi \Rightarrow \Phi' \qquad} \; 1\,(\Rightarrow I)$$

The two-dimensional format of natural deduction proofs may seem nice, but it is hard to study it meta-mathematically, since it is not always clear which auxiliary assumptions are needed to prove a formula in derivations: whenever we get a derivation of some formula, we have to go up the derivation and search it for non-discharged assumptions. For instance, here is a proof of $\Phi \vee \neg\Phi$ in natural deduction:

$$
\cfrac{
\cfrac{
[\neg(\Phi \vee \neg\Phi)]_3 \qquad
\cfrac{
\cfrac{[\neg\Phi]_1}{\Phi \vee \neg\Phi}(\vee I_2)
}{}
}{
\cfrac{\mathbf{F}}{\neg\neg\Phi}{}_1(\neg I)
}(\neg E) \qquad
\cfrac{\cfrac{\mathbf{F}}{\Phi}(\neg\neg E)}{}
\qquad
\cfrac{
[\neg(\Phi \vee \neg\Phi)]_3 \qquad
\cfrac{[\Phi]_2}{\Phi \vee \neg\Phi}(\vee I_1)
}{
\cfrac{\mathbf{F}}{\neg\Phi}{}_2(\neg I)
}(\neg E)
}{
\cfrac{
\cfrac{\mathbf{F}}{\neg\neg(\Phi \vee \neg\Phi)}{}_3(\neg I)
}{\Phi \vee \neg\Phi}(\neg\neg E)
}(\neg E)
$$

where $\mathbf{F}$ is some formula of the form $\Psi \wedge \neg\Psi$ (we don't care which). A hint: read the above proof top-down, from the left to the right, without the discharging brackets first; you should have to discharge the formulas indexes 1 first, then 2, then 3.

A more formalized presentation of natural deduction, which shows formulas together with the set of auxiliary assumptions, is *natural deduction in sequent form*. Instead of just deriving formulas Φ, we derive *sequents* $\Gamma \longrightarrow \Phi$, where Γ is a set of formulas, called the *context* of the sequent. This is precisely the set of auxiliary assumptions that we have while trying to prove Φ. The deduction rules are those of Figure 2.2, where $\mathbf{F}$ is an abbreviation for any formula of the form $\Phi' \wedge \neg\Phi'$. Notice how the context manipulations are less intricate than with the $[\ ]_i$ notation, as auxiliary assumptions are explicitly described in every sequent.

The $\longrightarrow$ entailment sign must not be confused with implication $\Rightarrow$, or with deducibility $\vdash$. Implication is a connective, $\longrightarrow$ builds sequents (couples of a set of formulas and a formula), and $\vdash$ is a mathematical relation. However, again, these notions are strongly correlated, as we now show.

The following definition parallels Definition 2.12:

Definition 2.19 *A natural deduction derivation is an inverted tree whose nodes are sequents connected by one of the deduction rules shown in Figure 2.2. The leaves are instances of the (Ax) rule, also called* axioms.

A proof π of a sequent $\Gamma \longrightarrow \Phi$ is a natural deduction derivation whose root is decorated by $\Gamma \longrightarrow \Phi$. We also say that π is a proof of Φ from Γ. If there is a proof of $\Gamma \longrightarrow \Phi$, we say that $\Gamma \longrightarrow \Phi$ is provable, or that Φ is provable from Γ, and we write $\vdash^{\mathcal{ND}} \Gamma \longrightarrow \Phi$, or $\Gamma \vdash^{\mathcal{ND}} \Phi$.

A theorem *is a formula that is provable from the empty set of hypotheses.*

$$\frac{}{\Gamma, \Phi \longrightarrow \Phi}\,(Ax)$$

$$\frac{\Gamma \longrightarrow \Phi \qquad \Gamma \longrightarrow \Phi'}{\Gamma \longrightarrow \Phi \wedge \Phi'}\,(\wedge I) \qquad \frac{\Gamma \longrightarrow \Phi_1 \wedge \Phi_2}{\Gamma \longrightarrow \Phi_i}\,(\wedge E_i) \quad (i = 1, 2)$$

$$\frac{\Gamma, \Phi \longrightarrow \Phi'}{\Gamma \longrightarrow \Phi \Rightarrow \Phi'}\,(\Rightarrow I) \qquad \frac{\Gamma \longrightarrow \Phi \Rightarrow \Phi' \qquad \Gamma \longrightarrow \Phi}{\Gamma \longrightarrow \Phi'}\,(\Rightarrow E)$$

$$\frac{\Gamma \longrightarrow \Phi_i}{\Gamma \longrightarrow \Phi_1 \vee \Phi_2}\,(\vee I_i) \quad (i = 1, 2) \qquad \frac{\Gamma \longrightarrow \Phi \vee \Phi' \qquad \Gamma, \Phi \longrightarrow \Phi'' \qquad \Gamma, \Phi' \longrightarrow \Phi''}{\Gamma \longrightarrow \Phi''}\,(\vee E)$$

$$\frac{\Gamma, \Phi \longrightarrow \mathbf{F}}{\Gamma \longrightarrow \neg\Phi}\,(\neg I) \qquad \frac{\Gamma \longrightarrow \Phi \qquad \Gamma \longrightarrow \neg\Phi}{\Gamma \longrightarrow \Phi'}\,(\neg E)$$

$$\frac{\Gamma \longrightarrow \Phi}{\Gamma \longrightarrow \neg\neg\Phi}\,(\neg\neg I) \qquad \frac{\Gamma \longrightarrow \neg\neg\Phi}{\Gamma \longrightarrow \Phi}\,(\neg\neg E)$$

Figure 2.2. Natural Deduction in sequent form

A few preliminary lemmas are in order:

Lemma 2.20 (Weakening) *Let* $\Gamma \longrightarrow \Phi$ *be a derivable sequent. Then for every* Γ' *such that* $\Gamma \subseteq \Gamma'$, $\Gamma' \longrightarrow \Phi$ *is derivable.*

Proof: By structural induction on the proof of $\Gamma \longrightarrow \Phi$ (this is a tree), we consistently add all elements of $\Gamma' \setminus \Gamma$ to the contexts of all sequents appearing in the proof. □

We have all the expected meta-theorems for this system of deduction:

Theorem 2.21 (Deduction Theorem) *Let* Γ *be a set of formulas,* Φ *and* Φ' *be formulas. Then* $\Gamma, \Phi \vdash^{\mathcal{ND}} \Phi'$ *if and only if* $\Gamma \vdash^{\mathcal{ND}} \Phi \Rightarrow \Phi'$.

Proof: If $\Gamma, \Phi \vdash^{\mathcal{ND}} \Phi'$, then we have a proof of $\Gamma, \Phi \longrightarrow \Phi'$. Append one application of the $(\Rightarrow I)$ rule to get a proof of $\Gamma \longrightarrow \Phi \Rightarrow \Phi'$.

Conversely, if $\Gamma \vdash^{\mathcal{ND}} \Phi \Rightarrow \Phi'$, then we have a proof of $\Gamma \longrightarrow \Phi \Rightarrow \Phi'$. By Lemma 2.20, we transform this into a proof of $\Gamma, \Phi \longrightarrow \Phi \Rightarrow \Phi'$. On the other hand, we produce $\Gamma, \Phi \longrightarrow \Phi$ by (Ax), then we apply $(\Rightarrow E)$ and the latter two to derive $\Gamma, \Phi \longrightarrow \Phi'$. □

To make a parallel with System $\mathcal{P}$, notice that we also have modus ponens: the $(\Rightarrow E)$ rule is precisely the (MP) rule of Section 3.1.

Theorem 2.22 (Soundness) *If* $\Gamma \vdash^{\mathcal{ND}} \Phi$, *then* $\Gamma \models \Phi$.

Proof: By structural induction on the derivation of $\Gamma \vdash^{\mathcal{ND}} \Phi$, the theorem follows immediately by inspection of each of the possible rules. □

We shall now proceed to prove that this system of natural deduction is complete. We first need the following lemma:

Lemma 2.23 *Let* ρ *be an assignment, and* Φ *be a formula. Define* Φ^ρ *as* Φ *if* $[\![\Phi]\!]\rho = \top$, *and as* $\neg\Phi$ *if* $[\![\Phi]\!]\rho = \bot$.

Let $A_1, \ldots, A_n$ *be the free variables in* Φ. *Then* $\Gamma, A_1^\rho, \ldots, A_n^\rho \vdash^{\mathcal{ND}} \Phi^\rho$ *for every context* Γ.

Proof: By structural induction on Φ:

- If Φ is a variable A, $\Gamma, A^\rho \longrightarrow A^\rho$ is an instance of (Ax).

- If Φ is a conjunction $\Phi' \wedge \Phi''$, by induction hypothesis, we have two derivations of $\Gamma, A_1'^\rho, \ldots, A_{m'}'^\rho \longrightarrow \Phi'^\rho$ and $\Gamma, A_1''^\rho, \ldots, A_{m''}''^\rho \longrightarrow \Phi''^\rho$ respectively, where $A_1', \ldots, A_{m'}'$ are the free variables of Φ' and $A_1'', \ldots, A_{m''}''$ are the free variables of Φ''. Let $A_1, \ldots, A_m$ be the free variables of Φ: they form a superset of the latter, so by the weakening Lemma 2.20, we get two derivations (π') and (π'') of $\Gamma, A_1^\rho, \ldots, A_m^\rho \longrightarrow \Phi'^\rho$ and $\Gamma, A_1^\rho, \ldots, A_m^\rho \longrightarrow \Phi''^\rho$ respectively. There are now three cases:

– if both Φ' and Φ'' are true under ρ, then $\Phi'^\rho = \Phi'$, $\Phi''^\rho = \Phi''$ and $\Phi^\rho = \Phi$, and we derive $\Gamma \longrightarrow \Phi$ by:

$$
\dfrac{
\begin{array}{cc}
\begin{array}{c} (\pi') \\ \vdots \\ \Gamma, A_1^\rho, \ldots, A_m^\rho \longrightarrow \Phi' \end{array}
&
\begin{array}{c} (\pi'') \\ \vdots \\ \Gamma, A_1^\rho, \ldots, A_m^\rho \longrightarrow \Phi'' \end{array}
\end{array}
}{\Gamma, A_1^\rho, \ldots, A_m^\rho \longrightarrow \Phi' \wedge \Phi''} \ (\wedge I)
$$

– if Φ' is false under ρ, notice that $\Phi'^\rho = \neg\Phi'$ and that $\Phi^\rho = \neg(\Phi' \wedge \Phi'')$. We first transform the proof (π') of $\Gamma, A_1^\rho, \ldots, A_m^\rho \longrightarrow \neg\Phi'$ into a proof (π_1') of $\Gamma, A_1^\rho, \ldots, A_m^\rho, \Phi' \wedge \Phi'' \longrightarrow \neg\Phi'$ by weakening, and write:

$$
\dfrac{
\dfrac{
\dfrac{\rule{3cm}{0.4pt}}{\Gamma_1, \Phi' \wedge \Phi'' \longrightarrow \Phi' \wedge \Phi''} \ (Ax)
}{\Gamma_1, \Phi' \wedge \Phi'' \longrightarrow \Phi'} \ (\wedge E_1)
\qquad
\begin{array}{c} (\pi_1') \\ \vdots \\ \Gamma_1, \Phi' \wedge \Phi'' \longrightarrow \neg\Phi' \end{array}
}{
\dfrac{\Gamma_1, \Phi' \wedge \Phi'' \longrightarrow \mathbf{F}}{\Gamma_1 \longrightarrow \neg(\Phi' \wedge \Phi'')} \ (\neg I)
} \ (\neg E)
$$

where $\Gamma_1 = \Gamma, A_1^\rho, \ldots, A_m^\rho$.

– the case where Φ'' is false under ρ is symmetric, with $(\wedge E_2)$ replacing $(\wedge E_1)$ and (π'') replacing (π').

• If Φ is a disjunction $\Phi' \vee \Phi''$, by induction hypothesis, we have two derivations (π') and (π'') of $\Gamma_1 \longrightarrow \Phi'^\rho$ and $\Gamma_1 \longrightarrow \Phi''^\rho$ respectively, as in the previous case, where Γ_1 denotes $\Gamma, A_1^\rho, \ldots, A_m^\rho$. There are three cases:

– if both Φ' and Φ'' are false under ρ, then $\Phi'^\rho = \neg\Phi'$, $\Phi''^\rho = \neg\Phi''$ and $\Phi^\rho = \neg\Phi$. Weaken (π') to a proof (π_1') of $\Gamma_1, \Phi', \Phi' \vee \Phi'' \longrightarrow \neg\Phi'$, and weaken (π'') to a proof (π_1'') of $\Gamma_1, \Phi'', \Phi' \vee \Phi'' \longrightarrow \neg\Phi''$. Let Γ' be $\Gamma_1, \Phi', \Phi' \vee \Phi''$, and Γ'' be $\Gamma_1, \Phi'', \Phi' \vee \Phi''$. Then, we derive $\Gamma, A_1^\rho, \ldots, A_m^\rho \longrightarrow \neg\Phi$ by:

$$
\dfrac{
\dfrac{\rule{3cm}{0.4pt}}{\Gamma_1, \Phi' \vee \Phi'' \longrightarrow \Phi' \vee \Phi''} \ (Ax)
\qquad
\begin{array}{cc}
\begin{array}{c} (\pi_2') \\ \vdots \\ \Gamma' \longrightarrow \mathbf{F} \end{array}
&
\begin{array}{c} (\pi_2'') \\ \vdots \\ \Gamma'' \longrightarrow \mathbf{F} \end{array}
\end{array}
}{
\dfrac{\Gamma_1, \Phi' \vee \Phi'' \longrightarrow \mathbf{F}}{\Gamma_1 \longrightarrow \neg(\Phi' \vee \Phi'')} \ (\neg I)
} \ (\vee E)
$$

where (π_2') is:

$$\frac{\dfrac{}{\Gamma_1,\Phi',\Phi'\vee\Phi''\longrightarrow\Phi'}\,(Ax) \qquad \begin{array}{c}(\pi_1')\\ \vdots\\ \Gamma_1,\Phi',\Phi'\vee\Phi''\longrightarrow\neg\Phi'\end{array}}{\Gamma_1,\Phi',\Phi'\vee\Phi''\longrightarrow\mathbf{F}}\,(\neg E)$$

and (π_2'') is:

$$\frac{\dfrac{}{\Gamma_1,\Phi'',\Phi'\vee\Phi''\longrightarrow\Phi''}\,(Ax) \qquad \begin{array}{c}(\pi_1')\\ \vdots\\ \Gamma_1,\Phi'',\Phi'\vee\Phi''\longrightarrow\neg\Phi''\end{array}}{\Gamma_1,\Phi'',\Phi'\vee\Phi''\longrightarrow\mathbf{F}}\,(\neg E)$$

- if Φ' is true under ρ, then we append to (π') one application of the $(\vee I_1)$ rule to get a proof of $\Gamma, A_1^\rho, \ldots, A_m^\rho \longrightarrow \Phi' \vee \Phi''$.

- similarly, if Φ'' is true under ρ, then we append to (π'') one application of the $(\vee I_2)$ rule to get a proof of $\Gamma, A_1^\rho, \ldots, A_m^\rho \longrightarrow \Phi' \vee \Phi''$.

- if Φ is an implication, by induction hypothesis and weakening, we have two derivations (π') and (π'') of $\Gamma_1 \longrightarrow \Phi'^\rho$ and $\Gamma_1 \longrightarrow \Phi''^\rho$ respectively, where $\Gamma_1 = \Gamma, A_1^\rho, \ldots, A_m^\rho$. There are three cases:

  - Φ' is true under ρ, Φ'' is false under ρ, so Φ is false under ρ. Weaken (π') and (π'') to get proofs (π_1') and (π_1'') of $\Gamma_1, \Phi \longrightarrow \Phi'$ and of $\Gamma_1, \Phi \longrightarrow \neg\Phi''$ respectively, then produce:

$$\frac{\dfrac{\dfrac{}{\begin{array}{c}\Gamma_1,\Phi\longrightarrow\\ \Phi'\Rightarrow\Phi''\end{array}}\,(Ax) \qquad \begin{array}{c}(\pi_1')\\ \vdots\\ \Gamma_1,\Phi\longrightarrow\Phi'\end{array}}{\Gamma_1,\Phi\longrightarrow\Phi''}\,(\Rightarrow E) \qquad \begin{array}{c}(\pi_1'')\\ \vdots\\ \Gamma_1,\Phi\longrightarrow\neg\Phi''\end{array}}{\dfrac{\Gamma_1,\Phi\longrightarrow\mathbf{F}}{\Gamma_1\longrightarrow\neg\Phi}\,(\neg I)}\,(\neg E)$$

  - if Φ' is false under ρ, weaken (π') to a proof (π_1') of $\Gamma_1, \Phi' \longrightarrow \neg\Phi'$ and produce:

$$\frac{\dfrac{\begin{array}{c}(\pi_1')\\ \vdots\\ \Gamma_1,\Phi'\longrightarrow\neg\Phi'\end{array} \qquad \dfrac{}{\Gamma_1,\Phi'\longrightarrow\Phi'}\,(Ax)}{\Gamma_1,\Phi'\longrightarrow\Phi''}\,(\neg E)}{\Gamma_1\longrightarrow\Phi'\Rightarrow\Phi''}\,(\Rightarrow I)$$

   – if Φ'' is true under ρ, weaken (π'') to a proof (π''_1) of $\Gamma_1, \Phi' \longrightarrow \Phi''$, and produce:

$$(\pi''_1)$$
$$\vdots$$
$$\frac{\Gamma, \Phi' \longrightarrow \Phi''}{\Gamma \longrightarrow \Phi' \Rightarrow \Phi''} \ (\Rightarrow I)$$

- if Φ is a negation $\neg\Phi'$, and by induction hypothesis we have a proof (π') of $\Gamma_1 \longrightarrow \Phi'^\rho$, with $\Gamma_1 = \Gamma, A_1^\rho, \ldots, A_m^\rho$ as above, then:

   – either Φ' is true under ρ, then $\Phi^\rho = \neg\neg\Phi'$, and we append one application of $(\neg\neg I)$ to (π'),

   – or Φ' is false under ρ, and (π') is already a proof of $\Gamma \longrightarrow \Phi^\rho$.

□

    The fact that there are only two truth-values is already captured in the following sense by the deduction system:

Lemma 2.24 *Let Γ be a set of formulas, Φ and Φ' be formulas. If $\Gamma, \Phi' \vdash^{\mathcal{ND}} \Phi$ and $\Gamma, \neg\Phi' \vdash^{\mathcal{ND}} \Phi$, then $\Gamma \vdash^{\mathcal{ND}} \Phi$.*

Proof: By Lemma 2.20, we have proofs (π_1) and (π_2) of $\Gamma, \Phi', \neg\Phi \longrightarrow \Phi$ and of $\Gamma, \neg\Phi', \neg\Phi \longrightarrow \Phi$, and we deduce $\Gamma \longrightarrow \Phi$ by:

□

    As we said in Section 3.1, to prove the Completeness Theorem in the general case, we usually need to prove the Compactness Theorem first. (This is not needed if the set of hypotheses is already assumed to be finite.)

Theorem 2.25 (Compactness) *If $\Gamma \models \Phi$, then there is a finite subset Δ of Γ such that $\Delta \models \Phi$.*

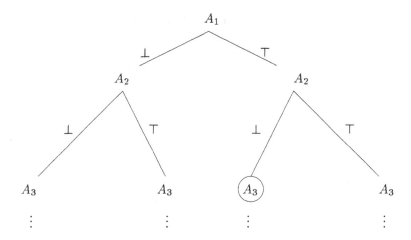

Figure 2.3. A decision tree

Proof: We shall use a topological argument, justifying the fact that this is indeed a compactness property. The interested reader might want to consult Appendix B for a brief introduction to topological concepts. There are other ways of proving this, and we shall see some others (notably using König's Lemma) in later chapters.

Assume that $\Gamma \models \Phi$, and let P be the set of all assignments on the set $\mathcal{X}$ of all propositional variables. P is exactly the product $\mathbb{B}^{\mathcal{X}}$, and $\mathbb{B}$ is compact under the discrete topology. So by Tychonoff's Theorem, P is compact.

With each formula Ψ, associate the set $P(\Psi)$ of all assignments that satisfy Ψ. For all variables A, $P(A)$ is both open and closed (open as it is the set of all ρ such that $\rho(A) = \top$, closed as it is the complement of the set of all ρ such that $\rho(A) = \bot$). By structural induction on Ψ, and from the fact that $P(\Psi_1 \wedge \Psi_2) = P(\Psi_1) \cap P(\Psi_2)$, $P(\Psi_1 \vee \Psi_2) = P(\Psi_1) \cup P(\Psi_2)$, $P(\neg \Psi_1) = P \setminus P(\Psi_1)$, $P(\Psi_1 \Rightarrow \Psi_2) = P \setminus (P(\Psi_1) \cap (P \setminus P(\Psi_2)))$, we infer that $P(\Psi)$ is both open and closed.

As every assignment ρ that satisfies Γ also satisfies Φ, every assignment ρ satisfies some formula in $\{\neg \Psi \mid \Psi \in \Gamma\} \cup \{\Phi\}$, i.e. P is included in the union of the open sets $P(\neg \Psi)$ for $\Psi \in \Gamma$, and $P(\Phi)$. Since P is compact, there is a finite subset Δ of Γ such that P is included in the union of the open sets $P(\neg \Psi)$, $\Psi \in \Delta$, and $P(\Phi)$, i.e. such that $\Delta \models \Phi$. □

Theorem 2.26 (Completeness) *If* $\Gamma \models \Phi$, *then* $\Gamma \vdash^{\mathcal{ND}} \Phi$.

Proof: We shall prove the theorem when Γ is empty first, then when Γ is finite, and we shall generalise to any set Γ.

- Case $\Gamma = \varnothing$:

  Let $A_1, \ldots, A_n$ be the free variables of Φ. We construct the *decision tree* for Φ as follows. (See Figure 2.3.) As in every tree, nodes are in bijection with the paths from the root leading to them, so a node represents a sequence of choices for the truth-values of the propositional variables. For example, the node $A_1, \neg A_2$ represents a state where we have assumed A_1 to be $\top$, and A_2 to be $\bot$, while we have assumed nothing of the other variables. (This is the circled node on Figure 2.3.)

  Formally, the nodes of the tree at depth i (from the root), $0 \leq i \leq n$, are sequences $A_1^\rho, \ldots, A_i^\rho$ for all assignments ρ. If $A_1^\rho, \ldots, A_i^\rho$ is a node, either $i = n$ and this is a *leaf*, or $i < n$, and it has two *sons* $A_1^\rho, \ldots, A_i^\rho$, $\neg A_{i+1}$ and $A_1^\rho, \ldots, A_i^\rho, A_{i+1}$. Alternatively, the two sons are $A_1^{\rho_1}, \ldots,$ $A_{i+1}^{\rho_1}$ and $A_1^{\rho_2}, \ldots, A_{i+1}^{\rho_2}$, where ρ_1 and ρ_2 are two assignments such that $\rho_1(A_j) = \rho_2(A_j) = \rho(A_j)$ for $1 \leq j \leq i$, and $\rho_1(A_{i+1}) = \bot, \rho_2(A_{i+1}) = \top$.

  Assume that $\models \Phi$. We prove by structural induction on the decision tree that, for every node $A_1^\rho, \ldots, A_i^\rho$, we have $A_1^\rho, \ldots, A_i^\rho \vdash^{\mathcal{ND}} \Phi$.

  This is true of the leafs: since Φ is valid, $\rho \models \Phi$, so $\Phi^\rho = \Phi$ by definition. By Lemma 2.23, $A_1^\rho, \ldots, A_i^\rho \vdash^{\mathcal{ND}} \Phi$.

  Let $A_1^\rho, \ldots, A_i^\rho$ be any non-leaf node, $0 \leq i \leq n-1$, with two sons $A_1^\rho, \ldots,$ $A_i^\rho, \neg A_{i+1}$ and $A_1^\rho, \ldots, A_i^\rho, A_{i+1}$, and assume that they both prove Φ, i.e. $A_1^\rho, \ldots, A_i^\rho, \neg A_{i+1} \vdash^{\mathcal{P}} \Phi$ and $A_1^\rho, \ldots, A_i^\rho, A_{i+1} \vdash^{\mathcal{P}} \Phi$. By Lemma 2.24, $A_1^\rho, \ldots, A_i^\rho \vdash^{\mathcal{ND}} \Phi$.

  This completes the proof that every node proves Φ. In particular, the root node (the empty sequence of nodes) proves Φ, i.e. $\vdash^{\mathcal{ND}} \Phi$.

- Case Γ finite:

  If Γ is some finite set $\Phi_1, \ldots, \Phi_m$, $m \in \mathbb{N}$, and $\Gamma \models \Phi$, then $\models \Phi_1 \Rightarrow \ldots \Rightarrow$ $\Phi_m \Rightarrow \Phi$. This is proved by induction on m, moving all Φ_is from the left to the right of $\models$ as in the proof of Theorem 2.16.

  Now, we apply the previous case ($\Gamma = \varnothing$), and infer that $\vdash^{\mathcal{ND}} \Phi_1 \Rightarrow \ldots \Rightarrow$ $\Phi_m \Rightarrow \Phi$. By applying the Deduction Theorem m times, we infer that $\Gamma \vdash^{\mathcal{ND}}$ Φ.

- General case:

  By the Compactness Theorem, if $\Gamma \models \Phi$, then there exists a finite subset Δ of Γ such that $\Delta \models \Phi$. By the previous case, $\Delta \vdash^{\mathcal{ND}} \Phi$, so by Lemma 2.20, $\Gamma \vdash^{\mathcal{ND}} \Phi$.

$\square$

⑤ Although we shall use it in later chapters, the case where Γ is infinite is rarely of great value in computer science. The lesson to be remembered in the proof is the use of decision trees as a means of inductively producing a proof from a statement of validity.

From the statement of completeness, it follows:

Theorem 2.27 (Decidability) *There is an algorithm which, given a finite set Γ of formulas and a formula Φ, decides whether $\Gamma \vdash^{\mathcal{ND}} \Phi$.*

Proof: The algorithm checks whether $\Gamma \models \Phi$, as in the proof of Theorem 2.18. □

⑤ Such an algorithm is said to be sound and complete, but you should not confuse these new notions of soundness and completeness (of procedures) with those for deduction systems! A proof procedure is *sound* provided that, given a formula as argument, if it returns true, then the formula is provable. And it is complete provided that, given any provable formula as argument, the procedure eventually stops and returns true. This distinction may seem like nitpicking, but look at the following example. To make things simpler, assume that formulas are built only from variables, implication and negation (no conjunction, no disjunction). We define a simpler proof procedure, deduced directly from Figure 2.2, as follows: on inputs Γ and Φ, the procedure returns true if $\Phi \in \Gamma$ (use rule (Ax)); otherwise, if Φ is an implication $\Phi_1 \Rightarrow \Phi_2$, then it calls itself recursively with Γ, Φ_1 and Φ_2 as arguments (apply rule $(\Rightarrow I)$ bottom-up); if Φ is a double negation $\neg\neg\Phi_1$, then it calls itself recursively on Γ and Φ_1 (rule $(\neg\neg I)$); otherwise, if $\Phi = \neg\Phi_1$, then it calls itself recursively on Γ, Φ_1 and **F** (rule $(\neg I)$); otherwise, it tries to apply some elimination rule: if Γ contains a formula $\neg\neg\Phi_1$, then replace it by Φ_1 in Γ (rule $(\neg\neg E)$) and start again; if it contains both Φ_1 and $\neg\Phi_1$, then return true (rule $(\neg E)$); if it contains both $\Phi_1 \Rightarrow \Phi_2$ and Φ_1, then the procedure calls itself recursively on Γ, Φ_2 and Φ (rule $(\Rightarrow E)$). Otherwise, the procedure returns false (unprovable). Now, is this procedure complete? It certainly explores a space of proof candidates in a complete system, namely natural deduction. But it does not explore all possible proof candidates, and is therefore incomplete. For example, apply the procedure on $\Gamma = \neg(A \Rightarrow B)$ and $\Phi = B \Rightarrow A$: it will get stuck trying to prove $\neg(A \Rightarrow B), B \longrightarrow A$, although there *is* a proof, for instance:

$$
\cfrac{
\cfrac{}{\neg(B \Rightarrow A), A, \neg B \longrightarrow \neg(B \Rightarrow A)}\ (Ax)
\qquad
\cfrac{
\cfrac{
\cfrac{}{\neg(B \Rightarrow A), A, \neg B, B \longrightarrow A}\ (Ax)
}{\neg(B \Rightarrow A), A, \neg B \longrightarrow B \Rightarrow A}\ (\Rightarrow I)
}{}\ (\neg E)
}{
\cfrac{
\cfrac{
\cfrac{\neg(B \Rightarrow A), A, \neg B \longrightarrow \mathbf{F}}{\neg(B \Rightarrow A), A \longrightarrow \neg\neg B}\ (\neg I)
}{\neg(B \Rightarrow A), A \longrightarrow B}\ (\neg\neg E)
}{\neg(B \Rightarrow A) \longrightarrow A \Rightarrow B}\ (\Rightarrow I)
}
$$

▶ **EXERCISE 2.16**

Show that the $(\neg\neg I)$ rule can be deduced from the other rules, i.e. show that, given a derivation of $\Gamma \longrightarrow \Phi$, we can produce a derivation of $\Gamma \longrightarrow \neg\neg\Phi$ without using $(\neg\neg I)$.

3.3 Gentzen Sequents

Although natural deduction is nicer than Hilbert-style systems for writing proofs, it is still awkward to find whether some proposition has a proof, apart from checking indirectly whether it is valid. If it is indeed valid, then the Completeness Theorem asserts that it has a proof. It even does so in a constructive manner, so that we can actually build a computer program to rebuild a proof from the knowledge that the formula is valid. (Examine the proof in detail, and don't forget that decidability assumes that the context is finite.)

Gerhard Gentzen invented what we now call *sequent systems* to represent proofs in classical logic. These systems use the notion of sequents that we introduced as the proper defining tool for natural deduction, but in a slightly different way. In particular, the only nasty rule from the proof-search perspective will be a special rule called Cut. Fortunately, it turns out that if a formula is provable at all, then it is provable without the Cut-rule: this is the so-called *cut-elimination* theorem, which has many interesting consequences. First, it provides a syntactic proof of consistency. Then, *cut-free proofs*, i.e. proofs where all cuts have been eliminated, span a much narrower space of proofs. This will in particular enable us to prove decidability of deducibility without referring to semantics. Cut-elimination is a fundamental tool of logic, whether pure or applied to computer science, and has more consequences, which we shall explore in chapters to come.

We now define the propositional version of Gentzen's **LK**$_0$ system.

Definition 2.28 *A* Gentzen sequent *is a couple of sets of formulas* Γ, Δ, *written* $\Gamma \longrightarrow \Delta$.

The main difference with natural deduction sequents is that we allow several formulas, or even no formula on the right of the sequents. This makes the calculus more symmetric, in particular, and will allow us to avoid the detours we sometimes had to take to prove certain formulas in natural deduction.

As far as semantics is concerned, $\Gamma \longrightarrow \Delta$ represents the assertion that the conjunction of formulas in Γ entails the *disjunction* of formulas in Δ, i.e. that some formula in Δ follows from all formulas of Γ.

Then **LK**$_0$ is the deduction system defined by the rules of Figure 2.4, where Γ, Γ', Δ and Δ' are sets of formulas and Φ, Φ' are formulas. We find axioms Ax, as in natural deduction systems, then left (L) and right (R) rules for each connective, and

$$\frac{}{\Gamma, \Phi \longrightarrow \Delta, \Phi} \text{ Ax}$$

$$\frac{\Gamma, \Phi, \Phi' \longrightarrow \Delta}{\Gamma, \Phi \wedge \Phi' \longrightarrow \Delta} \wedge\text{L} \qquad \frac{\Gamma \longrightarrow \Delta, \Phi \quad \Gamma \longrightarrow \Delta, \Phi'}{\Gamma \longrightarrow \Delta, \Phi \wedge \Phi'} \wedge\text{R}$$

$$\frac{\Gamma, \Phi \longrightarrow \Delta \quad \Gamma, \Phi' \longrightarrow \Delta}{\Gamma. \Phi \vee \Phi' \longrightarrow \Delta} \vee\text{L} \qquad \frac{\Gamma \longrightarrow \Delta, \Phi, \Phi'}{\Gamma \longrightarrow \Delta, \Phi \vee \Phi'} \vee\text{R}$$

$$\frac{\Gamma \longrightarrow \Phi, \Delta \quad \Gamma, \Phi' \longrightarrow \Delta}{\Gamma, \Phi \Rightarrow \Phi' \longrightarrow \Delta} \Rightarrow\text{L} \qquad \frac{\Gamma, \Phi \longrightarrow \Delta, \Phi'}{\Gamma \longrightarrow \Delta, \Phi \Rightarrow \Phi'} \Rightarrow\text{R}$$

$$\frac{\Gamma \longrightarrow \Delta, \Phi}{\Gamma, \neg\Phi \longrightarrow \Delta} \neg\text{L} \qquad \frac{\Gamma, \Phi \longrightarrow \Delta}{\Gamma \longrightarrow \Delta, \neg\Phi} \neg\text{R}$$

$$\frac{\Gamma \longrightarrow \Delta, \Phi \quad \Gamma', \Phi \longrightarrow \Delta'}{\Gamma, \Gamma' \longrightarrow \Delta, \Delta'} \text{ Cut}$$

Figure 2.4. Gentzen's System **LK**$_0$

the special Cut rule. We can see cuts as being applications of lemmata: from a proof of Φ (in disjunction with Δ) in the context Γ, and from a proof of a lemma asserting that if Φ is true, then Δ' follows from Γ', we get a proof of Δ' (in disjunction with Δ) in the context Γ, Γ'.

We define the intended semantics of sequents:

Definition 2.29 *Let* $\Gamma \longrightarrow \Delta$ *be a Gentzen sequent, and* ρ *be an assignment. We say that* ρ *satisfies* $\Gamma \longrightarrow \Delta$, *and we write* $\rho \models \Gamma \longrightarrow \Delta$ *if and only if either there is a formula* Φ *in* Δ *such that* $\rho \models \Phi$, *or there is a formula* Φ *in* Γ *such that* $\rho \not\models \Phi$.

We say that $\Gamma \longrightarrow \Delta$ *is* provable *in* **LK**$_0$ *if there is a derivation in* **LK**$_0$ *ending in* $\Gamma \longrightarrow \Delta$. *Then, we write* $\vdash^{\text{LK}_0} \Gamma \longrightarrow \Delta$, *or* $\Gamma \vdash^{\text{LK}_0} \Delta$.

We then have the usual theorems:

Theorem 2.30 (Soundness) *If* $\vdash^{\text{LK}_0} \Gamma \longrightarrow \Delta$, *then* $\models \Gamma \longrightarrow \Delta$.

Proof: By structural induction on the proof, checking each rule in turn.   □

Before we prove completeness, we give the analogue of Lemma 2.20:

Lemma 2.31 (Weakening) *Let* $\Gamma \longrightarrow \Delta$ *be a derivable sequent. Then for every* Γ' *such that* $\Gamma \subseteq \Gamma'$, *and every* Δ' *such that* $\Delta \subseteq \Delta'$, $\Gamma' \longrightarrow \Delta'$ *is derivable.*

Proof: By structural induction on the proof of $\Gamma \longrightarrow \Delta$, we consistently add all elements of $\Gamma' \setminus \Gamma$ to the left, and all elements of $\Delta' \setminus \Delta$ to the right of all sequents appearing in the proof.   □

Theorem 2.32 (Completeness) *If* $\models \Gamma \longrightarrow \Delta$, *then* $\vdash^{\mathbf{LK_0}} \Gamma \longrightarrow \Delta$.

Furthermore, if this is the case, we can find a proof in **LK_0** *that does not use the Cut rule. (We say that this is a cut-free proof.)*

Proof: We shall prove this directly. First, assume that Γ and Δ are finite. Let S be the set of sequents $\Gamma' \longrightarrow \Delta'$ such that $\Gamma' \cap \Delta' = \varnothing$ and $\Gamma' \cup \Delta' \not\subseteq \mathcal{X}$, and let f be a function mapping sequents $\Gamma' \longrightarrow \Delta'$ in S to some non-variable formula in $\Gamma' + \Delta'$, where $+$ denotes the direct sum. f will be called the *strategy*.

Build the following tree T whose root is $\Gamma \longrightarrow \Delta$, and whose nodes are sequents. If $\Gamma' \longrightarrow \Delta'$ is not in S, we say that $\Gamma' \longrightarrow \Delta'$ is a *leaf* of the tree. Otherwise, let Φ be $f(\Gamma' \longrightarrow \Delta')$. Then the sons of $\Gamma' \longrightarrow \Delta'$ are the premises of the rule whose conclusion is $\Gamma' \longrightarrow \Delta'$, acting on Φ.

More formally, either $\Phi \in \Gamma'$ or $\Phi \in \Delta'$. If $\Phi \in \Gamma'$, let Γ'' be $\Gamma' \setminus \{\Phi\}$, and let the sons of $\Gamma' \longrightarrow \Delta'$ be $\Gamma'' \longrightarrow \Phi', \Delta'$ if Φ has the form $\neg\Phi'$; $\Gamma'', \Phi', \Phi'' \longrightarrow \Delta'$ if Φ is $\Phi' \wedge \Phi''$; $\Gamma'', \Phi' \longrightarrow \Delta'$ and $\Gamma'', \Phi'' \longrightarrow \Delta'$ if Φ is $\Phi' \vee \Phi''$; $\Gamma'' \longrightarrow \Phi', \Delta'$ and $\Gamma'', \Phi'' \longrightarrow \Delta'$ if Φ is $\Phi' \Rightarrow \Phi''$. On the other hand, if $\Phi \in \Delta'$, let Δ'' be $\Delta' \setminus \{\Phi\}$, and let the sons of $\Gamma' \longrightarrow \Delta'$ be $\Gamma', \Phi' \longrightarrow \Delta''$ if Φ is $\neg\Phi'$; $\Gamma' \longrightarrow \Phi', \Delta''$ and $\Gamma' \longrightarrow \Phi'', \Delta''$ if Φ is $\Phi' \wedge \Phi''$; $\Gamma' \longrightarrow \Phi', \Phi'', \Delta''$ if Φ is $\Phi' \vee \Phi''$; $\Gamma', \Phi' \longrightarrow \Phi'', \Delta''$ if Φ is $\Phi' \Rightarrow \Phi''$. Notice that this tree is almost a cut-free proof in **LK_0**; indeed, the eight conditions defining the sons of non-leaf nodes are respectively $\neg$L, $\wedge$L, $\vee$L, $\Rightarrow$L, $\neg$R, $\wedge$R, $\vee$R, $\Rightarrow$R. (It helps to draw the tree upside-down, with the root at the bottom, to see that it really is similar to a proof in **LK_0**.) To show that T really is a proof, we have to check that it is a *finite* tree, and that its leaves are provable in **LK_0** (by the Ax rule).

We prove that T is a finite tree. Define the *length* of a term as follows: the length of a propositional variable is 1, the length of $\neg\Phi'$ is one plus the length of Φ', the length of $\Phi' \circ \Phi''$ is one plus the length of Φ' plus the length of Φ', for all binary connectives $\circ$. Then, by induction on the sum L of the lengths of the terms in Γ, Δ, we see that the depth of T does not exceed L, and that the number of nodes in it does not exceed 2^L.

We now show that the leaves are instances of the Ax rule. By construction, a non-leaf node of T is valid (as a sequent) if and only if its sons are valid. By structural induction on T, we conclude that the leaves are valid sequents. But if $\Gamma' \longrightarrow \Delta'$ is a leaf, by definition either $\Gamma' \cap \Delta' \neq \varnothing$ (so this is an instance of Ax), or Γ' and Δ' consist of propositional variables only. In the latter case, if we had $\Gamma' \cap \Delta' = \varnothing$, then we could define an assignment ρ by letting all variables in Γ' be $\top$, and all variables in Δ' be $\bot$, which would contradict the fact that $\Gamma' \longrightarrow \Delta'$ is valid. Hence, in both cases, the leaves are provable by the Ax rule. This terminates the proof in the finite case.

In the general case, if $\models \Gamma \longrightarrow \Delta$, then $\Gamma, \neg\Delta \models \mathbf{F}$, where $\neg\Delta$ is the set of negated formulas of Δ and $\mathbf{F}$ is $\Phi' \wedge \neg\Phi'$. By the Compactness Theorem, there is a finite subset Γ' of Γ, and a finite subset Δ' of Δ such that $\Gamma', \neg\Delta' \models \mathbf{F}$, hence such that $\models \Gamma' \longrightarrow \Delta'$. But then $\vdash^{\mathbf{LK_0}} \Gamma' \longrightarrow \Delta'$, so by Lemma 2.31, $\vdash^{\mathbf{LK_0}} \Gamma \longrightarrow \Delta$. (Notice that the construction of Lemma 2.31 does not introduce new cuts, so again the proof we build is cut-free.) $\qquad\square$

This theorem not only states that $\mathbf{LK_0}$ is complete, but that we actually do not need the Cut rule, i.e. if we can prove a proposition, we can also prove it without proving or using any auxiliary lemma. We express it, and extend it in the following way:

Theorem 2.33 (Cut Elimination) *The deduction system* $\mathbf{LK_0}$ *without the Cut rule is sound and complete.*

More precisely, there is a terminating algorithm that takes a proof in $\mathbf{LK_0}$, *and turns it into a cut-free proof of the same sequent.*

The first part of the theorem is a consequence of the Soundness and the Completeness Theorems. We state the second part of the theorem without proof (we shall prove a more general result in Theorem 6.26). This part says that translating a proof into a cut-free proof of the same theorem is actually an effective process (we can program it on a machine). Intuitively, cuts allow us to reuse previously proved lemmas. If, instead of using these lemmas, we replay their proofs, we end up with a proof without auxiliary lemmas, i.e. without the Cut rule. This terminates, because in essence a simple consistent deduction system such as $\mathbf{LK_0}$ cannot contain circular dependencies between lemmas.

A cut-free proof might be much bigger than a proof with cuts. The main interest of cut-elimination in automated deduction, however, is that the search space for cut-free proofs is much smaller than the space of all possible proofs; the latter is infinite, while the former is finite. To be more precise, any choice of a strategy f (see the proof of the Completeness Theorem) yields a terminating proof-search algorithm. The usual encoding of these algorithms are known as tableaux methods. (See Section 4.1.)

We have already stated that we should not confuse the notions of completeness of a deduction system and of a proof search procedure. This is a case where we have both.

Another important property of cut-free proofs that makes them interesting is the following:

Lemma 2.34 (Subformula Property) *In a cut-free proof of* $\Gamma \longrightarrow \Delta$, *all sequents are composed of subformulas of formulas in* Γ, Δ *only.*

Proof: By structural induction on the cut-free proof. We examine the last rule that we used: this is true of Ax; for all other rules (except Cut, which is not used), the set of subformulas of sequents on top of the rules is a subset of the set of subformulas of the sequent on the bottom. (The rules have actually been designed to have this property.) □

This result is important, as it means that we don't have to pull the right sequents from a magic hat to get a proof. It is enough to build sequents from subformulas of the formula to prove.

▶ **EXERCISE 2.17**

Prove that there is an algorithm that transforms every $\mathcal{ND}$ proof of a sequent $\Gamma \longrightarrow \Phi$ into an $\mathbf{LK_0}$ proof of the same sequent.

▶ **EXERCISE 2.18**

Deduce from Exercise 2.17 and from the Completeness Theorem for $\mathcal{ND}$ that $\mathbf{LK_0}$ is complete, without using Theorem 2.32. (You may use the Cut rule.)

4 AUTOMATED PROOF METHODS

Having a better understanding of propositional logic, its properties, and some of its prominent deduction systems, we turn to applying it to find automated proof methods. The question is the following: given a formula Φ, is Φ valid? (Or equivalently, is Φ, or $\longrightarrow \Phi$, provable?)

Traditionally, this problem is presented by first negating Φ, and asking whether $\neg\Phi$ is unsatisfiable. We won't adopt this point of view, as it is somehow unnatural. Its justification is mainly historical.

4.1 Tableaux

The method of tableaux to prove propositional formulas goes back to Beth, Hintikka and Smullyan in the late fifties and the early sixties. This method can be understood semantically, as a means of systematically trying to construct an interpretation that does not satisfy Φ by structural recursion on Φ (this is why the method is sometimes called *semantic tableaux*). It turns out that this way of finding what we shall call *counter-models* of Φ follows exactly the proof of Theorem 2.32. This is to say, tableaux are actually better understood syntactically, as a systematic search for cut-free proofs in $\mathbf{LK_0}$ (with a snag related to structural rules, which we shall come back to in Chapter 3).

We first present the method informally, by following a semantic intuition; then, we shall relate the construction to Gentzen sequents.

Take for example the formula $((A \Rightarrow B) \Rightarrow A) \Rightarrow A$. We want to show that it is valid, i.e. to show that the set of assignments that make it true is the set of all assignments. This formula is an implication, so we know that it is true whenever $(A \Rightarrow B) \Rightarrow A$ is false or A is true. Alternatively, the set of assignments that make

$((A \Rightarrow B) \Rightarrow A) \Rightarrow A$ true is the union of the set of assignments that make $(A \Rightarrow B) \Rightarrow A$ false, and of the set of assignments that make A true.

To represent our intentions of finding the assignments that make a formula true (resp. false), we prefix it with a *sign* + (resp. −). Therefore, we start off with $+((A \Rightarrow B) \Rightarrow A) \Rightarrow A$, and decompose it into $-(A \Rightarrow B) \Rightarrow A$ and $+A$. Graphically, this means we have expanded the initial tableau:

$$+((A \Rightarrow B) \Rightarrow A) \Rightarrow A$$

into:

$$+((A \Rightarrow B) \Rightarrow A) \Rightarrow A \ (*)$$
$$-(A \Rightarrow B) \Rightarrow A$$
$$+A$$

where the union of the sets of assignments corresponding to $-(A \Rightarrow B) \Rightarrow A$ and $+A$ is represented vertically by convention. Notice that we have marked $(*)$ expanded formulae, indicating that they do not need to be expanded again.

Then, $-(A \Rightarrow B) \Rightarrow A$ means that we are interested in the set of assignments that make $(A \Rightarrow B) \Rightarrow A$ false. This is the intersection of the set of assignments that make $A \Rightarrow B$ true and the set of assignments that make A false, i.e. we decompose it into $+(A \Rightarrow B)$ and $-A$. This yields:

$$+((A \Rightarrow B) \Rightarrow A) \Rightarrow A \ (*)$$
$$-(A \Rightarrow B) \Rightarrow A \ (*)$$
$$+A$$
$$+(A \Rightarrow B) \mid \qquad -A$$

where the conjunction of $+(A \Rightarrow B)$ and $-A$ is represented horizontally. Then, we decompose $+(A \Rightarrow B)$, and we get the following fully expanded tableau:

$$+((A \Rightarrow B) \Rightarrow A) \Rightarrow A \ (*)$$
$$-(A \Rightarrow B) \Rightarrow A \ (*)$$
$$+A$$
$$+(A \Rightarrow B) \ (*) \mid \qquad -A$$
$$+B$$
$$-A$$

Roughly speaking, tableaux are trees where unmarked formulas are the nodes, and where vertical bars are used to separate different paths. With each path B, we can associate the set $I(B)$ of assignments that make some positive formula in B true, or that make some negative formula in B false. By construction, the set of assignments that satisfy the original formula is the intersection of the sets $I(B)$ when B ranges over all paths of a tableau for the formula.

The last tableau above has a remarkable property: on all its paths, there is a formula Φ such that both $+\Phi$ and $-\Phi$ occur on the path. We then say that such a path is *closed*. A closed path B has the property that $I(B)$ is the set of all assignments. Hence, if all paths of a tableau for a formula Φ are closed, then Φ is valid. The converse holds: if Φ is valid, it has an expanded tableau where all paths are closed.

The proof of the latter is surprisingly similar to that of the completeness of $\mathbf{LK_0}$. The main reason is that tableaux are actually an implementation of a search for cut-free proofs in $\mathbf{LK_0}$. Indeed, take the example above. To prove it in $\mathbf{LK_0}$, we have to find a cut-free proof of $\longrightarrow ((A \Rightarrow B) \Rightarrow A) \Rightarrow A$:

$$\pi_1$$
$$\vdots$$
$$\longrightarrow ((A \Rightarrow B) \Rightarrow A) \Rightarrow A$$

where π_1 is the unknown. The only non-variable formula in the end sequent is $((A \Rightarrow B) \Rightarrow A) \Rightarrow A$, so to prove it without Cut, we must use $\Rightarrow$R as last proof rule. This translates the problem to finding a cut-free proof of $(A \Rightarrow B) \Rightarrow A \longrightarrow A$:

$$\pi_2$$
$$\vdots$$
$$\frac{(A \Rightarrow B) \Rightarrow A \longrightarrow A}{\longrightarrow ((A \Rightarrow B) \Rightarrow A) \Rightarrow A} \Rightarrow \text{R}$$

where π_2 is now the unknown. Notice that the formula on the left of the sequent $(A \Rightarrow B) \Rightarrow A \longrightarrow A$ is precisely the one with the $-$ sign, and that the formula on the right is the one with the $+$ sign in the second tableau above. The only non-variable formula in this sequent is $(A \Rightarrow B) \Rightarrow A$, so we must use $\Rightarrow$L as the last proof rule to derive $(A \Rightarrow B) \Rightarrow A \longrightarrow A$. This means that we now have to prove both $\longrightarrow A \Rightarrow B, A$ and $A \longrightarrow A$:

$$\frac{\displaystyle \mathop{\vphantom{\frac{A}{A}}}^{\textstyle \pi_3}_{\textstyle \vdots} \quad \frac{}{\longrightarrow A \Rightarrow B, A} \qquad \frac{}{A \longrightarrow A} \text{Ax}}{\dfrac{(A \Rightarrow B) \Rightarrow A \longrightarrow A}{\longrightarrow (A \Rightarrow B) \Rightarrow A \Rightarrow A} \Rightarrow \text{R}} \Rightarrow \text{L}$$

There still remains an unknown π_3, but we immediately recognise that $A \longrightarrow A$ is an instance of Ax. Again, π_3 must end in an application of $\Rightarrow$R, yielding:

$$\frac{\dfrac{\dfrac{\overline{A \longrightarrow B, A}\,\text{Ax}}{\longrightarrow A \Rightarrow B, A}\Rightarrow\text{R} \qquad \dfrac{}{A \longrightarrow A}\,\text{Ax}}{(A \Rightarrow B) \Rightarrow A \longrightarrow A}\Rightarrow\text{L}}{\longrightarrow (A \Rightarrow B) \Rightarrow A \Rightarrow A}\Rightarrow\text{R}$$

Notice how tableaux paths correspond to Gentzen sequents, where unmarked positive formulas are put on the right, and unmarked negative formulas are put on the left. Notice also how the expansion of a tableau actually builds an $\mathbf{LK}_0$ proof by guessing it from the bottom up, and how recognising instances of the Ax rule parallels the closing of tableaux paths. (To help see the correspondence, put the above sequent proofs upside-down, and compare with the tableaux above.)

We now make an important observation, which justifies the invention of *linear logic* (Chapter 3). Let's come back to the way how we solved the proof for the unknown π_2 above. We have claimed that, to get the sequent $(A \Rightarrow B) \Rightarrow A \longrightarrow A$, we had to apply the $\Rightarrow$L rule, with premises $A \longrightarrow A$ and $\longrightarrow A \Rightarrow B, A$. But look at it more closely: in general the premises of $\Rightarrow$L should be $\Gamma, A \longrightarrow A, \Delta$ and $\Gamma \longrightarrow A \Rightarrow B, A, \Delta$, with Γ and Δ such that $\Gamma, (A \Rightarrow B) \Rightarrow A \longrightarrow A, \Delta$ equals the conclusion $(A \Rightarrow B) \Rightarrow A \longrightarrow A$. Taking Γ and Δ to be empty is one solution, but not the only one! Take for instance $\Gamma = \{(A \Rightarrow B) \Rightarrow A\}$ and $\Delta = \{A\}$. Doing so results in the alternative incomplete proof:

$$
\cfrac{
\cfrac{
\begin{array}{c} \pi'_3 \\ \vdots \end{array} \\
(A \Rightarrow B) \Rightarrow A \longrightarrow A \Rightarrow B, A
\qquad
\cfrac{}{(A \Rightarrow B) \Rightarrow A, A \longrightarrow A} \ \text{Ax}
}{
(A \Rightarrow B) \Rightarrow A \longrightarrow A
} \ \Rightarrow\text{L}
}{
\longrightarrow (A \Rightarrow B) \Rightarrow A \Rightarrow A
} \ \Rightarrow\text{R}
$$

In tableau notation, this would mean that we are not forced to cross out the expanded formula $(A \Rightarrow B) \Rightarrow A$. That we can cross all expanded formulas out in classical propositional logic in fact means that we can eliminate the implicit rule of *contraction*:

$$
\cfrac{\Gamma, \Phi, \Phi \longrightarrow \Delta}{\Gamma, \Phi \longrightarrow \Delta}
\qquad\qquad
\cfrac{\Gamma \longrightarrow \Phi, \Phi, \Delta}{\Gamma \longrightarrow \Phi, \Delta}
$$

about which we shall speak more in Chapter 3. (This rule is implicit in our convention that Γ and Δ were sets. To be meaningful, this rule needs a change of conventions, and we have to consider Γ, Δ as multisets, i.e. "sets with multiplicities".) This rule is called *duplication* in tableaux, because in tableaux we look at proofs bottom-up. Contraction — or duplication — can be eliminated from $\mathbf{LK}_0$ proofs, but this shows how particular this system of classical logic can be: we won't be able to eliminate contraction from intuitionistic logic proofs (Chapter 3), from first-order proofs (because of necessary quantifier duplications, see Chapter 6), or in fact from any other system in this book.

Apart from contraction issues, we may be content with defining the tableaux method by the algorithm that attempts to build a Gentzen proof from the bottom up,

(α-rules)		
α	α_1	α_2
$+(X \wedge Y)$	$+X$	$+Y$
$-(X \vee Y)$	$-X$	$-Y$
$-(X \Rightarrow Y)$	$+X$	$-Y$
$+(\neg X)$	$-X$	
$-(\neg X)$	$+X$	

(β-rules)		
β	β_1	β_2
$-(X \wedge Y)$	$-X$	$-Y$
$+(X \vee Y)$	$+X$	$+Y$
$+(X \Rightarrow Y)$	$-X$	$+Y$

Figure 2.5. Tableaux rules

closing the proofs by instances of Ax, and failing when trying to derive a sequent consisting of propositional variables only that is not an instance of Ax. However, people working with tableaux often use a whole new vocabulary, which we now define.

The rules of $\mathbf{LK}_0$ can be classified into two categories (see Figure 2.5). Some of them, like $\wedge$R, have possibly several premises, but none is longer than the conclusion: they yield the so-called α-*rules*, which correspond respectively to $\wedge$R, $\vee$L, $\Rightarrow$L, $\neg$R and $\neg$L. All the others have a unique premise that is longer than their conclusion: these rules correspond to the so-called β-*rules*; they correspond respectively to $\wedge$L, $\vee$R, $\Rightarrow$R. Then tableau methods work on the set of paths through signed formulas:

Definition 2.35 (Paths) *The set of paths in Φ is the smallest set of sets of signed formulas such that:*

- *$\{+\Phi\}$ is a path;*

- *if C is a path, and α is a signed formula in C of type α, then $(C \setminus \{\alpha\}) \cup \{\alpha_1\}$ and $(C \setminus \{\alpha\}) \cup \{\alpha_2\}$ are paths (where α_1 and α_2 are defined as in Figure 2.5; in the negation case, the second path is omitted;)*

- *if C is a path, and β is a signed formula in C of type β, then $(C \setminus \{\beta\}) \cup \{\beta_1, \beta_2\}$ is a path (where β_1 and β_2 are defined as on Figure 2.5.)*

An expansion strategy f is a function from paths C with at least one non-atomic signed formula to such a formula. That is, $f(C)$ is a formula, $f(C) \in C$, and $f(C)$ is not atomic.

A path C is closed if there is a formula $+\Phi'$ in C such that $-\Phi'$ is in C, too. A tableau is a set of paths. It is closed if and only if all its paths are closed.

Paths are sequents, where the negative formulas are on the left of the sequent, and the positive formulas are on the right of the sequent. The expansion rules (α and β) correspond to all eight rules of $\mathbf{LK}_0$ except Ax and Cut. And a closed path is an instance of Ax. Notice that the choice of which formula to choose in a path is arbitrary, i.e. any strategy for choosing such a formula is able to prove all theorems. (See the use of the strategy f in the proof of completeness of $\mathbf{LK}_0$.)

The distinction between the two categories of rules comes from the following fact. α-rules are *ramification* rules, which may increase the number of paths to close, but never increase the number of formulas in paths; whereas β-rules are *prolongation* rules, which do not increase the number of paths to close, but expand the current path.

A classical implementation of the tableaux method is the following. Assume that a list is either the empty list [] or a couple of an object x and a list l, which we write $x :: l$. Let $[x_1, \ldots, x_n]$ be the list consisting of n objects $x_1, \ldots, x_n$, i.e. $x_1 :: \ldots :: x_n :: []$. Define a function prove taking as arguments a list of signed formulas representing the sequent we wish to prove (alternatively, the path we are on, and which we are trying to expand into closed paths) as follows. $\mathsf{prove}(l)$ is:

- if l contains both $+\Phi$ and $-\Phi$ for some Φ, return "yes";

- otherwise, if l contains only signed variables, return "no";

- otherwise, apply the expansion strategy to choose a non-variable signed formula Φ in l, and let l' be l with Φ removed.

  - if Φ is an α-type formula (with n α_i-formulas $\Phi_1, \ldots, \Phi_n$, where $n = 1$ or $n = 2$), then return "yes" if

$$\mathsf{prove}(\Phi_1 :: l'), \ldots, \mathsf{prove}(\Phi_n :: l')$$

  are all true; otherwise return "no".

  - if Φ is a β-type formula (yielding formulas Φ_1, Φ_2), then return $\mathsf{prove}(\Phi_1 :: \Phi_2 :: l')$.

Then, to find out whether Φ is a theorem, we call $\mathsf{prove}([+\Phi])$, and read off the answer.

The expansion strategy is used when we ask to choose a non-variable signed formula Φ from l in the third point above. Notice that this is a slight abuse, as the strategy is now a function from lists, not sets of signed formulas to signed formulas. (This abuse is justified by the fact that contraction can be eliminated.)

The tableaux method has the following important property:

Theorem 2.36 (Soundness, Completeness) *Tableaux are sound and complete for any strategy. That is, for any strategy, the tableaux expansion starting from $+\Phi$ terminates on a closed tableaux if and only if Φ is valid.*

Proof: Follows from Theorem 2.30 and the same arguments as in Theorem 2.32, working on lists rather than sets, for completeness. □

The terms "sound" and "complete" qualify proof search procedures, not deduction systems, here. Recall that a proof search method is sound if it returns "yes" only when its input proposition is a tautology. It is complete if it returns "yes" on all tautologies. Notice how subtle the distinction is: a proof search method is an *algorithm*, whereas a deduction system is a mere collection of rules, without any a priori computational interpretation. More concretely, saying that $\mathbf{LK_0}$ without Cut is complete means that every tautology has some proof, but we don't know in which order we must apply the rules, i.e. which strategy should be used. Saying that `prove` is complete tells us more: it states that the particular order in which `prove` unfolds a derivation will eventually produce a proof if the goal is a tautology; here, this follows from the even more general fact that *any* order in which we apply the proof rules (any strategy) is guaranteed to do so.

> Another possible definition is to say that a proof search method is sound *with respect to a deduction system* if it returns "yes" only when its input is a theorem of the system; and it is complete if it returns "yes" on all theorems. Tableaux, for example, are sound and complete with respect to $\mathbf{LK_0}$, and as $\mathbf{LK_0}$ is sound and complete as a deduction system for propositional logic, then tableaux are sound and complete (as a proof method) for propositional logic. These distinctions are important in incomplete deduction systems: by Gödel's incompleteness theorem, for example, there exist sound and complete proof methods for first-order Peano arithmetic (a deduction system for arithmetic), but not for arithmetic (as a model).

> Another subtlety is the following: a sound and complete proof search method returns "yes" on all tautologies, and only on tautologies. This definition allows us to consider as sound and complete a method that would *not* terminate when submitted non-tautologies. This will be necessary in Chapter 7. Here, we have a form of strong completeness, in that our proof search methods for propositional logic *always* terminate.

Theorem 2.37 (Termination) *Let Φ be a propositional formula. The expansion of a tableau from $+\Phi$ terminates, whatever expansion strategy we choose.*

Proof: The expansion of a tableaux builds a new node of the tree we defined in the proof of Theorem 2.32 at each step. As this tree is finite, the procedure terminates.

□

Again, we have to raise a subtle difference between the *termination* of a proof search method and the *decidability* of validity (resp. theoremhood). The latter only states that there is a decision method for validity (resp. theoremhood) that terminates on all inputs. The former says that a particular decision method (here, tableaux) terminates.

The value of tableaux is that they are simple to implement, and the basic computational steps can be made quite fast. Moreover, its space requirements are modest: we only need a stack to handle recursion in `prove`, and its depth is at most the size of the formula to prove. Another advantage is illustrated in Exercise 2.19. However, as the size of a fully expanded tableau is usually an exponential of the size of the formula to prove, it may take quite a while to prove a theorem by a tableau method. (This problem in fact plagues all proof methods.)

The main weakness of tableaux is, more specifically, that it develops every path of a tableau completely independently of the others. In particular, if we choose to expand the same formula in two different paths, we do exactly the same work twice, thus wasting some precious time. Such seems to be the price to pay for using little memory. (An approach that does quite the opposite in this respect will be presented in Section 4.4.)

▶ **EXERCISE 2.19**

Let Φ be a tautology, and σ be a propositional substitution. Using the function `prove`, show that proving $\Phi\sigma$ does not take more time than proving Φ, under reasonable assumptions.

What happens if Φ is not a tautology?

▶ **EXERCISE 2.20**

Let Φ be a propositional formula, and S be the set of unclosed fully expanded paths produced by a tableau expansion. Show that S represents (in a sense to be made precise) the set of all counter-models to Φ. (A *counter-model* of Φ is an interpretation ρ such that $[\![\Phi]\!]\rho = \mathbf{F}$.)

▶ **EXERCISE 2.21**

Let Ψ_n be the formula $A_n \Leftrightarrow (A_{n-1} \Leftrightarrow \ldots \Leftrightarrow (A_2 \Leftrightarrow A_1)\ldots)$, where A_1, $\ldots$, A_n are distinct variables. Let Φ_n be the formula $A_n \Leftrightarrow (A_{n-1} \Leftrightarrow \ldots \Leftrightarrow (A_1 \Leftrightarrow \Psi_n)\ldots)$. (Recall that $\Phi \Leftrightarrow \Phi'$ abbreviates $(\Phi \Rightarrow \Phi') \wedge (\Phi' \Rightarrow \Phi)$.)

Let the size of a formula be the number of distinct subformulas in it, or equivalently the number of nodes in its tree-like representation (where all identical formulas share the same node). What is the size of Φ_n?

Show that Φ_n is a tautology. (Hint: use a parity argument on assignments satisfying chains of equivalences.)

Convince yourself that any tableau proof of Φ_n requires an exponential amount of time in terms of n.

4.2   *Propositional Resolution*

The resolution method (invented in 1965 by J. Alan Robinson in the more general case of first-order proof search) and the Davis-Putnam method (invented in 1960 by Martin Davis and Hillary Putnam, also to help finding proofs of first-order theorems) both require a preliminary step. To prove Φ (or to refute $\neg\Phi$), we first put $\neg\Phi$ in conjunctive normal form. We shall explain it shortly. One way of explaining propositional resolution is then to interpret the conjunctive normal form for $\neg\Phi$ as a set of sequents, from which we strive to derive the absurd (empty) sequent $\longrightarrow$ with the rules of $\mathbf{LK}_0$.

Definition 2.38 *A formula* Φ *is in* negation normal form *(nnf) if and only if it is built with the connectives* $\wedge$, $\vee$, $\neg$ *only, and if any negated subformula* $\neg\Phi'$ *is such that* Φ' *is a propositional variable. (Negations do not govern composite formulas.)*

An atom *is another name for a propositional variable. A* literal *is a formula of the form* A *or* $\neg A$, *where* A *is an atom.*

A formula Φ *is in* disjunctive normal form *(dnf) if and only if it is in nnf and no disjunction occurs as argument of a conjunction. (That is,* Φ *is an n-ary disjunction of conjunctive clauses, where conjunctive clauses are m-ary conjunctions of literals.)*

Symmetrically, a formula Φ *is in* conjunctive normal form *(cnf) if and only if it is in nnf and no conjunction occurs as argument of a disjunction. (That is,* Φ *is an n-ary conjunction of disjunctive clauses, where disjunctive clauses are m-ary disjunctions of literals.)*

The empty disjunctive clause is written $\Box$.

For short, we shall say *clause* instead of *disjunctive clause*. A clause is then a disjunction of literals. We can interpret clauses as sequents by putting on the right all positive literals A, and putting on the left all atoms A coming from negative literals $\neg A$, and eliminating all duplicates. This establishes a correspondence between clauses and *atomic sequents*, i.e. sequents consisting of atoms only. (And $\Box$ then corresponds to $\longrightarrow$.) From now on, *clause* will always mean sequent, written with $\vee$ and $\neg$ instead of the previous arrow notation. Strictly speaking, this is an abuse of language, all the more so as contractions (replacing $\Phi \vee \Phi$ by Φ) are understood to occur implicitly in clauses. We use this notation because it is traditional, and because it conveys some useful semantic intuitions.

An algorithm for putting a formula Φ in nnf is the following: first rewrite all subformulas $\Phi' \Rightarrow \Phi''$ as $\neg\Phi' \vee \Phi''$, so that we get an equivalent formula using only $\wedge$, $\vee$ and $\neg$. Then use *de Morgan's identities*, namely:

$$\neg(\Phi' \wedge \Phi'') \text{ is logically equivalent to } (\neg\Phi') \vee (\neg\Phi'')$$
$$\neg(\Phi' \vee \Phi'') \text{ is logically equivalent to } (\neg\Phi') \wedge (\neg\Phi'')$$

to push negations inward. More precisely, while there is some subformula of Φ which is a negation of a non-atomic formula, do the following rewriting steps: $\neg\neg\Phi' \longrightarrow \Phi'$, $\neg(\Phi' \wedge \Phi'') \longrightarrow \neg\Phi' \vee \neg\Phi''$, $\neg(\Phi' \vee \Phi'') \longrightarrow \neg\Phi' \wedge \neg\Phi''$.

This process terminates: define the *length* $|\Phi|$ of any formula as its size as a tree, i.e., $|A| = 1$, $|\Phi' \wedge \Phi''| = |\Phi' \vee \Phi''| = |\Phi'| + |\Phi''| + 1$, $|\neg\Phi'| = 1 + |\Phi'|$; let the degree $d(\Phi)$ of any formula Φ be the sum of the lengths $|\Phi'|$ of the formulas Φ' such that $\neg\Phi'$ is (an occurrence of) a subformula of Φ, i.e. $d(\neg\Phi') = |\Phi'| + d(\Phi')$, $d(A) = 0$, $d(\Phi' \wedge \Phi'') = d(\Phi' \vee \Phi'') = d(\Phi') + d(\Phi'')$; $d(\Phi)$ is a non-negative integer that decreases strictly at each rewriting step. Moreover, the process yields a nnf that is logically equivalent to the original formula Φ.

Now, let Φ be in nnf. We can also rewrite it into a cnf by rewriting all subformulas $\Phi' \vee (\Phi_1 \wedge \Phi_2)$ into $(\Phi' \vee \Phi_1) \wedge (\Phi' \vee \Phi_2)$ and $(\Phi_1 \wedge \Phi_2) \vee \Phi'$ into

$(\Phi_1 \vee \Phi') \wedge (\Phi_2 \vee \Phi')$. This terminates again, but the argument is more complicated. One possible proof is the following. Let the and-degree $a(\Psi)$ of a formula
Ψ be defined as follows: let $a(A)$ be 1, $a(\neg A)$ be 1, $a(\Psi_1 \vee \Psi_2) = a(\Psi_1) \times a(\Psi_2)$,
and $a(\Psi_1 \wedge \Psi_2) = a(\Psi_1) + a(\Psi_2)$. Notice that the and-degree of any formula is
left unaffected by the rewriting steps. For every disjunction $\Psi_1 \vee \Psi_2$, let its or-
degree $o(\Psi_1 \vee \Psi_2)$ be $a(\Psi_1)^2 a(\Psi_2)^2$. Let now the degree $d(\Phi)$ of Φ be the sum
of the or-degrees of all occurrences of disjunction subformulas (i.e., $d(\Psi_1 \vee \Psi_2) =$
$o(\Psi_1 \vee \Psi_2) + d(\Psi_1) + d(\Psi_2)$, $d(\Psi_1 \wedge \Psi_2) = d(\Psi_1) + d(\Psi_2)$, $d(A) = d(\neg A) =$
0). The rewriting steps above change one occurrence of a disjunction of or-degree
$a(\Phi')^2(a(\Phi_1) + a(\Phi_2))^2$ into two occurrences of disjunctions, whose sum of or-
degrees is $a(\Phi')^2 a(\Phi_1)^2 + a(\Phi')^2 a(\Phi_2)^2$, which is lower by $2a(\Phi')^2 a(\Phi_1) a(\Phi_2)$.
The latter is at least 2, since all and-degrees are at least one, so the degree must
decrease by at least 2. The degree is also a non-negative integer, so the rewriting
process must terminate, and the end result is a cnf that is logically equivalent to the
original formula.

We then have the following:

Theorem 2.39 (Resolution) *Let Φ be a formula in cnf. We view Φ as a set S of
atomic sequents. Then Φ is unsatisfiable if and only if the empty sequent $\longrightarrow$ can
be deduced from S and the Cut rule only.*

Proof: Soundness (i.e., that if we can deduce $\longrightarrow$ from S and Cut, then Φ is unsatisfiable) is due to the following observation: if $\Gamma \longrightarrow \Phi', \Delta$ and $\Gamma', \Phi' \longrightarrow \Delta'$ are
cut on Φ' to get $\Gamma, \Gamma' \longrightarrow \Delta, \Delta'$, then every interpretation ρ satisfying the first two
must satisfy the latter (if $[\![\Phi']\!]\rho = \top$, then $\rho \models \Gamma' \longrightarrow \Delta'$, hence $\rho \models \Gamma, \Gamma' \longrightarrow$
Δ, Δ'; if $[\![\Phi']\!]\rho = \bot$, then $\rho \models \Gamma \longrightarrow \Delta$, hence $\rho \models \Gamma, \Gamma' \longrightarrow \Delta, \Delta'$ again).
Hence, if Φ were satisfiable, by induction on the length of the derivation of $\longrightarrow$
from S we would prove $\longrightarrow$ satisfiable, which is absurd.

Let's prove completeness, i.e. the converse.

We build a decision tree as in the proof of Theorem 2.26. (See Figure 2.3, page
33.) Let $A_1, \ldots, A_n$ be the free variables of Φ. Recall that the nodes of the tree at
depth i (from the root), $0 \leq i \leq n$, are sequences $A_1^\rho, \ldots, A_i^\rho$ for all assignments ρ.
If $A_1^\rho, \ldots, A_i^\rho$ is a node, either $i = n$ and this is a *leaf*, or $i < n$, and it has two *sons*
$A_1^\rho, \ldots, A_i^\rho, \neg A_{i+1}$ and $A_1^\rho, \ldots, A_i^\rho, A_{i+1}$. We call such a sequence $A_1^\rho, \ldots, A_i^\rho$ a
partial interpretation. If $i = n$, this is a *full interpretation*. (Full interpretations
define interpretations in the intuitive sense, mapping A_i to $\top$ if $A_i^\rho = A_i$, to $\bot$ if
$A_i^\rho = \neg A_i$.)

Assume the set S of sequents unsatisfiable as a conjunction, i.e. for each interpretation ρ, there is a $C \in S$ such that $\rho \not\models C$. For each full interpretation, the
corresponding interpretation ρ fails to satisfy at least one sequent in S. We call
failure node for a sequent C in S any highest node such that all full interpretations
going through this node fail to satisfy C. We then define the *closed decision tree*
for S as the initial decision tree, cut off at failure nodes (i.e. failure nodes become

leaves). Notice that every full interpretation must go through a failure node, so that the leaves of the closed tree are exactly the failure nodes of the initial tree.

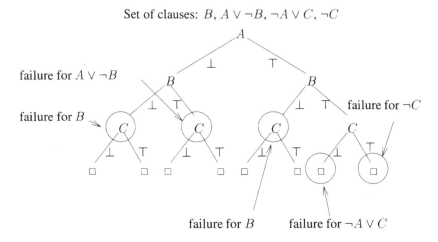

Figure 2.6. A decision tree, with failure nodes

As an example, look at Figure 2.6, where S consists of $\longrightarrow B, B \longrightarrow A, A \longrightarrow C, C \longrightarrow$ (or equivalently, as clauses, $B, A \vee \neg B, \neg A \vee C, \neg C$). Our enumeration of atoms is $A_1 = A$, $A_2 = B$, $A_3 = C$. The leftmost circled node (the partial interpretation $\neg A, \neg B$) is a failure node for the sequent $\longrightarrow B$ (i.e. the clause B) because B is falsified at this node, but not at the node above it. The failure nodes on this example are precisely the circled ones in Figure 2.6, and the closed tree is the subtree that lies at and above these failure nodes.

We now define *inference nodes* as those nodes in the closed tree that have two failure nodes as successors. We claim that if the closed tree has more than one node (i.e., if the root node is not a failure node), then there must be at least one inference node.

Indeed, let n be the number of nodes in the closed tree (including leaves), and prove the claim by induction on n. As the root is not a failure node, we must have $n \geq 3$. If $n = 3$, then the root is an inference node. Otherwise, let $n > 3$ and assume the claim proved for all closed trees with at least 3 nodes, but at most $n - 1$ nodes. Take a closed tree with n nodes, $n > 3$: the root is not a failure node, and one of its sons is not a failure node either; the subtree rooted at this non-failure son therefore has at least 3 nodes, and strictly less than n nodes, so by induction hypothesis it contains an inference node. Therefore the whole closed tree contains an inference node.

By definition, if an inference node N is labelled with a propositional variable A (i.e, A is A_i and N is a partial interpretation $A_1^\rho, \ldots, A_{i-1}^\rho$), then there must exist a sequent in S for which the left son of the node is a failure node (and this sequent must have the form $\Gamma \longrightarrow A, \Delta$), and symmetrically there must exist a sequent in S for which the right son of the node is a failure node (and it must have the form $\Gamma', A \longrightarrow \Delta'$).

For example, on Figure 2.6, the inference nodes are the leftmost B node (with left clause B, right clause $A \vee \neg B$), and the rightmost C node (with left clause $\neg A \vee C$, and right clause $\neg C$).

Then, produce the new clause $\Gamma, \Gamma' \longrightarrow \Delta, \Delta'$, which is deduced by applying Cut on these two sequents, and add it to S to get a new unsatisfiable set S' of sequents. Notice that every failure node for S is below or at the same place as some failure node in S'. Notice also that the inference node N is now a failure node for S', so that the closed tree for S' is strictly smaller than that for S. (In the example, assume we cut between B and $A \vee \neg B$ on the leftmost inference node, this produces a new clause A, and yields the decision tree of Figure 2.7.)

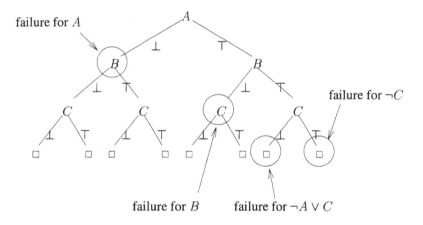

Figure 2.7. A decision tree, after applying Cut once

Therefore, if we take as initial value for S the set of clauses representing Φ, by applying Cut between clauses at inference nodes we generate larger and larger unsatisfiable sets of clauses with smaller and smaller closed trees. After finitely many such operations, the closed tree will be reduced to the root of the decision tree. But, if the root is a failure node, some clause must be satisfied by the empty partial assignment. As all clauses we generate are atomic, such an unsatisfiable clause must be the empty clause. This shows that we have been able to generate the empty clause in finitely many applications of the Cut rule from S.   □

This theorem is a bit surprising: if a formula Φ is provable, then let S be the set of atomic sequents corresponding to a cnf for $\neg\Phi$. By using the Cut rule, we can show that Φ is provable from $\mathbf{LK}_0$ if and only if $\longrightarrow$ is derivable from $\mathbf{LK}_0 + S$ (the deduction system consisting of all rules of $\mathbf{LK}_0$, plus all sequents of S taken as axioms). Then Theorem 2.39 is somehow the converse of Theorem 2.33: instead of eliminating the Cut rule, we eliminate all rules of $\mathbf{LK}_0$ but Cut.

> The solution to this riddle is given by the way we do cut elimination syntactically. This will be explained in Chapter 7, Section 5, but the idea is that we transform proofs with cuts by bubbling up instances of the Cut rule through the proof. In $\mathbf{LK}_0$, all cuts eventually apply to an Ax rule, and a cut between Ax and a proof just vanishes, yielding a weakened version of the latter proof. If we add new axioms from S, there will only remain cuts with sequents in S. But cuts between atomic sequents yield other atomic sequents, so we never need any of the non-Cut rules to derive the special atomic sequent $\longrightarrow$.

> In short, while the semantic proof of Theorem 2.39 is a relatively easy proof of completeness of resolution, the syntactic method of cut elimination (Theorem 6.26) gives us more insight. Cut elimination also gives rise to the functional interpretation of Chapter 3, a fundamental observation relating logics and the semantics of programming languages.

To apply Theorem 2.39 in practice, Robinson originally proposed a slight variant of the following algorithm. Given a set S of clauses, do:

1. Let S_0 be S, and n be 0 (n will be called the *level*, and S_n the clause set at level n).

2. If S_n contains the empty clause, then stop on "yes".

3. Otherwise, compute the set of all *resolvents* of clauses in S_n (a resolvent is the result of a Cut rule on two clauses), such that not both cut clauses are in S_{n-1} if $n \geq 1$; (this would only generate clauses that are already in S_n;) add them to S_n to yield a new set S_{n+1}.

4. If $S_{n+1} = S_n$, then stop on "no".

5. Otherwise, set n to $n + 1$, and go back to step 2.

This algorithm is called resolution by *level saturation search*. This is because S_n is the set of all consequences that we can get from S by applying the Cut rule at most n times.

This algorithm is grossly inefficient. The problem does not lie so much in the comparison between the sets S_n and S_{n+1}, but in the fact that the resolvents that are computed at step 3 may contain much redundancy: lots of tautologies (Ax sequents) and of clauses that were already generated are generated again and again. (See Exercise 2.24.) Actually, resolution is a very poor method for propositional proof search, but it is more interesting in the first-order case. Chapter 7 will be devoted to resolution and its improvements for first-order logic.

▶ **EXERCISE 2.22**

Let Φ be a propositional formula, and S be the set of unclosed fully expanded paths produced by a tableau expansion from Φ, viewed as sequents. Show how we can derive a cnf for Φ in a natural way from S. (Use Exercise 2.20.) Look at Exercise 2.21 and conclude.

▶ **EXERCISE 2.23**

(Short cnfs) Let Φ be a formula. With each non-variable subformula Φ' of Φ, associate a fresh propositional variable $\xi(\Phi')$ as follows (fresh means that $\Phi' \mapsto \xi(\Phi')$ is an injective function). With each non-variable subformula Φ' of Φ, associate the set $S(\Phi')$ of clauses defined as follows:

- if Φ' is a conjunction $\Psi \wedge \Psi'$, generate the clauses $\neg\xi(\Phi') \vee \xi(\Psi)$, $\neg\xi(\Phi') \vee \xi(\Psi')$, and $\neg\xi(\Psi) \vee \neg\xi(\Psi') \vee \xi(\Phi')$;

- if Φ' is a disjunction $\Psi \vee \Psi'$, generate the clauses $\neg\xi(\Psi) \vee \xi(\Phi')$, $\neg\xi(\Psi') \vee \xi(\Phi')$, and $\neg\xi(\Phi') \vee \xi(\Psi) \vee \xi(\Psi')$;

- if Φ' is a negation $\neg\Psi$, generate the clauses $\xi(\Phi') \vee \xi(\Psi)$ and $\neg\xi(\Psi) \vee \neg\xi(\Phi')$;

(We assume that Φ' does not use $\Rightarrow$.) Let S be the union of the sets $S(\Phi')$, plus the clause $\xi(\Phi)$.

Show that Φ is satisfiable if and only if S is satisfiable. Show that the size of S is linear in that of Φ. Finally, show that Φ is not in general equivalent to S.

▶ **EXERCISE 2.24**

Let S be the following set of clauses: $A \vee B$, $\neg A \vee B$, $A \vee \neg B$, $\neg A \vee \neg B$. Apply the level-saturation search strategy to derive the empty clause. What is the least level at which the empty clause $\square$ is derived? How many cut operations have you used? Discuss the redundancy in the generated clauses.

4.3   The Davis–Putnam Method

We now come to semantic methods. These methods are, in practice, among the fastest terminating, sound and complete procedures that we know as of today. Somewhat surprisingly, the Davis-Putnam method is at the root of some contest-winning programs for deciding whether a formula is satisfiable, although it was invented in 1960. (To be fair, what we call the Davis-Putnam method is the Davis-Logemann-Loveland method, an improvement over the previous one due to Martin Davis, George Logemann and Donald Loveland in 1964, where the resolution rule was replaced by the splitting rule—see below.)

The idea of the Davis-Putnam method is the following. We again take as input a set S of clauses representing $\neg\Phi$, where Φ is the formula to prove. Φ is valid if and only if S is unsatisfiable. If S is empty, then it is satisfiable, hence Φ is invalid.

If S contains the empty clause, then it is unsatisfiable, hence Φ is valid. Otherwise, choose a free propositional variable A in S, and basically replace A by true (resp. false) in S, then simplify to get a new set S_1 (resp. S_2) of clauses: then S is satisfiable if and only if S_1 or S_2 is satisfiable.

What makes the Davis-Putnam method efficient is that it also identifies a certain number of special cases where we don't have to split S between S_1 and S_2 on the value of a variable A. Indeed, splitting multiplies the number of subproblems to solve by 2 for each propositional variable, and is therefore responsible for the exponential explosion of the number of subproblems to solve.

The rules are the following:

Definition 2.40 (Tautologies) *A clause is said to be a* tautology *if and only if it contains A and $\neg A$ for some variable A.*

Notice indeed that a clause is valid if and only if it contains both A and $\neg A$, for some variable A.

Definition 2.41 (Splitting) *Let S be a set of non-tautological clauses, and A be a propositional variable.*
We let $S[\mathbf{T}/A]$ be S where all clauses of the form $C \vee A$ (containing A as a positive literal) have been removed, and where all clauses $C \vee \neg A$ have been replaced by C.
We let $S[\mathbf{F}/A]$ be S where all clauses of the form $C \vee A$ have been replaced by C, and where all clauses $C \vee \neg A$ have been removed.

Notice that an assignment ρ such that $\rho(A) = \top$ (resp. $\bot$) satisfies S if and only if it satisfies $S[\mathbf{T}/A]$ (resp. $S[\mathbf{F}/A]$), and that the latter two do not have A as a free variable any longer.

Definition 2.42 (Unit Clause) *A clause C is said to be a* unit clause *if it only consists of one literal.*

Unit clauses are interesting because if S contains a unit clause, say A, then S is satisfiable if and only if A is, i.e. any assignment satisfying S must make A true, so that S is satisfiable if and only if $S[\mathbf{T}/A]$ is. (If the unit clause is of the form $\neg A$, S is satisfiable if and only if $S[\mathbf{F}/A]$ is.) In particular, if S contains a unit clause, we don't need to split on A, as there is only one possible course of action.

Definition 2.43 (Pure Literals) *A literal A (resp. $\neg A$) is called* pure *in a set of clauses S if and only if $\neg A$ (resp. A) does not occur in any of the clauses in S.*
A clause is pure *in S if it contains a literal that is pure in S.*

Assume A (resp. $\neg A$) pure in S. Notice that $S[\mathbf{T}/A]$ (resp. $S[\mathbf{F}/A]$) is then a subset of S, so if S is satisfiable, then $S[\mathbf{T}/A]$ (resp. $S[\mathbf{F}/A]$) is, too. Conversely, if $S[\mathbf{T}/A]$ (resp. $S[\mathbf{F}/A]$) is satisfiable, let ρ be an assignment satisfying all clauses in it, and ρ' be the assignment mapping A to $\top$ (resp. $\bot$) and all other variables B to $\rho(B)$. Then ρ' satisfies all clauses in S; indeed, for each clause in C, either A does not occur in C and C is in $S[\mathbf{T}/A]$ (resp. $S[\mathbf{F}/A]$), so ρ' satisfies C, or C is of the form $C' \vee A$ (resp. $C' \vee \neg A$), and by definition of ρ' on A, ρ satisfies C again.

Another way of putting it is the following: if S contains a pure clause, then S without this pure clause is satisfiable if and only if S is satisfiable. We can therefore safely eliminate all pure clauses from S without altering the satisfiability status.

The Davis-Putnam procedure then proceeds as follows. Initially, S is a finite set of clauses:

1. if S is empty, return "satisfiable".

2. if S contains the empty clause, return "unsatisfiable".

3. if S contains a tautology (as defined in Definition 2.40) or a pure clause, remove it from S, and go to step 1.

4. if S contains a unit clause A (resp. $\neg A$), then replace S by $S[\mathbf{T}/A]$ (resp. $S[\mathbf{F}/A]$), and go to step 1.

5. otherwise, choose a free variable A in S. Apply the Davis-Putnam procedure recursively on $S[\mathbf{T}/A]$ and on $S[\mathbf{F}/A]$. Return "satisfiable" if one of the results was "satisfiable", and return "unsatisfiable" otherwise.

There is still some room for improvement in the last step. First, we improve it by, say, applying the Davis-Putnam procedure on $S[\mathbf{T}/A]$, then immediately returning "satisfiable" if the result is "satisfiable", otherwise applying the Davis-Putnam procedure on $S[\mathbf{F}/A]$. We might also decide to apply Davis-Putnam on $S[\mathbf{F}/A]$ first, then possibly on $S[\mathbf{T}/A]$, or launch both in parallel. . .

We need a strategy to decide which to test first. We also need a strategy to choose the variable A in the last step. The choices of strategy may heavily influence the speed at which the algorithm can conclude. Good heuristics are typically to choose A so that the number of clauses in which A appears times the number of clauses in which $\neg A$ appears is the greatest, (if we have to split, let's at least split on a variable that has the strongest influence on the set of clauses,) and to first test $S[\mathbf{T}/A]$ if there are more clauses containing A than clauses containing $\neg A$, and $S[\mathbf{F}/A]$ otherwise. (If we have to split, first test the set of clauses that has the fewest clauses.) A more sophisticated strategy is the Jeroslow-Wang rule: choose a literal L that maximizes $\sum_i 2^{-n_i}$, where i ranges over clauses containing L and n_i is the length of clause i. See (Hooker and Vinay, 1995) for a discussion.

Another important point is how to implement the Davis-Putnam method efficiently. The traditional implementation does not erase or modify clauses, but looks for a satisfying assignment. Let ρ denote a *partial assignment*, namely a set of literals such that not both A and $\neg A$ are in ρ. (The usual notion of assignment is recovered when, for each A, either A or $\neg A$ is in ρ.) A literal L is satisfied by a partial assignment ρ if $L \in \rho$; a clause C is satisfied by ρ if $C \cap \rho \neq \varnothing$; and a clause set S is satisfied by ρ if all its clauses are satisfied by ρ. The algorithm then tries to find a (partial) assignment ρ satisfying S.

It organizes clauses in a sparse matrix, where rows are clauses and columns are indexed by the propositional variables: if A is a propositional variable, then the column for A is the list of all clauses where A or $\neg A$ appears. In addition, each clause has integer attributes counting the number of literals that are true (satisfied), resp. false (whose negation is satisfied), resp. unassigned under the current partial assignment ρ. To augment ρ with a literal L (A or $\neg A$), as needed in splitting, unit propagation or pure clause elimination, go through the column for A, and update the counters accordingly. If the counter of true literals and the counter of unassigned literals are both false in some clause, then fail and backtrack to the last time that we applied splitting. To deal with unit clauses, maintain a stack of unit literals: each time the number of unassigned literals in a clause becomes one and the number of true literals is still zero, push the only remaining literal on the stack, unless it or its negation is already on the stack. (In the latter case, backtrack.) Pure clauses can be dealt with by maintaining numbers of positive, resp. negative occurrences of variables in clauses, associated with each column.

4.4   Binary Decision Diagrams

Another successful idea for proving propositional formulas that comes from semantic ideas is that of *binary decision diagrams*, or *BDDs*. We might say that they are a recent invention, as the originator of BDDs as we know them today was Randall E. Bryant in 1986. However, BDDs are just decision trees with a few well-known tricks thrown in, and decision trees go back at least to George Boole (1854), if not earlier. The ideas are straightforward, but implementations are usually more complex than with previous methods.

Roughly, the tricks that make BDDs work are the following: first, instead of representing decision trees as trees in memory (where there is a unique node from the root to any node), we represent them as *directed acyclic graphs* or *DAGs*, i.e. we share any two identical subtrees. Second, we use the following simplification rule: if some subtree is such that the two sons of its root are the same, replace it by this son; this is because, in essence, this subtree says "if A is true, then use the right subtree; if A is false, use the left subtree": as the left and the right subtree coincide, there really is no choice on the value of A. Third, we order the variables with respect to a given total ordering $<$, and insist that, if we go down the BDD along any path, we encounter variables in increasing order. This final property will ensure that BDDs are *canonical representations* of formulas modulo logical equivalence, i.e. if Φ and Φ' are two equivalent formulas, then their BDDs with respect to any given ordering is the same.

Essentially, BDDs are either **T** (true), **F** (false) or "if A then Φ_+ else Φ_-", where

Φ_+ and Φ_- are distinct BDDs. We shall represent the latter as $A \longrightarrow \Phi_+; \Phi_-$, or as the tree

This simple representation allows us to build BDDs piecewise. For example, to build the BDD for $\Phi \wedge \Phi'$, it is enough to combine the BDDs of Φ and Φ'. Say that the BDD of Φ is "if A then Φ_+ else Φ_-", and that of Φ' is "if A then Φ'_+ else Φ'_-"; then the BDD of $\Phi \wedge \Phi'$ is just "if A then $\Phi_+ \wedge \Phi'_+$ else $\Phi_- \wedge \Phi'_-$", where the two conjunctions $\Phi_+ \wedge \Phi'_+$ and $\Phi_- \wedge \Phi'_-$ are computed by recursive calls to the BDD conjunction function. The same principle applies equally well to disjunctions, implications, negations, and so on, and is called Shannon's decomposition principle (see below).

It follows that we can compute BDDs of formulas bottom-up: for example, to compute the BDD for $(A \wedge B) \vee C$, we compute the BDD of A, namely $A \longrightarrow \mathbf{T}; \mathbf{F}$, the BDD of B, namely $B \longrightarrow \mathbf{T}; \mathbf{F}$, then we compute their conjunctions, and then the disjunction of the latter with the BDD of C. The fact that BDDs are canonical forms ensure us that the original formula is valid if and only if its BDD is exactly $\mathbf{T}$.

We now proceed to define BDDs more formally. Let $\mathbf{T}$ (truth) and $\mathbf{F}$ (falsehood) be two new symbols. (Not to be confused with the values $\top$ and $\bot$, although this is what they are meant to represent.)

Lemma 2.44 (Shannon) *Define the semantics of $\mathbf{T}$ by $[\![\mathbf{T}]\!] = \top$, and that of $\mathbf{F}$ by $[\![\mathbf{F}]\!] = \bot$.*

For every formula Φ, and every propositional variable A, Φ is equivalent to $(A \Rightarrow \Phi[\mathbf{T}/A]) \wedge (\neg A \Rightarrow \Phi[\mathbf{F}/A])$, where A does not occur free in either $\Phi[\mathbf{T}/A]$ or $\Phi[\mathbf{F}/A]$. This is called Shannon's decomposition principle.

Proof: For any assignment ρ, if $\rho(A) = \top$, then $[\mathbf{T}/A]\rho = \rho$ (see Theorem 2.10 for the notation), so $[\![\Phi]\!]\rho = [\![\Phi]\!]([\mathbf{T}/A]\rho) = [\![\Phi[\mathbf{T}/A]]\!]\rho$ by Theorem 2.10; and if $\rho(A) = \bot$, then $[\![\Phi]\!]\rho = [\![\Phi]\!]([\mathbf{F}/A]\rho) = [\![\Phi[\mathbf{F}/A]]\!]\rho$ by Theorem 2.10 again.

So $\rho \models A$ if and only if, if $\rho \models A$ then $\rho \models \Phi[\mathbf{T}/A]$, and if $\rho \not\models A$ then $\rho \models \Phi[\mathbf{F}/A]$. That is, $\rho \models A$ if and only if $\rho \models (A \Rightarrow \Phi[\mathbf{T}/A]) \wedge (\neg A \Rightarrow \Phi[\mathbf{F}/A])$. As this holds for all ρ, the formulas are equivalent. $\qquad\square$

For example, look at the formula $(((A \Rightarrow B) \wedge C) \Rightarrow A) \vee \neg C$. Then, this is equivalent to $(A \Rightarrow (((\mathbf{T} \Rightarrow B) \wedge C) \Rightarrow \mathbf{T}) \vee \neg C) \wedge (\neg A \Rightarrow (((\mathbf{F} \Rightarrow B) \wedge C) \Rightarrow \mathbf{F}) \vee \neg C)$. Intuitively, we can simplify this formula by using the analogue of truth-tables: for example, $\mathbf{T} \Rightarrow \Phi$ simplifies to Φ, $\Phi \Rightarrow \mathbf{T}$ simplifies to $\mathbf{T}$, $\mathbf{F} \Rightarrow \Phi$ simplifies to $\mathbf{T}$, and so on. In the example, this means that the original formula can be rewritten as $(A \Rightarrow \mathbf{T}) \wedge (\neg A \Rightarrow \neg C)$. We seek to automate this process.

Definition 2.45 (Shannon Graphs) *A* Shannon graph *is a formula built on the two constants* **T**, **F** *with the only ternary connective _ ⟶ _; _ More precisely, the set of Shannon graphs is the smallest set such that* **T** *and* **F** *are Shannon graphs, and such that if A is a variable, and* Φ_+, Φ_- *are two Shannon graphs, then* $A \longrightarrow \Phi_+ ; \Phi_-$ *("if A then* Φ_+*, else* Φ_- *") is a Shannon graph.*

$A \longrightarrow \Phi_+ ; \Phi_-$ *will also be noted graphically as*

(Notice that the "A true" branch is on the right, while the "A false" branch is on the left, thus reversing the order of Φ_+ *and* Φ_-*.)*

The usual definition of BDDs proceeds by first defining Shannon trees. To be consistent with our definition of formulas (Definition 2.1), we prefer to define Shannon graphs. The difference, again, is that any two identical subtrees are not distinguished in the graph, whereas they could be distinguished by their occurrences in the tree. Although the difference between trees and graphs was of small significance until now, it will be most important with BDDs, as this is this coalescing of identical subtrees located at different occurrences—what we call *sharing* of identical subBDDs—which makes BDDs efficient.

We need to define the semantics. We have already defined $[\![\mathbf{T}]\!]\rho$ as $\top$, $[\![\mathbf{F}]\!]\rho$ as $\bot$. Then, $A \longrightarrow \Phi_+ ; \Phi_-$ is meant to represent the formula $(A \Rightarrow \Phi_+) \wedge (\neg A \Rightarrow \Phi_-)$ of Shannon's decomposition principle, so we let $[\![A \longrightarrow \Phi_+ ; \Phi_-]\!]\rho$ be $[\![\Phi_+]\!]\rho$ if $\rho(A) = \top$, and $[\![\Phi_-]\!]\rho$ if $\rho(A) = \bot$.

Then, we can represent formulas built with $\wedge$, $\vee$, $\neg$, $\Rightarrow$, etc. by logically equivalent Shannon graphs. For example, the formula $(((A \Rightarrow B) \wedge C) \Rightarrow A) \vee \neg C$ can be represented as the following Shannon graph:

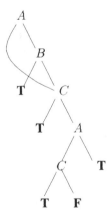

where we have represented multiple **T** nodes, instead of showing a single one, so as to avoid unnecessary clutter. (Check that this is indeed equivalent to the original formula, by enumerating all possible interpretations on A, B, C.) This is not the only representation of the formula as a Shannon graph, however.

Shannon graphs (or trees) have the following important property, which makes computing logical operations easy:

Theorem 2.46 (Orthogonality) *Let* Φ *be* $A \longrightarrow \Phi_+; \Phi_-$, Φ' *be* $A \longrightarrow \Phi'_+;$ Φ'_-.

The negation of Φ, *which we write* $\neg\Phi$, *is* $A \longrightarrow \neg\Phi_+; \neg\Phi_-$. *More formally, the latter graph is satisfied exactly by those interpretations that don't satisfy* Φ.

For any binary connective $\circ$ ($\wedge$, $\vee$, $\Rightarrow$, $\Leftrightarrow$, *respectively*), $\Phi \circ \Phi' = A \longrightarrow$ $(\Phi_+ \circ \Phi'_+); (\Phi_- \circ \Phi'_-)$; *more formally, the latter is satisfied exactly by those interpretations that satisfy both* Φ *and* Φ', *resp.* Φ *or* Φ', *resp.* Φ' *or* $\neg\Phi$, *resp. both* Φ *and* Φ' *or neither* Φ *nor* Φ'.

Proof: Immediate from the definition of the semantics of the $\_ \longrightarrow \_; \_$ connective.
□

We can do a bit better than Shannon graphs, and define BDDs:

Definition 2.47 (BDDs) *Let* Φ *be a Shannon graph, and* $<$ *be a strict total ordering of the variables in* $\mathrm{fv}(\Phi)$.

We say that Φ *is a* binary decision diagram, *or BDD, ordered by* $<$ *if and only if:*

- *(Reduced)* Φ *contains no subgraph of the form* $A \longrightarrow \Phi'; \Phi'$ *(i.e., with two identical sons),*

- *(Ordered) all subgraphs of* Φ *of the form* $A \longrightarrow \Phi_+; \Phi_-$ *are such that* A *is strictly less (by* $<$) *than all the variables in* $\mathrm{fv}(\Phi_+) \cup \mathrm{fv}(\Phi_-)$.

We can see a BDD Φ as an automaton for deciding whether a given interpretation ρ satisfies Φ. Indeed, start from the root of Φ and go down the BDD in the following way: when we are at a node labelled A (i.e., we are looking at a subgraph $A \longrightarrow \Phi_+; \Phi_-$), take the right branch (Φ_+ in the example) if $\rho(A) = \top$, and the left branch (Φ_-) if $\rho(A) = \bot$. This process terminates when we get to a leaf: if this leaf is **T**, then ρ satisfies Φ, whereas if it is **F**, ρ does not satisfy Φ. All this holds because, basically, BDDs are compact representations of decision trees.

BDDs are then canonical forms for formulas modulo logical equivalence (which Shannon graphs weren't):

Theorem 2.48 (Canonicity) *Let* Φ *and* Φ' *be two BDDs built on the same ordering* $<$.

Then Φ *and* Φ' *are logically equivalent if and only if they are equal.*

Proof: If they are equal, their logical equivalence is clear.

Conversely, assume Φ and Φ' logically equivalent. We show that they are equal by induction on the cardinality n of $\mathrm{fv}(\Phi) \cup \mathrm{fv}(\Phi')$.

If $n = 0$, then Φ and Φ' are in $\{\mathbf{T}, \mathbf{F}\}$; since Φ and Φ' are logically equivalent, they are either both $\mathbf{T}$ or both $\mathbf{F}$.

If $n \geq 1$, let A be the smallest variable in $\mathrm{fv}(\Phi) \cup \mathrm{fv}(\Phi')$ with respect to the $<$ ordering (i.e., the highest variable). Without loss of generality, assume that A is free in Φ. Because A is the smallest variable in $\mathrm{fv}(\Phi)$, we must have $\Phi = A \longrightarrow \Phi_+; \Phi_-$ for some Φ_+ and Φ_-.

If A is not free in Φ', then Φ' is logically equivalent to Φ_+ and Φ_-. Indeed, if $\rho \models \Phi'$, then $\rho' \models \Phi'$, where ρ' maps A to $\top$ and all other variables B to $\rho(B)$; since Φ' is logically equivalent to Φ, $\rho' \models \Phi$; since $\rho'(A) = \top$, $\rho' \models \Phi_+$. Conversely, if $\rho \models \Phi_+$, then $\rho' \models \Phi_+$ where ρ' is as above, so $\rho' \models \Phi$, and by logical equivalence $\rho' \models \Phi'$; since A is not free in Φ', $\rho \models \Phi'$. So Φ' is logically equivalent to Φ_+; and symmetrically, Φ' is logically equivalent to Φ_-. By induction hypothesis (the cardinalities of $\mathrm{fv}(\Phi_+) \cup \mathrm{fv}(\Phi')$ and of $\mathrm{fv}(\Phi_-) \cup \mathrm{fv}(\Phi')$ are indeed strictly less than n), Φ' equals both Φ_+ and Φ_-. But this is impossible, because Φ is reduced ($\Phi_+ \neq \Phi_-$).

So A is also free in Φ', and in fact $\Phi' = A \longrightarrow \Phi'_+; \Phi'_-$ for some Φ'_+ and Φ'_-. By similar arguments as above, Φ_+ is logically equivalent to Φ'_+ and Φ_- to Φ'_-, so by induction hypothesis $\Phi_+ = \Phi'_+$ and $\Phi_- = \Phi'_-$. Therefore $\Phi = \Phi'$. $\qquad\square$

Moreover, BDDs for formulas are usually compact. If a formula is valid or unsatisfiable, its BDD is $\mathbf{T}$ or $\mathbf{F}$, which is indeed quite compact. (Observe however that the BDDs of its subformulas, which we build while we compute the BDD of the whole formula, may be much larger.) But even if it is invalid and satisfiable, BDDs usually remain small. There are exceptions, and in general BDDs can get as large as an exponential of the number of free variables in them. (To be precise, if n is this number, the number of nodes in a BDD never exceeds $2^n/n$.)

On the example of the formula $(((A \Rightarrow B) \wedge C) \Rightarrow A) \vee \neg C$, if we choose the ordering $A < B < C$, then we can build its BDD in the following top-down manner: use Shannon's decomposition principle on variable A to see that it must have the form:

where Φ_1 is the BDD for $(((\mathbf{F} \Rightarrow B) \wedge C) \Rightarrow \mathbf{F}) \vee \neg C$, and Φ_2 is the BDD for $(((\mathbf{T} \Rightarrow B) \wedge C) \Rightarrow \mathbf{T}) \vee \neg C$. Use Shannon's decomposition principle on variable B, on both Φ_1 and Φ_2, to get:

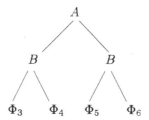

where Φ_3 through Φ_6 are BDDs for $(((\mathbf{F} \Rightarrow \mathbf{F}) \wedge C) \Rightarrow \mathbf{F}) \vee \neg C$, $(((\mathbf{F} \Rightarrow \mathbf{T}) \wedge C) \Rightarrow \mathbf{F}) \vee \neg C$, $(((\mathbf{T} \Rightarrow \mathbf{F}) \wedge C) \Rightarrow \mathbf{T}) \vee \neg C$ and $(((\mathbf{T} \Rightarrow \mathbf{T}) \wedge C) \Rightarrow \mathbf{T}) \vee \neg C$ respectively. Applying Shannon's decomposition principle again, and evaluating the now closed formulas at the leaves, we get:

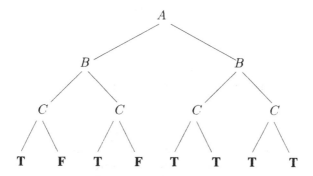

or equivalently, to make sharing more apparent:

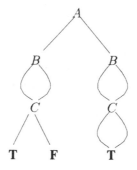

This is an ordered Shannon graph, by construction, but it is not reduced: to make it reduced, we have to replace the nodes that have two identical sons by their common son. This yields the final BDD:

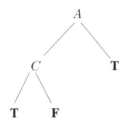

Notice that from this BDD, we can immediately conclude that the original formula is neither valid nor unsatisfiable (because this BDD is not reduced to **T** or **F**). We can also conclude that the original formula did not in fact depend on B, as B does not appear in the final BDD.

This top-down method for building BDDs is too naive to be useful. Indeed, the input formula contains n variables, we already need to build the full tree with 2^n nodes before we reduce it. Although the final BDD may be quite small, the tree might be too big to be built in reasonable time with a reasonable amount of memory. We therefore proceed to show how BDDs are really built in computer implementations, by a bottom-up method.

We first implement a function that allocates and shares triples $A \longrightarrow \Phi'; \Phi''$ as records in memory with three fields for A, Φ' and Φ'' respectively. To do this, we use a global hash-table (see (Knuth, 1973)) that memorises all previously built triples. Hash-tables are such that, given A, Φ', Φ'', either a $A \longrightarrow \Phi'; \Phi''$ triple is in the table and we return its address rapidly, or we quickly report its absence. A simple hash-table algorithm that works well in practice is the following: fix a large enough integer N (it is better to take N prime), and allocate an array with N entries (called *slots*), initially empty. These slots will contain linked lists of previously built triples.

To find whether $A \longrightarrow \Phi'; \Phi''$ is in the table, we first compute a *hash-function* $h(A, \Phi', \Phi'')$ on A, Φ', Φ'', which must be an integer between 0 and $N-1$ (i.e., an index in the table). Because all identical BDDs are shared, they are described in a unique way by their address in memory: a good choice for $h(A, \Phi', \Phi'')$ is the sum of the addresses of A, Φ' and Φ'', modulo N. The invariant of the table is that if $A \longrightarrow \Phi'; \Phi''$ is in the table, it is in the list at slot i in the hash-table, where $i = h(A, \Phi', \Phi'')$.

To find whether $A \longrightarrow \Phi'; \Phi''$ is already in the table, then, compute $i = h(A, \Phi', \Phi'')$; go through the list in slot i, and compare each element with the record having A, Φ' and Φ'' as fields. If we come to the expected triple, we return its address. Otherwise, we return on failure (the triple is not in the table). Symmetrically, to allocate and share $A \longrightarrow \Phi'; \Phi''$, look for it in the table, and return it if we found it; otherwise, allocate a new record containing A, Φ', Φ'', tack it in front of the list at slot i, and return its address.

The expected time to find or build and share a triple $A \longrightarrow \Phi'; \Phi''$ is of the order of n/N, where n is the number of previously allocated triples. If n is not too much larger than N, finding or sharing a triple is almost instantaneous.

To build BDDs, we first define the function BDDmake as follows: if $\Phi' = \Phi''$ (note that we can compare BDDs by address instead of recursively, by content, as they are shared), then BDDmake(A, Φ', Φ'') returns Φ'; otherwise, BDDmake(A, Φ', Φ'') allocates and shares the triple $A \longrightarrow \Phi'; \Phi''$ and returns its address. We have the following:

Lemma 2.49 *Let Φ' and Φ'' be two BDDs, and A a variable less than all variables in Φ' or Φ''. Then* BDDmake(A, Φ', Φ'') *returns a BDD equivalent to* $(A \Rightarrow \Phi') \wedge (\neg A \Rightarrow \Phi'')$.

Proof: The equivalence is trivial. Now, let Φ be the result returned by BDDmake(A, Φ', Φ''). We have to prove that it is a BDD. It is ordered by construction. And it is reduced: if $\Phi' = \Phi''$, Φ is exactly Φ', where no node has two identical sons; if $\Phi' \neq \Phi''$, then the nodes of Φ are either its root (which has two different sons by assumption), or nodes of Φ' or of Φ'', which always have distinct sons. $\square$

Then we can adapt the orthogonality principle for Shannon graphs to provide Boolean functions on BDDs. The negation BDDneg(Φ) of a BDD Φ is defined by structural recursion as follows:

$$\begin{aligned}
\text{BDDneg}(\mathbf{T}) &= \mathbf{F} \\
\text{BDDneg}(\mathbf{F}) &= \mathbf{T} \\
\text{BDDneg}(A \longrightarrow \Phi_+; \Phi_-) &= \text{BDDmake}(A, \text{BDDneg}(\Phi_+), \text{BDDneg}(\Phi_-))
\end{aligned}$$

If $\circ$ is any binary operator, we define the corresponding operation as follows. We take the example of disjunction, which we realize as a function BDDor:

- BDDor($\mathbf{T}, \Phi$) = $\mathbf{T}$, BDDor($\mathbf{F}, \Phi$) = Φ;

- BDDor($\Phi, \mathbf{T}$) = $\mathbf{T}$, BDDor($\Phi, \mathbf{F}$) = Φ;

- if $\Phi = A \longrightarrow \Phi_+; \Phi_-$, and $\Phi' = A' \longrightarrow \Phi'_+; \Phi'_-$, then:

  - if $A < A'$, then
    BDDor(Φ, Φ') = BDDmake(A, BDDor(Φ_+, Φ'), BDDor(Φ_-, Φ'));
  - if $A > A'$, then
    BDDor(Φ, Φ') = BDDmake(A', BDDor(Φ, Φ'_+), BDDor(Φ, Φ'_-));
  - if $A = A'$, then
    BDDor(Φ, Φ') = BDDmake(A, BDDor(Φ_+, Φ'_+), BDDor(Φ_-, Φ'_-)).

The problem, if we implement these functions naively (again), is that they run in time roughly proportional to the number of *paths* in the resulting BDD, not to the number of *nodes*. This is important, as because of sharing, BDDs may have many fewer nodes than they have paths (this is astronomical, see Exercise 2.27).

Therefore, we implement BDDneg and BDDor by *memo-functions*: a memo-function is a function f associated with a hash-table T that records all previously computed results of f. To be more precise, T maps arguments of f to the results of f on these arguments. A memo-function first consults the hash-table T, and looks for an entry matching its arguments. If it finds one, it returns the recorded result. Otherwise, it calls the regular function f, stores the couple (argument, result) in the hash-table and returns the result. We now assume that BDDneg and BDDor have been modified to become memo-functions.

Take again the example $(((A \Rightarrow B) \wedge C) \Rightarrow A) \vee \neg C$. The BDDs for A, B and C are:

We then build the BDD for $A \Rightarrow B$ by negating that for A (this produces a new BDD with the **T** leaves exchanged with the **F** leaves), and taking its disjunction with that for B, yielding:

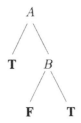

Then, we take its conjunction with the BDD for C (this is similar in operation to BDDor), so as to get:

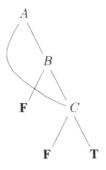

Then, we compute the implication between the latter and A, and get :

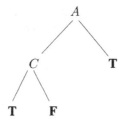

Finally, we take the disjunction with the BDD for $\neg C$, and get the same BDD, which is therefore the final answer.

▶ **EXERCISE 2.25**

From an efficiency perspective, discuss the pros and cons of implementing functions for disjunction and conjunction of BDDs separately, vs. implementing conjunction in terms of BDDor and BDDneg.

▶ **EXERCISE 2.26**

(Dual of Shannon's principle) Show that for every formula Φ, and every propositional variable A, Φ is equivalent to $(A \wedge \Phi[\mathbf{T}/A]) \vee (\neg A \wedge \Phi[\mathbf{F}/A])$.

▶ **EXERCISE 2.27**

Let Ψ_n be the formula of Exercise 2.21, i.e. $A_n \Leftrightarrow (A_{n-1} \Leftrightarrow \ldots (A_1 \Leftrightarrow A_1) \ldots)$. Draw its BDD for the ordering $A_n < A_{n-1} < \ldots < A_1$. How many nodes are there? How much time do we need to build it? How many paths are there in it?

▶ **EXERCISE 2.28**

(TDGs, due to Jean-Paul Billon, 1987) Typed Decision Graphs (TDGs) are a variant of BDDs, where pointers to nodes are decorated with a sign bit, $+$ or $-$: a TDG $+\Phi$ has $-\Phi$ as negation. Formally, TDG *vertices* are either the special leaf $\mathbf{T}$ or triples $A \longrightarrow \Phi_+ ; \Phi_-$, where Φ_+ and Φ_- are TDGs, and *TDGs* are couples (s, v) where $s \in \{+, -\}$ is a sign, and v is a TDG vertex (we also write $+v$ and $-v$ respectively). We also impose that TDGs are reduced, ordered and *normalised* in the following sense: for every TDG vertex $A \longrightarrow \Phi_+ ; \Phi_-$, the sign of Φ_+ is $+$.

Show how to convert a TDG to a BDD and conversely. Conclude that TDGs are canonical forms for formulas modulo logical equivalence.

Show that logical operations on TDGs are at least as fast as on BDDs, while negation can be done in constant time. What can you say on the question of Exercise 2.25 if we use TDGs?

(The aim of this exercise is to introduce this optimised version of BDDs. It is customary to say "BDD" when we really mean "TDG", as far as implementations are concerned.)

5   DIGRESSIONS

The following sections are somehow technical, and are best understood with some knowledge of complexity theory or of lattice theory. The uninclined reader might want to skip them on first reading.

5.1   *Expressiveness of Propositional Classical Logic*

After all we have said about validity and the art of proving propositional formulas, we discuss the expressiveness of propositional classical logic.

First, propositional logic is a formalism which we can implement in hardware: we can build circuits by connecting gates, each gate implementing a logical connective. (For example, a NAND gate is a small piece of circuit with two inputs and one output; in positive implementations of logic, when the voltage of both inputs is at the low level, representing $\bot$, the output is high, and if some input is at the high voltage, the output is low.) It cannot implement circuits with looping connections (as is customarily used in hardware to create unstable devices, like clocks), but almost all circuits in a contemporary computer are actually superpositions of logic gates. To simplify, a real-world computer is a big propositional formula implemented in hardware, plus a clock (which sends sequences of $\top$ and $\bot$ at a predefined rate) and interface components. Therefore, propositional logic is mostly expressive enough to describe any real-world computer.

The previous argument is a bit misleading, not only because the propositional formula representing a computer is tremendously large, but also because real-world computers are an approximation of what we would like a computer to be. The main defect in real computers is that an ideal computer should have an infinite amount of memory, and indeed virtual memory and secondary storage devices are all attempts to make computers closer to this ideal.

There are several mathematical models of ideal computers, and clearly the simpler it is, the easier it will be to work with. One of the simplest model for an ideal computer is the *Turing machine*. In a few words, a Turing machine is a finite-state machine (an automaton, with finitely many states and transitions between these states), with a read/write head scanning an infinitely long tape. The tape is a linear sequence of symbols taken from an alphabet (having at least two distinct letters), extending infinitely far in both directions. The Turing machine executes by reading the letter under its read/write head, uses it to decide of a transition to follow from its current state to a new state, of a letter to write in place of the letter read, and of a direction of movement of the head (one slot forward, or one slot backward). It does so until it reaches a specially marked *final state*. Turing machines represent partial functions from words to words in the following way: write the word on the tape, enclosed between special delimiting letters, the first delimiter being right under the head; then execute the Turing machine, and if it terminates, read the output from the tape.

Turing machines are universal, in that the set of functions that they compute is exactly the set of all computable functions. In practice, we are more interested in what functions we can compute on an idealised computer in some realistic amount of time, i.e. what functions are *tractable*. Research has shown that a good enough criterion for "realistic amount of time" was "time bounded by some low-degree polynomial of the size of the input". In particular, if the least upper bound we need to solve a problem is an exponential of the size of the input, we consider the problem as being intractable. A fundamental result is that a problem is tractable on a Turing machine if and only if it is tractable on most other idealisations, including our ideal real-world-computer-with-infinitely-large-memory. Studying the tractability or intractability of problems is the topic of abstract complexity theory (Garey and Johnson, 1979).

The expressiveness of propositional logic is then captured by the following theorem, due to Stephen Cook in 1971. We assume our Turing machines have two final states called "yes" and "no", and we will submit it questions such as "is the formula Φ satisfiable?", and read the answer in the state that the machine eventually reaches when it terminates. Moreover, Turing machines might progress either deterministically (the same choice of state and letter under the head yields the same next state, letter to write and direction of movement) or non-deterministically (if there may be several possible choices). A deterministic Turing machine accepts on some input if it terminates in the "yes" state; a non-deterministic Turing machine accepts on some input if there is an execution path that terminates in the "yes" state. Notice that we define "decision problem" as meaning "Boolean-valued function":

Theorem 2.50 (Cook) *Let SAT be the following decision problem:*
INSTANCE: a propositional formula Φ,
QUESTION: is Φ satisfiable?
or, equivalently, this is the function mapping Φ to its satisfiability status.

Let NP be the class of decision problems f such that $f(x)$ is true if and only if there is a non-deterministic Turing machine that accepts on input x in time polynomial in the size of x. (For short, NP is the class of problems solvable "in non-deterministic polynomial time".)

Then SAT is NP-complete, i.e.:

- *SAT is in NP,*

- *and for every problem f in NP, there is a deterministic polynomial-time algorithm that translates the input x of f into an input Φ for SAT, such that Φ is satisfiable if and only if $f(x)$ is true.*

This theorem can be read in two ways. The first way is the following. NP is (probably; this has not yet been proved or disproved) much larger than the class P of tractable problems. In particular, NP contains a lot of problems that we cannot, as of today, solve in less than exponential time. Being NP-hard (the second condition in the definition above of NP-completeness) means that SAT is at least as hard as all those problems.

The second reading is the following. Given the representation of a non-determ-inistic Turing machine M with pre-coded input on its tape (i.e., a non-deterministic program with its input data), we want to find out whether M accepts. We might execute M to find it out, but the proof of Cook's theorem actually builds a propositional formula Φ that is satisfiable if and only if M accepts. In short, the satisfiability problem for propositional logic has exactly the expressiveness of polynomial-time non-deterministic Turing machines.

This allows us to quantify the expressiveness of propositional logic more precisely: NP is quite probably more expressive than P (i.e., we can express more problems than just the tractable ones in propositional logic); but on the other hand, NP is also quite probably much smaller than DEXPTIME, the class of all problems solvable in time an exponential of some polynomial of the size of the input on a deterministic Turing machine (so, there are a lot of intractable problems, even working in exponential time like all the methods we know for SAT, that we probably cannot express as the satisfiability of a propositional formula).

"NP-complete" is yet another phrase where "complete" occurs. Here, "complete" just denotes some condition that we could roughly describe as follows: a complete problem for a complexity class (NP, here) is among the most complicated problems of this class. This has really nothing to do with notions of completeness of deduction systems or of proof-search procedures.

5.2   Boolean Algebras

Instead of using deduction systems to find out whether a propositional formula is valid (or satisfiable), it is sometimes practical to use an equational theory for logical equivalence, and try to derive **T** by replacing equals by equals in this theory.

For example, consider the formula $((A \Rightarrow B) \Rightarrow A) \Rightarrow A$. Rewrite $A \Rightarrow B$ into $\neg A \vee B$, then $(\neg A \vee B) \Rightarrow A$ into $\neg(\neg A \vee B) \vee A$. Then, by distributing negation over disjunction, get $(\neg\neg A \wedge \neg B) \vee A$. The double negation can be cancelled, yielding $(A \wedge \neg B) \vee A$. Now, the negation of the latter can be rewritten to $(\neg A \vee B) \wedge \neg A$, so $((A \Rightarrow B) \Rightarrow A) \Rightarrow A$ can be rewritten to $((\neg A \vee B) \wedge \neg A) \vee A$ by the same techniques. Distributing the inner conjunction over disjunction, we get $(\neg A \vee (B \wedge \neg A)) \vee A$, then by associativity and commutativity of $\vee$, $A \vee \neg A \vee (B \wedge \neg A)$. But $A \vee \neg A$ is valid, so the whole formula is.

The laws we used in the example are basically those of a *Boolean algebra*. We first introduce Boolean lattices:

Definition 2.51 *A lattice is a non-empty set L, with a partial order $\leq$, such that any finite subset S of L has a least upper bound $\bigvee S$ and a greatest lower bound $\bigwedge S$.*

We write $\top = \bigwedge \varnothing$ *(the greatest element of L),* $\bot = \bigvee \varnothing$ *(the least element of L),* $a \wedge b$ *for* $\bigwedge\{a, b\}$, $a \vee b$ *for* $\bigvee\{a, b\}$.

We say that L is a distributive lattice *if and only if* $\wedge$ *distributes over* $\vee$, *and* $\vee$ *distributes over* $\wedge$, *i.e.* $a \vee (b \wedge c) = (a \vee b) \wedge (a \vee c)$ *and* $a \wedge (b \vee c) = (a \wedge b) \wedge (a \wedge c)$.

We say that a distributive lattice L is a Boolean lattice *if there is a* complementation operation $\neg$ *such that* $a \vee \neg a = \top$, $a \wedge \neg a = \bot$, *and* $\neg\neg a = a$.

Johnstone (1992, Page 7), nicely captures the very heart of Boolean algebras when saying that, loosely, their axioms are "everything you can say about a two-element set [in the language of universal algebra]". Indeed:

Definition 2.52 *A* Boolean algebra *is a non-empty set A with elements* $\top$, $\bot$, *and operations* $\wedge$, $\vee$, $\neg$ *satisfying the following equations:*

- $a \wedge b = b \wedge a$, $a \wedge (b \wedge c) = (a \wedge b) \wedge c$, $a \wedge a = a$, $a \wedge \top = a$, $a \wedge \bot = \bot$,

- $a \vee b = b \vee a$, $a \vee (b \vee c) = (a \vee b) \vee c$, $a \vee a = a$, $a \vee \top = \top$, $a \vee \bot = a$,

- $a \wedge (b \vee c) = (a \wedge b) \vee (a \wedge c)$, $a \vee (b \wedge c) = (a \vee b) \wedge (a \vee c)$,

- $a \wedge \neg a = \bot$, $a \vee \neg a = \top$, $\neg\neg a = a$,

- $\neg(a \vee b) = \neg a \wedge \neg b$, $\neg(a \wedge b) = \neg a \vee \neg b$.

Any Boolean lattice trivially defines a Boolean algebra. Conversely, any Boolean algebra defines a Boolean lattice by letting $a \leq b$ if and only if $\neg a \vee b = \top$ (or equivalently, $a \wedge b = a$, or equivalently, $a \vee b = b$).

The fundamental completeness result (which we won't prove) is that a formula Φ is valid if and only if we can transform Φ into $\top$ in finitely many steps involving replacing one side of an equation of Boolean algebras by the other side. This is what we did on the example above.

As Boolean algebras are equivalent to Boolean lattices, we can also prove formulas by applying the laws of Boolean lattices instead. This sometimes yields faster proofs of validity by hand. To our knowledge, this has not been pursued in automated deduction.

Another idea is to use *Boolean rings* instead of Boolean algebras. This consists in axiomatising $\wedge$, $\oplus$ (the exclusive or, i.e. the negation of $\Leftrightarrow$), **T** (true) and **F** (false), and using them instead of $\wedge$, $\vee$ and $\neg$. The axioms can then be oriented as rewrite rules (Hsiang, 1985), modulo associativity and commutativity of $\wedge$ and $\oplus$. The translation rules are:

- $\neg\Phi \longrightarrow \Phi \oplus \mathbf{T}$,

- $\Phi \vee \Phi' \longrightarrow (\Phi \wedge \Phi') \oplus \Phi \oplus \Phi'$,

- $\Phi \Rightarrow \Phi' \longrightarrow (\Phi \wedge \Phi') \oplus \Phi \oplus \mathbf{T}$,

- $\Phi \Leftrightarrow \Phi' \longrightarrow \Phi \oplus \Phi' \oplus \mathbf{T}$.

And the simplification rules are:

- $\Phi \oplus \mathbf{F} \longrightarrow \Phi$,

- $\Phi \oplus \Phi \longrightarrow \mathbf{F}$,

- $\Phi \wedge (\Phi' \oplus \Phi'') \longrightarrow (\Phi \wedge \Phi') \oplus (\Phi \wedge \Phi'')$,

- $(\Phi \oplus \Phi') \wedge \Phi'' \longrightarrow (\Phi \wedge \Phi'') \oplus (\Phi' \wedge \Phi'')$,

- $\Phi \wedge \mathbf{T} \longrightarrow \Phi$,

- $\Phi \wedge \mathbf{F} \longrightarrow \mathbf{F}$,

- $\Phi \wedge \Phi \longrightarrow \Phi$.

The resulting normal form (modulo associativity and commutativity of $\wedge$ and $\oplus$) is called the *Reed-Muller form* of the formula. As for BDDs, it is canonical, and can be built bottom-up in the following way.

The Reed-Muller form of a formula is encoded as a set of sets of variables. (Given an ordering of the variables, we may represent a set of variables as an ordered list of these variables without duplicates, while a set of sets of variables is a list of these lists, ordered lexicographically, for instance.) The inner sets of variables are understood as the conjunctions of their variables, and the outer set is understood as the exclusive or of its elements.

Then, $\mathbf{F}$ is encoded as the empty set $\varnothing$, and $\mathbf{T}$ is encoded as the singleton set $\{\varnothing\}$. A variable A is coded as $\{\{A\}\}$. The exclusive or $\Phi \oplus \Phi'$ of two Reed-Muller forms Φ and Φ' is computed as the symmetric difference $(\Phi \setminus \Phi') \cup (\Phi' \setminus \Phi)$ of the two sets. (In the list representation, merge the two lists Φ and Φ' as a sorted list, then erase all adjacent pairs of identical sublists.) The conjunction $\Phi \wedge \Phi'$ of two Reed-Muller forms Φ and Φ' is computed as the symmetric difference of all sets $\{a \cup a'\}$, where $a \in \Phi$ and $a' \in \Phi'$. (In the list representation, produce the sorted list of all $a \cup a'$, then erase all adjacent couples of identical sublists.)

The Reed-Muller normal form can then be coded very much like BDDs, by using Shin-Ichi Minato's *zero-suppressed BDDs* (Minato, 1993), which are compact representations of sets of sets. We won't develop it here, but here is the basic idea. We use Shannon trees, but with a different semantics: $\mathbf{F}$ denotes the empty set $\varnothing$, $\mathbf{T}$ denotes the set containing only the empty set $\{\varnothing\}$, and $A \longrightarrow \Phi_+ ; \Phi_-$ denotes the set $\{\{A\} \cup s \mid s \in S_+\} \cup S_-$, where S_+ is the denotation of Φ_+ and S_- that of Φ_-. Informally, paths in the zero-suppressed BDD denotes sets of atoms A, where taking the Φ_+ branch means including A in the set, and taking the Φ_- means leaving it out. We can then share identical subtrees of these trees again, and order atoms along paths, as for BDDs. However, the new semantics imposes that we reduce them differently: $A \longrightarrow \Phi_+ ; \Phi_-$ rewrites to Φ_-, not when $\Phi_+ = \Phi_-$, but when $\Phi_+ = \mathbf{F}$. ($\mathbf{F}$ is sometimes written 0, hence the name "zero-suppressed".) Zero-suppressed BDDs are BDDs built using this alternative reduction rule, and are canonical forms for sets of sets of atoms—hence also for Reed-Muller forms.

How BDDs and Reed-Muller forms compare is somewhat tricky. Some problems are easier to deal with in one form, some are easier in the other form. There is no clear-cut advantage, although BDDs have traditionally been more fashionable than Reed-Muller forms.

We finish off this section by mentioning the following fact. While we have defined the semantics of classical logic by using the set of Booleans $\mathbb{B}$, we might as well have defined it by using a generic Boolean algebra A. Assignments would then map variables to elements of A, conjunction would be defined as the $\wedge$ operation in the algebra, disjunction as $\vee$, negation as complementation, and implication $a \Rightarrow b$ as $\neg a \vee b$. It is a remarkable fact that a formula is valid in the usual sense if and only if it is valid in the extended sense (i.e., if Φ evaluates to the top element $\top$ of A under all assignments). This idea, extended to first-order or even some second-order theories, has been successfully used by Dana Scott and Robert Solovay in the 1960s to give a simpler proof of the independence of the Generalised Continuum Hypothesis (GCH) from ZFC set theory, by the so-called method of *Boolean-valued models*. The GCH states that $2^{\aleph_\alpha} = \aleph_{\alpha+1}$ for all ordinals α; the Continuum Hypothesis (CH, i.e. $2^{\aleph_0} = \aleph_1$, or equivalently, all infinite subsets of $\mathbb{R}$ are either in bijection with $\mathbb{R}$ or in bijection with $\mathbb{N}$) was conjectured by Georg Cantor in the early twentieth century. While he could not prove or disprove it, Cohen proved in the early 1960s that not only CH, but GCH, was independent of set theory (i.e., neither it or its negation contradicts set theory). Cohen's proof is intricate and quite syntactic. Boolean-valued models provide a probably more readable semantic proof; see (Manin, 1977).

5.3 Quantified Propositional Logic

To the language of propositional logic, it is sometimes interesting to add quantifiers. This does not add any expressive power, but it can significantly shorten some propositional formulas.

If Φ is a formula, and A a propositional variable, then we may define $\forall A \cdot \Phi$ ("for all A, Φ") as $\Phi[\mathbf{T}/A] \wedge \Phi[\mathbf{F}/A]$, where $\mathbf{T}$ and $\mathbf{F}$ are two new constants such that $[\![\mathbf{T}]\!]\rho = \top$ and $[\![\mathbf{F}]\!]\rho = \bot$, as we did to introduce BDDs. Symmetrically, $\exists A \cdot \Phi$ ("for some A, Φ") can be defined as $\Phi[\mathbf{T}/A] \vee \Phi[\mathbf{F}/A]$.

We might as well define $\forall A \cdot \Phi$ and $\exists A \cdot \Phi$ directly by giving them a semantics (derived from the formulas above), and by extending deduction systems to handle the new operators. For example, we might extend $\mathbf{LK}_0$ by adding the following deduction rules:

$$\frac{\Gamma, \Phi[\Phi'/A] \longrightarrow \Delta}{\Gamma, \forall A \cdot \Phi \longrightarrow \Delta} \forall L \qquad \frac{\Gamma \longrightarrow \Delta, \Phi}{\Gamma \longrightarrow \Delta, \forall A \cdot \Phi} \forall R \text{ (if } A \notin \text{fv}(\Gamma) \cup \text{fv}(\Delta))$$

$$\frac{\Gamma, \Phi \longrightarrow \Delta}{\Gamma, \exists A \cdot \Phi \longrightarrow \Delta} \exists L \text{ (if } A \notin \text{fv}(\Gamma) \cup \text{fv}(\Delta)) \qquad \frac{\Gamma \longrightarrow \Delta, \Phi[\Phi'/A]}{\Gamma \longrightarrow \Delta, \exists A \cdot \Phi} \exists R$$

and this would yield exactly the same logic.

Quantified propositional logic is also implemented nicely on BDDs: substituting A for **T** (resp. **F**) means replacing all subBDDs of the form $A \longrightarrow \Phi_+; \Phi_-$ by Φ_+ (resp. Φ_-), and reducing the BDD above. We then implement $\forall$ and $\exists$ by taking the conjunction or the disjunction of the resulting BDDs.

The use of quantified propositional logic is common in hardware verification, for instance. As an example, we might represent a circuit in the following way: the inputs of the circuit are coded as propositional variables $A_1, \ldots, A_n$, while the outputs are propositional formulas on these variables. Assume for now that there is only one output Φ. Asking whether the output is valid means asking whether, whatever the inputs are, the output will always be $\top$. This means asking whether Φ is valid, or whether $\forall A_1 \cdot \ldots \forall A_n \cdot \Phi$ holds. Similarly, satisfiability (is there a case where Φ holds?) means asking whether $\exists A_1 \cdot \ldots \exists A_n \cdot \Phi$ holds. But some verification problems require more sophisticated quantifications than just checking for validity or satisfiability. It might be required to check that for any values of the first m inputs, there are values for the last $n - m$ inputs which make the output true. (For example, we need this to prove that there is no deadlock in some implementations of protocols, i.e. there is no state of the first m inputs such that we cannot make the output true with the right combination of the remaining inputs.) Checking this means verifying that $\forall A_1 \cdot \ldots \forall A_m \cdot \exists A_{m+1} \cdot \ldots \exists A_n \cdot \Phi$ holds. We shall see more on this in Chapter 5, Section 4.

In general, we are not limited to only one alternation of the quantifiers, and we may express propositions of the form $Q_1 A_1 \cdot \ldots Q_n A_n \cdot \Phi$, where the Q_is are either $\forall$ or $\exists$. If we had to express quantifications as conjunctions or disjunctions, this would expand the formula Φ by an exponential factor in general, meaning that it would already be intractable to *write down* the formula without quantifiers. (On BDDs, we can usually do that, because in good cases, sharing and reducing help keeping the BDD small enough, although it does not always work.)

Referring to the notion of expressiveness we sketched in Section 5.1, quantified propositional logic is as expressive as *polynomial-space bounded Turing machines*. More precisely, checking a quantified propositional formula is PSPACE-complete, where PSPACE is the class of decision problems that we can check in finite time on a (deterministic or non-deterministic, it actually does not matter) Turing machine using polynomial space (i.e., only a polynomial amount of slots on the tape). There are very probably many more problems in PSPACE than there are in NP, leading us to think that PSPACE-complete problems present more difficulties than problems in NP. However, PSPACE is still inside DEXPTIME, so we can still solve them in exponential time.

CHAPTER 3

OTHER PROPOSITIONAL LOGICS

1 INTRODUCTION

In the first chapter we mentioned several logical systems (classical, intuitionistic, linear, modal, etc.), each one having specific characteristics that make it well adapted for a particular class of applications. In this chapter we are going to study in detail two of these systems which can be thought of as refinements (or alternatives) to classical propositional logic. The systems that we will focus on are *intuitionistic* and *linear* logic. The first gives a different perspective on both the proof theory and model theory of logic: it is a *constructive* logic. By constructivity we mean that in contrast with classical logic we are interested not just in the truth value of a formula, but in the actual proof objects themselves. The second, linear logic, gives a new perspective on the *use* of *assumptions* (or *resources*) in derivations, which provides a deeper insight into the structure of a derivation and the process of cut elimination.

1.1 Constructivity

Intuitionistic logic is perhaps one of the most significant alternatives to classical logic since it brings a new way of thinking about logic and mathematics, especially with respect to proofs. It can be regarded as a subsystem of classical logic since the only difference is that certain laws are forbidden in the proof formation. The double negation elimination, $\neg\neg\Phi\Rightarrow\Phi$, or equivalently the law of excluded middle, $\Phi\vee\neg\Phi$, are not provable in intuitionistic logic.

As we have seen, classical logic is based on a notion of *truth* — each formula can be assigned a truth value. This is the essence of Tarski semantics (cf. truth tables for classical logic). There is a different school of thought that claims that the right notion is not about the *truth* value of a formula, but rather about the actual *proof* — a formula is true just when there is a *proof object* corresponding to that formula; this is essence of *Heyting semantics*.

A common example used to express the differences between classical and intuitionistic logic is the formula $\Phi\vee\neg\Phi$. This is trivially true in truth table semantics, but what constitutes a proof of this formula? A proof of $\Phi\vee\Phi'$ must take the form of either a proof of Φ *or* a proof of Φ', together with some indication of which of the two disjuncts has been proved. Hence, for $\Phi \vee \neg\Phi$, we must give a proof of Φ or a proof of $\neg\Phi$, which is not possible in general. To give a concrete example of this kind of reasoning, consider that you are my theorem prover, and I ask the question:

$$raining \vee \neg raining$$

There are many ways that you can "prove" this. The really lazy method is that you notice that the question is of the form $\Phi \vee \neg\Phi$, and you can tell me *true* straight away. Alternatively, you can go and look to see if it is raining or not, in which case you can evaluate the sentence using truth tables, and reply *true because it is raining*, for example. A third possible option is that you look to see if it is raining or not, evaluate the sentence, and tell me *true*, but without telling me why. The point is that I don't know what you decided to do. The following diagram shows some of the possibilities, but of course there are many others.

In any case, knowing the result *true* does not tell me anything about the weather. What would be much more useful is to have a reason (a proof or some evidence) why the sentence is *true*. For example the sentence is *true* because it is raining.

Equivalently, the formula $\neg\neg\Phi \Rightarrow \Phi$ is excluded from intuitionistic logic. $\neg\neg\Phi$ expresses a proof that $\neg\Phi$ is not provable, and from this information there is no reasonable way to believe, for the same reasons that we discussed above, that Φ is provable just because we cannot prove $\neg\Phi$.

In general, logics that follow this school of thought are called *constructive*. The origins of this kind of thinking date back to the work of Brouwer and Heyting under the name *intuitionism*. The ideas are somehow simple and obvious, but give a fundamentally new perspective on mathematical and logical reasoning. Their ideas are based on the fact that mathematics deals with mental *constructions* as opposed to formal manipulations of syntax—the mathematical language is secondary and is only used to communicate ideas. More specifically, it doesn't make sense to talk about truth or falsity of mathematical sentences independently of our knowledge about *why* it is true or false (our mental construction). This gives the following scenario:

- Φ is true if we can prove it.

- Φ is false if we can show that under the assumption that we have a proof of Φ we get a contradiction.

Therefore, for a suitable Φ, the statement $\Phi \vee \neg\Phi$ cannot be deemed to be true or false since there is no reasonable way of constructing a proof. In the next chapter we will see a computational application of these ideas, in the form of a type system for functional languages.

1.2 Resources

Linear logic is a further refinement of classical logic in that every assumption must be used *exactly once*. When we write the rule for modus ponens:

$$\frac{\Phi \qquad \Phi \Rightarrow \Phi'}{\Phi'}$$

we generally do not think about how many times the assumption Φ is used to deduce Φ'. There are basically three obvious possibilities, consisting of using Φ exactly once, more than once, or not at all. If Φ is used exactly once then we say it is used in a *linear* way. If Φ is used more than once, then the assumption needs to be *duplicated*. In the final case, where Φ is not used at all, the assumption needs to be *erased*. Below are three concrete examples, where we show, respectively, a linear and two non-linear deductions corresponding to duplicating and erasing.

$$\frac{\Phi \qquad \Phi \Rightarrow \Phi}{\Phi} \qquad\qquad \frac{\Phi \qquad \Phi \Rightarrow \Phi' \qquad \Phi \Rightarrow \Phi''}{\Phi' \wedge \Phi''} \qquad\qquad \frac{\Phi \qquad \Phi'}{\Phi'}$$

With this perspective, it makes sense to ask if Φ is still valid after the rule has been applied. More precisely, in deducing Φ' was the assumption Φ *consumed* by the rule? In real life, the notion of resource is of course taken very seriously, and examples are not hard to find. Take for example flour as a basic resource. The process of baking bread $F \Rightarrow B$ *consumes* a certain quantity of flour, which cannot be too easily re-used to bake a cake. The maintenance of resources, or assumptions, seems a very natural property that one should ask from a logic, especially if we want to use the logic to reason about physical quantities.

However, there are other scenarios where consumption of resources is not capturing the desired behaviour. Let us consider programs, written in the C programming language for example, as a basic resource. The process of compilation $C \Rightarrow O$ transforms the source code into some object code O, but just as important, we would hope that we still had the source code — we would not be very happy if the C compiler consumed our wonderful programs!

From this second example it is clear that a logic that allows only linear derivations, where everything is consumed exactly once is a rather weak logic. Linear logic starts from this restricted framework, but then builds in the power of non-linearity by introducing duplicating and erasing of resources in a controlled way. This is achieved by marking assumptions Φ that are allowed to be used in a non-linear way as $!\Phi$; only formulae of this shape can be duplicated and erased.

It is then easy to see that if we have $!F$ (as much flour as we want) then we can make as many cakes and loafs of bread as we wish. The same idea also applies to C programs. Interpreting a program as $!C$, which can be thought of as a file containing the source, the compiler can extract a copy of the program which is then consumed in the production of the object code.

This extra control over the way assumptions are used gives greater information about a derivation, and a more controlled cut-elimination procedure. This is one reason why it could be claimed that linear logic seems well adapted to being a logic for the dynamics of computation.

1.3   Semantic Paradigms

There are two semantic paradigms that are used to explain, or give a meaning to, the logical connectives, which reflects the difference between non-constructive and constructive logics.

Model-theoretic (Tarski) Tarski's idea was to fix a mathematical universe, say $\mathcal{M}$, where we can model formulas. Let us write $\mathcal{M} \models \Phi$ if Φ is valid in the model, otherwise $\mathcal{M} \not\models \Phi$. Assume that atomic sentences are assigned a truth value, then compound formulae are assigned a denotation as follows:

$$\mathcal{M} \models \Phi \wedge \Phi' \quad \text{iff} \quad \mathcal{M} \models \Phi \text{ and } \mathcal{M} \models \Phi'$$
$$\mathcal{M} \models \Phi \vee \Phi' \quad \text{iff} \quad \mathcal{M} \models \Phi \text{ or } \mathcal{M} \models \Phi'$$
$$\mathcal{M} \models \Phi \Rightarrow \Phi' \quad \text{iff} \quad \mathcal{M} \not\models \Phi \text{ or } \mathcal{M} \models \Phi'$$
$$\mathcal{M} \models \neg\Phi \quad \text{iff} \quad \mathcal{M} \not\models \Phi$$

Since this is just another formulation of the truth tables that we saw in Chapter 2, Section 2, it is clear that the approach fits very well with classical logic.

This approach gives a *static* view of the world in that sentences are either *true* or *false* (according to their denotation). The approach is reminiscent of traditional denotational semantics for programming languages.

It appears that there is very little foundation in this paradigm, for instance there is no real reason to believe that "and" is any more primitive than $\wedge$, etc. However, for classical logic the paradigm is certainly powerful, as we have demonstrated in the previous chapter.

Proof-theoretic (Brouwer-Heyting-Kolmogorov) In contrast to giving a denotation of a formula in some mathematical universe, the proof theoretic approach is to model *not* the denotation, but the *proofs* themselves. One way to do this is to interpret proofs using sets and functions. Atomic sentences are assumed to be proved by external means, and then formulae are interpreted as *data types* in the following way:

$$\begin{aligned}
\Phi \wedge \Phi' &= \Phi \times \Phi' = \{(a,b) : a \in \Phi, b \in \Phi'\} \\
\Phi \vee \Phi' &= \Phi + \Phi' = \{(0,a) : a \in \Phi\} \cup \{(1,b) : b \in \Phi'\} \\
\Phi \Rightarrow \Phi' &= \Phi \rightarrow \Phi' = \{f : \text{for all } a \in \Phi \text{ implies } f(a) \in \Phi'\} \\
\neg\Phi &= \Phi \rightarrow \bot \\
\bot &= \varnothing
\end{aligned}$$

To give an example, $\Phi \Rightarrow \Phi$ is interpreted as a function f which transforms any proof $a \in \Phi$ to a proof $a \in \Phi$, for example, the identity function.

Again, one could say that there is very little foundational content in this paradigm, for example the function space $\rightarrow$ is explained in terms of *for all* and *implies*. However, this paradigm has been very useful for understanding constructive logics, and in generating a connection between logic and computation, the Curry-Howard correspondence, that we will study in Chapter 4.

This second semantic paradigm is what we shall focus on. It supports intuitionistic logic (more generally constructive logics), not classical logic; the example of *raining* $\vee$ $\neg raining$ illustrates the point.

1.4   Notation

In contrast to the treatment of classical logic of the last chapter, here we will mostly develop the proof theory, and only hint at the model theory of the logics and theorem proving techniques. Throughout this chapter we will make use of the sequent presentation of logics, whether *sequent calculus* or *natural deduction in sequent form*. This will facilitate the closer analysis of the usage of assumptions in derivations. Since we are going to be looking at the use of resources in derivations, we prefer to work with lists of contexts rather than sets of the previous chapter. As a consequence, we make explicit the manipulation of the contexts.

To avoid introducing many different logical symbols, we adopt the same logical connectives for intuitionistic logic as we used for classical logic. This of course needs to be understood with care, but should be clear from the context. For linear logic, however, there will be a new set of connectives, since they are decompositions of both classical and intuitionistic ones.

To expose further the structure of the proof systems we are going to use, we divide the rules into three groups which are described below.

Identity group There are two rules that come under the heading of identities: the *axiom* and the *cut*.

The *axiom* of a logic is required to "get things off the ground" — it is a rule with no premises. For both the sequent calculus and natural deduction in sequent form we always have the axiom.

$$\frac{}{\Phi \longrightarrow \Phi} \, (Ax)$$

Compare this to the axiom rules presented previously; here we have the axiom cut down to its essential content. In the presence of the structural rules (see below) we can derive the axiom rules that we have been using up to now. There are several perspectives that one can put on this rule, all are fruitful. The most basic understanding of the axiom is "Φ *is* Φ", but we can also think of it as given Φ as input, Φ is the only possible output (of some process).

For the sequent calculus, we often use the *cut* rule, although as we have seen for classical logic, it can be eliminated and is thus redundant. However, it is often convenient to have the rule in the sequent calculus, and this gives us our second identity in the logic.

$$\frac{\Gamma, \Phi \longrightarrow \Delta \qquad \Gamma' \longrightarrow \Phi, \Delta'}{\Gamma, \Gamma' \longrightarrow \Delta, \Delta'} \, (Cut)$$

We have seen this rule many times already in the context of classical logic. However, we never mentioned that this was an identity before. In fact it can be understood as the dual of the axiom in the sense that it relates an output formula with the *same* input formula Φ.

Structural group The structural (or bookkeeping) rules are normally hidden from the logical system. They are used to make the usage of assumptions explicit in a derivation.

There are three structural rules that we normally use. The first one, called *exchange*, says that the order of assumptions is not important. In classical logic (either sequent calculus or natural deduction in sequent from) we would write this as:

$$\frac{\Gamma, \Phi, \Phi', \Gamma' \longrightarrow \Delta}{\Gamma, \Phi', \Phi, \Gamma' \longrightarrow \Delta} \, (XL)$$

where there is a symmetric rule for (XR).

The second structural rule, called *weakening*, says that the system is *monotonic* — if we can prove Δ from Γ, then we can also prove it with an extra assumption Φ. In classical logic we would write:

$$\frac{\Gamma \longrightarrow \Delta}{\Gamma, \Phi \longrightarrow \Delta} \, (WL)$$

where there is a symmetric rule for (WR). Logics without this rule are called *affine*, or *relevance* logics.

Finally, the third structural rule, called *contraction*, states that the number of times we are given an assumption is not important; if we can prove Δ using Φ twice, then we can also prove it from one Φ. Again, for classical logic we write:

$$\frac{\Gamma, \Phi, \Phi \longrightarrow \Delta}{\Gamma, \Phi \longrightarrow \Delta} \, (CL)$$

where again there is a symmetric rule for (CR).

Of course, once we make these structural operations explicit the proofs will become much larger. To keep the proofs to reasonable size, we adopt the following abbreviation for the multiple application of a specific structural rule (here shown for weakening)

$$\frac{\Gamma, \Gamma' \longrightarrow \Delta}{\Gamma \longrightarrow \Delta} \, (WL)$$

which shows that the whole context Γ' has been weakened.

Logical group The logical group gives the logical rules for manipulating the connectives. For the sequent calculus, they take the form of left and right introduction rules, and for natural deduction, they take the form of introduction and elimination rules.

For the purpose of this chapter only, where we compare intuitionistic logic with classical logic, we will use a slight variant of the system $\mathbf{LK}_0$ (classical propositional sequent calculus), as defined in Chapter 2, Section 3.3, which uses explicit structural rules as described above. The translation of $\mathbf{LK}_0$ into this variant is quite straightforward; see Exercise 3.1 for a first hint at how to do this.

2   INTUITIONISTIC LOGIC

In this section we give various presentations of intuitionistic logic. We begin with the natural deduction system in sequent form which we take as the main formulation of the logic. We then give the sequent calculus and Hilbert-style axioms for the logic, showing how they relate to the natural deduction system.

The main characteristic of intuitionistic logic in our presentations is the restriction of sequents $\Gamma \longrightarrow \Delta$ so that Δ has *at most one* formula. Hence, for all the presentations below, sequents are either $\Gamma \longrightarrow \Phi$ or $\Gamma \longrightarrow$. For the latter, we sometimes write $\Gamma \longrightarrow \bot$, where $\bot$ represents a *distinguished* formula which has no proof. Our natural deduction systems for intuitionistic logic will be characterised by the lack of the double negation elimination rule.

We refer the reader to the articles by van Dalen (1986), Ryan and Sadler (1992) for additional material on intuitionistic logic. Troelstra and van Dalen (1988a; 1988b) provide an encyclopedia for constructivism in mathematics, hence a great deal of material on intuitionistic logic.

2.1   *Intuitionistic Natural Deduction* ($\mathbf{NJ_0}$)

The natural deduction system for intuitionistic logic, which we shall call $\mathbf{NJ_0}$, is undoubtedly the most well-known formulation of this logic. Natural deduction systems are well adapted for intuitionistic logic, and the rules are related in a direct way with the *proof theoretic approach* to semantics. We therefore take it as the first, and main, presentation. Our only slight variant is that since we are taking the use of assumptions seriously, we give the natural deduction system in sequent form, which, for the purpose of this chapter, we also call $\mathbf{NJ_0}$.

Intuitionistic natural deduction can be obtained from the natural deduction system for classical logic by excluding the rule for double negation elimination:

$$\frac{\Gamma \longrightarrow \neg\neg\Phi}{\Gamma \longrightarrow \Phi} \ (\neg\neg E)$$

As we shall see, this has quite a lot of ramifications on the class of provable sequents.

The rules for $\mathbf{NJ_0}$ are given in Figure 3.1, and we invite the reader to compare the system with that of classical logic. With this system we derive sequents of the form $\Gamma \longrightarrow \Phi$. A proof π of a sequent $\Gamma \longrightarrow \Phi$ is then defined by a derivation using the rules given with root $\Gamma \longrightarrow \Phi$. Following our usual notation, we write $\vdash^{\mathbf{NJ_0}} \Gamma \longrightarrow \Phi$, or $\Gamma \vdash^{\mathbf{NJ_0}} \Phi$ to mean that $\Gamma \longrightarrow \Phi$ is provable in $\mathbf{NJ_0}$.

The reader will convince himself that the Deduction Theorem (cf. Theorem 2.21, page 29) also holds for this system.

There are several remarks that can be made about this logical system. For the structural group, all the manipulations are on the left, there is no way of applying *exchange* and *contraction* to the formula on the right, since there is at most one. As with classical natural deduction, all the logical rules for each connective come in pairs: introduction (I) and elimination (E) rules, and remark that the ($\vee E$) rule introduces a new formula Φ'' which had nothing to do with the original formula (sometimes called a *parasitic* formula). Finally, it is also worth noting that for natural deduction we do not actually need to label the nodes of the tree since the applied

Identity group

$$\frac{}{\Phi \longrightarrow \Phi} \, (Ax)$$

Structural group

$$\frac{\Gamma, \Phi, \Phi', \Delta \longrightarrow \Phi''}{\Gamma, \Phi', \Phi, \Delta \longrightarrow \Phi''} \, (X) \qquad \frac{\Gamma \longrightarrow \Phi'}{\Gamma, \Phi \longrightarrow \Phi'} \, (W) \qquad \frac{\Gamma, \Phi, \Phi \longrightarrow \Phi'}{\Gamma, \Phi \longrightarrow \Phi'} \, (C)$$

Logical group

$$\frac{\Gamma \longrightarrow \Phi \qquad \Delta \longrightarrow \Phi'}{\Gamma, \Delta \longrightarrow \Phi \wedge \Phi'} \, (\wedge I)$$

$$\frac{\Gamma \longrightarrow \Phi \wedge \Phi'}{\Gamma \longrightarrow \Phi} \, (\wedge E_1) \qquad \frac{\Gamma \longrightarrow \Phi \wedge \Phi'}{\Gamma \longrightarrow \Phi'} \, (\wedge E_2)$$

$$\frac{\Gamma, \Phi \longrightarrow \Phi'}{\Gamma \longrightarrow \Phi \Rightarrow \Phi'} \, (\Rightarrow I) \qquad \frac{\Gamma \longrightarrow \Phi \Rightarrow \Phi' \qquad \Delta \longrightarrow \Phi}{\Gamma, \Delta \longrightarrow \Phi'} \, (\Rightarrow E)$$

$$\frac{\Gamma \longrightarrow \Phi}{\Gamma \longrightarrow \Phi \vee \Phi'} \, (\vee I_1) \qquad \frac{\Gamma \longrightarrow \Phi'}{\Gamma \longrightarrow \Phi \vee \Phi'} \, (\vee I_2)$$

$$\frac{\Gamma \longrightarrow \Phi \vee \Phi' \qquad \Delta, \Phi \longrightarrow \Phi'' \qquad \Delta, \Phi' \longrightarrow \Phi''}{\Gamma, \Delta \longrightarrow \Phi''} \, (\vee E)$$

Figure 3.1. Intuitionistic Natural Deduction (**NJ**$_0$)

rule (introduction or elimination of a connective) is clear from the formula because all the *action* appears on the right hand side of the $\longrightarrow$. However, we prefer to keep this extra information for clarity when talking about the rules.

To be precise, what we have actually presented here is *minimal* intuitionistic logic. The full system can be obtained by adding the following rule to the natural deduction system. We mention this rule as an aside here since it will only play a "minimal" rôle in what follows, where we show how intuitionistic logic relates to classical logic.

$$\frac{\Gamma \longrightarrow \bot}{\Gamma \longrightarrow \Phi}\,(\bot)$$

where $\bot$ means contradiction; a formula for which there is no proof ($\longrightarrow \bot$ is *not* derivable). This rule says that if we can derive a contradiction, then we can deduce *anything* (cf. truth tables for $\Rightarrow$ in classical logic). Intuitionistic negation is then *defined* as $\neg\Phi = \Phi \Rightarrow \bot$. $\neg\Phi$ is provable in the system if under the assumption that if we have a proof of Φ, we get a contradiction. It is easy now to see that negation is no longer an involution, since $\neg\neg\Phi = (\Phi \Rightarrow \bot) \Rightarrow \bot$ is not the same as Φ. It is worth noting some consequences of this negation with respect to classical logic. Recall that every connective can be explained in terms of one other and negation. For example we had the following equivalences: $\neg(\Phi \wedge \Phi') \Leftrightarrow \neg\Phi \vee \neg\Phi'$, $\neg(\Phi \vee \Phi') \Leftrightarrow \neg\Phi \wedge \neg\Phi'$, $\Phi \Rightarrow \Phi' \Leftrightarrow \neg\Phi \vee \Phi'$ and also $\neg\neg\Phi \Leftrightarrow \Phi$, where $\Phi \Leftrightarrow \Phi'$ abbreviates $(\Phi \Rightarrow \Phi') \wedge (\Phi' \Rightarrow \Phi)$. None of these equivalences hold in intuitionistic logic: de Morgan duals no longer hold, implication ($\Rightarrow$) is no longer defined in terms of other connectives, and negation is not an involution.

In Exercise 3.3 we give the reader the opportunity to verify that one direction of the above holds. Here we give an example, to show that a restricted form of double negation elimination $\neg\neg\neg\Phi \Leftrightarrow \neg\Phi$ does hold.

$$\cfrac{\cfrac{\cfrac{\cfrac{\cfrac{\overline{\Phi \longrightarrow \Phi}\,(Ax) \qquad \overline{\neg\Phi \longrightarrow \neg\Phi}\,(Ax)}{\Phi, \neg\Phi \longrightarrow \bot}\,(\Rightarrow E)}{\Phi \longrightarrow \neg\neg\Phi}\,(\Rightarrow I) \qquad \overline{\neg\neg\neg\Phi \longrightarrow \neg\neg\neg\Phi}\,(Ax)}{\neg\neg\neg\Phi, \Phi \longrightarrow \bot}\,(\Rightarrow E)}{\neg\neg\neg\Phi \longrightarrow \neg\Phi}\,(\Rightarrow I)}{\longrightarrow \neg\neg\neg\Phi \Rightarrow \neg\Phi}\,(\Rightarrow I)$$

(The proof may be clearer if the reader replaces $\neg\Phi$ by $\Phi \Rightarrow \bot$.) It is easy to see that the other direction holds, since we built a proof of $\Phi \longrightarrow \neg\neg\Phi$ as part of the above derivation.

▶ **EXERCISE 3.1**

The structural rules presented above for intuitionistic logic are a subset of the structural rules for classical logic. In Section 1.4 we introduced structural rules using classical logic, but only gave the left rules. Give the missing rules to complete the structural group for classical logic.

Show that the usual axiom of classical logic can then be derived from the axiom $\Phi \longrightarrow \Phi$, together with these structural rules.

Remark also that this holds for both sequent calculus and natural deduction in sequent form.

▶ **EXERCISE 3.2**

Build derivations for the following (closed) formulae, i.e. for each formula Φ, show that $\longrightarrow \Phi$ is derivable.

$$
\begin{array}{ll}
(i) & A \Rightarrow B \Rightarrow A \\
(ii) & (A \Rightarrow B \Rightarrow C) \Rightarrow (A \Rightarrow B) \Rightarrow (A \Rightarrow C) \\
(iii) & (A \wedge B \Rightarrow C) \Rightarrow (A \Rightarrow B \Rightarrow C) \\
(iv) & (A \Rightarrow B \Rightarrow C) \Rightarrow (A \wedge B \Rightarrow C) \\
(v) & ((A \vee B) \wedge C) \Rightarrow ((A \wedge C) \vee (B \wedge C)) \\
(vi) & (A \Rightarrow (B \wedge C)) \Rightarrow ((A \Rightarrow B) \wedge (A \Rightarrow C))
\end{array}
$$

▶ **EXERCISE 3.3**

This question concerns defining connectives for intuitionistic logic in terms of the others. Build derivations to show which of the following hold, in the case that only one direction holds, give the derivation.

$$
\begin{array}{lll}
\neg(\Phi \wedge \Phi') & \Leftrightarrow & \neg\Phi \vee \neg\Phi' \\
\neg(\Phi \vee \Phi') & \Leftrightarrow & \neg\Phi \wedge \neg\Phi' \\
\Phi \Rightarrow \Phi' & \Leftrightarrow & \neg\Phi \vee \Phi' \\
\Phi \wedge \neg\Phi & \Leftrightarrow & \bot
\end{array}
$$

▶ **EXERCISE 3.4**

Show that the weakening rule can be deduced from the other rules. [Hint: use $\wedge$ introduction and elimination rules.]

2.2 Intuitionistic Sequent Calculus ($\mathbf{LJ}_0$)

In this section we give a sequent calculus formulation of intuitionistic logic, and briefly examine some of the relations between this and the natural deduction presentation. Recall that the sequent calculus presentation of a logic uses left and right rules rather than introduction and elimination rules for the logical connectives. The axiom and the structural rules are the same as for natural deduction (cf. Figure 3.1), but we need the additional (Cut) rule:

$$\frac{\Gamma \longrightarrow \Phi \qquad \Delta, \Phi \longrightarrow \Phi'}{\Gamma, \Delta \longrightarrow \Phi'} \ (Cut)$$

Thus, in Figure 3.2 we just give the logical group. Again, we invite the reader to compare intuitionistic sequent calculus with the classical sequents in Chapter 2.

Logical group

$$\frac{\Gamma \longrightarrow \Phi \qquad \Delta \longrightarrow \Phi'}{\Gamma, \Delta \longrightarrow \Phi \wedge \Phi'} \ (\wedge R)$$

$$\frac{\Gamma, \Phi \longrightarrow \Phi''}{\Gamma, \Phi \wedge \Phi' \longrightarrow \Phi''} \ (\wedge L_1) \qquad \frac{\Gamma, \Phi' \longrightarrow \Phi''}{\Gamma, \Phi \wedge \Phi' \longrightarrow \Phi''} \ (\wedge L_2)$$

$$\frac{\Gamma \longrightarrow \Phi}{\Gamma \longrightarrow \Phi \vee \Phi'} \ (\vee R_1) \qquad \frac{\Gamma \longrightarrow \Phi'}{\Gamma \longrightarrow \Phi \vee \Phi'} \ (\vee R_2)$$

$$\frac{\Gamma, \Phi \longrightarrow \Phi'' \qquad \Delta, \Phi' \longrightarrow \Phi''}{\Gamma, \Delta, \Phi \vee \Phi' \longrightarrow \Phi''} \ (\vee L)$$

$$\frac{\Gamma, \Phi \longrightarrow \Phi'}{\Gamma \longrightarrow \Phi \Rightarrow \Phi'} \ (\Rightarrow R) \qquad \frac{\Gamma \longrightarrow \Phi \qquad \Delta, \Phi' \longrightarrow \Phi''}{\Gamma, \Delta, \Phi \Rightarrow \Phi' \longrightarrow \Phi''} \ (\Rightarrow L)$$

Figure 3.2. Intuitionistic Sequent Calculus ($\mathbf{LJ_0}$)

In addition to the sequents given, there are the following two rules for negation:

$$\frac{\Gamma \longrightarrow \Phi}{\Gamma, \neg \Phi \longrightarrow \bot} \ (\neg L) \qquad \frac{\Gamma, \Phi \longrightarrow \bot}{\Gamma \longrightarrow \neg \Phi} \ (\neg R)$$

together with an additional structural rule which allows weakening on the right in a restricted way:

$$\frac{\Gamma \longrightarrow \bot}{\Gamma \longrightarrow \Phi} \ (\bot)$$

Again, we derive sequents of the form $\Gamma \longrightarrow \Phi$. A proof π of the sequent is a derivation ending with $\Gamma \longrightarrow \Phi$ at the root. We write $\Gamma \vdash^{\mathbf{LJ_0}} \Phi$ if $\Gamma \longrightarrow \Phi$ is provable in $\mathbf{LJ_0}$.

We remark that the two rules $\wedge L_i$ $i \in \{1, 2\}$ are different from the single rule that was used for classical sequents. One might have expected to see the following single rule for $\wedge L$:

$$\frac{\Gamma, \Phi, \Phi' \longrightarrow \Phi''}{\Gamma, \Phi \wedge \Phi' \longrightarrow \Phi''} \ (\wedge L)$$

However, they are equivalent in $\mathbf{LJ_0}$. We prefer this version since it reflects some symmetry between the rules for $\vee$ and $\wedge$. We will look more into this distinction later, in Section 6, where we also compare different presentations of contexts.

The way negation works here is a little different from what we had for $\mathbf{NJ_0}$. First, the rule ($\perp$) has now become a structural rule, and there are explicit rules that allow formulae to move across the entailment. There is of course nothing deep in this, and as we shall see shortly, both systems prove the same formulae. Before giving the reader the opportunity to prove some sequents, we give a simple example derivation showing how the negation rules work.

$$\frac{\dfrac{\dfrac{\rule{1.5cm}{0.4pt}}{\neg\Phi \longrightarrow \neg\Phi} \ (Ax)}{\dfrac{\neg\Phi, \neg\neg\Phi \longrightarrow \perp}{\dfrac{\neg\Phi \longrightarrow \neg\neg\neg\Phi}{\longrightarrow \neg\Phi \Rightarrow \neg\neg\neg\Phi} \ (\Rightarrow I)} \ (\neg R)} \ (\neg L)}$$

▶ **EXERCISE 3.5**

Which of the following are derivable in intuitionistic sequent calculus?

(i) $\neg A \Rightarrow A \longrightarrow A$ (ii) $A \Rightarrow \neg A \longrightarrow \neg A$

(iii) $A \Rightarrow B \longrightarrow \neg B \Rightarrow \neg A$ (iv) $\neg A \Rightarrow \neg B \longrightarrow B \Rightarrow A$

(v) $\longrightarrow \neg\perp$

▶ **EXERCISE 3.6**

Show that $\vdash^{\mathbf{LJ_0}} \Phi \vee \neg\Phi \Leftrightarrow \neg\neg\Phi \Rightarrow \Phi$, thus showing that if we have either $\Phi \vee \neg\Phi$ or $\neg\neg\Phi \Rightarrow \Phi$, then we have the other.

2.3   *Relating Sequent Calculus and Natural Deduction*

In this section we ask if it is possible to translate between sequent calculus and natural deduction presentations of intuitionistic logic. It should come as no surprise that both systems prove the same sequents. More formally we have the following result.

Theorem 3.1 $\Gamma \vdash^{\mathbf{LJ_0}} \Phi$ *iff* $\Gamma \vdash^{\mathbf{NJ_0}} \Phi$.

A proof can be obtained by exhibiting translations between the logical systems. Here we look at one half of the theorem, by giving a translation of $\mathbf{LJ_0}$ into $\mathbf{NJ_0}$.

(Note that the other direction has already been studied in Exercise 2.17 for classical logic, which is easily converted into the intuitionistic case.)

To give a little motivation into the usefulness of this translation, we mention the following "bad feature" of the sequent calculus (whether intuitionistic or classical). There are many ways of deriving a proof in the sequent calculus, even for *cut-free* proofs, which is due to the possibility of *permuting* various rules. However, in natural deduction there is a canonical notion of proof, the so-called normal proofs which we shall define shortly. To give an example of permuting rules in the sequent calculus we can apply the rules (left or right) "non-deterministically". A classic example of this is shown below where we give two proofs of the sequent $\Phi \wedge \Phi' \longrightarrow \Phi' \wedge \Phi$.

The first derivation, reading upwards, applied all the rules on the left-hand side first:

$$
\cfrac{
  \cfrac{
    \cfrac{
      \cfrac{
        \cfrac{}{\Phi' \longrightarrow \Phi'}(Ax)
        \qquad
        \cfrac{}{\Phi \longrightarrow \Phi}(Ax)
      }{\Phi, \Phi' \longrightarrow \Phi' \wedge \Phi}(\wedge R)
    }{\Phi, \Phi \wedge \Phi' \longrightarrow \Phi' \wedge \Phi}(\wedge L)
  }{\Phi \wedge \Phi', \Phi \wedge \Phi' \longrightarrow \Phi' \wedge \Phi}(\wedge L)
}{\Phi \wedge \Phi' \longrightarrow \Phi' \wedge \Phi}(C)
$$

whereas the second derivation applies a right rule first, thus duplicating the $(\wedge L)$ rule:

$$
\cfrac{
  \cfrac{
    \cfrac{\cfrac{}{\Phi' \longrightarrow \Phi'}(Ax)}{\Phi \wedge \Phi' \longrightarrow \Phi'}(\wedge L)
    \qquad
    \cfrac{\cfrac{}{\Phi \longrightarrow \Phi}(Ax)}{\Phi \wedge \Phi' \longrightarrow \Phi}(\wedge L)
  }{\Phi \wedge \Phi', \Phi \wedge \Phi' \longrightarrow \Phi' \wedge \Phi}(\wedge R)
}{\Phi \wedge \Phi' \longrightarrow \Phi' \wedge \Phi}(C)
$$

In natural deduction, however, there is no choice in the derivation. In fact the proofs are *syntax directed*, meaning that the proof is determined uniquely from the sequent being proved. Hence, for this example, there is only one (*normal form*) derivation in **NJ**$_0$:

$$
\cfrac{
  \cfrac{
    \cfrac{\cfrac{}{\Phi \wedge \Phi' \longrightarrow \Phi \wedge \Phi'}(Ax)}{\Phi \wedge \Phi' \longrightarrow \Phi'}(\wedge E_2)
    \qquad
    \cfrac{\cfrac{}{\Phi \wedge \Phi' \longrightarrow \Phi \wedge \Phi'}(Ax)}{\Phi \wedge \Phi' \longrightarrow \Phi}(\wedge E_1)
  }{\Phi \wedge \Phi', \Phi \wedge \Phi' \longrightarrow \Phi' \wedge \Phi}(\wedge I)
}{\Phi \wedge \Phi' \longrightarrow \Phi' \wedge \Phi}(C)
$$

This gives some kind of evidence that a natural deduction proof is more "primitive" than a sequent proof in intuitionistic logic.

We define a function $\mathcal{N}$ from proofs in $\mathbf{LJ}_0$ to proofs in $\mathbf{NJ}_0$ by induction on the structure of a sequent proof tree, denoted by π. To save writing the translation out in full, we state the general idea, and just show the interesting cases. The summary of the translation is given in the following table:

π	$\mathcal{N}(\pi)$
Axiom	Axiom
Cut	Subs
Right rules	Introduction rules
Left rules	Elimination rules+Subs

The translation that we give uses a rule Subs (for substitution) which is *not* part of the natural deduction system presented. This additional rule is defined as:

$$\frac{\Gamma \longrightarrow \Phi \qquad \Delta, \Phi \longrightarrow \Phi'}{\Gamma, \Delta \longrightarrow \Phi'} \; (Subs)$$

Although it is not part of the natural deduction system, it is a *derivable extension*.

$$\frac{\vdots \qquad \dfrac{\begin{array}{c}\vdots\\ \Delta, \Phi \longrightarrow \Phi'\end{array}}{\Delta \longrightarrow \Phi \Rightarrow \Phi'} \; (\Rightarrow I)}{\Gamma, \Delta \longrightarrow \Phi'} \; (\Rightarrow E)$$

where we have a $(\Rightarrow I)$ followed by a $(\Rightarrow E)$. Later, in Section 3, we see that this rule can be eliminated. We see in the above table that the right rules of $\mathbf{LJ}_0$ are translated trivially into introduction rules, that is to say that the right rules are of the same form as the introduction rules.

Let us assume that $\Gamma \vdash^{\mathbf{LJ}_0} \Phi$. By induction over the proof rules we build a proof in $\mathbf{NJ}_0$ as follows, where we only give the translations of axiom, cut and the left rules.

- The axiom $\Phi \longrightarrow \Phi$ becomes the corresponding axiom of natural deduction $\Phi \longrightarrow \Phi$.

- The (*Cut*) rule:

$$\frac{\begin{array}{c}\pi\\ \overline{\Gamma \longrightarrow \Phi}\end{array} \qquad \begin{array}{c}\pi'\\ \overline{\Delta, \Phi \longrightarrow \Phi'}\end{array}}{\Gamma, \Delta \longrightarrow \Phi'} \; (Cut)$$

is mapped to the following derivation:

$$\frac{\begin{array}{c}\mathcal{N}(\pi)\\ \overline{\Gamma \longrightarrow \Phi}\end{array} \qquad \begin{array}{c}\mathcal{N}(\pi')\\ \overline{\Delta, \Phi \longrightarrow \Phi'}\end{array}}{\Gamma, \Delta \longrightarrow \Phi'} \; (Subs)$$

- The $(\wedge L_1)$ rule:

$$\dfrac{\dfrac{\pi}{\Gamma, \Phi \longrightarrow \Phi''}}{\Gamma, \Phi \wedge \Phi' \longrightarrow \Phi''} \, (\wedge L_1)$$

is mapped to the derivation:

$$\dfrac{\dfrac{\overline{\Phi \wedge \Phi' \longrightarrow \Phi \wedge \Phi'}}{\Phi \wedge \Phi' \longrightarrow \Phi} \, (\wedge E_1) \qquad \dfrac{\mathcal{N}(\pi)}{\Gamma, \Phi \longrightarrow \Phi''}}{\Gamma, \Phi \wedge \Phi' \longrightarrow \Phi''} \, (Subs)$$

The $(\wedge L_2)$ rule is similar.

- The $(\Rightarrow)L$ rule:

$$\dfrac{\dfrac{\pi}{\Gamma \longrightarrow \Phi} \qquad \dfrac{\pi'}{\Delta, \Phi' \longrightarrow \Phi''}}{\Gamma, \Delta, \Phi \Rightarrow \Phi' \longrightarrow \Phi''} \, (\Rightarrow L)$$

is mapped to the derivation:

$$\dfrac{\dfrac{\dfrac{\overline{\Phi \Rightarrow \Phi' \longrightarrow \Phi \Rightarrow \Phi'}}{} \, (Ax) \quad \dfrac{\mathcal{N}(\pi)}{\Gamma \longrightarrow \Phi}}{\Gamma, \Phi \Rightarrow \Phi' \longrightarrow \Phi'} \, (\Rightarrow E) \qquad \dfrac{\mathcal{N}(\pi')}{\Delta, \Phi' \longrightarrow \Phi''}}{\Gamma, \Delta, \Phi \Rightarrow \Phi' \longrightarrow \Phi''} \, (Subs)$$

- The $(\vee L)$ rule:

$$\dfrac{\dfrac{\pi}{\Gamma, \Phi \longrightarrow \Phi''} \qquad \dfrac{\pi'}{\Delta, \Phi' \longrightarrow \Phi''}}{\Gamma, \Delta, \Phi \vee \Phi' \longrightarrow \Phi''} \, (\vee L)$$

is mapped to the derivation:

$$\dfrac{\dfrac{}{\Phi \vee \Phi' \longrightarrow \Phi \vee \Phi'} \, (Ax) \quad \dfrac{\dfrac{\mathcal{N}(\pi)}{\Gamma, \Phi \longrightarrow \Phi''}}{\Gamma, \Delta, \Phi \longrightarrow \Phi''} \, (W) \quad \dfrac{\dfrac{\mathcal{N}(\pi')}{\Delta, \Phi' \longrightarrow \Phi''}}{\Gamma, \Delta, \Phi' \longrightarrow \Phi''} \, (W)}{\Gamma, \Delta, \Phi \vee \Phi' \longrightarrow \Phi''} \, (\vee E)$$

One problem of this translation is that we used the (*Subs*) rule, but the *real* problem is that a *cut-free* proof is translated into a proof with (Subs). For example the sequent proof

$$\frac{\displaystyle \frac{}{\Phi \longrightarrow \Phi}\ (Ax)}{\Phi \wedge \Phi' \longrightarrow \Phi}\ (\wedge L)$$

becomes the following natural deduction derivation under the translation:

$$\frac{\displaystyle \frac{\overline{\Phi \wedge \Phi' \longrightarrow \Phi \wedge \Phi'}\ (Ax)}{\Phi \wedge \Phi' \longrightarrow \Phi}\ (\wedge E_1) \qquad \frac{}{\Phi \longrightarrow \Phi}\ (Ax)}{\Phi \wedge \Phi' \longrightarrow \Phi}\ (Subs)$$

There are several ways to avoid this problem by being more careful in the translation. However, here we take a "corrective" approach and show that the (Subs) rule can be eliminated. More generally, any pair of rules of the form introduction followed by an elimination of the same connective can be eliminated from the proof. This will be the subject of Section 3.1.

2.4 Hilbert-style Presentation

For completeness of our presentation of intuitionistic logic, we next present the Hilbert-style axioms *in sequent form* and show how it relates to $\mathbf{NJ_0}$. In the same way as for classical logic, the Hilbert system $\mathbf{H}$ for intuitionistic logic consists of a collection of *axiom schemes* and the rule modus ponens. These are given below, and note in particular that we need to present more axioms—several for each connective—whereas for classical logic this is not necessary since we can reduce the number of axioms by half by defining some connectives in terms of others by de Morgan dualities. Compare with the Hilbert system presented for classical logic (Chapter 2, Section 3.1).

- Axiom schemata:

    1. $\Phi \Rightarrow \Phi' \Rightarrow \Phi$, for all formulas Φ, Φ'
    2. $(\Phi \Rightarrow \Phi' \Rightarrow \Phi'') \Rightarrow (\Phi \Rightarrow \Phi') \Rightarrow (\Phi \Rightarrow \Phi'')$, for all formulas Φ, Φ', Φ''
    3. $\Phi \Rightarrow \Phi' \Rightarrow \Phi \wedge \Phi'$, for all formulas Φ, Φ'
    4. $\Phi \wedge \Phi' \Rightarrow \Phi$, for all Φ, Φ'
    5. $\Phi \wedge \Phi' \Rightarrow \Phi'$, for all Φ, Φ'
    6. $\Phi \Rightarrow \Phi \vee \Phi'$, for all Φ, Φ'
    7. $\Phi' \Rightarrow \Phi \vee \Phi'$, for all Φ, Φ'

8. $(\Phi \Rightarrow \Phi'') \Rightarrow ((\Phi' \Rightarrow \Phi'') \Rightarrow (\Phi \vee \Phi' \Rightarrow \Phi''))$, for all Φ, Φ', Φ''

9. $\perp \Rightarrow \Phi$, for all Φ

- Rule: (modus ponens)

$$\frac{\Delta \longrightarrow \Phi \Rightarrow \Phi' \qquad \Gamma \longrightarrow \Phi}{\Delta, \Gamma \longrightarrow \Phi'} \; (MP)$$

Axioms 1 and 2 capture implication, axioms 3–5 capture the introduction and elimination of $\wedge$, axioms 6–8 the introduction and elimination of $\vee$, and the last axiom captures negation. It is worth remarking that the first two rules also express *weakening* and *contraction* respectively.

We write $\Gamma \longrightarrow \Phi$ if we can deduce Φ from the axioms using assumptions Γ. We also adopt a convention and write just Φ if the context Γ is empty, rather than $\longrightarrow \Phi$. A proof π of a sequent $\Gamma \longrightarrow \Phi$ is then a derivation ending with $\Gamma \longrightarrow \Phi$ at the root, and axioms at the leaves. We write $\Gamma \vdash^{H} \Phi$ if $\Gamma \longrightarrow \Phi$ is provable in **H**.

It is important to remark that, with respect to presentation details, the Hilbert style system given here is based on deriving sequents. The relationship between *Hilbert systems* and *Hilbert systems in sequent form* is in the same spirit as the connection between natural deduction and natural deduction in sequent form, see Chapter 2, Section 3.2.

A common example of the axioms shows that we can deduce $\Phi \Rightarrow \Phi$ from instances of axioms 1 and 2, as shown below, where we abbreviate $\Phi \Rightarrow \Phi$ as I.

$$\frac{\dfrac{(\Phi \Rightarrow I \Rightarrow \Phi) \Rightarrow (\Phi \Rightarrow I) \Rightarrow I \qquad \Phi \Rightarrow I \Rightarrow \Phi}{(\Phi \Rightarrow I) \Rightarrow I} \; (MP) \qquad \Phi \Rightarrow I}{\Phi \Rightarrow \Phi} \; (MP)$$

Reading from the top, we have replaced Φ' by $\Phi \Rightarrow \Phi$ and Φ'' by Φ for axiom 2, and replaced Φ' by $\Phi \Rightarrow \Phi$ for axiom 1 and applied the (MP) rule. Next, replacing Φ' by Φ for axiom 1 and again applying (MP) gives the required result.

When we have proved a *theorem* from the axioms, it is very useful to memoise it, and use it as a lemma for further proofs. The previous example in fact turns out to be very useful, and is often used many times in Hilbert-style proofs. There are also alternative sets of axioms from what is shown above, that includes $\Phi \Rightarrow \Phi$, for all Φ, as an axiom. As we have just demonstrated, this axiom is derivable from the others.

Next we present an important result connecting the Hilbert-style system **H** and the deduction system NJ_0. It goes without saying that there are also similar connections between **H** and LJ_0.

Theorem 3.2 $\Gamma \vdash^{\mathbf{NJ_0}} \Phi$ *iff* $\Gamma \vdash^{\mathbf{H}} \Phi$.

Proof: Right to left is quite straightforward — we just need to show that all the axioms are derivable in $\mathbf{NJ_0}$. In Exercise 3.2 the we already saw some of these cases, and we leave the reader to verify the rest (see Exercise 3.7).

To show the other direction we must show how to translate a deduction in $\mathbf{NJ_0}$ into a deduction in $\mathbf{H}$. It is convenient to do this in two steps. First, we define an intermediate system $\mathbf{H'}$ which is just $\mathbf{H}$, but where we allow an additional rule corresponding to $(\Rightarrow I)$. Hence we first translate $\mathbf{NJ_0}$ into $\mathbf{H'}$, then show how we can eliminate all the occurrences of $(\Rightarrow I)$ thus completing the proof.

The first part of the proof proceeds by an induction on the height of the $\mathbf{NJ_0}$ deduction. We define a function $H'(\cdot)$ taking proofs from $\mathbf{NJ_0}$ to proofs of $\mathbf{H'}$. Assume $\Gamma \vdash^{\mathbf{NJ_0}} \Phi$, then we show how to transform the proof into the Hilbert style axioms as follows. The function is defined by cases depending on the rule applied at the root.

- The Axiom rule (Ax) $\Phi \longrightarrow \Phi$ simply becomes the corresponding axiom rule $\Phi \longrightarrow \Phi$ of $\mathbf{H'}$.

- The weakening (W) rule:

$$\frac{\begin{array}{c} \pi \\ \Gamma \longrightarrow \Phi \end{array}}{\Gamma, \Phi' \longrightarrow \Phi} \, (W)$$

becomes the following, using Axiom 1,

$$\frac{\dfrac{\Phi \Rightarrow \Phi' \Rightarrow \Phi \quad \dfrac{H'(\pi)}{\Gamma \longrightarrow \Phi}}{\Gamma \longrightarrow \Phi' \Rightarrow \Phi} \quad \dfrac{}{\Phi' \longrightarrow \Phi'} \, (Ax)}{\Gamma, \Phi' \longrightarrow \Phi} \, (MP)$$

where an (Ax) rule has been used to introduce the assumption Φ'.

- The contraction (C) rule:

$$\frac{\begin{array}{c} \pi \\ \Gamma, \Phi, \Phi \longrightarrow \Phi' \end{array}}{\Gamma, \Phi \longrightarrow \Phi'}$$

becomes the following. First we build a proof of $\Gamma \longrightarrow (\Phi \Rightarrow \Phi) \Rightarrow (\Phi \Rightarrow \Phi')$, using the hypothesis and Axiom 2, as follows:

$$\dfrac{(\Phi \Rightarrow \Phi \Rightarrow \Phi') \Rightarrow (\Phi \Rightarrow \Phi) \Rightarrow \Phi \Rightarrow \Phi' \qquad \dfrac{\dfrac{\dfrac{\dfrac{H'(\pi)}{\Gamma, \Phi, \Phi \longrightarrow \Phi'}\,(\Rightarrow I)}{\Gamma, \Phi \longrightarrow \Phi \Rightarrow \Phi'}\,(\Rightarrow I)}{\Gamma \longrightarrow \Phi \Rightarrow \Phi \Rightarrow \Phi'}}{}}{\Gamma \longrightarrow (\Phi \Rightarrow \Phi) \Rightarrow (\Phi \Rightarrow \Phi')}\,(MP)$$

Since we know already how to build a proof of $\Phi \Rightarrow \Phi$, we can then apply (MP) to get $\Gamma \longrightarrow \Phi \Rightarrow \Phi'$. Finally we complete the proof by:

$$\dfrac{\Gamma \longrightarrow \Phi \Rightarrow \Phi' \qquad \dfrac{}{\Phi \longrightarrow \Phi}\,(Ax)}{\Gamma, \Phi \longrightarrow \Phi'}\,(MP)$$

Note also that this is the first time that we have used the $(\Rightarrow I)$ rule.

- The $(\wedge I)$ rule:

$$\dfrac{\dfrac{\pi_1}{\Gamma \longrightarrow \Phi} \qquad \dfrac{\pi_2}{\Delta \longrightarrow \Phi'}}{\Gamma, \Delta \longrightarrow \Phi \wedge \Phi'}\,(\wedge I)$$

becomes, using Axiom 3,

$$\dfrac{\dfrac{\Phi \Rightarrow \Phi' \Rightarrow \Phi \wedge \Phi' \qquad \dfrac{H'(\pi_1)}{\Gamma \longrightarrow \Phi}}{\Gamma \longrightarrow \Phi' \Rightarrow \Phi \wedge \Phi'}\,(MP) \qquad \dfrac{H'(\pi_2)}{\Delta \longrightarrow \Phi'}}{\Gamma, \Delta \longrightarrow \Phi \wedge \Phi'}\,(MP)$$

- The $(\wedge E_1)$ rule:

$$\dfrac{\dfrac{\pi}{\Gamma \longrightarrow \Phi \wedge \Phi'}}{\Gamma \longrightarrow \Phi}\,(\wedge E_1)$$

becomes, using Axiom 4,

$$\dfrac{\Phi \wedge \Phi' \Rightarrow \Phi \qquad \dfrac{H'(\pi)}{\Gamma \longrightarrow \Phi \wedge \Phi'}}{\Gamma \longrightarrow \Phi}\,(MP)$$

The case for $\wedge E_2$ is entirely similar, using Axiom 5.

- The $(\vee I_1)$ rule:

$$\frac{\dfrac{\pi}{\Gamma \longrightarrow \Phi}}{\Gamma \longrightarrow \Phi \vee \Phi'} \, (\vee I_1)$$

becomes, using Axiom 6,

$$\frac{\Phi \Rightarrow \Phi \vee \Phi' \qquad \dfrac{H'(\pi)}{\Gamma \longrightarrow \Phi}}{\Gamma \longrightarrow \Phi \vee \Phi'} \, (MP)$$

The case for $(\vee I_2)$ is entirely similar, using Axiom 7.

- The $(\vee E)$ rule:

$$\frac{\dfrac{\pi_1}{\Gamma \longrightarrow \Phi \vee \Phi'} \qquad \dfrac{\pi_2}{\Delta, \Phi \longrightarrow \Phi''} \qquad \dfrac{\pi_3}{\Delta, \Phi' \longrightarrow \Phi''}}{\Gamma, \Delta \longrightarrow \Phi''} \, (\vee E)$$

becomes the following. First we build a proof of $\Delta \longrightarrow (\Phi' \Rightarrow \Phi'') \Rightarrow (\Phi \vee \Phi' \Rightarrow \Phi'')$ using the hypothesis and Axiom 8:

$$\frac{(\Phi \Rightarrow \Phi'') \Rightarrow ((\Phi' \Rightarrow \Phi'') \Rightarrow (\Phi \vee \Phi' \Rightarrow \Phi'')) \qquad \dfrac{\dfrac{H'(\pi_2)}{\Delta, \Phi \longrightarrow \Phi''}}{\Delta \longrightarrow \Phi \Rightarrow \Phi''} \, (\Rightarrow I)}{\Delta \longrightarrow (\Phi' \Rightarrow \Phi'') \Rightarrow (\Phi \vee \Phi' \Rightarrow \Phi'')} \, (MP)$$

Also, from $H'(\pi_3)$ we can build a proof of $\Delta \longrightarrow \Phi' \Rightarrow \Phi''$ using $(\Rightarrow I)$. Now, using the rule (MP) we get a proof of $\Delta, \Delta \longrightarrow \Phi \vee \Phi' \Rightarrow \Phi''$.

We can now complete the proof in the following way:

$$\frac{\dfrac{\Delta, \Delta \longrightarrow \Phi \vee \Phi' \Rightarrow \Phi'' \qquad \dfrac{H'(\pi_1)}{\Gamma \longrightarrow \Phi \vee \Phi'}}{\Gamma, \Delta, \Delta \longrightarrow \Phi''} \, (MP)}{\Gamma, \Delta \longrightarrow \Phi''} \, (C)$$

where we have used both $(\Rightarrow I)$, and contraction (C) that we defined earlier.

- The ($\bot$) rule:

$$\frac{\begin{array}{c}\pi\\ \hline \Gamma \longrightarrow \bot\end{array}}{\Gamma \longrightarrow \Phi}\ (\bot)$$

becomes, using Axiom 9,

$$\frac{\bot \Rightarrow \Phi \qquad \dfrac{H'(\pi)}{\Gamma \longrightarrow \bot}}{\Gamma \longrightarrow \Phi}\ (MP)$$

Finally, the cases for ($\Rightarrow I$) and ($\Rightarrow E$) are straightforward since we use the same rules in both systems. This completes the first part of the translation. Next, we have to show how to eliminate the ($\Rightarrow I$) rules from the system. Assume that we have a derivation ending in:

$$\frac{\begin{array}{c}\pi\\ \hline \Gamma, \Phi \longrightarrow \Phi'\end{array}}{\Gamma \longrightarrow \Phi \Rightarrow \Phi'}\ (\Rightarrow I)$$

We define a function $H(\cdot)$ by induction on the height of proofs in $\mathbf{H}'$ which removes all occurrences of the ($\Rightarrow I$) rule. Thus we obtain a proof of the form:

$$\frac{\begin{array}{c}H(\pi)\\ \hline \Gamma, \Phi \longrightarrow \Phi'\end{array}}{\Gamma \longrightarrow \Phi \Rightarrow \Phi'}\ (\Rightarrow I)$$

All we have to show is how to remove the last occurrence of the ($\Rightarrow I$) rule.

We do this by showing how to translate each sub-derivation of π ending in $\Gamma \longrightarrow \Phi'$ into a derivation ending in $\Gamma' \longrightarrow \Phi \Rightarrow \Phi'$, where Γ' is obtained from Γ by possibly discharging an assumption Φ.

There are three base cases to consider depending on which kind of axiom is at the leaf. If π is the axiom

$$\frac{}{\Phi \longrightarrow \Phi}\ (Ax)$$

then $H(\pi)$ is given by $\Phi \Rightarrow \Phi$, which we have already shown to be derivable from axioms in $\mathbf{H}$. If π is the axiom

$$\frac{}{\Phi' \longrightarrow \Phi'}\ (Ax)$$

where Φ' is different from Φ, then $H(\pi)$ is given by the following:

$$\frac{\Phi' \Rightarrow \Phi \Rightarrow \Phi' \qquad \dfrac{\overline{\quad}}{\Phi' \longrightarrow \Phi'}(Ax)}{\Phi' \longrightarrow \Phi \Rightarrow \Phi'}(MP)$$

The final possibility is that there is an axiom Φ'' of **H** at the leaf, in which case it is just a slight variation of the previous case, but there is no context.

$$\frac{\Phi'' \Rightarrow \Phi \Rightarrow \Phi'' \qquad \Phi''}{\Phi \Rightarrow \Phi''}(MP)$$

Next for the inductive case, for which there is only one case to consider

$$\frac{\overset{\pi_1}{\Gamma \longrightarrow \Phi'' \Rightarrow \Psi} \qquad \overset{\pi_2}{\Delta \longrightarrow \Phi''}}{\Gamma, \Delta \longrightarrow \Psi}(MP)$$

becomes the following. First we build a proof of $\Gamma' \longrightarrow (\Phi \Rightarrow \Phi'') \Rightarrow \Phi \Rightarrow \Psi$ using the hypothesis and Axiom 2:

$$\frac{(\Phi \Rightarrow \Phi'' \Rightarrow \Psi) \Rightarrow (\Phi \Rightarrow \Phi'') \Rightarrow \Phi \Rightarrow \Psi \qquad \overset{H(\pi_1)}{\Gamma' \longrightarrow \Phi \Rightarrow \Phi'' \Rightarrow \Psi}}{\Gamma' \longrightarrow (\Phi \Rightarrow \Phi'') \Rightarrow \Phi \Rightarrow \Psi}(MP)$$

Now, using the hypothesis again we can build a proof of $\Delta' \longrightarrow \Phi \Rightarrow \Phi''$ and use (MP) to give a proof of the required sequent.

Hence, starting from the top of the proof, we have eliminated all $(\Rightarrow I)$ in **H′**, thus giving a system only in terms of the Hilbert-style axioms for intuitionistic logic. $\qquad\qquad\square$

▶ **EXERCISE 3.7**
Show each of the Hilbert axioms in $\mathbf{NJ_0}$, thus showing that the axioms are sound.

3   NORMALISATION AND CUT ELIMINATION

In this section we study a useful notion of proof transformation for both the natural deduction and sequent calculus systems. The notion of proof transformation that we are concerned with is replacing part of a proof by another proof of the same sequent, but which is in some sense simpler. We can then keep iterating this process until the proof cannot be simplified further, at which point we have a proof in *normal form*. The key results of the section is that this process of proof simplification terminates.

The process of simplifying proofs in sequent and natural deduction systems are similar and related. The nomenclature used for the natural deduction is *normalisation*, whereas for sequent systems, the word *cut elimination* is used. A natural deduction proof is said to be a *normal derivation* if no further normalisation can be done; a sequent proof is said to be a *cut free* derivation if no further cut elimination can be done. The word *cut* is often overloaded to refer to both systems.

We now look at normalisation and cut elimination in turn.

3.1   Normalisation

In natural deduction, a *cut* arises when we have an introduction rule of some connective, followed by an elimination rule that eliminates precisely the occurrence of the connective that has just been introduced (this will become clearer when we specify the normalisation steps below). This is called a *detour* by Dag Prawitz (1965). Intuitively, this seems to be an inefficiency in a proof, and therefore we should be able to eliminate both rules and still obtain an equivalent proof. The process of eliminating these cuts in natural deduction is called normalisation. In this section we give a recipe for normalisation and show that the process terminates giving a normal form for proofs. This process therefore also eliminates the (Subs) derived rule which we (temporarily) introduced in the last section.

The basic process is to systematically eliminate parts of proofs where there is an introduction of a connective immediately followed by an elimination of the same connective, as described above. Assuming that we have a well formed proof, we define the main rules that we are interested in eliminating as one step normalisation transformations.

Definition 3.3 (One Step Normalisation)   • $(\wedge I)$ *followed by* $(\wedge E_1)$:

$$
\cfrac{
  \cfrac{\dfrac{\pi_1}{\Gamma \longrightarrow \Phi} \qquad \dfrac{\pi_2}{\Delta \longrightarrow \Phi'}}{\Gamma, \Delta \longrightarrow \Phi \wedge \Phi'}\ (\wedge I)
}{\Gamma, \Delta \longrightarrow \Phi}\ (\wedge E_1)
$$

becomes the following derivation where the proof of Φ' *has been deleted.*

$$
\cfrac{\dfrac{\pi_1}{\Gamma \longrightarrow \Phi}}{\Gamma, \Delta \longrightarrow \Phi}\ (W)
$$

There is a similar case for $(\wedge I)$ *followed by* $(\wedge E_2)$.

- $(\Rightarrow I)$ *followed by* $(\Rightarrow E)$:

$$
\cfrac{\cfrac{\pi_1}{\Gamma \longrightarrow \Phi} \qquad \cfrac{\cfrac{\pi_2}{\Delta, \Phi \longrightarrow \Phi'}}{\Delta \longrightarrow \Phi \Rightarrow \Phi'}(\Rightarrow I)}{\Gamma, \Delta \longrightarrow \Phi'}(\Rightarrow E)
$$

becomes the proof π_2', *which is the proof* π_2 *where all axioms* $\Phi \longrightarrow \Phi$ *are replaced (substituted) by a proof of* π_1.

$$
\cfrac{\pi_2'}{\Gamma, \Delta \longrightarrow \Phi'}
$$

- $(\vee I_1)$ *followed by* $(\vee E)$:

$$
\cfrac{\cfrac{\cfrac{\pi_1}{\Gamma \longrightarrow \Phi}}{\Gamma \longrightarrow \Phi \vee \Phi'}(\vee I_1) \qquad \cfrac{\pi_2}{\Delta, \Phi \longrightarrow \Phi''} \qquad \cfrac{\pi_3}{\Delta, \Phi' \longrightarrow \Phi''}}{\Gamma, \Delta \longrightarrow \Phi''}(\vee E)
$$

becomes the proof π_2' *derived from substitution as described for the previous case, and* π_3 *has been deleted.*

$$
\cfrac{\pi_2'}{\Gamma, \Delta \longrightarrow \Phi''}
$$

There is a similar case for $(\vee I_2)$ *followed by* $(\vee E)$.

If we can replace a proof π by a proof π' by applying one of the above transformations then we say that π' is the contractum of π. In other words, each transformation is a *contraction step*.

We have used the word *contraction* for two entirely different purposes. The reader should not confuse contraction steps *on proofs* with the rule of contraction *inside proofs*.

There are several important results that can be stated about the system $\mathbf{NJ_0}$. None of the proofs will be given here, since we will look into these kinds of results in the following chapter for an equivalent system where the objects for manipulation will be a little simpler.

Theorem 3.4 (Normal form) *If* $\Gamma \vdash^{\mathbf{NJ_0}} \Phi$ *then there is a normal derivation of* Φ *from assumptions* Γ.

This result formalises the intuition that we can always write proofs without redundant steps, i.e. without using an introduction followed by an elimination of a connective. Moreover, we can do much better as the following sequence of results shows. We first define a notion of *reduction sequence*.

Definition 3.5 *A reduction sequence is given by a sequence of deductions* $\pi_0, \ldots, \pi_{n-1}$, *where, for all* $i < n - 1$, π_{i+1} *is obtained from* π_i *by applying a contraction step. We write* $\pi \Rightarrow \pi'$ *to mean that there is a reduction sequence from* π *to* π'.

Our next result states that it is possible to transform *any* proof in $\mathbf{NJ}_0$ into a normal derivation.

Theorem 3.6 (Normalisation) *If* π_0 *is a proof of* $\Gamma \longrightarrow \Phi$, *then there exists a finite reduction sequence* $\pi_0, \ldots, \pi_n$ *such that* π_n *is a normal derivation.*

In fact there is an even stronger result which states that *any* reduction sequence terminates with a normal derivation.

Theorem 3.7 (Strong Normalisation) *If* π_0 *is a proof of* $\Gamma \longrightarrow \Phi$, *then there is a number* n *such that all reduction sequences starting from* π_0 *are of length less than* n.

The final result gives the uniqueness of the normal derivation obtained by application of a reduction sequence. First an important theorem which states that the order of the reduction sequence is not important.

Theorem 3.8 (Church-Rosser) *Given two reduction sequences* $\pi \Rightarrow \pi'$ *and* $\pi \Rightarrow \pi''$, *then there is a proof* π''' *such that* $\pi' \Rightarrow \pi'''$ *and* $\pi'' \Rightarrow \pi'''$.

From this, one almost immediately obtains the following.

Corollary 3.9 (Uniqueness of normal form) *A proof* π *has at most one normal form.*

As a consequence of the above results we have a way of writing proofs in canonical form (normal form), and we know this is unique.

3.2   Cut Elimination

Our aim here is to eliminate the (Cut) rule from the sequent calculus presentation of intuitionistic logic.

$$\frac{\Gamma \longrightarrow \Phi \qquad \Delta, \Phi \longrightarrow \Phi'}{\Gamma, \Delta \longrightarrow \Phi'} \, (Cut)$$

The intuition is that we are eliminating the lemmas in a proof—putting the proof of the lemmas *in-line*. As with normalisation, the idea is to transform the proof to an equivalent one, this time without the (Cut) rule. The idea is essentially to push a cut rule up through a proof. The result that we will show is that this process will in fact terminate, leaving a proof without cuts, thus showing that the cut rule is redundant in intuitionistic sequent calculus.

We begin by showing the main cases of transformation which are defined as follows:

Definition 3.10 (One Step Cut Elimination)

- *A cut of a proof π against an axiom:*

$$
\cfrac{\cfrac{}{\Phi \longrightarrow \Phi}\ (Ax) \qquad \cfrac{\pi}{\Gamma, \Phi \longrightarrow \Phi'}}{\Gamma, \Phi \longrightarrow \Phi'}\ (Cut)
$$

becomes simply the proof π:

$$
\cfrac{\pi}{\Gamma, \Phi \longrightarrow \Phi'}
$$

- *A cut of a proof π against a contraction:*

$$
\cfrac{\cfrac{\pi}{\Gamma \longrightarrow \Phi} \qquad \cfrac{\Delta, \Phi, \Phi \longrightarrow \Phi'}{\Delta, \Phi \longrightarrow \Phi'}\ (C)}{\Gamma, \Delta \longrightarrow \Phi'}\ (Cut)
$$

becomes two copies of the proof π in the following configuration:

$$
\cfrac{\cfrac{\pi}{\Gamma \longrightarrow \Phi} \qquad \cfrac{\cfrac{\pi}{\Gamma \longrightarrow \Phi} \qquad \Delta, \Phi, \Phi \longrightarrow \Phi'}{\Gamma, \Delta, \Phi \longrightarrow \Phi'}\ (Cut)}{\Gamma, \Delta \longrightarrow \Phi'}\ (Cut)
$$

- *A cut of a proof π against a weakening:*

$$
\cfrac{\cfrac{\pi}{\Gamma \longrightarrow \Phi} \qquad \cfrac{\Delta \longrightarrow \Phi'}{\Delta, \Phi \longrightarrow \Phi'}\ (W)}{\Gamma, \Delta \longrightarrow \Phi'}\ (Cut)
$$

becomes the following simplification where the proof of π has been removed.

$$\frac{\Delta \longrightarrow \Phi'}{\Gamma, \Delta \longrightarrow \Phi'} \, (W)$$

Note that the resulting proof includes a weakening on the context Γ.

- *A cut $(\wedge R)$ against $(\wedge L_1)$:*

$$\frac{\dfrac{\Gamma \longrightarrow \Phi \quad \Delta \longrightarrow \Phi'}{\Gamma, \Delta \longrightarrow \Phi \wedge \Phi'} \, (\wedge R) \qquad \dfrac{\Theta, \Phi \longrightarrow \Phi''}{\Theta, \Phi \wedge \Phi' \longrightarrow \Phi''} \, (\wedge L_1)}{\Gamma, \Delta, \Theta \longrightarrow \Phi''} \, (Cut)$$

becomes the following derivation:

$$\frac{\dfrac{\Gamma \longrightarrow \Phi \quad \Theta, \Phi \longrightarrow \Phi''}{\Gamma, \Theta \longrightarrow \Phi''} \, (Cut)}{\Gamma, \Delta, \Theta \longrightarrow \Phi''} \, (W)$$

Again, there is a need to complete the proof with a weakening on the context Δ.

- *The $(\Rightarrow L)$ and $(\Rightarrow R)$ cut:*

$$\frac{\dfrac{\Gamma \longrightarrow \Phi \quad \Delta, \Phi' \longrightarrow \Phi''}{\Gamma, \Delta, \Phi \Rightarrow \Phi' \longrightarrow \Phi''} \, (\Rightarrow L) \qquad \dfrac{\Theta, \Phi \longrightarrow \Phi'}{\Theta \longrightarrow \Phi \Rightarrow \Phi'} \, (\Rightarrow R)}{\Gamma, \Delta, \Theta \longrightarrow \Phi''} \, (Cut)$$

becomes the following derivation:

$$\frac{\dfrac{\Gamma \longrightarrow \Phi \quad \Theta, \Phi \longrightarrow \Phi'}{\Gamma, \Theta \longrightarrow \Phi'} \, (Cut) \qquad \Delta, \Phi' \longrightarrow \Phi''}{\Gamma, \Delta, \Theta \longrightarrow \Phi''} \, (Cut)$$

- *The $(\vee L)$ cut against $(\vee R_1)$:*

$$\frac{\dfrac{\Gamma, \Phi \longrightarrow \Phi'' \quad \Delta, \Phi' \longrightarrow \Phi''}{\Gamma, \Delta, \Phi \vee \Phi' \longrightarrow \Phi''} \, (\vee L) \qquad \dfrac{\Theta \longrightarrow \Phi}{\Theta \longrightarrow \Phi \vee \Phi'} \, (\vee R_1)}{\Gamma, \Delta, \Theta \longrightarrow \Phi''} \, (Cut)$$

becomes the following derivation:

$$\frac{\dfrac{\Gamma, \Phi \longrightarrow \Phi'' \qquad \Theta \longrightarrow \Phi}{\Gamma, \Theta \longrightarrow \Phi''} \ (Cut)}{\Gamma, \Delta, \Theta \longrightarrow \Phi''} \ (W)$$

The $(\vee L)$ cut against $(\vee R_2)$ is symmetric.

The following definition gives a sequence of measures on the complexity of formulae and proofs that give a hint to the proof of the forthcoming theorem.

Definition 3.11   *1. The* degree $\deg(\Phi)$ *of a formula Φ is defined as:*

$$
\begin{aligned}
\deg(A) &= 1 \\
\deg(\Phi \wedge \Phi') &= \max\{\deg(\Phi), \deg(\Phi')\} + 1 \\
\deg(\Phi \Rightarrow \Phi') &= \deg(\Phi \vee \Phi') = \deg(\Phi \wedge \Phi') \\
\deg(\neg\Phi) &= \deg(\Phi) + 1
\end{aligned}
$$

(The degree is, intuitively, the height of the formula drawn as a tree.)

2. The degree of a cut is defined to be the degree of the cut formula.

3. The height $h(\pi)$ of a proof π is defined as:

- *if π is an axiom, $h(\pi) = 0$*

- *if π is a proof built from π' and structural rules, then $h(\pi) = h(\pi')$.*

- *if π is a proof built from sub-proofs π_1 and π_2 and a binary logical rule, then $h(\pi) = \max\{h(\pi_1) + 1, h(\pi_2) + 1\}$.*

- *The cut-rank of a proof π is defined as $r(\pi) = \max\{\deg(\Phi_i) + h(\pi_i)\}$ where the π_i range over all subproofs of π ending in a cut on formula Φ_i between two cut-free subproofs. We define $r(\pi) = 0$ if π is cut-free.*

Theorem 3.12 (Gentzen, 1934) *If π is a proof with $r(\pi) = d$ $(d > 0)$ then we can construct a proof π' with $r(\pi') < d$ of the same sequent.*

We will not give the details of the proof here, since we will cover this proof for a more complicated logic in Chapter 6. The important point to note here is that this process of pushing a cut up to the leaves of a proof is a terminating process, and thus shows that all cuts in a proof can be eliminated.

One can wonder why the cut rule is there in the first place since it is redundant. The cut rule provides a way of *compressing* proofs. Following the lemma analogy, it states that we can write proofs by factoring out lemmas. Hence, the *cut-formula* Φ only needs to be proved once, but can be used many times in the proof of Φ'.

4   SEMANTICS OF INTUITIONISTIC LOGIC

Thus far we have mainly been concerned with the proof theory of intuitionistic logic, more specifically, derivability of sequents in a particular proof system. In this section we will give a semantics of intuitionistic logic, so that we can then see that the proof systems we have been using do in fact capture validity in a model.

As a first thought, we might try to give a truth table semantics for intuitionistic logic. However the following theorem, due to Gödel, quickly puts an end to this idea if we would like to have a sound and complete model.

Theorem 3.13 *There is no finite truth table semantics for intuitionistic logic that* is *sound* and *complete.*

Proof: Assume that there is a model with n truth values. We show that the formula:

$$\bigvee_{1 \leq i < j \leq n+1} (A_i \Leftrightarrow A_j)$$

is valid in the model, but not provable in intuitionistic logic.

To show that it is valid in the model it suffices to see that if there are n truth values then the formula above will have one such disjunct $(A_i \Leftrightarrow A_j)$ which has to be true in the model. To see this, take $n = 2$ as in the case of classical logic. Now, since there are only two truth values, the formula $(A_1 \Leftrightarrow A_2) \vee (A_1 \Leftrightarrow A_3) \vee (A_2 \Leftrightarrow A_3)$ must have an assignment ρ such that $\llbracket A_i \rrbracket \rho = \llbracket A_j \rrbracket \rho$, $(i \neq j)$. Thus the whole formula is valid in the model.

There is however no way of giving a proof of the above formula in intuitionistic logic, since an attempt to build a derivation would require the use of the law of excluded middle.

$\square$

Indeed, the truth values for classical logic provide a sound model of intuitionistic logic (the reader will easily verify that if $\Gamma \longrightarrow \Phi$ is derivable, then $\Gamma \models \Phi$.) A complete model can be obtained in fact if we introduce a notion of *partiality*. This is the essence of Kripke semantics, which is the first semantics that we present.

4.1   Kripke Semantics

Kripke models provide a very intuitive idea about what is provable in intuitionistic logic. Before presenting the semantics basic order-theoretic notion:

Definition 3.14 *A partially ordered set (often abbreviated to* poset*) is a mathematical structure* $(P, \leq)$ *which consists of a set P equipped with a binary relation $\leq$ (pronounced "less than") satisfying the following laws:*

   1. Reflexivity: $x \leq x$, for all $x \in P$

2. *Transitivity: if $x \leq y$ and $y \leq z$ then $x \leq z$, for all $x, y, z \in P$*

3. *Antisymmetry: if $x \leq y$ and $y \leq x$ then $x = y$, for all $x, y \in P$*

We say that two elements $x, y \in P$ are comparable *whenever $x \leq y$ or $y \leq x$, otherwise x and y are said to be* incomparable.

Instances of partial orders are not hard to find. Below are several examples that the reader will easily verify to be posets.

- $(\mathbb{N}, \leq)$, where $\mathbb{N}$ is the set of natural numbers and $\leq$ is the usual 'less than or equal to' ordering.

- $(\mathcal{P}(X), \subseteq)$, where $\mathcal{P}(X)$ is the powerset of some set X and $\subseteq$ is the usual subset inclusion ordering.

- $(\{0, 1\}^*, \leq)$, where $\{0, 1\}^*$ are finite binary strings and $\leq$ is the *prefix* ordering. For example $0 \leq 00, 0 \leq 01$, etc., and the elements 00 and 01 are incomparable.

The last example given will be the one that we use as a running example, but *any* poset do. A useful way of drawing a poset is by the use of *Hasse diagrams*, a diagrammatic representation of posets of which we will see several examples in this section. Using a partially ordered structure allows us to build the basic model of intuitionistic logic. We begin with the definition of a Kripke frame, then show how the partial order structure is used.

Definition 3.15 (Kripke frame) *A Kripke frame $\mathcal{F}$ is a structure $(X, \leq, \models)$, where:*

- $(X, \leq)$ *is a poset, and*

- $\models$ *is a binary* forcing *relation such that if $x \models A$ and $x \leq y$ then $y \models A$, for all $x, y \in X$.*

  We write $x \models A$ to mean that the formula A is forced at x. The notation $x \not\models A$ is used for x does not force A.

The semantics of the intuitionistic connectives is then given by extending the forcing relation to all the connectives in the following way:

$$
\begin{array}{lll}
x \models \Phi \wedge \Phi' & \textit{iff} & x \models \Phi \textit{ and } x \models \Phi' \\
x \models \Phi \vee \Phi' & \textit{iff} & x \models \Phi \textit{ or } x \models \Phi' \\
x \models \Phi \Rightarrow \Phi' & \textit{iff} & \textit{for all } y, x \leq y, y \models \Phi \textit{ implies } y \models \Phi' \\
x \models \neg \Phi & \textit{iff} & \textit{for all } y, x \leq y, y \not\models \Phi \\
x \not\models \bot & & \textit{for every } x
\end{array}
$$

Forcing is then extended to sequents in the following way: $x \models \Gamma \longrightarrow \Phi$ iff (for all $\Psi \in \Gamma, x \models \Psi$) implies $x \models \Phi$, that is, iff whenever $x \models \Psi$ for every $\Psi \in \Gamma$, then $x \models \Phi$. We will write $\Gamma \models_x \Phi$ as an abbreviation for $x \models \Gamma \longrightarrow \Phi$.

The interpretation of intuitionistic negation in a frame given above is in fact a redundant case, since it is derivable from the other cases together with the fact that $\neg \Phi$ is equivalent to $\Phi \Rightarrow \bot$. See Exercise 3.8.

Definition 3.16 *A formula Φ is said to be* forced *in the frame $\mathcal{F}$ if every x in X forces Φ. Φ is* intuitionistically valid *if it is forced in* every *frame. This notion extends to sequents in the natural way.*

Intuitively, think of the forcing relation $x \models \Phi$ as meaning that Φ is *true* at x. The partial order $(X, \leq)$ provides a way of talking about a *possible world*, or *state* of knowledge. We think of $x \leq y$ as meaning y is the future of x, or y extends x. To give a little intuition about how the partial order information is used, we consider the example of whether $A \vee \neg A$ is forced or not, i.e. $0 \models A \vee \neg A$. We start from no knowledge, and increase our information as we go up the partial order. The following frame gives us a starting point, which we draw as a *Hasse diagram*.

$$00 \models A$$

$$0$$

Here, we see that at 0, nothing is forced. However, we cannot eliminate the possibility of A being proven in the future, so we postulate that A might be forced later, shown as $00 \models A$. Now, at 0 we cannot assume $\neg A$ since we already said that A might hold in the future, and we assume (see below) that we never get inconsistent information in a frame. We cannot assume A because we know nothing about it. Hence the above diagram shows a frame that does not force $A \vee \neg A$ at 0, thus it is not intuitionistically valid. We will see in the next subsection that there is a systematic way of building frames, using the tableaux proof technique.

The following lemma states that the forcing relation is preserved by the ordering. This is called *monotonicity* and holds for the kind of logic that we are dealing with. We remark that there are other logics, in particular the so-called *non-monotonic* logics, that do not satisfy this property.

Lemma 3.17 (Monotonicity) *If $x \models \Phi$ and $x \leq y$, then $y \models \Phi$.*

Proof: The proof proceeds by an induction on the structure of Φ. See Exercise 3.9.

$\square$

Here are three examples of using frames to show if a formula is intuitionistically valid or not. These examples demonstrate just how useful the monotonicity lemma is.

1. We show first that $\Phi \Rightarrow \Phi$ is intuitionistically valid. Now, to show $x \models \Phi \Rightarrow \Phi$ we must show that for all $y, x \leq y, y \models \Phi$ implies $y \models \Phi$, which is immediate.

2. Next we show that $\Phi \Rightarrow \neg\neg\Phi$ is intuitionistically valid. To show $x \models \Phi \Rightarrow \neg\neg\Phi$ we must show that for all y, $x \leq y$, $y \models \Phi$ implies $y \models \neg\neg\Phi$. Now, $y \models \neg\neg\Phi$ if and only if for all u, $y \leq u$, there exists v, $u \leq v$ such that $v \models \Phi$ (see comment below). The result follows by transitivity of $\leq$, and the monotonicity Lemma 3.17.

3. Next we give a formula that we know not to be valid. $x \models \neg\neg\Phi \Rightarrow \Phi$ if and only if for all y, $x \leq y$, $y \models \neg\neg\Phi$ implies $y \models \Phi$. Now, from the previous example, using transitivity again, we are left to show that this is valid if and only if for all u, $y \leq u$ there is a v, $u \leq v$ such that $v \models \Phi$ implies $y \models \Phi$. Putting in some specific values shows that this does not hold, and hence the formula is not valid. We can see this by giving the following frame (which incidentally, is the same as the frame used to show that $A \vee \neg A$ is not valid):

$$00 \models A$$

$$0$$

In the second example above we used the following *fact*:

$$x \models \neg\neg\Phi \text{ iff for all } y, x \leq y, \text{ there exists } u, y \leq u, \text{ such that } u \models \Phi.$$

To justify this, we have to use *classical* reasoning in our meta-language, rather than intuitionistic reasoning. See Troelstra and van Dalen (1988a, page 78) for more discussion on this point.

It is worth remarking that all the formulas in the above examples that are intuitionistically valid are in fact provable in the proof systems that we have presented, and the example that shows a non-valid formula is known *not* to be a derivable sequent. This gives a hint that the proof systems that we have been using for intuitionistic logic do in fact capture intuitionistic validity. We show the results here for natural deduction, which we already know is equivalent to sequent calculus and Hilbert-style axioms.

Theorem 3.18 (Soundness) *If* $\Gamma \longrightarrow \Phi$ *is a derivable sequent in* **NJ**$_0$ *then* $\Gamma \longrightarrow \Phi$ *is intuitionistically valid.*

Proof: A straightforward induction over the height of a proof in **NJ**$_0$. We just give a selection of cases.

- Identity group. To show that the axiom rule (Ax) is valid we have to show is that $x \models \Phi$ implies $x \models \Phi$, which is immediate. Equally simple are the cases for the structural group.

- The $(\Rightarrow I)$ rule. By hypothesis $\Gamma, \Phi \models_x \Phi'$, and we must show that $\Gamma \models_x$ $\Phi \Rightarrow \Phi'$. Unpacking the hypothesis yields that for all Ψ in Γ, Φ, (for all y, $x \leq$ $y, y \models \Psi$) implies $y \models \Phi'$. Unpacking what we are required to prove and using monotonicity and transitivity quickly gives the required result.

- The $(\Rightarrow E)$ rule. By hypothesis we have $\Gamma \models_x \Phi \Rightarrow \Phi'$ and $\Gamma \models_x \Phi$, and we must show that $\Gamma, \Delta \models_x \Phi'$. Again expanding these, using transitivity of $\leq$ and monotonicity of $\models$ gives the result.

The remaining cases follow in a similar way. $\qquad\qquad\qquad\qquad\square$

The above result shows in fact that all the derivations that we have made for intuitionistic logic are in fact valid. It is also possible to show that the systems we have been using are also complete, in the following sense.

Theorem 3.19 (Completeness) *If* $\Gamma \quad \longrightarrow \quad \Phi$ *is intuitionistically valid, then* $\Gamma \vdash^{\mathrm{NJ_0}} \Phi$.

We will not enter into the details of proving completeness here. In the following section we turn to a theorem proving technique for intuitionistic logic which corresponds *exactly* with Kripke semantics, and the connection with proof systems comes out much clearer in that system.

Finally, we remark that there is a degenerate form of the Kripke semantics in which we have only one state, say $x = 0$. One can easily see that with just one state we get *classical* validity.

Additional material on Kripke semantics for intuitionistic logic can be found in Troelstra and van Dalen (1988a; 1988b).

▶ **EXERCISE 3.8**
Prove that $x \models \Phi \Rightarrow \bot$ iff $x \models \neg\Phi$, thus showing that, in the definition of forcing, the case for negation is redundant.

▶ **EXERCISE 3.9**
Prove the Monotonicity Lemma 3.17: If $x \models \Phi$ and $x \leq y$, then $y \models \Phi$.

4.2   *Intuitionistic Tableaux*

Intuitionistic tableaux provide an automated proof method for intuitionistic logic, which are a simple variant of the tableaux proof method for classical logic that we discussed in Chapter 2, Section 4.1. There is a remarkably strong connection between Kripke semantics and the tableaux proof method for intuitionistic logic. What we give here is a direct re-casting of the Kripke semantics in a tableaux framework. Indeed, it is instructive to understand the Kripke semantics in terms of this theorem proving method—it also gives a simple way of constructing counter-examples for non-valid formulae.

This "connection" is no accident. In fact there are many variants of the Kripke semantics, and there is always a corresponding tableaux proof method. The particular variant that we present here follows that of Nerode (1990). Here we will just present the proof system, and give some examples and elementary results.

We build intuitionistic tableaux in a similar way to the classical case, presented in Chapter 2, but with several important differences. First, to show an alternative method for building tableaux, we follow the traditional approach, and negate the formula that we are trying to prove, $\neg\Phi$, and ask whether $\neg\Phi$ is unsatisfiable. Second, the tableaux will be labelled with *forcing relations* $+x \models \Phi$ (resp. $-x \models \Phi$) with the intended meaning that x forces (resp. does not force) the formula Φ (x being nothing more than a partial order that we already introduced for forcing relations in the previous section).

We build tableaux by placing $-0 \models \Phi$ at the root, and using the tableaux rules that we will shortly define.

Definition 3.20 *A path is said to be* open *if both* $+x \models \Phi$ *and* $x \models \Phi$ *do not* occur *on the same path, for all partial order elements* $x \in X$ *and formulas* Φ.

If $+x \models \Phi$ *and* $-x \models \Phi$ *are on the same path, then the path is declared* closed, *which we will indicate as '(closed)'.*

The rules for building intuitionistic tableaux are given in Figure 3.3. Several comments are in order. Rules $\vee$ and $\wedge$ are quite straightforward. However, the rules for $\Rightarrow$ and $\neg$ contain restrictions on the partial order structure. The rule $+x \models A \Rightarrow B$ is decomposed and creates a branching in the tableaux with entries $-x' \models A$ and $+x' \models B$ for any x', $x \leq x'$. By *for any* x' we mean specifically any x' that already appears in the branch developed so far. Similarly for the $-x \models A\Rightarrow$, we extend the tableaux by appending the forcing relations $+x' \models A$ and $-x' \models B$, *for some new* x', $x \leq x'$. By *for some new* x' we mean specifically that x' is a value that doesn't already appear in the branch being developed so far, and is not comparable with any element between x and x'. The rules for negation are similar. Finally, the monotonicity rule says that we can propagate forced formulae along paths in a tableaux. This rule turns out to be very useful in closing paths in the tableaux.

Suppose that we want to show that the formula $A \Rightarrow B \Rightarrow A$ is intuitionistically valid. We begin the process of building the proof by writing $-0 \models A \Rightarrow B \Rightarrow A$ at the root. Applying the rule for $\Rightarrow$ from Figure 3.3 allows us to start building the tableaux:

$$-0 \models A \Rightarrow B \Rightarrow A$$

$$+00 \models A$$

$$-00 \models B \Rightarrow A$$

where 00 ($0 \leq 00$) is new. We again apply the $\Rightarrow$ rule, giving the following tableaux:

Rules for $\vee$:

$$+x \models A \vee B$$

$$+x \models A \quad\bigg|\quad +x \models B$$

$$-x \models A \vee B$$
$$-x \models A$$
$$-x \models B$$

Rules for $\wedge$:

$$+x \models A \wedge B$$
$$+x \models A$$
$$+x \models B$$

$$-x \models A \wedge B$$
$$-x \models A \quad\bigg|\quad -x \models B$$

Rules for $\Rightarrow$:

$$+x \models A \Rightarrow B$$
$$-x' \models A \quad\bigg|\quad +x' \models B$$

$$-x \models A \Rightarrow B$$
$$+x' \models A$$
$$-x' \models B$$

For any x', $x \leq x'$ For some new x', $x \leq x'$

Rules for $\neg$:

$$+x \models \neg A$$
$$-x' \models A$$

$$-x \models \neg A$$
$$+x' \models A$$

For any x', $x \leq x'$ For some new x', $x \leq x'$

Rule for $\perp$:

$$+x \models \perp$$

(closed)

Monotonicity rule:

$$+x \models A$$

$$+x' \models A$$
For any x', $x \leq x'$

Figure 3.3. Intuitionistic Tableaux

$$-0 \models A \Rightarrow B \Rightarrow A$$

$$+00 \models A$$

$$-00 \models B \Rightarrow A$$

$$+000 \models B$$

$$-000 \models A$$

where 000 ($00 \leq 000$) is new. The only rule that can now apply is Monotonicity, which allows us to enter $+000 \models A$. We now have a path with $+000 \models A$ and $-000 \models A$, which is a *closed path*, thus showing that the formula $A \Rightarrow B \Rightarrow A$ is intuitionistically valid. It is very clear from this example that proof search in intuitionistic tableaux is exactly like proving a formula valid using Kripke frames.

To see how tableaux relate to frames (and thus see how the partial order structure is used) we examine the tableaux and write all the forced formulae ($+x \models A$) as a Hasse diagram. Consider the tableaux for $-0 \models (A \Rightarrow B) \vee (B \Rightarrow A)$ which we can build as follows.

$$-0 \models (A \Rightarrow B) \vee (B \Rightarrow A)$$

$$+0 \models A \Rightarrow B$$

$$+0 \models B \Rightarrow A$$

$$+00 \models A$$

$$-00 \models B$$

$$+01 \models B$$

$$-01 \models A$$

There is no way of closing the tableax, hence this is not a valid intuitionistic formula. We generate the frame by selecting all the forced elements of the tableaux, giving elements $+00 \models A$ and $+01 \models B$, with nothing forced at 0.

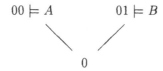

$$00 \models A \qquad 01 \models B$$

$$0$$

Here we see a splitting in the frame structure.

It is also worth remarking that if we set the state x to be the partially ordered set with just one element, then we get a variant of the classical tableaux presented in Chapter 2, Section 4.1.

We can in fact use this proof method, in a similar way as we have shown for classical tableaux, to build a derivation of a formula in intuitionistic sequent calculus. Remark also that since intuitionistic tableaux and Kripke frames coincide, we have a hint at how to prove completeness of Kripke semantics for intuitionistic logic. Further details of intuitionistic tableaux, and also Kripke frames, can be found in (Nerode, 1990).

▶ **EXERCISE 3.10**
Construct intuitionistic tableaux for the following formulae:

$$(i) \quad A \vee \neg A \qquad (ii) \quad \neg\neg A \Rightarrow A \qquad (iii) \quad A \Rightarrow \neg\neg A$$

4.3 Functional Interpretation

To complete this section on semantics of intuitionistic logic, we investigate, in a little more detail, the *proof theoretic* paradigm hinted at in the introduction. The functional interpretation, also known as Heyting, BHK[1] or Realisability semantics, models *proofs* of formulae as *functions*. This kind of semantics will also provide the starting point for the material in the next chapter where we study the Curry-Howard isomorphism.

Definition 3.21 (Heyting proofs) *To each formula Φ we define the corresponding notion of a proof object by induction on the structure of Φ.*

1. *A proof π of an atomic formula A is given by the context (in other words we assume we have already atomic proofs).*

2. *A proof π of $\Phi \wedge \Phi'$ is a pair (π_1, π_2) such that π_1 is a proof of Φ, and π_2 is a proof of Φ'.*

3. *A proof π of $\Phi \vee \Phi'$ is a pair (i, π') such that:*

   - *either $i = 0$ and π' is a proof of Φ;*
   - *or $i = 1$ and π' is a proof of Φ'.*

   In other words, we require some evidence of which side of the $\vee$ we have proved.

4. *A proof π of $\Phi \Rightarrow \Phi'$ is a function f such that if π' is a proof of Φ, the application $f(\pi')$ is a proof of Φ'.*

5. *A proof π of $\neg\Phi$ is a function f such that if π' is a proof of Φ, the application $f(\pi')$ is a proof of $\bot$. In other words, a function with an empty domain since there are no proofs of $\bot$.*

---

[1] Brouwer-Heyting-Kolmogorov.

It is instructive to see an example so that the structure of a proof becomes more evident. Consider the formula F defined as $A \Rightarrow B \Rightarrow A$, which we know is intuitionistically valid. Following the definition above, we see that a proof of the formula F is a function f such that when f is applied to a proof a of A, $f(a)$ is a proof of $B \Rightarrow A$. Iterating the process again, we see that a proof of $B \Rightarrow A$ is a function f' such that if b is a proof of B the application $f'(b)$ is a proof of A, i.e. $f'(b) = a$. Putting all this together we have:

$$f(a)(b) \quad = \quad f'(b) \quad = \quad a$$

So a proof of $A \Rightarrow B \Rightarrow A$ is a function f of two arguments, and returns the first as result.

With practice, one can write functions straight away from the formula. For example a proof f of $A \wedge B \Rightarrow B \wedge A$ is a function which takes a pair (a, b), where a is a proof of A and b is a proof of B, and gives a pair (b, a). Thus, $f(a, b) = (b, a)$.

Now let us try to find the interpretation corresponding to the formula F defined by $A \vee \neg A$, which we know is *not* intuitionistically valid. A proof of F is a pair (i, π) where π is either a proof of A or a proof of $\neg A$. If we try to give a proof of A then we fail immediately: there is no context here. Let us then try to provide a proof of $\neg A$. A proof of $\neg A$, if it exists, is a function f with an empty domain; in other words f exists only if there is no proof of A which we cannot say. Hence there is no proof of the formula $A \vee \neg A$.

As a final example, we give the interpretation of $\neg \bot$; an intuitionistically valid formula. Now, a proof of $\neg \bot$ is a function f such that if a is a proof of $\bot$ then $f(a)$ is a proof of $\bot$. Hence the interpretation can be given by the identity function. However, remark that this function only exists only if there is no proof of $\bot$, which is indeed the case, thus the function will be empty.

In the next chapter we will be more precise about what these proof objects are by defining a specific calculus (*the λ-calculus*) to represent proofs.

▶ **EXERCISE 3.11**
Show the functional interpretations of the following:

(i)	$A \Rightarrow A$	(ii)	$\bot \Rightarrow A$
(iii)	$A \Rightarrow B \longrightarrow \neg B \Rightarrow \neg A$	(iv)	$A \Rightarrow \bot$
(v)	$A \Rightarrow A \vee \neg A$	(vi)	$\neg\neg A \Rightarrow A$

5   RELATING INTUITIONISTIC AND CLASSICAL LOGIC

When we have different systems of logic, we are often interested in how they compare. In particular, we might be interested in knowing whether one logic is "stronger" than another; whatever we mean by that! By a simple analysis of the logical rules for classical logic and intuitionistic logic we see that the latter has a rule missing—the double negation elimination for the natural deduction presentation, and sequents restricted to at most one formula for the sequent calculus presentation. Hence the rules for intuitionistic logic are a proper subset of the rules for classical logic. Therefore we can deduce the following result, which states that intuitionistic logic is "weaker" than classical logic.

Theorem 3.22 $\Gamma \vdash^{\mathbf{LJ}_0} \Phi$ *implies* $\Gamma \vdash^{\mathbf{LK}_0} \Phi$

The proof of this theorem is almost trivial, since the rules for intuitionistic logic are included in classical logic. However, this is not the end of the story. What is much more interesting is to ask whether classical logic can be encoded into intuitionistic logic. At first sight, it looks quite impossible to encode all, since we cannot prove $\Phi \vee \neg\Phi$ in the latter. However, there are many encodings that allows all of classical logic to be encoded into intuitionistic logic. The translation that we give here is one of the simplest, and we will mention several other translations at the end of the section.

These results hold equally for both sequent and natural deduction presentations. Here we will show the results for the sequent calculus.

Definition 3.23 *We define a translation function which takes a classical sequent* $\Gamma \longrightarrow \Delta$ *into an intuitionistic sequent* $[\Gamma], \neg[\Delta] \longrightarrow \perp$ *where:*

$$
\begin{aligned}
{[A]} &= A \\
{[\neg\Phi]} &= \neg[\Phi] \\
{[\Phi \wedge \Phi']} &= \neg\neg[\Phi] \wedge \neg\neg[\Phi'] \\
{[\Phi \vee \Phi']} &= \neg\neg[\Phi] \vee \neg\neg[\Phi'] \\
{[\Phi \Rightarrow \Phi']} &= \neg[\Phi'] \Rightarrow \neg[\Phi]
\end{aligned}
$$

Theorem 3.24 (Embedding LK$_0$ into LJ$_0$)   *1.* $\vdash^{\mathbf{LK}_0} (\Phi \Leftrightarrow [\Phi])$

2. $\Gamma \vdash^{\mathbf{LK}_0} \Phi$ *iff* $[\Gamma], \neg[\Phi] \vdash^{\mathbf{LJ}_0} \perp$

Proof:

1. This follows trivially by the fact that $\neg\neg\Phi \Leftrightarrow \Phi$ and the de Morgan dualities are classically valid.

2. ($\Leftarrow$) follows from Theorem 3.22. ($\Rightarrow$) is proved by an induction over the length of the proof.

To verify the last implication, we define the translation function $L(\cdot)$ from proofs in $\mathbf{LK}_0$ to proofs in $\mathbf{LJ}_0$ such that a proof of a sequent $\Gamma \longrightarrow \Delta$ in $\mathbf{LK}_0$ becomes a proof of the sequent $[\Gamma], \neg[\Delta] \longrightarrow \perp$ in $\mathbf{LJ}_0$. $L(\cdot)$ is defined by induction over the structure of the classical sequent calculus proofs. We assume $\Gamma \vdash^{\mathbf{LK}_0} \Delta$, and define the translation of proofs by cases depending on the rule applied at the root.
Identity group

- The axiom $\Phi \longrightarrow \Phi$ is translated in the following way:

$$\cfrac{\cfrac{}{[\Phi] \longrightarrow [\Phi]}\,(Ax)}{[\Phi], \neg[\Phi] \longrightarrow \bot}\,(\neg L)$$

- The (Cut) rule

$$\cfrac{\cfrac{\pi_1}{\Gamma \longrightarrow \Phi, \Delta} \qquad \cfrac{\pi_2}{\Gamma', \Phi \longrightarrow \Delta'}}{\Gamma, \Gamma' \longrightarrow \Delta, \Delta'}\,(Cut)$$

becomes the following derivation:

$$\cfrac{\cfrac{L(\pi_1)}{[\Gamma], \neg[\Phi], \neg[\Delta] \longrightarrow} \qquad \cfrac{\cfrac{L(\pi_2)}{[\Gamma'], [\Phi], \neg[\Delta'] \longrightarrow}{[\Gamma'], \neg[\Delta'] \longrightarrow \neg[\Phi]}\,(\neg L)}{[\Gamma], [\Gamma'], \neg[\Delta], \neg[\Delta'] \longrightarrow}\,(Cut)$$

Structural group

All the structural rules of $\mathbf{LK}_0$ translate in a straightforward way, and become the corresponding proof in $\mathbf{LJ}_0$, applied on the left hand side. Here we just show an exchange and a weakening to illustrate the point.

$$\cfrac{\cfrac{\pi}{\Gamma \longrightarrow \Delta, \Phi, \Phi', \Delta'}}{\Gamma \longrightarrow \Delta, \Phi', \Phi, \Delta'}\,(RX) \quad \Longrightarrow \quad \cfrac{\cfrac{L(\pi)}{[\Gamma], \neg[\Delta], \neg[\Phi], \neg[\Phi'], \neg[\Delta'] \longrightarrow}}{[\Gamma], \neg[\Delta], \neg[\Phi'], \neg[\Phi], \neg[\Delta'] \longrightarrow}\,(X)$$

$$\cfrac{\cfrac{\pi}{\Gamma \longrightarrow \Delta}}{\Gamma \longrightarrow \Phi, \Delta}\,(RW) \quad \Longrightarrow \quad \cfrac{\cfrac{L(\pi)}{[\Gamma], \neg[\Delta] \longrightarrow}}{[\Gamma], \neg[\Phi], \neg[\Delta] \longrightarrow}\,(W)$$

Logical Group

The logical rules require a little care to keep within $\mathbf{LJ}_0$. All the cases are shown below in full, however we encourage the reader to reconstruct them.

- The $(\Rightarrow L)$ rule

$$\cfrac{\cfrac{\pi_1}{\Gamma \longrightarrow \Phi, \Delta} \qquad \cfrac{\pi_2}{\Gamma', \Phi' \longrightarrow \Delta'}}{\Gamma, \Gamma', \Phi \Rightarrow \Phi' \longrightarrow \Delta, \Delta'}\,(\Rightarrow L)$$

becomes the following derivation:

$$
\frac{
L(\pi_1)
\qquad
\dfrac{
\dfrac{L(\pi_2)}{[\Gamma'],[\Phi'],\neg[\Delta'] \longrightarrow}
}{[\Gamma'],\neg[\Delta'] \longrightarrow \neg[\Phi']}\,(\neg R)
}{
[\Gamma],[\Gamma'],\neg[\Phi'] \Rightarrow \neg[\Phi],\neg[\Delta],\neg[\Delta'] \longrightarrow
}\,(\Rightarrow L)
$$

$$
\frac{[\Gamma],\neg[\Phi],\neg[\Delta] \longrightarrow}{}
$$

- The $(\Rightarrow R)$ rule

$$
\frac{
\dfrac{\pi}{\Gamma,\Phi \longrightarrow \Phi',\Delta}
}{
\Gamma \longrightarrow \Phi \Rightarrow \Phi',\Delta
}\,(\Rightarrow R)
$$

becomes the following derivation:

$$
\frac{
\dfrac{
\dfrac{
\dfrac{L(\pi)}{[\Gamma],[\Phi],\neg[\Phi'],\neg[\Delta] \longrightarrow}
}{[\Gamma],\neg[\Phi'],\neg[\Delta] \longrightarrow \neg[\Phi]}\,(\neg R)
}{[\Gamma],\neg[\Delta] \longrightarrow \neg[\Phi'] \Rightarrow \neg[\Phi]}\,(\Rightarrow R)
}{
[\Gamma],\neg(\neg[\Phi'] \Rightarrow \neg[\Phi]),\neg[\Delta] \longrightarrow
}\,(\neg L)
$$

- The $(\wedge L)$ rule

$$
\frac{
\dfrac{\pi}{\Gamma,\Phi \longrightarrow \Delta}
}{
\Gamma,\Phi \wedge \Phi' \longrightarrow \Delta
}\,(\wedge L_1)
$$

becomes the following derivation:

$$
\frac{
\dfrac{
\dfrac{
\dfrac{L(\pi)}{[\Gamma],[\Phi],\neg[\Delta] \longrightarrow}
}{[\Gamma],\neg[\Delta] \longrightarrow \neg[\Phi]}\,(\neg R)
}{[\Gamma],\neg\neg[\Phi],\neg[\Delta] \longrightarrow}\,(\neg L)
}{
[\Gamma],\neg\neg[\Phi] \wedge \neg\neg[\Phi'],\neg[\Delta] \longrightarrow
}\,(\wedge L)
$$

The case for $(\wedge L_2)$ is similar.

- The $(\wedge R)$ rule

$$
\frac{
\dfrac{\pi_1}{\Gamma \longrightarrow \Phi,\Delta}
\qquad
\dfrac{\pi_2}{\Gamma' \longrightarrow \Phi',\Delta'}
}{
\Gamma,\Gamma' \longrightarrow \Phi \wedge \Phi',\Delta,\Delta'
}\,(\wedge R)
$$

becomes the following derivation:

$$
\cfrac{
  \cfrac{
    \cfrac{L(\pi_1)}{[\Gamma], \neg[\Phi], \neg[\Delta] \longrightarrow}
    {[\Gamma], \neg[\Delta] \longrightarrow \neg\neg[\Phi]} \; (\neg R)
  \qquad
  \cfrac{
    \cfrac{L(\pi_2)}{[\Gamma'], \neg[\Phi'], \neg[\Delta'] \longrightarrow}
    {[\Gamma'], \neg[\Delta'] \longrightarrow \neg\neg[\Phi']} \; (\neg R)
  }
  {
    \cfrac{[\Gamma], [\Gamma'], \neg[\Delta], \neg[\Delta'] \longrightarrow \neg\neg[\Phi] \wedge \neg\neg[\Phi']}{[\Gamma], [\Gamma'], \neg(\neg\neg[\Phi] \wedge \neg\neg[\Phi']), \neg[\Delta], \neg[\Delta'] \longrightarrow} \; (\neg L)
  } \; (\wedge R)
}
$$

- The $(\vee L)$ rule

$$
\cfrac{
  \cfrac{\pi_1}{\Gamma, \Phi \longrightarrow \Delta}
  \qquad
  \cfrac{\pi_2}{\Gamma', \Phi' \longrightarrow \Delta'}
}
{\Gamma, \Gamma', \Phi \vee \Phi' \longrightarrow \Delta, \Delta'} \; (\vee L)
$$

becomes the following derivation:

$$
\cfrac{
  \cfrac{
    \cfrac{
      \cfrac{L(\pi_1)}{[\Gamma], [\Phi], \neg[\Delta] \longrightarrow}
      {[\Gamma], \neg[\Delta] \longrightarrow \neg[\Phi]} \; (\neg R)
    }
    {[\Gamma], \neg\neg[\Phi], \neg[\Delta] \longrightarrow} \; (\neg L)
  \qquad
  \cfrac{
    \cfrac{
      \cfrac{L(\pi_2)}{[\Gamma'], [\Phi'], \neg[\Delta'] \longrightarrow}
      {[\Gamma'], \neg[\Delta'] \longrightarrow \neg[\Phi']} \; (\neg R)
    }
    {[\Gamma'], \neg\neg[\Phi'], \neg[\Delta'] \longrightarrow} \; (\neg L)
  }
}
{[\Gamma], [\Gamma'], \neg\neg[\Phi] \vee \neg\neg[\Phi'], \neg[\Delta], \neg[\Delta'] \longrightarrow} \; (\vee L)
$$

- The $(\vee R)$ rule

$$
\cfrac{
  \cfrac{\pi}{\Gamma \longrightarrow \Phi, \Delta}
}
{\Gamma \longrightarrow \Phi \vee \Phi', \Delta} \; (\vee R_1)
$$

becomes the following derivation:

$$
\cfrac{
  \cfrac{
    \cfrac{
      \cfrac{L(\pi)}{[\Gamma], \neg[\Phi], \neg[\Delta] \longrightarrow}
      {[\Gamma], \neg[\Delta] \longrightarrow \neg\neg[\Phi]} \; (\neg R)
    }
    {[\Gamma], \neg[\Delta] \longrightarrow \neg\neg[\Phi] \vee \neg\neg[\Phi']} \; (\vee R_1)
  }
}
{[\Gamma], \neg(\neg\neg[\Phi] \vee \neg\neg[\Phi']), \neg[\Delta] \longrightarrow} \; (\neg L)
$$

There is a similar case for $(\vee R_2)$.

- The rules for negation are quite trivial. $(\neg L)$ becomes no rule at all:

$$\frac{\dfrac{\pi}{\Gamma \longrightarrow \Phi, \Delta}}{\Gamma, \neg\Phi \longrightarrow \Delta}\,(\neg L) \qquad\Longrightarrow\qquad \frac{L(\pi)}{[\Gamma], \neg[\Phi], \neg[\Delta] \longrightarrow}$$

and $(\neg R)$ requires that we double negate a formula by applying the $(\neg R)$ followed by the $(\neg L)$ rule, as shown below.

$$\frac{\dfrac{\pi}{\Gamma, \Phi \longrightarrow \Delta}}{\Gamma \longrightarrow \neg\Phi, \Delta}\,(\neg R) \qquad\Longrightarrow\qquad \frac{\dfrac{\dfrac{L(\pi)}{[\Gamma], [\Phi], \neg[\Delta] \longrightarrow}}{[\Gamma], \neg[\Delta] \longrightarrow \neg[\Phi]}\,(\neg R)}{[\Gamma], \neg\neg[\Phi], \neg[\Delta] \longrightarrow}\,(\neg L)$$

$\square$

We end this section with by stating several other translations of classical logic into intuitionistic logic. There are a wealth of translations, and some are better than others. Of course, all of these translations work equally well for both natural deduction and sequent calculus presentations of the logics. First, we mention the translation due to Gödel which is perhaps one of the best known translations.

$$\begin{aligned}
A^\circ &= \neg\neg A \\
(\neg\Phi)^\circ &= \neg\Phi^\circ \\
(\Phi \wedge \Psi)^\circ &= \Phi^\circ \wedge \Psi^\circ \\
(\Phi \vee \Psi)^\circ &= \neg(\neg\Phi^\circ \wedge \neg\Psi^\circ) \\
(\Phi \Rightarrow \Psi)^\circ &= \neg(\Phi^\circ \wedge \neg\Psi^\circ)
\end{aligned}$$

Another translation is due to Kolmogorov which prefixes every formulae by a double negation.

$$\begin{aligned}
A^\circ &= \neg\neg A \\
(\neg\Phi)^\circ &= \neg\Phi^\circ \\
(\Phi \wedge \Psi)^\circ &= \neg\neg(\Phi^\circ \wedge \Psi^\circ) \\
(\Phi \vee \Psi)^\circ &= \neg\neg(\Phi^\circ \vee \Psi^\circ) \\
(\Phi \Rightarrow \Psi)^\circ &= \neg\neg(\Phi^\circ \Rightarrow \Psi^\circ)
\end{aligned}$$

▶ EXERCISE 3.12

Derive the following in $\mathbf{LJ}_0$ using the translation given.

- $\longrightarrow [\Phi \vee \neg\Phi]$

- $\longrightarrow [\neg\neg\Phi \Rightarrow \Phi]$

▶ EXERCISE 3.13

There is a problem with this translation in that certain properties that we would expect from a translation do not hold. In particular the $\vee$ connective is not associative under this translation.

Verify that the following does *not hold*:

$$\vdash^{\mathbf{LJ_0}} [\Phi \vee (\Phi' \vee \Phi'')] \Leftrightarrow [(\Phi \vee \Phi') \vee \Phi'']$$

6   ADDITIVE AND MULTIPLICATIVE CONNECTIVES

In this section we study two different ways of presenting sequent systems which brings out a decomposition of the usual logical connectives. This takes us in the direction of linear logic.

In our presentations of logical systems (both classical and intuitionistic) whether sequent calculus or natural deduction in sequent form we were not so precise about the manipulation of contexts in logical rules. For example the $\wedge I$ rule can be presented as either of the following:

$$\frac{\Gamma \longrightarrow \Phi \qquad \Gamma \longrightarrow \Phi'}{\Gamma \longrightarrow \Phi \wedge \Phi'} (\wedge I) \qquad\qquad \frac{\Gamma \longrightarrow \Phi \qquad \Delta \longrightarrow \Phi'}{\Gamma, \Delta \longrightarrow \Phi \wedge \Phi'} (\wedge I)$$

In the first one, we *merge* the contexts Γ, and in the second, we *concatenate* the contexts Γ and Δ. The reason that we were not so formal is that in the presence of the structural rules *weakening* and *contraction* they are equivalent. The first can be simulated in the second by the use of contraction

$$\frac{\dfrac{\Gamma \longrightarrow \Phi \qquad \Gamma \longrightarrow \Phi'}{\Gamma, \Gamma \longrightarrow \Phi \wedge \Phi'} (\wedge I)}{\Gamma \longrightarrow \Phi \wedge \Phi'} (C)$$

The second can be simulated in the first by the use of weakening

$$\frac{\dfrac{\Gamma \longrightarrow \Phi}{\Gamma, \Delta \longrightarrow \Phi} (W) \qquad \dfrac{\Delta \longrightarrow \Phi'}{\Gamma, \Delta \longrightarrow \Phi'} (W)}{\Gamma, \Delta \longrightarrow \Phi \wedge \Phi'} (\wedge I)$$

However, if we take away the structural rules, the above rules are no longer equivalent, and in fact lead us to *two* kinds of $\wedge$ connective.

- The first is when we concatenate contexts. This is the *multiplicative* connective $\otimes$, called the *tensor product*:

$$\frac{\Gamma \longrightarrow \Phi \qquad \Delta \longrightarrow \Phi'}{\Gamma, \Delta \longrightarrow \Phi \otimes \Phi'} (\otimes I)$$

In the absence of weakening and contraction, there is *no* possibility of projections: $\Phi \otimes \Phi' \Rightarrow \Phi$ or diagonals: $\Phi \Rightarrow \Phi \otimes \Phi$.

- The second is when we merge contexts. This is the *additive* connective &, called *with*:

$$\frac{\Gamma \longrightarrow \Phi \qquad \Gamma \longrightarrow \Phi'}{\Gamma \longrightarrow \Phi \& \Phi'} \, (\& I)$$

With this connective we do have the possibility of projections. It behaves very much like we would expect $\wedge$ (conjunction), however we only get one chance to project.

Next we have the linear versions of the "or" connective ($\vee$). Again this is decomposed into two operators: the multiplicative one $\mathbin{\bindnasrepma}$ and the additive one $\oplus$ for similar reasons to those described above. For example, consider the $\vee L$ rule which can be presented as either of the following:

$$\frac{\Gamma, \Phi \longrightarrow \Phi'' \qquad \Gamma, \Phi' \longrightarrow \Phi''}{\Gamma, \Phi \vee \Phi' \longrightarrow \Phi''} \qquad \frac{\Gamma, \Phi \longrightarrow \Phi'' \qquad \Delta, \Phi' \longrightarrow \Phi''}{\Gamma, \Delta, \Phi \vee \Phi' \longrightarrow \Phi''}$$

Again, it is possible to show that the two are equivalent in the presence of structural rules. Dropping these rules leads to a splitting of $\vee$ into two distinctive connectives. The first gives us the additive connective $\oplus$, called *sum*:

$$\frac{\Gamma, \Phi \longrightarrow \Phi'' \qquad \Gamma, \Phi' \longrightarrow \Phi''}{\Gamma, \Phi \oplus \Phi' \longrightarrow \Phi''}$$

and the second gives the multiplicative $\mathbin{\bindnasrepma}$ connective, called *par*:

$$\frac{\Gamma, \Phi \longrightarrow \Phi'' \qquad \Delta, \Phi' \longrightarrow \Phi''}{\Gamma, \Delta, \Phi \mathbin{\bindnasrepma} \Phi' \longrightarrow \Phi''}$$

Observe that the information carried around in additive sequents is much larger than that of the multiplicative case. This can be demonstrated by building and comparing the proof of $\Phi \wedge \Phi' \Rightarrow \Phi' \wedge \Phi$ in both the additive and multiplicative styles.

Decomposing the disjunction and conjunction into two separate components leaves the question as to what happens with the units of these connectives. In classical logic, true is the unit for $\wedge$ and false is the unit for $\vee$. As expected, true and false are split into two giving both multiplicative and additive versions.

Multiplicative units $(1, \perp)$

$$\frac{\Gamma \longrightarrow \Delta}{\Gamma, 1 \longrightarrow \Delta} \, (1L) \qquad \frac{}{\longrightarrow 1} \, (1R) \qquad \frac{}{\perp \longrightarrow} \, (\perp L) \qquad \frac{\Gamma \longrightarrow \Delta}{\Gamma \longrightarrow \perp, \Delta} \, (\perp R)$$

Additive units $(0, \top)$

$$\frac{}{\Gamma \longrightarrow \top, \Delta}\,(\top) \qquad \frac{}{\Gamma, 0 \longrightarrow \Delta}\,(0)$$

It is now easy to see that **1** is the unit for $\otimes$, $\bot$ is the unit for $\invamp$, 0 for $\oplus$ and $\top$ for $\&$. This may be verified by proving various equivalences, for example $\Phi \otimes 1 \multimapboth \Phi$, etc., where $\Phi \multimapboth \Phi'$ is an abbreviation for $(\Phi \multimap \Phi') \& (\Phi \multimap \Phi')$. Remark that there are *no* rules for $(\top L)$ and $(\mathbf{0R})$.

It is worth remarking at this point that additive formulations of logics is better for backward proof search, whereas the multiplicative formulation is better suited for forward proof search.

▶ **EXERCISE 3.14**

Build derivations of $\neg \Phi \vee \Phi$ in *classical logic*, taking first $\vee$ to be multiplicative, then additive.

▶ **EXERCISE 3.15**

Try to build a derivation in intuitionistic logic *without the structural rules* of $\Phi \Rightarrow \Phi \wedge \Phi$ first taking $\wedge$ to be the multiplicative connective $\otimes$, then as the additive connective $\&$.

▶ **EXERCISE 3.16**

Prove that the multiplicative and additive $\vee$ connectives $\invamp$ and $\oplus$ are equivalent in the presence of the structural rules.

▶ **EXERCISE 3.17**

We have shown the decomposition for the $\wedge$ and $\vee$ using intuitionistic sequent calculus. Show all the cases for classical sequent calculus. [Hint: There are eight rules altogether: for each of the $\wedge$-$\vee$ connectives there is the left, right, additive and multiplicative versions.]

7 LINEAR LOGIC

Linear logic is a relatively recent system of logic introduced by Jean-Yves Girard (1987a). It fits into the constructive paradigm in two ways. First, we can see it as a refinement of intuitionistic logic, where the maintenance of resources is taken more seriously — this leads to intuitionistic linear logic. Another perspective sees it as a refinement of classical logic, which is obtained not by translation as we used for intuitionistic logic, but by careful decomposition of the classical connectives as already described in the previous section.

Decomposing the operations gives a way to dig deeper into understanding the logical connectives, and to provide greater insight into the cut elimination process.

More significantly, this new insight has generated a *constructive* classical logic — linear logic is a constructive logic having negation as an involution: $\Phi^{\perp\perp} = \Phi$, where $\perp$ is the notation for linear negation. As expected, all connectives have duals, and are related by de Morgan-like dualities.

7.1   Syntax of Linear Logic

Linear logic has a rich collection of connectives. From a syntactical point of view, linear logic comes from a more careful presentation of the structural rules, as we saw in the previous section. In the absence of structural rules we get a splitting of the usual $\wedge$ and $\vee$ connectives. Of course, a logic without structural rules is a very weak one — all proofs will normalise in linear time. In linear logic, the idea then is to re-introduce the rules *contraction* and *weakening*, not as structural rules, but in a controlled way as first class logical rules. This will be achieved by the introduction of a *modality*. We have already seen most of the connectives for linear logic, below we recap what we know so far:

- The *linear negation* $(\cdot)^{\perp}$ (read 'perp') satisfies $\Phi^{\perp\perp} = \Phi$, but remains constructive.

- The *multiplicative* connectives are $\otimes$ (read 'tensor') and $\mathbin{\bindnasrepma}$ (read 'par'), together with their units 1 and $\perp$ respectively. 'tensor' and 'par' are de Morgan-like duals, as expressed by:

$$
\begin{array}{rclcrcl}
(\Phi \otimes \Psi)^{\perp} & = & \Phi^{\perp} \mathbin{\bindnasrepma} \Psi^{\perp} & \quad & (\Phi \mathbin{\bindnasrepma} \Psi)^{\perp} & = & \Phi^{\perp} \otimes \Psi^{\perp} \\
1^{\perp} & = & \perp & & \perp^{\perp} & = & 1
\end{array}
$$

- The *additive* connectives are $\&$ (read 'with') and $\oplus$ (read 'sum'), together with their units $\top$ and 0 respectively. 'with' and 'sum' are de Morgan duals, as expressed by:

$$
\begin{array}{rclcrcl}
(\Phi \& \Psi)^{\perp} & = & \Phi^{\perp} \oplus \Psi^{\perp} & \quad & (\Phi \oplus \Psi)^{\perp} & = & \Phi^{\perp} \& \Psi^{\perp} \\
\top^{\perp} & = & 0 & & 0^{\perp} & = & \top
\end{array}
$$

We now need to recover the ability to do the copying and discarding in a controlled form. This is done by explicit logical operations, rather than as "hidden" structural rules. To this end, we introduce the modality ! (read 'exponential', or 'of course'), which allows us to write formulae of the form $!\Phi$. A proof of $!\Phi$ can be used to produce any finite number of copies (including zero) of the proof of Φ. This connective, like all the others, has a dual which is ? (read 'why not'). We will give some intuition about these connectives below, but first we can now add the final item to our list of connectives:

- The *exponentials* are ! and ?. 'Of course' and 'why not' are de Morgan duals, as expressed by:

$$(!\Phi)^{\perp} \;=\; ?\Phi^{\perp} \qquad (?\Phi)^{\perp} \;=!\Phi^{\perp}$$

To summarise, we have the following connectives:

Multiplicatives	$\otimes$	$\wp$
Additives	&	$\oplus$
Exponentials	!	?

7.2 Classical Linear Sequent Calculus

There is a nice symmetry between the left and right rules for dual formulae. For example the $(\otimes L)$ is a mirror image of $(\wp R)$, etc. (cf. Exercise 3.17.)

$$\frac{\Gamma, \Phi, \Phi' \longrightarrow \Delta}{\Gamma, \Phi \otimes \Phi' \longrightarrow \Delta}\,(\otimes L) \qquad \frac{\Gamma \longrightarrow \Delta, \Phi, \Phi'}{\Gamma \longrightarrow \Delta, \Phi\wp\Phi'}\,(\wp R)$$

If we were to move all the formulae to the right-hand side of the $\longrightarrow$ (negating the formulas on the left), then we see that both rules are in fact the same. This allows us to make a more economic presentation of the logical system because we no longer need to have the left (L) rules; all the action is on the right-hand side. More precisely, this is made possible by the following fact:

$$\Gamma \longrightarrow \Delta \text{ is provable iff } \longrightarrow \Gamma^{\perp}, \Delta \text{ is provable}$$

We also adopt a convention and write Γ rather than $\longrightarrow \Gamma$, since there is no context to the left. We also take the negation as defined on atoms only, hence we enrich the language of atomic formulas with $A^{\perp}$, $B^{\perp}$, ..., and then define negation on compound formulae in the following way:

$$\begin{aligned}
(A)^{\perp} &= A^{\perp} & \Phi^{\perp\perp} &= \Phi \\
(\Phi \otimes \Psi)^{\perp} &= \Phi^{\perp}\wp\Psi^{\perp} & (\Phi\wp\Psi)^{\perp} &= \Phi^{\perp} \otimes \Psi^{\perp} \\
(!\Phi)^{\perp} &= ?\Phi^{\perp} & (?\Phi)^{\perp} &= !\Phi^{\perp} \\
(\Phi \& \Psi)^{\perp} &= \Phi^{\perp} \oplus \Psi^{\perp} & (\Phi \oplus \Psi)^{\perp} &= \Phi^{\perp} \& \Psi^{\perp}
\end{aligned}$$

Linear implication $\Phi \multimap \Psi$ can then be defined as an abbreviation for $\Phi^{\perp}\wp\Psi$ (cf. $\neg\Phi \vee \Psi$ in classical logic), so we do not need to have an explicit rule. Making negation defined only on atoms of course means that we don't have any explicit rules for negation in the system either.

We are now in a position to give the logical rules which will be explained in some detail below. The sequent calculus formulation for classical linear logic is given in Figure 3.4.

Identity Group

$$\frac{}{\Phi^{\perp}, \Phi}\,(Ax) \qquad \frac{\Gamma, \Phi \qquad \Delta, \Phi^{\perp}}{\Gamma, \Delta}\,(Cut)$$

Structural Group

$$\frac{\Gamma, \Phi, \Phi', \Delta}{\Gamma, \Phi', \Phi, \Delta}\,(X)$$

Multiplicative Group

$$\frac{\Gamma, \Phi \qquad \Delta, \Phi'}{\Gamma, \Delta, \Phi \otimes \Phi'}\,(\otimes) \qquad \frac{\Gamma, \Phi, \Phi'}{\Gamma, \Phi \,\8\, \Phi'}\,(\8)$$

Additive Group

$$\frac{\Gamma, \Phi}{\Gamma, \Phi \oplus \Phi'}\,(\oplus_1) \qquad \frac{\Gamma, \Phi'}{\Gamma, \Phi \oplus \Phi'}\,(\oplus_2) \qquad \frac{\Gamma, \Phi \qquad \Gamma, \Phi'}{\Gamma, \Phi \& \Phi'}\,(\&)$$

Exponential Group

$$\frac{\Gamma, \Phi}{\Gamma, ?\Phi}\,(D) \qquad \frac{\Gamma}{\Gamma, ?\Phi}\,(W) \qquad \frac{\Gamma, ?\Phi, ?\Phi}{\Gamma, ?\Phi}\,(C) \qquad \frac{?\Gamma, \Phi}{?\Gamma, !\Phi}\,(!)$$

Figure 3.4. Classical Linear Logic

Note first that the rules for Axiom and Cut capture both the usual rules together with negation. The structural, multiplicative and additive group should be quite clear from what has already been discussed. However, the rules for the exponential group require some explanation. There are four rules, $(D), (W),$ (C) and $(!)$ which are called, respectively, *dereliction, weakening, contraction*, and *promotion*.

Think of a formula $!\Phi$ as being a specially marked formula with a *box* around a proof of it. If we have a derivation ending in $!\Phi$ we think of the whole derivation being enclosed in this box. The parts of a proof that are enclosed in a box are precisely those proofs that can be used in a non-linear way. We then need to have operations to manipulate these boxes. In particular we need to be able to do all the non-linear operations on them, for example duplication and erasure. This is precisely the purpose of the rules contraction and weakening. We also need two additional operations on boxes — a way of *opening* them so that we can use the contents, and the dual operation of placing a box around the proof. Dereliction corresponds to opening a box, and promotion builds a box around a proof.

Promotion, however, has to be dealt with some care. By placing a box around a *linear* Φ gives the *non-linear* $!\Phi$. If we could do this anywhere in the proof, then nothing would be gained. However, promotion is restricted in that we insist that the context is built up of non-linear formulae, of the form $?\Phi_1, \ldots, ?\Phi_n$ (which we write as $?\Gamma$)

$$\frac{?\Gamma, \Phi}{?\Gamma, !\Phi} \, (!)$$

This rule in fact can be better understood if we resort back for a moment to the two-sided sequents. We see that the rule transforms a proof $!\Gamma \longrightarrow \Phi$ into $!\Gamma \longrightarrow !\Phi$, where $!\Gamma$ is a context built up of $!\Phi_1, \ldots, !\Phi_n$.

We can explain this intuition a little more concretely by looking at the *cut elimination* steps for the exponential group.

The first cut is between two proofs, one ending in a dereliction, with the cut formula $?\Phi^\perp$, and the other ending in a promotion with cut formula $!\Phi$.

$$\frac{\dfrac{\Gamma, \Phi^\perp}{\Gamma, ?\Phi^\perp} \, (D) \qquad \dfrac{?\Delta, \Phi}{?\Delta, !\Phi} \, (!)}{\Gamma, ?\Delta} \, (Cut)$$

As usual, the one step of cut elimination pushes the cut up through the proof to the sub-formulas, in this case eliminating both the dereliction and promotion rules:

$$\frac{\Gamma, \Phi^\perp \qquad ?\Delta, \Phi}{\Gamma, ?\Delta} \, (Cut)$$

Thus we see that the box created by the promotion rule has been opened by a dereliction during cut elimination.

The next cut that we show demonstrates the action of weakening. A cut between two proofs, one ending in a weakening and another ending in a promotion is shown below:

$$
\cfrac{\cfrac{\Gamma}{\Gamma, ?\Phi^{\perp}}\,(W) \qquad \cfrac{?\Delta, \Phi}{?\Delta, !\Phi}\,(!)}{\Gamma, ?\Delta}\,(Cut)
$$

Pushing the cut up through the proof yields the following:

$$
\cfrac{\Gamma}{\Gamma, ?\Delta}\,(W)
$$

where now we see that the box created by the promotion has been *erased* by the weakening — the whole proof of Φ inside the box has been deleted from the proof. Note, as with cut elimination for $\mathbf{LJ}_0$, we also create a sequence of weakenings on the context, here written as $?\Delta$. We think of this context $?\Delta$ as being the *auxiliary doors* to the box (we say that the formula $!\Phi$ is the *principal door*). The additional weakenings can then be understood as erasing all assumptions of the box, whereas the cut itself only deleted the box and the contents.

The next cut that we show demonstrates the action of contraction. A cut between two proofs, one ending in a contraction and another ending in a promotion is shown below:

$$
\cfrac{\cfrac{\Gamma, ?\Phi^{\perp}, ?\Phi^{\perp}}{\Gamma, ?\Phi^{\perp}}\,(C) \qquad \cfrac{?\Delta, \Phi}{?\Delta, !\Phi}\,(!)}{\Gamma, ?\Delta}\,(Cut)
$$

Again pushing the cut up through the proofs yields the following:

$$
\cfrac{\cfrac{\Gamma, ?\Phi^{\perp}, ?\Phi^{\perp} \qquad \cfrac{?\Delta, \Phi}{?\Delta, !\Phi}\,(!)}{\Gamma, ?\Delta, ?\Phi^{\perp}}\,(Cut) \qquad \cfrac{?\Delta, \Phi}{?\Delta, !\Phi}\,(!)}{\cfrac{\Gamma, ?\Delta, ?\Delta}{\Gamma, ?\Delta}\,(C)}\,(Cut)
$$

where this time the box created by the promotion has been duplicated, together with the contents. Note that we create a sequence of contractions on the context $?\Delta$. This can now be understood as the process of duplicating all the assumptions of the box, in the same way as they were erased in the case of weakening.

The key difference between linear logic and classical logic, on the syntactic side, is that for weakening, the formula introduced is of the form $?\Phi$ indicating that it is used in a non-linear way, and similarly only non-linear formulae $?\Phi$ can be contracted. Hence all formulae that are not marked are used linearly.

To give an example of a derivation in the linear sequent calculus we prove the formula $?(\Phi^\perp \oplus \Psi^\perp)\,\invamp\,(!\Phi\otimes!\Psi)$.

$$
\cfrac{
  \cfrac{
    \cfrac{
      \cfrac{
        \cfrac{\rule{2cm}{0.4pt}}{\Phi^\perp,\Phi}(Ax)
      }{\Phi^\perp \oplus \Psi^\perp,\Phi}(\oplus_2)
      \quad\quad
      \cfrac{
        \cfrac{\rule{2cm}{0.4pt}}{\Psi^\perp,\Psi}(Ax)
      }{\Phi^\perp \oplus \Psi^\perp,\Psi}(\oplus_1)
    }{}{}
  }{}{}
}{}
$$

$$
\cfrac{
\cfrac{
\cfrac{\dfrac{\dfrac{\dfrac{\rule{1.5cm}{0.4pt}}{\Phi^\perp,\Phi}(Ax)}{\Phi^\perp \oplus \Psi^\perp,\Phi}(\oplus_2)}{?(\Phi^\perp \oplus \Psi^\perp),\Phi}(D)}{?(\Phi^\perp \oplus \Psi^\perp),!\Phi}(!)
\quad
\cfrac{\dfrac{\dfrac{\dfrac{\rule{1.5cm}{0.4pt}}{\Psi^\perp,\Psi}(Ax)}{\Phi^\perp \oplus \Psi^\perp,\Psi}(\oplus_1)}{?(\Phi^\perp \oplus \Psi^\perp),\Psi}(D)}{?(\Phi^\perp \oplus \Psi^\perp),!\Psi}(!)
}{?(\Phi^\perp \oplus \Psi^\perp),?(\Phi^\perp \oplus \Psi^\perp),!\Phi\otimes!\Psi}(\otimes)
}{\cfrac{?(\Phi^\perp \oplus \Psi^\perp),!\Phi\otimes!\Psi}{?(\Phi^\perp \oplus \Psi^\perp)\,\invamp\,(!\Phi\otimes!\Psi)}(\invamp)}(C)
$$

▶ **EXERCISE 3.18**

Which of the following are provable in linear logic?

- Implicational fagment:

(i) $\quad A \multimap A$

(ii) $\quad A \multimap B \multimap A$

(iii) $\quad A \multimap (A \multimap B) \multimap B$

(iv) $\quad (A \multimap B \multimap C) \multimap (A \multimap B) \multimap A \multimap C$

- Multiplicative/Additive

(i) $\quad A \multimap A \& B$

(ii) $\quad A \multimap A \& A$

(iii) $\quad A \otimes B \multimap A \& B$

(iv) $\quad A \otimes (A \multimap B) \multimap B$

(v) $\quad A \& B \multimap B \& A$

(vi) $\quad A \oplus A^\perp$

- Exponentials:

(i) $\quad !A \multimap A$

(ii) $\quad !(A \multimap A)$

(iii) $\quad !(A \& B) \multimap !A \otimes !B$

(iv) $\quad A \multimap !B \multimap A$

(v) $\quad A \multimap !A$

(vi) $\quad !A \multimap !!A$

(vii) $\quad (!A \multimap B \multimap C) \multimap (!A \multimap B) \multimap !A \multimap C$

▶ **EXERCISE 3.19**

We have shown the main cut elimination steps for the exponential fragment in the text. Complete the cases by giving the steps for the identity, multiplicative and additive groups.

7.3   Coding Intuitionistic Logic

Linear logic *without* the exponentials is clearly weaker than both classical and intuitionistic logics. However, with the addition of the exponentials we would hope to be able to recover provability of the other logics we have been studying. In this section we will formally prove this, by giving an encoding of intuitionistic logic into linear sequent calculus. For this we define a translation $L(\cdot)$ which takes a derivation of intuitionistic natural deduction and generates a derivation in the linear sequent calculus. First we define the encoding of formulas.

Definition 3.25 *We define a function* $[\cdot]$ *which translates formulas in intuitionistic logic to linear logic.*

$$
\begin{aligned}
[A] &= A \\
[\Phi \Rightarrow \Phi'] &= ![\Phi] \multimap [\Phi'] \\
[\Phi \wedge \Phi'] &= [\Phi] \& [\Phi'] \\
[\Phi \vee \Phi'] &= ![\Phi] \oplus ![\Phi']
\end{aligned}
$$

This translation extends to sequents in the following way:

$$
[\Gamma \longrightarrow \Phi] = ?[\Gamma]^{\perp}, [\Phi]
$$

This translation in fact brings out one of the most interesting aspects of linear logic. The functional arrow $\Phi \Rightarrow \Phi'$ is now unveiled to be more than an atomic operation. In fact it constitutes two distinct operations: a linear function $\Phi \multimap \Phi'$, which means that Φ is used exactly once, and a non-linear operation $!\Phi$ which allows Φ to be used multiple times, thus giving $!\Phi \multimap \Phi'$.

We show results here for natural deduction, but of course, there is also a translation from the sequent calculus presentation of intuitionistic logic.

Theorem 3.26 *If* $\Gamma \vdash^{\mathbf{NJ}_0} \Phi$ *then there is a derivation in the linear sequent calculus ending in* $?[\Gamma]^{\perp}, [\Phi]$.

Proof: We prove this theorem by showing how to transform each $\mathbf{NJ}_0$ derivation into a linear derivation. We remark that this translation is working from a natural deduction system to a sequent calculus system, and hence there are similar techniques that we used for the translation of natural deduction to sequent calculus for classical logic (cf. Exercise 2.17). The translation is defined by cases depending on the rule applied at the root, and we omit the square brackets for clarity.

- The axiom becomes the following derivation:

$$\cfrac{\cfrac{}{\Phi^{\perp},\Phi}\,(Ax)}{?\Phi^{\perp},\Phi}\,(D)$$

- The $(\Rightarrow I)$ rule:

$$\cfrac{\cfrac{\pi}{\Gamma,\Phi \longrightarrow \Phi'}}{\Gamma \longrightarrow \Phi \Rightarrow \Phi'}$$

becomes the following derivation:

$$\cfrac{\cfrac{L(\pi)}{?\Gamma^{\perp},?\Phi^{\perp},\Phi'}}{?\Gamma^{\perp},?\Phi^{\perp}\mathbin{⅋}\Phi'}\,(⅋)$$

- The $(\Rightarrow E)$ rule:

$$\cfrac{\cfrac{\pi_1}{\Gamma \longrightarrow \Phi \Rightarrow \Phi'} \qquad \cfrac{\pi_2}{\Delta \longrightarrow \Phi}}{\Gamma,\Delta \longrightarrow \Phi'}$$

becomes the following derivation:

$$\cfrac{\cfrac{L(\pi_1)}{?\Gamma^{\perp},?\Phi^{\perp}\mathbin{⅋}\Phi'} \qquad \cfrac{\cfrac{\cfrac{L(\pi_2)}{?\Delta^{\perp},\Phi}}{?\Delta^{\perp},!\Phi}\,(!) \qquad \cfrac{}{\Phi'^{\perp},\Phi'}\,(Ax)}{?\Delta^{\perp},!\Phi \otimes \Phi'^{\perp},\Phi'}\,(\otimes)}{?\Gamma^{\perp},?\Delta^{\perp},\Phi'}\,(Cut)$$

- The $\wedge I$ rule:

$$\cfrac{\cfrac{\pi_1}{\Gamma \longrightarrow \Phi} \qquad \cfrac{\pi_2}{\Delta \longrightarrow \Phi'}}{\Gamma,\Delta \longrightarrow \Phi \wedge \Phi'}\,(\wedge I)$$

becomes the following derivation:

$$\cfrac{\cfrac{\cfrac{L(\pi_1)}{?\Gamma^\perp,\Phi}}{?\Gamma^\perp,?\Delta^\perp,\Phi}\,(W) \qquad \cfrac{\cfrac{L(\pi_2)}{?\Delta^\perp,\Phi'}}{?\Gamma^\perp,?\Delta^\perp,\Phi'}\,(W)}{?\Gamma^\perp,\Delta^\perp,\Phi\&\Phi'}\,(\&)$$

- The $(\wedge E_1)$ rule:

$$\cfrac{\cfrac{\pi}{\Gamma \longrightarrow \Phi \wedge \Phi'}}{\Gamma \longrightarrow \Phi}\,(\wedge E_1)$$

becomes the following derivation:

$$\cfrac{\cfrac{L(\pi)}{?\Gamma^\perp,\Phi\&\Phi'} \qquad \cfrac{\cfrac{}{\Phi^\perp,\Phi}\,(Ax)}{\Phi^\perp \oplus \Phi'^\perp,\Phi}\,(\oplus_1)}{?\Gamma^\perp,\Phi}\,(Cut)$$

- The $(\vee I_1)$ rule:

$$\cfrac{\cfrac{\pi}{\Gamma \longrightarrow \Phi}}{\Gamma \longrightarrow \Phi \vee \Phi'}\,(\vee I_1)$$

becomes the following derivation:

$$\cfrac{\cfrac{\cfrac{L(\pi)}{?\Gamma^\perp,\Phi}}{?\Gamma^\perp,!\Phi}\,(!)}{?\Gamma^\perp,!\Phi\oplus!\Phi'}\,(\oplus_1)$$

The case for $\vee I_2$ is entirely similar.

- The $(\vee E)$ rule:

$$\cfrac{\cfrac{\pi_1}{\Gamma \longrightarrow \Phi \vee \Phi'} \qquad \cfrac{\pi_2}{\Delta,\Phi \longrightarrow \Phi''} \qquad \cfrac{\pi_3}{\Delta,\Phi' \longrightarrow \Phi''}}{\Gamma,\Delta \longrightarrow \Phi''}\,(\vee E)$$

becomes the following derivation

$$
\cfrac{
L(\pi_1) \qquad
\cfrac{
\cfrac{L(\pi_2)}{?\Delta^{\perp},?\Phi^{\perp},\Phi''} \qquad
\cfrac{L(\pi_3)}{?\Delta^{\perp},?\Phi'^{\perp},\Phi''}
}{?\Delta^{\perp},\Phi'',?\Phi^{\perp}\&?\Phi'^{\perp}}(\&)
}{?\Gamma^{\perp},?\Delta^{\perp},\Phi''}(Cut)
$$

with $L(\pi_1)$ over $?\Gamma^{\perp},!\Phi\oplus!\Phi'$.

Weakening and contraction are quite straightforward.   □

▶ **EXERCISE 3.20**

Are there any other translations from intuitionistic logic to linear logic? Try to write at least one other. [Hint: translate the arrow as $\Phi \Rightarrow \Phi'$ as $!(\Phi \multimap \Phi')$.]

▶ **EXERCISE 3.21**

Write each of the formulas from Exercise 3.2 in linear logic under these translations and give the corresponding derivations of them. What do you notice about the structure of the proofs that are generated?

CHAPTER 4

THE CURRY-HOWARD CORRESPONDENCE

1 INTRODUCTION

In this chapter we study a remarkable correspondence between two independently defined systems: intuitionistic natural deduction and the typed λ-calculus. This correspondence is known as the *Curry-Howard isomorphism* and establishes a precise relation between logic and (functional) computation.

The λ-calculus can be regarded as the theoretical foundation of functional programming. Indeed, it is the canonical form of the pure fragment of such languages. It is a system consisting of *functional abstraction* and *application*, which are two universal features in programming languages: abstraction is the mechanism that corresponds to procedure definitions, and application corresponds to a procedure call. The λ-calculus has also contributed to the design of modern languages, for example the notion of polymorphism that we find presently in many languages was first developed for the λ-calculus. Also, the λ-calculus is used as the *meta-language* for defining the semantics (specifically denotational) of different kinds of programming languages.

The Curry–Howard isomorphism establishes a tight relationship between the *typed λ-calculus* (a restricted form of the λ-calculus) and intuitionistic propositional logic. The isomorphism can be explained by considering the *term formation* rules for the typed λ-calculus which can be given as a natural deduction system. We write $t : \sigma$ to represent that a *term* t of the λ-calculus has *type* σ. There are three term constructions in the λ-calculus, which are variables (denoted by $x, y, z, \ldots$), abstraction (denoted by $\lambda x.t$) and application (denoted by juxtaposition of terms, i.e. tu). The rules for building terms are given below:

$$x : \sigma \longrightarrow x : \sigma$$

$$\frac{\Gamma, x : \sigma \longrightarrow t : \tau}{\Gamma \longrightarrow (\lambda x.t) : \sigma \to \tau} \qquad \frac{\Gamma \longrightarrow t : \sigma \to \tau \qquad \Delta \longrightarrow u : \sigma}{\Gamma, \Delta \longrightarrow (tu) : \tau}$$

If the terms are taken out of the above presentation we are left with the following, which, if we replace the functional arrow ($\to$) by implication ($\Rightarrow$) and types by

131

formulas, is the natural deduction presentation of (the implicational fragment of) intuitionistic logic:

$$\overline{\Phi \longrightarrow \Phi}$$

$$\frac{\Gamma, \Phi \longrightarrow \Phi'}{\Gamma \longrightarrow \Phi \Rightarrow \Phi'} \qquad \frac{\Gamma \longrightarrow \Phi \Rightarrow \Phi' \quad \Delta \longrightarrow \Phi}{\Gamma, \Delta \longrightarrow \Phi'}$$

These rules are called Axiom (Ax), implication introduction ($\Rightarrow I$) and implication elimination ($\Rightarrow E$) respectively. From this presentation one sees immediately that there is a correspondence between types of the λ-calculus and formulae of intuitionistic logic: all we do is systematically replace each type σ, τ by formulae; here Φ, Φ'. Slightly more hidden is the fact that there is a correspondence between the terms of the calculus and the proofs of the logic: variables correspond to the rule (Ax), abstraction corresponds to the ($\Rightarrow I$) rule and application corresponds to the ($\Rightarrow E$) rule. Moreover, an analysis of the process of normalisation in intuitionistic logic and the normalisation process of the λ-calculus yields a further correspondence: the process of β-reduction in the λ-calculus corresponds to the normalisation procedure in logic.

If we put all this together, taking programs to be terms from the λ-calculus, the Curry-Howard isomorphism can be seen as:

programs	$\sim$	proofs
types	$\sim$	formulae
computation	$\sim$	normalisation

and is known under a series of other names including *formulae-as-types* and *proofs-as-programs*.

The most interesting aspect of the correspondence is the relationship between normalisation and computation. This gives a dynamical aspect of logic, and a logical side to operational semantics of programming languages. The correspondence permits results to be carried over from one framework to the other, for example normalisation theorems.

This kind of correspondence is by no means restricted to intuitionistic propositional logic and the typed λ-calculus. One of the most well-known examples of this correspondence is the term calculus for second order propositional intuitionistic logic, known as System F. Second order propositional intuitionistic logic extends the system $\mathbf{NJ}_0$ with the following:

$$\frac{\Gamma \longrightarrow \Phi}{\Gamma \longrightarrow \forall X.\Phi} \, (\forall I) \qquad \frac{\Gamma \longrightarrow \forall X.\Phi}{\Gamma \longrightarrow \Phi[\Phi'/X]} \, (\forall E)$$

where the ($\forall I$) rule has the side condition that X is not free in Γ. Later we shall see that we can annotate derivations with terms giving a calculus for this logic. Results obtained in the calculus, for example *strong normalisation, confluence* and *consistency*, can then be applied directly to the logic, once the isomorphism is established.

Throughout this chapter all the logical systems will be given using the multiplicative presentations (cf. Chapter 3, Section 6), but the results are by no means restricted to this presentation.

1.1 Functions

We are all familiar with the notion of a *function* which is an input/output relation. For example we can construct functions using names such as:

$$
\begin{aligned}
f &: \quad \mathbb{N} \rightarrow \mathbb{N} \\
f(x) &= \quad x + 1
\end{aligned}
$$

where $f : \mathbb{N} \rightarrow \mathbb{N}$ states that the function f has the set of natural numbers as domain and codomain, and $f(x) = x + 1$ gives the algorithm to compute the value of the function for each element of $\mathbb{N}$. In this terminology we can write integer expressions (e.g. $x + 1$) as functions as above, but there is a discrepancy between the status of functions and expressions. There is no reason at all why we shouldn't consider functions themselves as expressions, i.e. make them first class citizens. To facilitate this we use the λ-notation — a way of writing functions as first class data objects anonymously (without having to resort to giving it a name, such as f as above). To motivate this notation, we consider some examples:

$$
\begin{aligned}
succ &= \quad \lambda x.x + 1 \\
apply &= \quad \lambda f.\lambda x.fx
\end{aligned}
$$

Where we think of $\lambda x.t$ as a notation for:

$$
\text{function } (x) \\
t
$$

In other words, $\lambda x.e$ is a function that returns a value t depending on the argument x. For the examples above, $succ$ is quite obvious in that it is a function which takes an argument, and returns the successor. The function $apply$ is slightly more complicated in that it is a function that takes two arguments; the first is a function and the second is an argument for that function, and the result is the application of that function to the argument. Understanding that a function can take a function as an argument is a fundamental aspect of the λ-calculus. Before giving a formal definition of the calculus, let us see how we can use this notation. There is an evident notion of function application:

$$
(\lambda x.t)u
$$

Note that in this notation it is more common to have the parentheses around the function rather than the argument. Function application is *simply* substituting the occurrence of the variable x in the term t with the term u. (Compare with $f(x) = x + 1$ and the application $f(3)$ for example.) Hence we have a rewriting rule that we write like:

$$(\lambda x.t)u \rightarrow t[u/x]$$

where $t[u/x]$ is the notation for substitution that we will formalise shortly. Here is an example in a familiar setting:

$$(\lambda x.x + 1)3 \quad \rightarrow \quad (x + 1)[3/x] \quad = \quad 3 + 1$$

which can then be reduced to the final result 4.

To show how functions themselves can be used as arguments, here is the application of $(apply\ succ)0$ in this notation. To shorten the trace of the computation, we assume that substitutions are done immediately.

$$
\begin{aligned}
(\lambda f.\lambda x.fx)(\lambda x.x + 1)0 \quad &\rightarrow \quad (\lambda x.(\lambda x.x + 1)x)0 \\
&\rightarrow \quad (\lambda x.x + 1)0 \\
&\rightarrow \quad 0 + 1
\end{aligned}
$$

which again can be reduced to give 1 as the answer.

When using this notation, one has to take great care in understanding the *scope* of a variable. In the example above, we see that we have the variable x appearing several times in the term $(\lambda x.(\lambda x.x + 1)x)0$. Variables are either bound by a λ, or they are free. In the example, all variables are bound, but if we look at the subterm $(\lambda x.\underline{x} + 1)x$ there are two occurrences of x of which one is free, and the other (underlined) is bound by the λ. To overcome this confusion, we will adopt a convention that all free variables are named differently from bound variables.

We refer the reader to other texts for a more thorough introduction into the λ-calculus, see for example Hankin (1994) and Barendregt (1984; 1992).

2   TYPED λ-CALCULUS AND NATURAL DEDUCTION

Here we define the theory of the *typed λ-calculus* without recourse to the corresponding system of natural deduction. In the following subsection we will make the correspondence precise.

2.1   Typed λ-calculus

Definition 4.1 *We define a set of types T inductively as:*

1. *A collection of type variables:* $\alpha, \beta, \gamma, \ldots$

2. *If σ and τ are types, then:*

(a) $(\sigma \times \tau)$ *is a (product) type.*

(b) $(\sigma \rightarrow \tau)$ *is a (function) type.*

The set of typed λ-terms *are generated from the following. We write* $t : \sigma$ *to mean that term* t *has type* σ.

1. *A collection of* typed variables $x : \sigma, \ldots$

2. *If* $u : \sigma$ *and* $v : \tau$ *are typed terms, then the* product $\langle u, v \rangle : \sigma \times \tau$ *is a typed term.*

3. *If* $t : \sigma \times \tau$ *is a typed term, then:*

   (a) $\mathsf{fst}(t) : \sigma$

   (b) $\mathsf{snd}(t) : \tau$

   are typed terms.

4. *If* $v : \tau$ *is a term and* $x : \sigma$ *is a variable, then the* abstraction $(\lambda x^{\sigma}.v) : \sigma \rightarrow \tau$ *is a typed term.*

5. *If* $t : \sigma \rightarrow \tau$ *and* $u : \sigma$ *are terms, then the* application $(tu) : \tau$ *is also a typed term.*

There is also a system of λ-calculus without types, called the *untyped λ-calculus*, or simply *the* λ-calculus. The untyped version is exactly the same as the calculus that we have presented here, except that more terms are admitted since the term construction does not depend on the types.

When referring to the typed λ-calculus we shall often write abstractions $\lambda x^{\sigma}.\ t : \sigma \rightarrow \tau$ as just $\lambda x.t$ when we are not so interested in the type, and similarly for other terms.

There are several accepted syntactic conventions for both types and typed terms. For types, we assume that $\rightarrow$ associates to the right, and binds more strongly than $\times$, and outermost parentheses can be dropped. These are exactly the same conventions that we used for logical connectives $\Rightarrow$ and $\wedge$. For typed terms, we adopt the following conventions:

- Application associates to the *left* and we drop outermost parentheses, for example:

$$((xy)z) \quad \text{becomes} \quad xyz$$
$$(x(yz)) \quad \text{becomes} \quad x(yz)$$

- Multiple λ's can be abbreviated, for example:

$$\lambda x.\lambda y.\lambda z.t \text{ becomes } \lambda xyz.t$$

- There is a notion of renaming that allows us to write the terms using different variable names. For example, $\lambda x^\sigma.x$ and $\lambda z^\sigma.z$ are obviously defining the same function. This renaming operation is known as α-conversion (or α-renaming) and is not at all trivial. For instance, in the term $\lambda x.xy$, a renaming of the variable x to y would give $\lambda y.yy$ which is not at all the same function that we originally had. We refer the interested reader to the literature for more details on this point, and just remark that one should take care with variable renamings.

Before going on, we give a few examples of typed λ-terms that the reader can easily check are well formed from the definition.

$$\lambda x.x : \sigma \to \sigma \qquad\qquad \lambda x.\lambda y.x : \sigma \to \tau \to \sigma$$
$$\lambda fx.fx : (\sigma \to \tau) \to \sigma \to \tau \qquad \langle \lambda x.x, \lambda x.x \rangle : (\sigma \to \sigma) \times (\tau \to \tau)$$

In contrast, the term $\lambda x.xx$ does *not* have a type in the system given, since this would require that we have both $x : \sigma$ and $x : \sigma \to \sigma$ at the same time. The term $\lambda x.x : \sigma \times \tau$ is an example of a badly formed typed term, but remark that the term and the type are both well formed; they just don't correspond to each other.

A λ-term is said to be *closed* (or a *combinator*) if there are no *free variables* in the term, otherwise it is *open*. For example $\lambda x.y$ is an open term since the variable y is not bound to any λ. In contrast, the term $\lambda yx.y$ is closed since all variables are bound. Formally, we define the *free variables* (FV) of a λ-term as follows:

$$
\begin{aligned}
FV(x) &= \{x\} \\
FV(\lambda x.t) &= FV(t) \setminus \{x\} \\
FV(tu) &= FV(\langle t,u \rangle) = FV(t) \cup FV(u) \\
FV(\mathsf{fst}(t)) &= FV(\mathsf{snd}(t)) = FV(t)
\end{aligned}
$$

and the *bound variables* (BV) as:

$$
\begin{aligned}
BV(x) &= \varnothing \\
BV(\lambda x.t) &= BV(t) \cup \{x\} \\
BV(tu) &= BV(\langle t,u \rangle) = BV(t) \cup BV(u) \\
BV(\mathsf{fst}(t)) &= BV(\mathsf{snd}(t)) = BV(t)
\end{aligned}
$$

It is easy to see that the variables of a term are precisely $FV(t) \cup BV(t)$.

Definition 4.2 (Redex) *A redex (reducible expression) is a λ-term of one of the following forms:* $(\lambda x.t)u$, $\mathsf{fst}(\langle t,u \rangle)$, $\mathsf{snd}(\langle t,u \rangle)$.

The redex $\mathsf{fst}(\langle t,u \rangle)$ is a term constructed by building the pair $\langle t,u \rangle$ and then applying the projection fst. It seems rather evident that having both t and u and then applying a projection function on the first element of the pair should just be the same term t. Hence we would expect that we have the equality $\mathsf{fst}(\langle t,u \rangle) = t$.

Similarly, the redex $(\lambda x.t)u$ has been constructed by building the function $\lambda x.t$ (from the term t) and then building the application of this function with u. This process (building a function and applying it) can be eliminated in the term, giving just the term t where the variable x has been replaced by the term u. Again, we should expect the equality $(\lambda x.t)u = t[u/x]$.

It is *clear* that the right-hand sides of the equalities are in some way simpler than the terms on the left. Formally, we have the following *reduction rules* for the typed λ-calculus, which define a reduction relation written as $\rightarrow_\beta$ and called β-reduction.

$$
\begin{aligned}
(\lambda x.t)u &\quad\rightarrow\quad t[u/x] \\
\mathsf{fst}(\langle t, u \rangle) &\quad\rightarrow\quad t \\
\mathsf{snd}(\langle t, u \rangle) &\quad\rightarrow\quad u
\end{aligned}
$$

where *substitution* $t[u/x]$ is defined as:

$$
\begin{aligned}
x[v/x] &\;=\; v \\
y[v/x] &\;=\; y \\
(\lambda y.t)[v/x] &\;=\; \lambda y.(t[v/x]) \\
(tu)[v/x] &\;=\; (t[v/x])(u[v/x]) \\
\langle t, u \rangle[v/x] &\;=\; \langle t[v/x], u[v/x] \rangle \\
\mathsf{fst}(t)[v/x] &\;=\; \mathsf{fst}(t[v/x]) \\
\mathsf{snd}(t)[v/x] &\;=\; \mathsf{snd}(t[v/x])
\end{aligned}
$$

We adopt a common convention, called the variable convention, that states that all bound variables are different from free variables, so that it is assumed that for the third case there are no free occurrences of the variable y in term v that could become bound. (Thus eliminating the problems of α-conversion that we mentioned previously).

We also have the following reduction rules

$$
\begin{aligned}
\lambda x.tx &\quad\rightarrow\quad t \;(x \notin FV(t)) \\
\langle \mathsf{fst}(t), \mathsf{snd}(t) \rangle &\quad\rightarrow\quad t
\end{aligned}
$$

which we again can think of as a reduction from left to right. These reductions are the *duals* of the β-reductions and are called η-reductions, which we write as $\rightarrow_\eta$. The notation $\rightarrow_{\beta\eta}$ will be used for the union of $\rightarrow_\beta$ and $\rightarrow_\eta$, but we will often omit the subscripts and simply write $\rightarrow$, where the meaning will be clear from the context. $t \rightarrow u$ captures the fact that t rewrites to u in one step. The notation $\rightarrow^*$ is used for the transitive reflexive closure of $\rightarrow$, and $\rightarrow^+$ represents one or more steps of reduction. We can also define a notion of *conversion*, denoted by $=_\beta$, to be the symmetric closure of $\rightarrow^*_\beta$. The reduction relation applies in any *context*, which we formalise as follows.

Definition 4.3 *A context $C[\,]$ is defined inductively by:*

- *A variable x is a context.*

- *[ ] is a context, called a* hole.

- *The application $C_1[\ ]C_2[\ ]$ is a context if both $C_1[\ ]$ and $C_2[\ ]$ are.*

- *The abstraction $\lambda x.C_1[\ ]$ is a context if $C_1[\ ]$ is a context.*

- *If $C_1[\ ]$ and $C_2[\ ]$ are contexts then so are $\mathsf{fst}(C_1[\ ])$, $\mathsf{snd}(C_1[\ ])$ and $\langle C_1[\ ], C_2[\ ]\rangle$.*

If t is a λ-term and $C[\ ]$ is a context, then $C[t]$ denotes the context $C[\ ]$ where all the holes are replaced by the term t.

We can now formally state what it means for a reduction to take place in any context:

Lemma 4.4 *If $t \to u$ then, for all contexts $C[\ ]$, $C[t] \to C[u]$.*

Proof: A straightforward induction over the structure of contexts. $\square$

Finally, a remark on substitutions. When writing terms with multiple substitutions, we have to be a little careful. The following lemma gives one result about how we can reorder them.

Lemma 4.5 (Substitution) *If x and y are distinct variables and $x \notin FV(v)$, then*

$$t[u/x][v/y] = t[v/y][u[v/y]/x]$$

Proof: The proof is a straightforward induction over the structure of t. See Exercise 4.1. $\square$

We refer the reader to the literature cited for a more thorough treatment of the λ-calculus, where the reader can find additional material on both the syntactic and semantic aspects of the calculus.

▶ **EXERCISE 4.1**
Prove the Substitution Lemma 4.5: If x and y are distinct variables and $x \notin FV(v)$, then $t[u/x][v/y] = t[v/y][u[v/y]/x]$.

▶ **EXERCISE 4.2**
We said that is was clear that the reductions make the terms simpler. However, give an example of a term such that reduction makes the term bigger. In what sense can it be said that reductions make the terms simpler?

2.2   Term Assignment for Natural Deduction

There are other ways that we could have presented the λ-calculus term formation
rules. One of the most common is to present the rules as a deduction system where
we derive sequents of the form:

$$x_1 : \sigma_1, \ldots, x_n : \sigma_n \longrightarrow t : \tau$$

where the x_i are distinct variables, and t is a λ-term that depends on the variables x_i.
We can then define deduction rules that build terms of the right form. In Figure 4.1
we give such a set of rules. There are *three* kinds of rules to this system. The first is
for the *variable*, which allows us to construct typed variables. The next group contains three *structural rules* which allow manipulations on the context: exchange
(X), weakening (W) and contraction (C). Throughout this chapter we will work
modulo the exchange rule. The final group, *term formation rules*, gives all the possible ways that we can construct terms. Each term construction has a corresponding
rule, for example a function $\lambda x.t$ is built using the (ABS) rule, application is built
using the (APP) rule, etc. One can easily show that if there is a derivation ending
in $\Gamma \longrightarrow t : \sigma$, then t is a well typed term.

Remark that in this system we have the following property: If $\Gamma \longrightarrow t : \sigma$, then
$\Gamma, \Delta \longrightarrow t : \sigma$. This property holds simply by application of the weakening rule
(W).

We have also introduced three extra kinds of typed λ-term in the deduction system: inl, inr and case. These correspond to the *sum*, written as $+$, which is the
λ-calculus counterpart of the $\vee$ connective. In Exercise 4.3 we give the reader the
opportunity to define the rules for β and η-reduction to complete the definition of
the typed λ-calculus.

In this formulation of the typed λ-calculus a number of interesting connections
arise. There is a striking similarity between the term formation rules and the system
of natural deduction that we presented in Chapter 3, Section 2.1. For instance the
rule defining application (APP) and implication elimination $(\Rightarrow E)$ are essentially
the same rule. More precisely, the types of terms correspond to logical formulae:
$\rightarrow$ corresponds to $\Rightarrow$, $\times$ to $\wedge$ and $+$ to $\vee$, and the term forming rules correspond to
the logical rules, in full we have:

$$
\begin{array}{lcl}
(\text{ABS}) & \sim & (\Rightarrow I) \\
(\text{APP}) & \sim & (\Rightarrow E) \\
(\text{PAIR}) & \sim & (\wedge I) \\
(\text{FST}) & \sim & (\wedge E_1) \\
(\text{SND}) & \sim & (\wedge E_2) \\
(\text{INL}) & \sim & (\vee I_1) \\
(\text{INR}) & \sim & (\vee I_2) \\
(\text{CASE}) & \sim & (\vee E)
\end{array}
$$

Variable:

$$\frac{}{x : \sigma \longrightarrow x : \sigma} \; (\text{VAR})$$

Structural rules:

$$\frac{\Gamma, x : \sigma, y : \tau, \Delta \longrightarrow t : \gamma}{\Gamma, y : \tau, x : \sigma, \Delta \longrightarrow t : \gamma} \; (X)$$

$$\frac{\Gamma \longrightarrow t : \gamma}{\Gamma, x : \sigma \longrightarrow t : \gamma} \; (W)$$

$$\frac{\Gamma, x : \sigma, y : \sigma \longrightarrow t : \gamma}{\Gamma, z : \sigma \longrightarrow t[z/x, z/y] : \gamma} \; (C)$$

Term formation rules:

$$\frac{\Gamma \longrightarrow t : \sigma \qquad \Delta \longrightarrow u : \tau}{\Gamma, \Delta \longrightarrow \langle t, u \rangle : \sigma \times \tau} \; (\text{PAIR})$$

$$\frac{\Gamma \longrightarrow t : \sigma \times \tau}{\Gamma \longrightarrow \text{fst}(t) : \sigma} \; (\text{FST}) \qquad \frac{\Gamma \longrightarrow t : \sigma \times \tau}{\Gamma \longrightarrow \text{snd}(t) : \tau} \; (\text{SND})$$

$$\frac{\Gamma, x : \sigma \longrightarrow t : \tau}{\Gamma \longrightarrow \lambda x.t : \sigma \to \tau} \; (\text{ABS}) \qquad \frac{\Gamma \longrightarrow t : \sigma \to \tau \qquad \Delta \longrightarrow u : \sigma}{\Gamma, \Delta \longrightarrow tu : \tau} \; (\text{APP})$$

$$\frac{\Gamma \longrightarrow t : \sigma}{\Gamma \longrightarrow \text{inl}(t) : \sigma + \tau} \; (\text{INL}) \qquad \frac{\Gamma \longrightarrow u : \tau}{\Gamma \longrightarrow \text{inr}(u) : \sigma + \tau} \; (\text{INR})$$

$$\frac{\Gamma \longrightarrow t : \sigma + \tau \qquad \Delta, x : \sigma \longrightarrow u : \gamma \qquad \Delta, y : \tau \longrightarrow v : \gamma}{\Gamma, \Delta \longrightarrow \text{case } t \text{ of inl}(x) \Longrightarrow u \mid \text{inr}(y) \Longrightarrow v : \gamma} \; (\text{CASE})$$

Figure 4.1. Term Assignment for Natural Deduction

and it is also the case that the axiom and structural rules coincide.

With this perspective we can see the typed λ-calculus as a term assignment to natural deduction proofs, that is to say that we can decorate $\mathbf{NJ}_0$ derivations with λ-terms yielding the system of typed λ-calculus. Having two systems that are isomorphic in this way gives a new perspective on both systems. For example, we could take the *logical view* where the formulae and their proofs are primary, and the terms are nothing more than notations for proofs. We can also take the *computational view* where terms (functional programs) are primary, and the natural deduction system is a *type inference system* for terms. However, the Curry-Howard isomorphism states that these two views exist together; we can switch perspective as we wish.

We give an example. Consider the following derivation in $\mathbf{NJ}_0$.

$$
\cfrac{
\cfrac{
\cfrac{\rule{2cm}{0.4pt}}{\Phi \longrightarrow \Phi}\ (Ax)
}{\longrightarrow \Phi \Rightarrow \Phi}\ (\Rightarrow I)
\qquad
\cfrac{
\cfrac{\rule{2cm}{0.4pt}}{\Phi' \longrightarrow \Phi'}\ (Ax)
}{\longrightarrow \Phi' \Rightarrow \Phi'}\ (\Rightarrow I)
}{\longrightarrow (\Phi \Rightarrow \Phi) \wedge (\Phi' \Rightarrow \Phi')}\ (\wedge I)
$$

If we annotate this proof of $(\Phi \Rightarrow \Phi) \wedge (\Phi' \Rightarrow \Phi')$ with λ-terms throughout, we get the term $\langle \lambda x.x, \lambda y.y \rangle$ at the root. Replacing formulae by types and logical connectives with type constructors, we see that this term has type $(\sigma \rightarrow \sigma) \times (\tau \rightarrow \tau)$.

We have seen that there is a precise connection between formulae and types, and proofs and terms (programs). However, there is a third and most important part which connects the process of normalisation in natural deduction with reduction (computation) in the typed λ-calculus. The Curry-Howard isomorphism is then summarised by the following table:

formulae	$\sim$	types
proofs	$\sim$	programs
normalisation	$\sim$	computation

The first two parts of the isomorphism are clear from what we have said above. Here we focus on the third part where we see most of the applications of this connection. If we look at the calculus, we have ways of *constructing* terms, and ways of *destructing* them. The following table gives a summary:

	Constructors	Destructors
$\times$	(,)	fst, snd
$\rightarrow$	λ	apply
$+$	inl, inr	case

The reduction relation (β) that we have defined corresponds precisely to the case when a constructor meets a destructor. On the logical side, we can extract from the above table that constructors correspond to the *introduction* rules and destructors correspond to *elimination* rules. The third part of the Curry-Howard isomorphism states that each of the reduction rules for the λ-calculus corresponds to a proof reduction.

For example, the term $\mathsf{fst}(\langle t, u \rangle)$ is an occurrence of such a reduction: it is a product constructor followed by a product destructor. In the calculus we gave the equation $\mathsf{fst}(\langle t, u \rangle) \to t$. The derivations for $\mathsf{fst}(\langle t, u \rangle)$ and t are given by the following:

$$
\begin{array}{cc}
(\pi_1) \qquad\qquad (\pi_2) \\
\vdots \qquad\qquad\quad \vdots \\
\dfrac{\Gamma \vdash t : \sigma \qquad \Delta \vdash u : \tau}{\dfrac{\Gamma, \Delta \vdash \langle t, u \rangle : \sigma \times \tau}{\Gamma, \Delta \vdash \mathsf{fst}(\langle t, u \rangle) : \sigma}}
\end{array}
\qquad \Rightarrow \qquad
\begin{array}{c}
(\pi_1) \\
\vdots \\
\Gamma \vdash t : \sigma \\
\hline\hline
\Gamma, \Delta \longrightarrow t : \sigma
\end{array} (W)
$$

Removing the terms yields exactly the same reduction for natural deduction normalisation that we saw in Chapter 3. The same applies to $(\lambda x.t)u \to t[u/x]$:

$$
\begin{array}{cc}
(\pi_1) \\
\vdots \\
\dfrac{\Gamma, x : \sigma \longrightarrow t : \tau}{\dfrac{\Gamma \longrightarrow \lambda x.t : \sigma \to \tau \qquad \begin{array}{c}(\pi_2)\\ \vdots \\ \Delta \longrightarrow u : \sigma\end{array}}{\Gamma, \Delta \longrightarrow (\lambda x.t)u : \tau}}
\end{array}
\qquad \Rightarrow \qquad
\begin{array}{c}
(\pi_2) \cdots (\pi_2) \\
\vdots \\
(\pi_1) \\
\vdots \\
\Gamma, \Delta \longrightarrow t[u/x] : \tau
\end{array}
$$

This case makes a little clearer the notion of substituting the hypotheses of σ by proofs of σ, since we have the notation from the λ-calculus, i.e. $t[u/x]$.

η-reductions are also reflected in the logic. Here we show the case $\langle \mathsf{fst}(t), \mathsf{snd}(t) \rangle = t$.

$$
\dfrac{\dfrac{\Gamma \longrightarrow t : \sigma \times \tau}{\Gamma \longrightarrow \mathsf{fst}(t) : \sigma} \qquad \dfrac{\Gamma \longrightarrow t : \sigma \times \tau}{\Gamma \longrightarrow \mathsf{snd}(t) : \tau}}{\Gamma \longrightarrow \langle \mathsf{fst}(t), \mathsf{snd}(t) \rangle : \sigma \times \tau}
\quad \Rightarrow \quad
\dfrac{}{\Gamma \longrightarrow t : \sigma \times \tau}
$$

The remaining cases follow in the same style.

It should be emphasized that a proof π of $\Gamma \longrightarrow t : \sigma$ is given in terms of t, so that the proof π can be reconstructed, in a unique way (modulo applications of the structural rules), from the term t. This is because the term contains the history of the proof. There are various refinements that can be made to the λ-calculus which take into account the structural rules, thus capturing *all* the history of the proof. Such a calculus would capture the substitution process, thus a calculus of explicit substitutions.

▶ **EXERCISE 4.3**

Give the β and η-reductions for the case construct. [Hint: write the rules for normalisation in $\mathbf{NJ}_0$ first.]

2.3 Properties

Here we establish some of the key results for the typed λ-calculus. By the Curry-Howard isomorphism, all of these results apply also to intuitionistic natural deduction.

Subject reduction. If we apply one of the reduction rules of the typed λ-calculus to a typed λ-term, we would hope that we still have a term of the same type. The notion of subject reduction states that types are preserved under reduction. This gives a kind of semantic soundness for reduction; nothing bad will happen. The statement of the theorem and proof follow.

Theorem 4.6 *If* $\Gamma \longrightarrow t : \sigma$ *and* $t \rightarrow_{\beta\eta} t'$ *then* $\Gamma \longrightarrow t' : \sigma$.

In proving this theorem, it is useful to factor out a very general lemma.

Lemma 4.7 (Substitution) *If* $\Gamma, x : \sigma \longrightarrow t : \tau$ *and* $\Delta \longrightarrow u : \sigma$, *then* $\Gamma, \Delta \longrightarrow t[u/x] : \tau$.

Proof: By induction over the length of the derivation $\Gamma, x : \sigma \longrightarrow t : \tau$.   □

We are now in a position where we can prove the subject reduction theorem.
Proof: (of Theorem 4.6)

By induction over the derivation of $\Gamma \longrightarrow t : \sigma$. If the derivation ends in a structural rule, or the reduction $t \rightarrow t'$ takes place in a proper subterm of t, then the result follows by induction. We consider only the case where the reduction takes place at the root, and we prove the theorem for each kind of reduction.

1. Case: $(\lambda x.t)u \rightarrow_\beta t[u/x]$

   Without loss of generality, we assume that the derivation ends in

   $$\dfrac{\dfrac{\Gamma, x : \sigma \longrightarrow t : \tau}{\Gamma \longrightarrow \lambda x.t : \sigma \to \tau}\ (\text{ABS}) \qquad \Delta \longrightarrow u : \sigma}{\Gamma, \Delta \longrightarrow (\lambda x.t)u : \tau}\ (\text{APP})$$

   (The case where there are structural rules between APP and ABS follows in a similar way.)

   We can now apply Lemma 4.7 on the premises of the above to obtain $\Gamma, \Delta \longrightarrow t[u/x] : \tau$ as required.

2. Case: $\lambda x.tx \rightarrow_\eta t$

   As in the previous case, we can assume without loss of generality, that the derivation ends in

$$
\cfrac{
  \Gamma \longrightarrow t : \sigma \rightarrow \tau \qquad
  \cfrac{}{x : \sigma \longrightarrow x : \sigma}\ (\text{VAR})
}{
  \cfrac{
    \Gamma, x : \sigma \longrightarrow tx : \tau
  }{
    \Gamma \longrightarrow \lambda x.tx : \sigma \rightarrow \tau
  }\ (\text{ABS})
}\ (\text{APP})
$$

   Then $\Gamma \longrightarrow t : \sigma \rightarrow \tau$ follows from the first premise.

3. Case: $\text{fst}(\langle t, u\rangle) \rightarrow_\beta t$.

   Assume without loss of generality that the derivation ends in

$$
\cfrac{
  \cfrac{
    \Gamma \longrightarrow t : \sigma \qquad \Delta \longrightarrow u : \tau
  }{
    \Gamma, \Delta \longrightarrow \langle t, u\rangle : \sigma \times \tau
  }\ (\text{PAIR})
}{
  \Gamma, \Delta \longrightarrow \text{fst}(\langle t, u\rangle) : \sigma
}\ (\text{FST})
$$

   Then $\Gamma, \Delta \longrightarrow t : \sigma$ follows from the first premise, by using the weakening rule.

4. Case: $\text{snd}(\langle t, u\rangle) \rightarrow_\beta u$. This is only a slight variation from the previous case.

5. Case: $\langle \text{fst}(t), \text{snd}(t)\rangle \rightarrow_\eta t$.

   Assume, again without loss of generality, that the derivation ends in

$$
\cfrac{
  \cfrac{\Gamma \longrightarrow t : \sigma \times \tau}{\Gamma \longrightarrow \text{fst}(t) : \sigma}\ (\text{FST}) \qquad
  \cfrac{\Gamma \longrightarrow t : \sigma \times \tau}{\Gamma \longrightarrow \text{snd}(t) : \tau}\ (\text{SND})
}{
  \Gamma, \Gamma \longrightarrow \langle \text{fst}(t), \text{snd}(t)\rangle : \sigma \times \tau
}\ (\text{PAIR})
$$

   Then $\Gamma \longrightarrow t : \sigma \times \tau$ from one of the premises, together with the contraction rule.

$\square$

For intuitionistic natural deduction, the subject reduction theorem says that contracting a proof using one step normalisation gives a proof of the same formula.

Normalisation.   Normalisation results generally come in two flavours. First, there is a notion of the existence of a reduction strategy that terminates. This is the property that we prove here. There is also a stronger variety which states that all reduction sequences terminate. Here we just consider the β-reductions, but the results extend to η in a straightforward way. We begin with a definition of the class of terms that have already terminated.

Definition 4.8 *A term t is said to be in β-normal form if there are no sub-terms of the form $(\lambda x.t)u$, $\mathsf{fst}(\langle t, u \rangle)$ or $\mathsf{snd}(\langle t, u \rangle)$.*

We can now formalise two normalisation results.

1. Weak normalisation. There exists a finite sequence

$$t \rightarrow t_1 \rightarrow t_2 \rightarrow \cdots \rightarrow t_n$$

such that t_n is a β-normal form.

2. Strong normalisation. *All* reduction sequences are finite.

Before proving either of these, we define a number of measures on typed terms.

Definition 4.9   • *The degree $\deg(\sigma)$ of a type σ is defined as:*

$$
\begin{aligned}
\deg(\alpha) \quad &= \quad 1 \\
\deg(\sigma \times \tau) \quad &= \quad \deg(\sigma \rightarrow \tau) = \max\{\deg(\sigma), \deg(\tau)\} + 1
\end{aligned}
$$

• *The rank $R(r)$ of a redex r is defined as:*

   1. If $r = (\lambda x.t)u$, and $\lambda x.t : \sigma \rightarrow \tau$, then

$$R((\lambda x.t)u) = \deg(\sigma \rightarrow \tau)$$

   2. If $r = \mathsf{fst}(\langle t, u \rangle)$ or $r = \mathsf{snd}(\langle t, u \rangle)$, and $\langle t, u \rangle : \sigma \times \tau$, then

$$R(\mathsf{fst}(\langle t, u \rangle)) = R(\mathsf{snd}(\langle t, u \rangle)) = \deg(\sigma \times \tau)$$

• *The degree $d(t)$ of a term t is defined as the supremum of the ranks of the redexes in t:*

   $d(t) = \max\{R(r_i) : r_i \text{ is a redex in the term } t\}$.

   If t is in normal form, then $d(t) = 0$.

Lemma 4.10 *If $x : \sigma$ then $d(t[u/x]) \leq \max\{d(t), d(u), \deg(\sigma)\}$.*

Proof: If the substitution does not create any new redexes, the result trivially holds. The only problematic case is the creation of a new redex, which must be one of: $(\lambda x.t')u$, $\mathsf{fst}(u)$ or $\mathsf{snd}(u)$. In all three cases, the degree of the new redex is $\deg(\sigma)$. □

We are now ready to state the main result.

Theorem 4.11 (Weak Normalisation) *If there is a derivation ending in* $\Gamma \longrightarrow t_0 : \sigma$ *then there is a finite sequence of reductions* $t_0 \to t_1 \to \cdots \to t_n$ *such that* t_n *is in* β-normal form.

Proof: It suffices to show that given a reducible term t we can always find a u such that $t \to^+ u$ and $d(t) > d(u)$. For this we will show that reducing all of the innermost redexes of t with maximal rank r, we obtain a term u in which all the redexes have rank less than r.

We first assume that there is only one innermost redex with maximal rank, and show that by reducing this redex we decrease the degree of the term.

We can do this by an induction over the structure of terms. The cases for variables and abstractions are trivial. The only interesting case is for applications. There are three cases to consider. If $tu \to tu'$ then the result follows by hypothesis. The first interesting case is when $(\lambda x.t)u \to t[u/x]$, then the result follows by the previous Lemma 4.10:

$$d(t[u/x]) \le \deg(\sigma) < \deg(\sigma \to \tau) = d((\lambda x.t)u)$$

The final case is when $tu \to t'u$. If this reduction creates a redex, for example $t \to \lambda x.t''$, then we must show that the hypothesis holds, which is again straightforward.

Now, the result for a term with n innermost redexes with maximal rank follows by a simple induction. □

This result can indeed be strengthened to give the strong normalisation property for the typed λ-calculus, but we will not cover that here.

Of course, this result (and method of proof) also holds for the logic, thus providing a proof of the normalisation theorem for $\mathbf{NJ}_0$ (Theorem 3.6 of Chapter 3).

Church-Rosser. The Church-Rosser property tells us that the choice of reduction order is not important.

Theorem 4.12 (Church-Rosser) *If* $t \to^* t_1$ *and* $t \to^* t_2$ *then there exists a term* t_3 *such that* $t_1 \to^* t_3$ *and* $t_2 \to^* t_3$.

Corollary 4.13 (Uniqueness of Normal Form) *A term* $t : \sigma$ *has exactly one* β-*normal form.*

Proof: Assume $t : \sigma$, and that t has two β-normal forms, say u and v. By the Church-Rosser theorem there is a term w such that $u \to^* w$ and $v \to^* w$. By assumption u and v cannot be reduced further, hence $u = v = w$. □

Consistency.   From the Church-Rosser theorem, we obtain the fact that the calculus is consistent.

Corollary 4.14 *The λ-calculus is consistent — it is not possible to prove $u =_\beta v$ for arbitrary terms.*

Proof: Take two terms t and u. If $t \to^* u$ then $t =_\beta u$, and if $t =_\beta u$ then it is easy to show from Church-Rosser that there is a common term w such that $t \to^* w$ and $u \to^* w$. If we take two variables x and y, which are in normal form, then there is no such term w. Hence it is not possible to show $x =_\beta y$.   □

Again, all of these results apply to intuitionistic natural deduction, showing that the system is consistent.

Summary.   We have seen a sequence of results for the typed λ-calculus that are applicable to intuitionistic natural deduction proofs, thanks to the Curry-Howard isomorphism. In Section 4 we show some other applications of this isomorphism which do not rely on two pre-existing systems.

▸ **EXERCISE 4.4**
Build *normal* derivations of the following formulae:

$$
\begin{aligned}
(i) \quad & A \Rightarrow B \Rightarrow A \\
(ii) \quad & (A \Rightarrow B \Rightarrow C) \Rightarrow (A \Rightarrow B) \Rightarrow (A \Rightarrow C) \\
(iii) \quad & (A \wedge B \Rightarrow C) \Rightarrow (A \Rightarrow B \Rightarrow C) \\
(iv) \quad & (A \Rightarrow B \Rightarrow C) \Rightarrow (A \wedge B \Rightarrow C)
\end{aligned}
$$

Annotate each derivation with λ-terms, replace $\Rightarrow$ by $\to$, propositions by type variables, and write the corresponding λ-term for each formula.

3   COMBINATORY LOGIC AND HILBERT-STYLE AXIOMS

In this section we look at another formulation of the Curry-Howard correspondence, which applies to the Hilbert-style axioms of intuitionistic logic and a system called combinatory logic. We restrict our attention here to the implicational fragment of the logic.

3.1   *Combinatory Logic*

Combinatory logic is another formal system, like the λ-calculus, which is also regarded as being a foundation of functional programming.   Indeed, it has proved fruitful as an implementation technique for this kind of language, see for example Peyton Jones (1987).

Combinatory logic is built up from only two typed elements, called *combinators*:

$$\mathbf{K}_{\sigma,\tau} \quad : \quad \sigma \to \tau \to \sigma$$
$$\mathbf{S}_{\sigma,\tau,\gamma} \quad : \quad (\sigma \to \tau \to \gamma) \to (\sigma \to \tau) \to \sigma \to \gamma$$

together with a collection of typed variables, $x : \sigma$, etc. Terms can be combined using only the rule of application:

$$\frac{\Gamma \longrightarrow M : \sigma \to \tau \qquad \Delta \longrightarrow N : \sigma}{\Gamma, \Delta \longrightarrow (MN) : \tau}$$

Note that this is nothing more than modus ponens, or the $(\Rightarrow E)$ rule. As usual, we assume application associates to the left, and drop excessive parentheses.

This theory comes equipped with the following two axioms:

$$\begin{aligned} \mathbf{K}PQ &= P \\ \mathbf{S}PQR &= PR(QR) \end{aligned}$$

where we have dropped types for clarity (we will avoid writing types whenever possible). As with the λ-calculus we can think of this as a *reduction* from left to right, and use $\to_w$ instead of the equality and $\to_w^*$ for the reflexive and transitive closure of $\to_w$. These two rewrite rules are allowed in any context (which is the same as for the λ-calculus, Definition 4.3, except that there is no case for abstraction). The $\to_w$ notation is used to signify that reduction in combinatory logic is the so-called *weak* reduction. We will come back to this issue below.

A simple example of reduction is given by:

$$\begin{aligned} \mathbf{SKK}M \quad &\to_w \quad \mathbf{K}M(\mathbf{K}M) \\ &\to_w \quad M \end{aligned}$$

which shows that, for any M, we have $\mathbf{SKK}M \to_w^* M$, thus $\mathbf{SKK}$ behaves as an *identity* combinator. It is convenient to have the identity combinator as primitive in the theory, which is denoted by $\mathbf{I}_\sigma : \sigma \to \sigma$. The reader will easily verify that the type we assigned is in fact the correct one inferred from the application of $\mathbf{SKK}$.

Combinatory logic is very close in fact to the typed λ-calculus. The main difference between the two systems is the missing abstraction rule from the term construction in combinatory logic, and the missing two *combinators* from the λ-calculus. However, it is possible to translate between the two systems. Let us first look at coding combinatory logic in the λ-calculus, where all we are required to do is give a λ-term corresponding to each of the combinators $\mathbf{S}, \mathbf{K}$ and $\mathbf{I}$.

Definition 4.15 *We define a translation* $(\cdot)_\lambda$ *from (typed) combinators to the (typed) λ-calculus as follows:*

$$
\begin{aligned}
(x)_\lambda &= x \\
(\mathbf{I}_\sigma)_\lambda &= \lambda x.x : \sigma \to \sigma \\
(\mathbf{K}_{\sigma,\tau})_\lambda &= \lambda xy.x : \sigma \to \tau \to \sigma \\
(\mathbf{S}_{\sigma,\tau,\gamma})_\lambda &= \lambda xyz.xz(yz) : (\sigma \to \tau \to \gamma) \to (\sigma \to \tau) \to \sigma \to \gamma \\
(MN) &= (M)_\lambda (N)_\lambda
\end{aligned}
$$

We leave the reader to verify that the typed terms are indeed well formed. Next, we verify that the given λ-terms have the required behaviour.

Theorem 4.16 *If* $P \to_w^* Q$ *then* $(P)_\lambda \to^* (Q)_\lambda$.

Proof: A straightforward computation shows the result:

- $(\mathbf{I}x)_\lambda = (\lambda x.x)x \to x$.

- $(\mathbf{K}xy)_\lambda = (\lambda xy.x)xy \to (\lambda y.x)y \to x$.

- $(\mathbf{S}xyz)_\lambda = (\lambda xyz.xz(yz))xyz \to^* xy(yz)$ as required.

$\square$

Finally, as an example, we show the reduction of $(\mathbf{S}\mathbf{K}\mathbf{K}M)_\lambda$ in the λ-calculus. Here are a few snap-shots of the reduction:

$$
\begin{aligned}
(\mathbf{S}\mathbf{K}\mathbf{K}M)_\lambda &= (\lambda xyz.xz(yz))(\lambda xy.x)(\lambda xy.x)M \\
&\to (\lambda yz.(\lambda xy.x)z(yz))(\lambda xy.x)M \\
&\to (\lambda z.(\lambda xy.x)z((\lambda xy.x)z))M \\
&\to (\lambda xy.x)M((\lambda xy.x)M) \\
&\to^* (\lambda y.M)(\lambda y.M) \\
&\to M
\end{aligned}
$$

The reader will be able to add the types to the above to check that all the reductions were indeed well-formed.

Next, we turn to understanding the λ-calculus in combinatory logic.

Definition 4.17 *Let us define a translation of terms from the (implicational fragment of) λ-calculus into combinatory logic. We do this in two steps, first we define a translation* $(\cdot)_{CL}$ *into a system close to combinatory logic where we allow abstraction. In other words, we add to combinatory logic the following abstraction operation:* $[\cdot]$. *We now define the translation as follows (again, written without types for clarity):*

$$
\begin{aligned}
(x)_{CL} &= x \\
(tu)_{CL} &= (t)_{CL}(u)_{CL} \\
(\lambda x.t)_{CL} &= [x](t)_{CL}
\end{aligned}
$$

*Next, we must eliminate this additional abstraction operation to complete the cod-
ing into combinatory logic. This is achieved by defining $[\cdot]$ as the following function:*

$$
\begin{aligned}
[x]x &= \mathbf{I} \\
[x]t &= \mathbf{K}t \; (\text{if } x \notin FV(t)) \\
[x](tu) &= \mathbf{S}([x]t)([x]u)
\end{aligned}
$$

*where FV is the obvious free variable function for combinatory logic ($FV(\mathbf{S}) =
FV(\mathbf{K}) = \varnothing$, $FV(x) = x$, and $FV(MN) = FV(M) \cup FV(N)$). Note that
there is no rule of the form $[x]([y]t)$, in other words, the translation works from the
inside out.*

Let us look at an example of this translation. We will compute $(\lambda xy.x)_{CL}$ as
follows:

$$
\begin{aligned}
(\lambda xy.x)_{CL} &= [x]([y]x) \\
&= [x](\mathbf{K}x) \\
&= \mathbf{S}([x]\mathbf{K})([x]x) \\
&= \mathbf{S}(\mathbf{K}\mathbf{K})\mathbf{I}
\end{aligned}
$$

Again we leave the reader to check that the types are well behaved under this translation.

Corresponding to Theorem 4.16 we would like to prove an analogous results
stating that reduction in the λ-calculus is preserved under this translation. However, this is not the case since reduction in combinatory logic is *weak*. To illustrate
what we mean weak reduction, we consider an example. Take the term $\mathbf{KI}$ which
we can see by inspection of the rules for $\rightarrow_w$ is in normal form — no reduction can
be applied. However, if we look at the corresponding λ-term, using the translation
given previously, we get $(\mathbf{SK})_{\lambda} = (\lambda xy.x)(\lambda y.y)$ which is a λ-term not in normal
form.

We refer the interested reader to the literature, especially Barendregt (1984,
Chapter 7), for a discussion on how to add additional axioms to combinatory logic
so that β-reduction is preserved under the translation of λ-calculus into combinatory logic.

3.2   The Correspondence with Hilbert Axioms

For the λ-calculus we showed that if we remove the terms from typing rules, we
get exactly the natural deduction system for intuitionistic logic. Here we show that
we can do exactly the same trick for combinatory logic. If we take away the terms
from the following:

$$
\begin{aligned}
\mathbf{K}_{\sigma,\tau} &\;:\; \sigma \rightarrow \tau \rightarrow \sigma \\
\mathbf{S}_{\sigma,\tau,\gamma} &\;:\; (\sigma \rightarrow \tau \rightarrow \gamma) \rightarrow (\sigma \rightarrow \tau) \rightarrow \sigma \rightarrow \gamma
\end{aligned}
$$

and replace the functional arrow $\rightarrow$ by implication $\Rightarrow$, and types by formulas, as is familiar by now. We are then left with the Hilbert-style axiom schema for the implicational fragment for intuitionistic logic:

$$\Phi \Rightarrow \Phi' \Rightarrow \Phi$$
$$(\Phi \Rightarrow \Phi' \Rightarrow \Phi'') \Rightarrow (\Phi \Rightarrow \Phi') \Rightarrow \Phi \Rightarrow \Phi''$$

As we have already seen, the rule for application in combinatory logic is nothing more than modus ponens (MP). **S** and **K** correspond to the axioms of **H** (Chapter 3, Section 2.4), terms built out of the combinators correspond to Hilbert-style proofs, and types correspond to formulae.

As with all formulations of this style of correspondence, the main application is that we have two different view points on the *same* theory. The objects for manipulation in each case are quite different and lend themselves well to certain manipulations. To give an example of this, we show how we can give an alternative proof of Theorem 3.2 from Chapter 3. We are required to prove that the system **NJ$_0$** and Hilbert axioms are equivalent systems, in the sense that the same formulae are provable. Recasting this theorem into the corresponding formalisms requires that we prove that the λ-calculus and combinatory logic are equivalent systems, in other words, we need to show that we can translate any derivation (modulo the structural rules) in the λ-calculus $\Gamma \longrightarrow t : \sigma$ into a derivation in combinatory logic, and vice-versa. But this is precisely what we did in Definitions 4.17 and 4.15 above.

There is no doubt that the proof at the level of λ-terms and combinators is more manageable. Inductions over the structure of terms are not only easier to grasp, but also easier to write down than the corresponding inductions over the length of the derivation. Of course, the two proofs, the first relating **NJ$_0$** and the hilbert axioms **H** and the second relating λ-calculus with combinatory logic, are essentially the same proofs because of the Curry-Howard correspondence.

A second application of this correspondence is that we get another way of building proofs in Hilbert-style intuitionistic logic. We give an example and give a proof of $\Phi \Rightarrow \Phi$ from the axioms. To do this, we have to build a term **I** out of the combinators such that $\mathbf{I} : \sigma \rightarrow \sigma$ and satisfies $\mathbf{I}x = x$. Now, **SKK** is such a term, as we have already seen. The term **SKK** is considerably more manageable than the proof that we used for this in Chapter 3, Section 2.4.

We refer the reader to Howard (1980), and Girard *et al.* (1989) for further details on the Curry-Howard isomorphism.

▶ **EXERCISE 4.5**

Compute the following, using the translations $(\cdot)_{CL}$ and $(\cdot)_\lambda$ given. Where necessary, reduce the terms to normal form (after performing the translation!).

(i) $(\lambda x.x)_{CL}$ (ii) $(\lambda xy.x)_{CL}$

(iii) $(\lambda xy.y)_{CL}$ (iv) $((\lambda xy.x)(\lambda x.x))_{CL}$

(v) $(\mathbf{KI})_{\lambda}$ (vi) $(xy)_{\lambda}$

(vii) $(x\mathbf{I})_{\lambda}$ $(viii)$ $(\mathbf{S}(\mathbf{KS})\mathbf{K})_{\lambda}$

4   APPLICATIONS OF THE CURRY-HOWARD CORRESPONDENCE

We have already seen several applications of the Curry-Howard correspondence in the form of simpler proof methods for the typed λ-calculus (programming language) which carry over to the logic. In this section we will study some other applications, in particular to the development of new languages.

It is worth remarking that the correspondence between the λ-calculus and intuitionistic natural deduction, and combinatory logic and Hilbert axioms, are two of the simplest cases of the isomorphism. The ideas can in fact be generalised to many other logics. Probably one of the most well-known additions is Girard's (1972) System F, which is a term assignment to *second order intuitionistic logic*. By establishing an isomorphism between System F and the logic all results obtained in the calculus can be carried over. Girard also gave a proof of strong normalisation for System F using a proof technique called "Candidats de Reductibilité" (See Gallier (1990) for a good exposition), which, thanks to the Curry-Howard isomorphism, gives a strong normalisation result of second order intuitionistic logic in purely logical terms. These ideas were then generalised to the calculus F_{ω} giving a strong normalisation result for higher-order intuitionistic logic using only proof theory (a result previously conjectured by Takeuti; see for example (Girard, 1987b; Girard, 1972) for details).

Another example is the $\lambda\mu$-calculus developed by Michel Parigot (1992) as the term assignment of classical logic, thus giving a programming language based on classical logic. This work, starting from work of Griffin (1990) extends the Curry-Howard style isomorphism to classical logic, and allows us to get the computational content from a classical logic proof. Double negation elimination, $\neg\neg A \Rightarrow A$, turns out to be the type of Felleisen's (1988) *catch* operation, thus giving classical logic the status of a calculus to reason about control and *continuation* operations.

A further example is the search for a programming language based on linear logic giving a linear functional programming language. With linear logic we have a system that enables us to talk about the fine details of the reduction process of a program. In particular it brings out features of copying and discarding which are strongly acknowledged as the "expensive" parts of a computation. This computational-logic connection has been used to develop programming languages that keep a tighter control over resource management. See for example (Abramsky, 1993).

This list goes on, and is rather long! We now outline one of these applications in a little more detail.

4.1   System F

One of the most significant applications of the Curry-Howard isomorphism is the work on System F, which extends the isomorphism to second order intuitionistic logic.

In the same way as the λ-calculus allows functions to become first class citizens (they can be passed as arguments to other functions), the idea of System F can be understood intuitively as making *types* first class citizens. If we write a function $\lambda x.t : \sigma \to \tau$, then the type for this function is fixed. The idea is to extend the system to allow types to be passed as arguments. Hence we need a notion of *type abstraction* and *type application*; a way of making terms depend on types.

We write $\Lambda X.t$ as a term t that depends on a type X, to reflect this in the type, assuming $t : \sigma$, we write $\Lambda X.t : \forall X.\sigma$, where $\forall$ is the second order universal quantification. Here is an example of type abstraction: $\Lambda X.\lambda x.x : \forall X.X \to X$. This is a function that is the identity function at *all* types X. (A function working on *any* type is called *polymorphic*.)

To use this function, we have to apply it first to some type. For example, assume to have already defined types such as integers (N), Booleans (B), etc. Then the following application says that we want to use the identity function only on integers:

$$(\Lambda X.\lambda x.x : \forall X.X \to X)N$$

There is a corresponding reduction rule which is similar to β-reduction that allows us to reduce this to the expected $\lambda x.x : N \to N$.

We will now be a little more formal about the types and terms of System F.

Definition 4.18 (Type schemes) *The types of System F are given by the set T:*

- *Type variables, denoted by $X, Y, \ldots$.*

- *If $\sigma, \tau \in T$ then $\sigma \to \tau \in T$.*

- *If $\sigma \in T$ and X is a type variable, then $\forall X.\sigma \in T$.*

Note that these types are nothing more than propositions quantified at the second order (cf. Chapter 2, Section 5.3.)

Definition 4.19 (Terms) *The set of terms in System F are given by extending the λ-calculus with the following operations:*

- *Type abstraction: If $t : \sigma$ is a term, then $\Lambda X.t : \sigma$ is a term of type $\forall X.\sigma$ provided that X does not occur free in any of the free variables of t.*

- *Type application: If $t : \forall X.\sigma$ is a typed term, and τ is a type, then $t\tau$ is a term of type $\sigma[\tau/X]$.*

We can now give the term assignment for System F to second order propositional logic.

$$\frac{\Gamma \longrightarrow t : \sigma}{\Gamma \longrightarrow \Lambda X.t : \forall X.\sigma} \ (\forall I) \qquad \frac{\Gamma \longrightarrow t : \forall X.\sigma \quad \tau \in \mathcal{T}}{\Gamma \longrightarrow t\tau : \sigma[\tau/X]} \ (\forall E)$$

where there is the usual side condition on the $(\forall I)$ rule that X is not free in t or Γ.

There are two reductions in this calculus. The usual β-reduction as we saw for the λ-calculus, together with an additional rule for type application, denoted β^t. The two reductions are given by:

$$\begin{array}{rcl} \beta & : & (\lambda x.t)u \to t[u/x] \\ \beta^t & : & (\Lambda X.t)\tau \to t[\tau/X] \end{array}$$

where term substitution is already defined (Page 137), and type substitution is defined in the obvious way.

System F is a very powerful calculus. It captures all of second order arithmetic, and thus all of *current* mathematics. It can also be seen as a programming language: all the usual data-types (inductive types) that one uses in programming languages are indeed definable in System F. Note in particular that we restricted the calculus to the implication fragment. We can recover products and sums in a natural way. For example, define $\sigma \times \tau$ as $\forall X.(\sigma \to \tau \to X) \to X$, which gives the term $\langle u, v \rangle : \sigma \times \tau$ as $\Lambda X.\lambda x.xuv$. We leave the definition of the projections fst and snd as an exercise. The reader can find many more examples of this kind of data structure in Girard et al. (1989).

If we look at the reduction step in second order propositional logic, we see that there is an additional reduction in the case where we have an $(\forall I)$ followed by a $(\forall E)$. As usual, this can be eliminated from the derivation in the following way:

$$\frac{\dfrac{\Gamma \longrightarrow \Phi}{\Gamma \longrightarrow \forall X.\Phi} \ (\forall I)}{\Gamma \longrightarrow \Phi[\Phi'/X]} \ (\forall E)$$

becomes simply the derivation ending in $\Gamma \longrightarrow \Phi[\Phi'/X]$.

This calculus has been shown to be strongly normalising, and satisfies the Church–Rosser property. This gives a purely proof theoretic account of showing consistency of second order propositional calculus. Remark also that via the double negation translations of classical logic into intuitionistic logic given in Chapter 3, Section 5, this result covers classical logic too. However, it is also possible to work directly with classical logic and develop a corresponding calculus, this is what has been done by Parigot (1992) with the $\lambda\mu$-calculus.

▶ **EXERCISE 4.6**

Define in System F the operations fst and snd so that they satisfy the usual properties, i.e. $\mathsf{fst}(\langle t, u \rangle) \to^* t$ and $\mathsf{snd}(\langle t, u \rangle) \to^* u$.

▶ **EXERCISE 4.7**

Give a typing system for the following calculus, which is a restriction of the λ-calculus where all variables occur exactly once in the body of a term.

Term	Constraint	FV
x		x
$\lambda x.t$	$x \in FV(t)$	$FV(t) \setminus \{x\}$
tu	$FV(t) \cap FV(u) = \varnothing$	$FV(t) \cup FV(u)$

Then prove strong normalisation for this calculus, and hence also for the corresponding logic.

CHAPTER 5

MODAL AND TEMPORAL LOGICS

1 INTRODUCTION AND MOTIVATION

Modal logics, of which temporal logics are particular instances, were proposed at the beginning of the twentieth century by Clarence Lewis (Lewis, 1918). Lewis was dissatisfied by the material implication of classical propositional logic, and explored another variety of logics, where a new operator □ (read "box") is introduced. Originally, □Φ meant "necessarily, Φ", and ¬□¬Φ, also written ◇Φ, meant "possibly, Φ". Lewis then proposed several different sets of rules for the new connectives, and discussed their possible intuitive meanings.

Some of these interpretations are related to philosophical reflections on the nature of consequence and of implication. Consider for example the formula (□Φ) ⇒ Φ. If we interpret □Φ as "Mr. A knows that Φ holds", then □Φ ⇒ Φ means "If Mr. A knows that Φ holds, then Φ holds"; if it is true for all Φ, this formula asserts that Mr. A does not make mistakes. The resulting logic is called a logic of knowledge. These logics can also model temporal reasoning: if □Φ means "in all possible futures, Φ" (or "from now on, Φ"), then (□Φ) ⇒ Φ states that whenever Φ holds from some instant on, then it also holds at this precise instant.

Under these interpretations, the formula Φ ⇒ □Φ does not hold in general. For instance, in the interpretation by knowledge, Φ ⇒ □Φ means that either Φ is false or that Mr. A knows Φ; this excludes the case where Φ is true but Mr. A doesn't know it, a very likely situation. In fact, if Mr. A knew that Φ was true whenever Φ was true, then there would be no point in distinguishing Mr. A's state of knowledge and reality. In the temporal interpretation, Φ ⇒ □Φ would mean that if Φ holds at some point in time, then it will hold forever from then: this would force the world to be entirely static, negating the essence of time.

On the other hand, if we have proved Φ in any modal system, we shall be able to deduce □Φ. Intuitively, if Φ is proved, then it *always* holds. (This is why □ is sometimes read "necessary".) Or, if Φ is proved (by Mr. A), then Mr. A knows that Φ holds. Or, if Φ is proved, then it holds at all points in time, therefore if holds in every future of every instant. The surprising fact that we can deduce □Φ from Φ, but that Φ ⇒ □Φ is not valid means that implication ⇒ and consequence ⊢ won't coincide in modal logics, i.e. the Deduction Theorem won't hold.

Why then, are we interested in such logics? Apart from philosophical considerations, modal logics and in particular temporal logics have been shown to provide
an expressive enough, while still decidable, language in which to express specifications of sequential and parallel models of computation, and with which the verification of an implementation with respect to a specification is decidable with simple automaton-based techniques. From a logical point of view, it is also interesting
to explore the connections between modal and non-modal logics (like intuitionistic
logic), so as to understand better how all of them relate to each other. This has been
put to good use in the framework of *non-monotonic logics*, whose aim is to model
statements like "in general, Φ holds" that can survive the discovery of a particular
case where Φ does not hold.

We shall now study both aspects of modal logics. We first study the particular
modal logic S4 in Section 2, and relate it to non-monotonic logics. We choose S4
because it is one of the simplest modal logics of interest, and because it is also a
temporal logic. This will lead us naturally to the study of temporal logics for specifying properties of transition systems in Section 3, and of methods for checking
that a given transition system satisfies a given specification in these logics (this is
called *model-checking*).

The reader interested in logical aspects of modal logics is referred to (Chellas,
1980), or to (Hughes and Creswell, 1968; Hughes and Creswell, 1984). Texts on
temporal logics and model-checking include (Emerson, 1990; Kozen and Tiuryn,
1990).

2   S4 AND NON-MONOTONIC LOGICS

2.1   *Syntax*

A formula of S4 is any formula built on propositional variables with the usual propo-
sitional connectives $\wedge$, $\vee$, $\neg$, $\Rightarrow$ and the special *modality* $\square$: if Φ is a formula, then
$\square\Phi$ is a formula, too ("box Φ", or "necessarily, Φ; temporally speaking, "in every
future, Φ"). We assume that $\square$ binds tighter than any other logical operator, so that
$\square A \Rightarrow A$ means $(\square A) \Rightarrow A$, for example. We also define $\diamondsuit\Phi$ as an abbreviation
for $\neg\square\neg\Phi$ ("diamond Φ" or "possibly, A"; temporally speaking, "in some future,
A").

In addition to all proof rules for classical logic (see Chapter 2, Section 3.1), in
S4 we add the following axioms:

(K) $\square(\Phi \Rightarrow \Phi') \Rightarrow \square\Phi \Rightarrow \square\Phi'$

(T) $\square\Phi \Rightarrow \Phi$

(4) $\square\Phi \Rightarrow \square\square\Phi$

for all formulas Φ and Φ', and in addition to the modus ponens (MP) rule, we add the following *necessitation rule*:

(Nec) from Φ, deduce $\Box\Phi$, for every formula Φ.

Axioms (K) and (T) are common to most modal logics. Axiom (K) says that if $\Phi \Rightarrow \Phi'$ and Φ are necessary (or known, or true in every future), then Φ' is necessary (or known—so here Mr. A can deduce everything that is deducible from what he originally knew—; or true in every future, in the temporal interpretation). Axiom (T), as well as Rule (Nec), were discussed in the prologue of the chapter.

Together, Axioms (K) and (T) are also called (S). The logic S4 is then named after its axioms, and indeed it is sometimes referred to as the logic KT4.

What makes S4 particular is Axiom (4). In the interpretation of $\Box$ as knowledge, then (4) states that Mr. A is capable of introspection: if Mr. A knows something, then he knows that he knows it. (Axiom (5), which we shall see later, would be even more demanding, as it would ask that if A is consistent with what Mr. A knows, then the agent must know it is indeed consistent; in S5, that is, S4 plus the axiom (5), $\Box$ actually modelizes beliefs rather than knowledge.) The temporal logic interpretation is more straightforward: (4) expresses the transitivity of time, i.e. if Φ holds in every future, then it will again hold in every future of every future.

As we have already noticed in Chapter 2, Section 3.1, Hilbert-style systems are not easy to work with, and we shall define the semantics of S4 in Section 2.3, and sequent systems in Section 2.5.

2.2 Philosophical Logic and Non-Monotonic Logics

There is a flurry of other modal logics coming from the philosophical logic community. Most are based on Axiom (K), and as such they are called *normal* modal logics. Some of them use Axiom (T). Other possible axioms that have been used to define modal logics are:

(5) $\Diamond\Phi \Rightarrow \Box\Diamond\Phi$,

(D) $\Box\Phi \Rightarrow \Diamond\Phi$,

(R) $\Diamond\Box\Phi \Rightarrow (\Phi \Rightarrow \Box\Phi)$,

(F) $(\Box\Phi \Rightarrow \Phi') \vee (\Diamond\Box\Phi' \Rightarrow \Phi)$,

(2) $\Diamond\Box\Phi \Rightarrow \Box\Diamond\Phi$

(T) is also sometimes called (M), (5) is sometimes called (E) (for Euclidean), and (R) is sometimes called (W5).

We won't delve into what logics using any combination of these axioms might mean, and we refer the interested reader to Chellas' book.

In every modal logic L, we have a notion of deduction $\vdash^L$: if Γ is a set of formulas, and Φ is a formula, $\Gamma \vdash^L \Phi$ if and only Φ can be deduced from the set of hypotheses Γ and the axioms of L by using the only rules (MP) and (Nec). But, as we have already stated, the Deduction Theorem does not hold in modal logics, so $\Phi_1, \ldots, \Phi_n \vdash^L \Phi$ is not equivalent to $\vdash^L \Phi_1 \Rightarrow \ldots \Rightarrow \Phi_n \Rightarrow \Phi$. (Although the latter entails the former.)

This definition of deducibility is the one adopted by Rajeev Goré (Goré, 1991), and is called *strong deducibility*. *Weak deducibility* is defined as the relation $\vdash_L$ such that $\Gamma \vdash_L \Phi$ if and only if there is a finite subset $\{\Phi_1, \ldots, \Phi_n\}$ of Γ such that $\vdash^L \Phi_1 \Rightarrow \ldots \Rightarrow \Phi_n \Rightarrow \Phi$. (Notice that, somewhat paradoxically, weak deducibility entails strong deducibility, but not the converse.)

Weak deducibility and strong deducibility coincide with an empty set of hypotheses, and a formula Φ such that $\vdash^L \Phi$ (equivalently, $\vdash_L \Phi$) is called a *theorem* of L.

One of the main interests of modal logics is that strong deducibility plus a certain completion process are the essence of *non-monotonic logics*, modal or not. We quote Rajeev Goré (Goré, 1991):

> Propositional modal logics have been used to model epistemic notions like knowledge and belief for quite a while now where the formula $\Box A$ is read as "A is believed" or as "A is known". Given a (monotonic) modal logic **S** and and a set of formulae Γ, the formula A is a monotonic consequence of Γ in **S** if it is deducible in **S** from Γ, usually written as $\Gamma \vdash_S A$. The set Γ is usually called a theory and the monotonic consequences of Γ in **S** are all the formulae deducible from Γ in **S**; that is $Cn_S(\Gamma) = \{A \mid \Gamma \vdash_S A\}$. The system is "monotonic" in that if A is in $Cn_S(\Gamma)$ then it will be in $Cn_S(\Gamma')$ for any superset Γ' of Γ.
>
> To obtain non-monotonicity we assume $\neg\Box A$ ("A is not known") if there is no deduction of A in **S** from Γ and previous assumptions. More formally, the theory T is an **S**-expansion of theory Γ if it satisfies the equation $T = Cn_S(\Gamma \cup \{\neg\Box A \mid A \notin T\})$. Since T appears in the right hand side, the definition is circular, and consequently, a theory Γ may have zero, one or more **S**-expansions. To compensate for this phenomenon, the set of *non-monotonic* consequences of Γ in **S** is usually defined as the intersection of all **S**-expansions of Γ. The new system is "nonmonotonic" because, although A may be a nonmonotonic consequence of Γ, it may not be a non-monotonic consequence of a superset of Γ.

The interest in non-monotonic logics then arises in trying to capture the notion of common-sense reasoning, where *default* conclusions can be drawn from incomplete information, possibly to be retracted on presentation of contrary evidence. John McCarthy's *circumscription* was the first attempt to formalise common-sense reasoning, followed by Reiter's *default logic* or Halpern and Moses' *auto-epistemic logic*. Marek, Schwarz and Truszczyński then noticed that auto-epistemic logic was nothing else but non-monotonic K45D, and that S4F generalised both default logic and auto-epistemic logic.

▶ **EXERCISE 5.1**
Show that in all modal logics L with Axioms (K) and (T), $\Phi \vdash^L \Phi'$ if and only if $\vdash^L \Box\Phi \Rightarrow \Phi'$.

▶ **EXERCISE 5.2**
Show that in all normal modal logics L (i.e., with Axiom (K)), $\Phi \vdash^L \Phi'$ if and only if $\vdash^L (\Box\Phi \wedge \Phi) \Rightarrow \Phi'$.

2.3 Kripke Semantics

The semantics of modal logics like S4 are usually given in terms of *possible worlds interpretations*. Indeed, a formula does not just hold under an interpretation, but also at a given time point, or in a certain state of knowledge. We represent the set of all points in time or the set of all states of knowledge as a non-empty *universe* W, whose points are called *worlds*. In the case of S4, worlds are partially ordered: if w and w' are worlds, $w \leq w'$ may either mean that w is an instant that occurs before w', or that w is a subset of all the things that we know in w'. Then, a formula $\Box\Phi$ is true at some world w if and only if Φ is true at all subsequent worlds (at all later times, or in any expanded state of knowledge).

The definition generalises to all modal logics in the following way (this is due to Saul Kripke in 1963):

Definition 5.1 *A* Kripke frame *is a couple* (W, R)*, where* W *is a non-empty set called the* universe *and* R *is a binary relation on* W *called the* accessibility *relation. The elements of* W *are called* worlds.

An assignment *or* interpretation ρ *is a map from propositional variables to sets of worlds.*

The semantics of a modal formula Φ *is given by the Boolean valued evaluation function* $[\![\Phi]\!]w\rho$*, where* w *is a world and* ρ *is an assignment. We also write* $w, \rho \models \Phi$ *for* $[\![\Phi]\!]w\rho$*, and define it by:*

- $w, \rho \models A$ *if and only if* $w \in \rho(A)$*, where* A *is a propositional variable;*

- $w, \rho \models \Phi \wedge \Phi'$ *if and only if* $w, \rho \models \Phi$ *and* $w, \rho \models \Phi'$*;*

- $w, \rho \models \Phi \vee \Phi'$ *if and only if* $w, \rho \models \Phi$ *or* $w, \rho \models \Phi'$*;*

- $w, \rho \models \neg\Phi$ *if and only if* $w, \rho \not\models \Phi$*;*

- $w, \rho \models \Phi \Rightarrow \Phi'$ *if and only if* $w, \rho \not\models \Phi$ *or* $w, \rho \models \Phi'$*;*

- $w, \rho \models \Box\Phi$ *if and only if for every world w' such that wRw', then $w' \models \Phi$.*

When $w, \rho \models \Phi$, we also say that w, ρ satisfy Φ.

A Kripke frame is satisfied *by Φ, or is a* model *of Φ if and only if $w, \rho \models \Phi$ for some world w and assignment ρ. Φ is valid in the given frame if and only if $w, \rho \models \Phi$ for every world w and assignment ρ.*

If C is a class of frames, we say that Φ is valid in C, and we write $C \models \Phi$, if and only if Φ is valid in every frame of the class.

Note that, viewing R as a set of couples, a Kripke frame (W, R) is nothing but an *oriented graph* with W as set of vertices and R as set of edges.

It turns out that a formula of S4 is a theorem if and only if it is valid in the class of pre-ordered sets, that is, of frames (W, R) where R is a reflexive and transitive relation on W. Interpreting frames as graphs means that the theorems of S4 are actually theorems about the set of paths in collections of directed acyclic graphs (in fact, trees suffice). We make this precise:

Definition 5.2 *Let C be a class of frames, and L be a modal logic.*

L is sound *with respect to C if and only if for every formula Φ, $\vdash^L \Phi$ entails $C \models \Phi$.*

L is complete *with respect to C if and only if for every formula Φ, $C \models \Phi$ entails $\vdash^L \Phi$.*

We characterise frames by the properties of their accessibility relation R: they may be reflexive, symmetric, transitive, serial (from each world, there is an accessible world), convergent (for all worlds w, if w_1 and w_2 are accessible from w, then there is world accessible from both w_1 and w_2).

Then, in the case of S4:

Theorem 5.3 (Soundness, Completeness) *S4 is sound and complete with respect to the class of reflexive and transitive frames.*

Proof: That S4 is sound for reflexive and transitive frames is clear.

To prove completeness, instead of proving that $\models \Phi$ implies $\vdash^{S4} \Phi$ (or equivalently, $\vdash_{S4} \Phi$), we shall assume that $\nvdash_{S4} \Phi$, and prove $\not\models \Phi$.

Choose a formula **F** to denote false, i.e. a formula that provably implies all others. Call a set Γ of formulas *consistent* if $\Gamma \vdash_{S4} \mathbf{F}$ does not hold. (Or equivalently, such that **F** is not weakly deducible from Γ.) If $\nvdash_{S4} \Phi$, then $\neg\Phi$ is consistent. Indeed, if we had $\neg\Phi \vdash_{S4} \mathbf{F}$, then we would have $\vdash_{S4} \neg\Phi \Rightarrow \mathbf{F}$, therefore $\vdash_{S4} \Phi$ by classical reasoning. (That is, without using any modal rules or axioms.)

We now build a model for $\neg\Phi$ (i.e., a frame and an interpretation such that $\neg\Phi$ holds at some world). We define the particular Kripke frame $(W, \leq)$, where W is the set of all maximal consistent sets of propositions of S4, and where $w \leq w'$ if and only if for every formula of the form $\Box\Phi'$ in w, Φ' is in w'. We define the following interpretation ρ: for every variable A, $\rho(A)$ is the set of worlds w such that $A \in w$.

First, notice a few facts about maximally consistent sets w. Since w is consistent, it cannot contain both Φ and $\neg\Phi$, for any formula Φ. But for every formula Φ, w must contain either Φ or $\neg\Phi$. Indeed, assume it contains none. Then, $w \cup \{\Phi\}$ and $w \cup \{\neg\Phi\}$ are both inconsistent by maximality, so there is a finite set of formulas $\Phi_1, \ldots, \Phi_n$ in $w \cup \{\Phi\}$ such that $\vdash^{S4} \Phi_1 \Rightarrow \ldots \Rightarrow \Phi_n \Rightarrow \mathbf{F}$, and similarly there is a finite set of formulas $\Phi'_1, \ldots, \Phi'_{n'}$ in $w \cup \{\neg\Phi\}$ such that $\vdash^{S4} \Phi'_1 \Rightarrow \ldots \Rightarrow \Phi'_{n'} \Rightarrow \mathbf{F}$. Without loss of generality, we may assume that Φ is among the Φ_i (otherwise we add it to the finite set of the Φ_i's), and in fact that $\Phi = \Phi_n$, and similarly we may assume that $\neg\Phi = \Phi'_{n'}$. We may also assume, by adding enough formulas in each finite set, that the sets $\{\Phi_1, \ldots, \Phi_{n-1}\}$ and $\{\Phi'_1, \ldots, \Phi'_{n'-1}\}$ coincide. Then, we have $\vdash^{S4} \Phi_1 \Rightarrow \ldots \Rightarrow \Phi_{n-1} \Rightarrow \Phi \Rightarrow \mathbf{F}$ and $\vdash^{S4} \Phi_1 \Rightarrow \ldots \Rightarrow \Phi_{n-1} \Rightarrow \neg\Phi \Rightarrow \mathbf{F}$, hence $\vdash^{S4} \Phi_1 \Rightarrow \ldots \Rightarrow \Phi_{n-1} \Rightarrow (\Phi \vee \neg\Phi) \Rightarrow \mathbf{F}$, but the latter amounts to saying that $\vdash^{S4} \Phi_1 \Rightarrow \ldots \Rightarrow \Phi_{n-1} \Rightarrow \mathbf{F}$. Moreover $\Phi_1, \ldots, \Phi_{n-1}$ are all in w, so this proves that w is inconsistent, which is contradictory. Hence w must contain either Φ or $\neg\Phi$.

Observe also that any maximally consistent set is stable by weak deduction: if w is a maximal consistent set such that $w \vdash_{S4} \Phi$, and if $\neg\Phi$ was in w, then we could deduce $\mathbf{F}$ from w by deducing Φ first and taking the conjunction with $\neg\Phi$. So $\neg\Phi \notin w$, so $\Phi \in w$.

Our first claim is that $(W, \leq)$ is a reflexive and transitive frame, i.e. that $\leq$ is a preorder. It is indeed reflexive, because for each $\Box\Phi' \in w$, Φ' is weakly deducible from it by modus ponens with (T); but any maximal consistent set is closed under (weak) deducibility, so $\Phi' \in w$. It is also transitive: if $w_1 \leq w_2 \leq w_3$, then for every formula $\Box\Phi' \in w_1$, $\Box\Box\Phi' \in w_1$ by (4), hence $\Box\Phi' \in w_2$ since $w_1 \leq w_2$, hence $\Phi' \in w_3$ since $w_2 \leq w_3$; this entails $w_1 \leq w_3$.

Our second claim is that for every formula Φ', $w, \rho \models \Phi'$ if and only if $\Phi' \in w$. This is proved by structural induction on Φ':

- If Φ' is a variable, this follows from the definition of ρ.

- If Φ' is of the form $\Phi'' \vee \Phi'''$, then: if $w, \rho \models \Phi'$, then $w, \rho \models \Phi''$ or $w, \rho \models \Phi'''$, hence Φ'' or Φ''' is in w, by induction hypothesis; but if $\Phi'' \in w$ (resp. $\Phi''' \in w$), then $\Phi' = \Phi'' \vee \Phi'''$ is weakly deducible from w, so $\Phi' \in w$.

  Conversely, if $w, \rho \not\models \Phi'$, then $w, \rho \not\models \Phi''$ and $w, \rho \not\models \Phi'''$, hence $\Phi'' \notin w$ and $\Phi''' \notin w$ by induction hypothesis; therefore, by maximality of w, $\neg\Phi''$ and $\neg\Phi'''$ must both be in w, hence the weakly deducible formulas $\neg\Phi'' \wedge \neg\Phi'''$ and $\neg(\Phi'' \vee \Phi''')$ (that is, $\neg\Phi'$) are in w, so Φ' cannot be in w, by consistency.

- The case of conjunctions is similar.

- If Φ' is of the form $\neg\Phi''$, then: if $w, \rho \models \Phi'$, then $w, \rho \not\models \Phi''$, so by induction hypothesis, $\Phi'' \not\subseteq w$; by maximality of w, $w \cup \{\Phi''\}$ is not consistent, and in particular $\neg\Phi''$ is weakly deducible from $w \cup \{\Phi''\}$, hence from w; therefore $\Phi' = \neg\Phi'' \in w$. Conversely, if $\Phi' \in w$, then Φ'' is not in w as otherwise w would not be consistent, so by induction hypothesis $w, \rho \not\models \Phi''$, hence $w, \rho \models \Phi'$.

- If Φ' is of the form $\Box\Phi''$, then: if $\Phi' \in w$, then $\Phi'' \in w'$ for every $w' \geq w$ by definition of $\leq$, so by induction hypothesis $w', \rho \models \Phi''$ for every $w' \geq w$, hence $w, \rho \models \Phi'$.

  Conversely, if $w, \rho \models \Phi'$, then $w', \rho \models \Phi''$ for every $w' \geq w$, so by induction hypothesis, $\Phi'' \in w'$ for every $w' \geq w$. Now, consider the set $w_0' = \{\Psi \mid \Box\Psi \in w\}$. We claim that $w_0' \cup \{\neg\Phi''\}$ is inconsistent. Indeed, assume the contrary, i.e., assume $w_0' \cup \{\neg\Phi''\}$ consistent.

  We claim that there is a maximally consistent superset w_1' of $w_0' \cup \{\neg\Phi''\}$. To show this, we use Zorn's Lemma (another form of the Axiom of Choice), which states that every inductively ordered set S has a maximal element. Recall that a set is *inductive* if and only if every totally ordered subset has a least upper bound. In our case, let S be the set of all consistent sets, ordered by inclusion. Given a totally ordered set of consistent sets C_i, $i \in I$, let C be $\bigcup_{i \in I} C_i$. C is consistent, because otherwise there would be finitely many formulas $\Phi_1, \ldots, \Phi_n$ in C such that $\vdash^{S4} \Phi_1 \Rightarrow \ldots \Rightarrow \Phi_n \Rightarrow \mathbf{F}$. Each Φ_i is in some C_i, and since the C_i's are totally ordered, there is a C_i containing all of $\Phi_1, \ldots, \Phi_n$; but then C_i would be inconsistent, a contradiction. So S is inductive, and by Zorn's Lemma, there is a maximally consistent superset w_1' of $w_0' \cup \{\neg\Phi''\}$.

  By definition of w_0', w_1' and of $\leq$, $w \leq w_1'$. Because $\Phi'' \in w'$ for every $w' \geq w$, we must therefore have $\Phi'' \in w_1'$. But then, w_1' contains both Φ'' and $\neg\Phi''$ (by construction), and is inconsistent.

  So $w_0' \cup \{\neg\Phi''\}$ is inconsistent, and we can weakly derive $\mathbf{F}$ from $w_0', \neg\Phi''$, hence by classical reasoning we can weakly derive Φ'' from w_0'. (Notice that weak deduction satisfies the Deduction Theorem by definition.) But then, there are finitely many formulas $\Psi_1, \ldots, \Psi_n$ in w_0' such that $\vdash_{S4} \Psi_1 \Rightarrow \ldots \Rightarrow \Psi_n \Rightarrow \Phi''$. Hence $\vdash_{S4} \Box(\Psi_1 \Rightarrow \ldots \Rightarrow \Psi_n \Rightarrow \Phi'')$ by (Nec), then $\vdash_{S4} \Box\Psi_1 \Rightarrow \ldots \Rightarrow \Box\Psi_n \Rightarrow \Box\Phi''$ by n applications of Axiom (K) and classical reasoning. So $\Box\Phi''$, i.e. Φ', is weakly deducible from the $\Box\Psi_i$, $1 \leq i \leq n$, which are all in w by construction. Therefore, Φ' itself must be in w.

Finally, as $\neg\Phi$ is consistent, by Zorn's Lemma again it belongs to some maximal consistent set w, and by definition of W, $\leq$ and ρ, it follows that $w, \rho \models \neg\Phi$. $\Box$

This theorem means that Φ is provable in S4 if and only if it is true in every frame whose accessibility relation is a preorder. Actually, it is enough to consider frames where the accessibility relation is an ordering, i.e. where it is also antisymmetric. (This is because the models defined by filtration below will also be antisymmetric.)

Although we have proved the theorem for S4, similar theorems can be proved for other logics as well. For example, KD is sound and complete for all serial frames, S for all reflexive frames, K4 for all transitive frames, S5 for all equivalence relations (reflexive, symmetric, transitive), and S4.2 for all convergent preorders. The proof is exactly the same as the above. In particular, we build the accessibility relation $\leq$ by stating that $w \leq w'$ if and only if for every formula $\Box\Psi$ in w, Ψ is in w'. The only difference is that $\leq$ will then be something else: in the case of KD, it will be a serial relation, in the S case, it will be reflexive, and so on.

▶ **EXERCISE 5.3**

Let Φ be a propositional formula, and $\overline{\Phi}$ be the modal formula obtained by replacing implication subformulas $\Psi \Rightarrow \Psi'$ of Φ by $\Box(\Psi \Rightarrow \Psi')$, negated subformulas $\neg\Psi$ of Φ by $\Box(\neg\Psi)$ and variables A by $\Box A$.

Show that Φ is intuitionistically provable if and only if $\overline{\Phi}$ is a theorem of S4. (This is one of many translations of intuitionistic logic to S4. Others are due to Gödel and McKinsey-Tarski. Hint: use Kripke semantics for intuitionistic logic and S4.)

2.4 Decidability

To help us prove decidability for any modal logic L, we define an equivalence relation on worlds (an indistinguishability relation) and the quotient of frames by this relation. This definition is justified by the construction of Theorem 5.7.

Definition 5.4 (Filtration) *Let Φ be a formula of L, and ρ be an assignment. Let $S(\Phi)$ be the set of subformulas of Φ.*

The indistinguishability relation $\equiv_{\Phi,\rho}$ *on a frame (W, R) is defined by $w \equiv_{\Phi,\rho} w'$ if and only if for every formula $\Phi' \in S(\Phi)$, $w, \rho \models \Phi'$ if and only if $w', \rho \models \Phi'$.*

The filtration *of (W, R) by $S(\Phi)$, ρ is the frame $(W/\Phi\rho, R/\Phi\rho)$, where $W/\Phi\rho$ is the quotient of W by $\equiv_{\Phi,\rho}$ and if v, v' are elements of $W/\Phi\rho$, then $v(R/\Phi\rho)v'$ if and only if there are $w \in v$, $w' \in v'$ such that wRw'.*

We let $w_{\Phi,\rho}$ be the class of w modulo $\equiv_{\Phi,\rho}$, and we let ρ_Φ be the assignment in $(W/\Phi\rho, R/\Phi\rho)$ mapping each variable A to the quotient $\rho(A)/\Phi\rho = \{w_{\Phi,\rho} \mid w \in \rho(A)\}$.

The $\equiv_\Phi$ relation is, indeed, an equivalence relation on worlds. The filtration preserves models:

Lemma 5.5 *For every subformula Φ' of Φ, $(W, R, \rho) \models \Phi'$ if and only if $(W/\Phi\rho, R/\Phi\rho, \rho_\Phi) \models \Phi'$.*

Proof: We claim that, if $v \in W/\Phi\rho$, then $v, \rho_\Phi \models \Phi'$ if and only if any of the following equivalent conditions is met:

1. there is a $w \in W$ such that $v = w_{\Phi,\rho}$ and $w, \rho \models \Phi'$;

2. for every $w \in W$ such that $v = w_{\Phi,\rho}$, then $w, \rho \models \Phi'$.

These conditions are indeed equivalent: 2 implies 1 by the non-emptiness of equivalence classes, and 1 implies 2 by the indistinguishability of worlds in the same equivalence class and by the fact that Φ' is a subformula of Φ. The claim is then proved by a straightforward structural induction on Φ'.

To illustrate the process, we deal with the case where Φ' is of the form $\Box\Phi''$. If $v, \rho_\Phi \models \Box\Phi''$, then for every v' such that $v(R/\Phi\rho)v'$, $v', \rho_\Phi \models \Phi''$ holds; then, for every $w \in v$, for every w' such that $w \ R \ w'$, $w'/\Phi\rho, \rho_\Phi \models \Phi''$ by definition of $R/\Phi\rho$; by induction hypothesis, $w', \rho \models \Phi''$ for every such w', hence $w, \rho \models \Box\Phi''$. Conversely, if for every $w \in v$, $w, \rho \models \Box\Phi''$, then for every w' such that $w \ R \ w'$, $w', \rho \models \Phi''$; then, for every v' such that $v(R/\Phi\rho)v'$, there is a $w \in v$ and a $w' \in v'$ such that $w \ R \ w'$, so $w', \rho \models \Phi''$, and by induction hypothesis, $v', \rho_\Phi \models \Phi''$; as v' is arbitrary, it follows that $v, \rho_\Phi \models \Box\Phi''$. $\quad\Box$

Moreover, the filtration preserves most of the properties of the model:

Lemma 5.6 *If (W, R) is reflexive (resp. symmetric, serial), then $(W/\Phi\rho, R/\Phi\rho)$ is also reflexive (resp. symmetric, serial).*

Proof: If (W, R) is reflexive, then for every $v \in W/\Phi\rho$, there is a world $w \in v$, and $w \ R \ w$ by assumption, so $v(R/\Phi\rho)v$ by definition.

If (W, R) is symmetric, and $v(R/\Phi\rho)v'$, then there are $w, w' \in W$ such that $w \ R \ w'$, hence $w' \ R \ w$ by symmetry, therefore $v'(R/\Phi\rho)v$.

If (W, R) is serial, then for every $v \in W/\Phi\rho$, there is a w in v, hence there is a $w' \in W$ such that $w \ R \ w'$ by assumption, so $v(R/\Phi\rho)w'/\Phi\rho$, and $(W/\Phi\rho, R/\Phi\rho)$ is serial. $\quad\Box$

However, when (W, R) is transitive, $(W/\Phi\rho, R/\Phi\rho)$ need not be transitive. Indeed, if $v(R/\Phi\rho)v'$ and $v'(R/\Phi\rho)v''$, then there are worlds w, w_1', w_2', w'' such that $w \ R \ w_1'$ and $w_2' \ R \ w''$, but we cannot conclude from here, since w_1' may not be related to w_2'. Similarly, when (W, R) is convergent, $(W/\Phi\rho, R/\Phi\rho)$ need not be convergent.

Finally, the filtration is small enough:

Theorem 5.7 (Small Model) *Let Φ be a formula of size n. The cardinal of $W/\Phi\rho$ is less than or equal to 2^n.*

Proof: For every $\Phi' \in S(\Phi)$, let $f_0(\Phi')$ be $\{w \in W \mid w, \rho \models \Phi'\}$, and $f_1(\Phi')$ be $\{w \in W \mid w, \rho \not\models \Phi'\}$.

Now, let $\Phi_1, \ldots, \Phi_n$ be the n subformulas of Φ, and let $s_1, \ldots, s_n$ be any sequence of elements of $\{0, 1\}$. Then $\bigcap_{i=1}^{n} f_{s_i}(\Phi_i)$ is an equivalence class modulo $\equiv_{\Phi,\rho}$, that is, an element of $W/\Phi\rho$.

Conversely, every such equivalence class is determined by the Boolean value (true if $s_i = 0$, false if $s_i = 1$) of every subformula Φ_i, $1 \leq i \leq n$, of Φ. That is every such equivalence class has the form $\bigcap_{i=1}^{n} f_{s_i}(\Phi_i)$ for some sequence $s_1, \ldots, s_n$.

Therefore, there are at most as many worlds in $W/\Phi\rho$ than there are sequences of length n over $\{0, 1\}$, i.e. at most 2^n. $\qquad\square$

This establishes the so-called *small model* property for all those logics for which $(W/\Phi\rho, R/\Phi\rho, \rho_\Phi)$ is a model of Φ whenever (W, R, ρ) is. This works for any combination of reflexivity, symmetry and seriality.

When models are transitive, $(W/\Phi\rho, R/\Phi\rho, \rho_\Phi)$ may not be a model of Φ at all even though (W, R, ρ) is. In this case, we use the following instead of Lemma 5.5:

Lemma 5.8 *Assume R transitive. For every subformula Φ' of Φ, $(W, R, \rho) \models \Phi'$ if and only if $(W/\Phi\rho, (R/\Phi\rho)^+, \rho_\Phi) \models \Phi'$, where $(R/\Phi\rho)^+$ is the transitive closure of $R/\Phi\rho$.*

Proof: The proof follows the same lines as that of Lemma 5.5. $\qquad\square$

Therefore, when R is transitive, if (W, R, ρ) is a model of Φ, then $(W/\Phi\rho, (R/\Phi\rho)^+, \rho_\Phi)$ is another model of Φ, which is obviously small. A similar technique holds for convergent (resp. convergent and transitive) Kripke frames.

The result that makes the small model property useful is the following :

Theorem 5.9 (Decidability) *Let L be a modal logic with the small model property. Then the validity of a formula Φ in L is decidable.*

Proof: Let the size of Φ be n, and the size of a small model for Φ be bounded from above by $f(n)$. (We had $f(n) = O(2^n)$ above.) We enumerate all graphs with at most $f(n)$ vertices that fit in the class $\mathcal{C}$. To do this, we enumerate all graphs with exactly m vertices, for all $1 \leq m \leq f(n)$: it is enough to enumerate all graphs on $\{1, \ldots, m\}$, and there are at most 2^{m^2} of them. On a graph with m vertices, there are at most 2^{nm} possible assignments (we map each variable and world to a Boolean stating whether the variable is true at the world or not), and to check each we need at most $O(nm)$ time (tabulate every subformula at each world, starting from atomic formulas and going up in Φ). To sum up, the running time is bounded from above by

$$O(\sum_{m=1}^{f(n)} 2^{m^2} . 2^{nm} . nm) \leq nf(n)^2 . 2^{f(n)^2 + nf(n)}$$

$\qquad\square$

This is, as we have seen, in particular true of the class of frames satisfying any combination of reflexivity, symmetry, transitivity, seriality and convergence.

> In complexity theory jargon, the bound on the running time of the small model algorithm above is said to be doubly-exponential. Looking a bit more carefully to the algorithm, we see that finding whether Φ is invalid means guessing a frame and an assignment in this frame in single exponential time, then checking them in single exponential time. Therefore, the algorithm runs in *non-deterministic single exponential time*. However, most modal logics are solvable in *deterministic exponential time*, which is far less intractable, although still provably intractable (i.e., definitely not polynomial-time). The satisfiability problem in most of these logics, like S4, is in fact PSPACE-complete, and for S5 it is even "only" NP-complete.

> The trick that makes the small model property work is to consider an indistinguishability relation defined on the set $S(\Phi)$ of all subformulas of Φ. This is a semantical way of saying that the semantics of Φ only depends on its subformulas, and is therefore an analogue of the subformula property that we usually derive from the syntactic process of cut-elimination. There are modal logics, like PDL (see Section 3.3) where we have to extend this notion of subformula to something more complicated, called the Fischer-Ladner closure of the formula, essentially because the subformula property cannot be enforced.

2.5   Sequent Systems and Tableaux

Melvin Fitting (1983) was one of the first logicians to be interested in proof systems for modal logics. It turns out that most of the well-known proof methods for classical logic have failed to be adapted to modal logics. However, tableaux methods, i.e. cut-free Gentzen sequent systems, have been quite successful in the area. (We can still use resolution, provided that we first translate modal formulas into first-order logic formulas; first-order logic will be the subject of Chapter 6; for a survey of the translation techniques, see (Ohlbach, 1993).)

However, cut elimination or the subformula property may fail in some modal logics. S4R (Goré, 1991), for instance, has a cut-free sequent system, but it breaks the subformula property: we may need subformulas of $\Box\Phi$, not only Φ, during the search for a proof. This is not too serious, as decidability is preserved. Logics like S4.2 and S4F have Gentzen systems that not only break the subformula property (we need to consider all subformulas of $\Box\neg\Box\Phi$), but also need a so-called *analytical Cut rule*. This special form of Cut, is such that the cut formula (the one that disappears in the conclusion of Cut), is always a direct subformula of one of the formulas in the conclusion of the Cut rule, thus limiting the explosion of the search space; again, decidability is preserved, although at some cost.

Let's see a few concrete examples. A sequent system derived from what Goré proposes for S4 is $\mathbf{LK_0}$ minus Cut plus the following rules:

$$\frac{\Box\Gamma \longrightarrow \Phi}{\Box\Gamma, \Gamma' \longrightarrow \Box\Phi, \Delta} \ (S4) \qquad \frac{\Gamma, \Phi \longrightarrow \Delta}{\Gamma, \Box\Phi \longrightarrow \Delta} \ (T)$$

where if Γ is a set of formulas, $\Box\Gamma$ is the set $\{\Box\Psi \mid \Psi \in \Gamma\}$.

Interpreting these as tableau rules means the following. Assume we are on an unclosed path. α and β-formulas expand as in the classical case. But, if we decide to expand a formula of the form $+\Box\Phi$, we must do so by replacing it in the current path by $+\Phi$, but we must also shadow all non-boxed negative formulas and all positive formulas except Φ itself (rule (S4)). And, if we wish to expand a formula of the form $-\Box\Phi$, we must do so by adding by $-\Phi$ on the path. (The implicit contraction on the left of the sequents in rule (T) is *not* eliminable, just like in the intuitionistic system $\mathbf{LJ}_0$, see Chapter 3, Section 2.)

Notice that, contrarily to classical logic, but similarly to intuitionistic logic, we cannot choose to decompose any arbitrary formula in the path: if we decompose a right boxed formula using (S4), we won't be able to access formulas in Γ' or Δ any longer, and we may have to backtrack to choose another formula to decompose if the first decomposition failed. Because of rules like (S4), tableaux for modal logics are not as simple to use as in the classical case.

A question that comes to the mind is the following: are world jumps visible on proofs? The answer is yes, but to understand why, it is helpful to remember how we proved completeness for S4. We represented worlds as maximal consistent sets of formulas. Sequents $\Gamma \longrightarrow \Delta$ represent sets S of formulas (all formulas in Γ, and the negated versions of formulas in Δ). In turn, S determines the class $C(S)$ of all maximal consistent sets of formulas containing it. Then, whenever we use a rule of $\mathbf{LK}_0$ or (T), we don't modify $C(S)$. However, when going up through rule (S4), we erase all unboxed formulas from S, and retain only the formulas Φ where $\Box\Phi \in S$: this changes the class $C(S)$ in roughly the same way as the $\leq$ relation was defined in the proof of Theorem 5.3. This is where we jump to a next world.

Other modal logics have much more complicated sequent systems than S4, though. For example, here is a variant of Goré's system for S4.2. Apart from the rules of $\mathbf{LK}_0$ (without Cut), rules (T) and (S4) above, we include:

$$\frac{\Gamma, \Box\neg\Box\Phi \longrightarrow \Box\Phi, \Delta \quad \Gamma, \Box(\neg\Box\neg\Box\Phi)^* \longrightarrow \Box\Phi, \Delta}{\Gamma \longrightarrow \Box\Phi, \Delta} \ (S4.2)$$

where $(\ldots)^*$ is called a starred formula, and the rule is restricted to the case where $\Box\Phi$ is not starred in the conclusion. (Star marks are just annotations on formula occurrences, and do not participate in the structure of the formula; their purpose is to detect looping, i.e. situations where the same rule would be applied on the same formula twice.) To find a proof in K45, when we come to expand a formula of the form $+\Box\Phi$, we have the choice between applying rule (S4), or rule (K45); the latter actually enriches the current branch, producing two branches on which (S4) will produce more informative sequents.

But the (S4.2) rule is not sufficient to find all proofs, and we have to add the following analytical Cut rules:

$$\frac{\Gamma \longrightarrow \Phi, \Phi', \Delta \quad \Gamma, \Phi \longrightarrow \Phi', \Delta \quad \Gamma, \Phi' \longrightarrow \Phi, \Delta}{\Gamma \longrightarrow \Phi \wedge \Phi', \Delta}$$

$$\frac{\Gamma, \Phi \longrightarrow \Box\Phi, \Delta \quad \Gamma \longrightarrow \Phi, \Box\Phi, \Delta}{\Gamma \longrightarrow \Box\Phi, \Delta}$$

to which we must add corresponding rules for $\vee$ and the other connectives. (Goré only uses $\neg$, $\wedge$ and $\Box$ as basic connectives.) These rules add a non-negligible amount of non-determinism and of branching: finding proofs in such logics is in fact not easy.

3   OTHER MODAL LOGICS OF INTEREST IN COMPUTER SCIENCE

Examples of modal logics that are in even wider use in computer science are Hennessy–Milner logic or HML, which was invented to have a logical language for Milner's CCS language for concurrency; temporal logics like CTL, designed to have a richer language for specifying hardware circuits; or Propositional Dynamic Logic, which is basically Hoare's logic for reasoning about programs, in a smaller and more elegant clothing. We give a succinct description of each, but be aware that several hundred different modal logics exist to achieve similar goals, and that our presentation is a mere illustration. (See (Milner, 1989) for more about CCS and HML, (Emerson, 1990) for more about temporal logics, and (Kozen and Tiuryn, 1990) for more about dynamic logic.)

The starting idea for using these logics is that they provide a rich enough language for expressing useful properties on automata (a.k.a. finite state machines, a.k.a. labelled state-transition graphs). An automaton is a set S of *states*, together with a transition relation R between states. (We omit initial and final states in this definition.) A *transition* is a couple of states $(s, s') \in R$, and is assumed to be labelled by some letter taken from a finite alphabet $\mathcal{A}$. We write $s \xrightarrow{a} s'$ for the transition (s, s') labelled by a. Then, we can see S4 or more complicated temporal logics as expressing properties about automata, which they see as Kripke frames: for example, in S4, the fact that $\Box\Phi$ is satisfied at state s would mean that Φ is satisfied at all states that we can reach from s by following any finite number of transitions. (The paths that we may take represent all the possible futures from s.)

3.1   Hennessy-Milner Logic

Hennessy-Milner logic (HML, for short) is a multi-agent version of the basic modal logic K; instead of having only one $\Box$ (resp. $\Diamond$) modality, we have several $[s]$ (resp. $\langle s \rangle$) modalities, each corresponding to a possible set s of transitions in an automaton representing all possible executions of some CCS program. Transitions in such systems are meant to represent individual atomic actions, like sending a message or receiving a message. Transitions may be put in sequence, or in parallel. Moreover, transitions may synchronise, and this is how communication between parallel processes is achieved: for each transition i (say, "send some message"), there is a distinct matching transition $\bar{i}$ ("receive the above message"), such that any two processes, one of which is ready to fire i, and the other is ready to fire $\bar{i}$, may fire both at the same time and continue execution. Execution proceeds by following transitions from one state to the next. A state is often called an *agent* or a *process*. We write $A \xrightarrow{i} A'$ to state that agent A may transform into agent A' by following the transition i.

Hennessy-Milner logic represents the behaviours of processes by modal formulas. The fact that an agent A satisfies $[s]\Phi$, where s is a set of transitions, means: "for every transition $A \xrightarrow{i} A'$ with $i \in s$, A' satisfies Φ", and $\langle s \rangle \Phi$ (i.e., $\neg [s] \neg \Phi$) then means: "there is a transition $A \xrightarrow{i} A'$ with $i \in s$ such that A' satisfies Φ". Although this is not strictly necessary, we shall now restrict ourselves to *finite* sets of transitions, for simplicity.

Because of the definitions, we may assume that $[\{i_1, \ldots, i_n\}]\Phi$ is an abbreviation for $[i_1]\Phi \wedge \ldots \wedge [i_n]\Phi$ (or some valid formula $\mathbf{T}$ if $n = 0$), where $[i]\Phi$ means $[\{i\}]\Phi$. Then, the axioms are those of classical propositional logic plus:

(K) $[i](A \Rightarrow B) \Rightarrow [i]A \Rightarrow [i]B$,

with the modified necessitation rule:

(Nec) from Φ, deduce $[i]\Phi$, for every formula Φ and every transition i.

HML has the remarkable property that two processes are *strongly bisimilar* if and only if they satisfy exactly the same formulas in HML. We won't define the term "strongly bisimilar" (See (Milner, 1989)), suffice it to say roughly that two processes are strongly bisimilar if and only if they have the same observable behaviour (non-observable transitions are ignored). Therefore, HML captures exactly all observable properties of CCS processes, no less and no more.

This logic can be made more expressive by extending it by, for instance, a least fixed point operator μ and a greatest fixed point operator ν, yielding what we shall call the *strong modal μ-calculus*.

The idea is that, if A is a propositional variable that occurs only positively in Φ (assuming that $\Rightarrow$ has been replaced by $\neg$ and $\vee$, this means that A is wrapped up inside an even number of negations; this is a syntactic criterion ensuring that is $A \Rightarrow A'$ holds, then $\Phi \Rightarrow \Phi[A'/A]$ holds, too), then $\mu A \cdot \Phi$ and $\nu A \cdot \Phi$ are both formulas such that Φ' is logically equivalent to $\Phi[\Phi'/A]$: they are *fixed points* of the function $A \mapsto \Phi$, modulo logical equivalence. Moreover, whenever Φ' is a formula such that Φ' is equivalent to $\Phi[\Phi'/A]$, then $\mu A \cdot \Phi \Rightarrow \Phi'$ and $\Phi' \Rightarrow \nu A \cdot \Phi$. They obey the following axioms and rules:

- $(\mu A \cdot \Phi) \Leftrightarrow \Phi[(\mu A \cdot \Phi)/A], (\nu A \cdot \Phi) \Leftrightarrow \Phi[(\nu A \cdot \Phi)/A]$,

- $\dfrac{\Phi' \Leftrightarrow \Phi[\Phi'/A]}{(\mu A \cdot \Phi) \Rightarrow \Phi'} \qquad \dfrac{\Phi' \Leftrightarrow \Phi[\Phi'/A]}{\Phi' \Rightarrow (\nu A \cdot \Phi)}$

Intuitively, $\mu A \cdot \Phi$ is the least fixpoint of the function f mapping A to Φ, i.e. the disjunction of all the $\underbrace{f(f(\ldots f(\mathbf{F})\ldots)}_{n \text{ times}}$, $n \in \mathbb{N}$, where $\mathbf{F}$ is some formula denoting false. Conversely, $\nu A \cdot \Phi$ is intuitively the conjunction of all the $\underbrace{f(f(\ldots f(\mathbf{T})\ldots)}_{n \text{ times}}$, $n \in \mathbb{N}$, and $\mathbf{T}$ is some formula denoting true.

The μ and ν operators allow us to write properties that could not be expressed in HML, because μ and ν actually incorporate some form of induction, which we didn't have in HML. For example, liveness can be expressed as $\nu A \cdot \langle \mathcal{A} \rangle \mathbf{T} \wedge [\mathcal{A}]A$, where $\mathcal{A}$ is the set of all transition letters; this means that at each computation step, we can execute some transition, and whatever transition we choose, we can still execute some transition, and then whatever transition we choose, … We can also write the statement that a process can deadlock by writing $\mu A \cdot [\mathcal{A}]\mathbf{F} \vee \langle \mathcal{A} \rangle A$. In general, if Φ does not depend on A, $\mu A \cdot \Phi \vee \langle \mathcal{A} \rangle A$ states that there is an execution path such that eventually Φ holds, while $\mu A \cdot \Phi \vee [\mathcal{A}]A$ says that, on all execution paths, Φ eventually holds (i.e., Φ is *inevitable*). A more complicated formula like $\nu B \cdot (\mu A \cdot \Phi \vee [\mathcal{A}]A) \wedge \langle \mathcal{A} \rangle \mathbf{T} \wedge [\mathcal{A}]B$ says that Φ holds infinitely often on all execution paths. (The inner μ-formula means that Φ eventually holds, so the whole formula says that Φ eventually holds at some state, then we can follow some transition, then whatever transition we follow, Φ eventually holds again, and so on.)

▶ **EXERCISE 5.4**

Let Φ be a formula, and let Φ' be $\neg \Phi[\neg A / A]$. Show that A occurs only positively in Φ if and only if A occurs only positively in Φ'.

Show that $\mu A \cdot \Phi$ is logically equivalent to $\neg \nu A \cdot \Phi'$, and that $\nu A \cdot \Phi$ is logically equivalent to $\neg \mu A \cdot \Phi'$.

Show also that any variable that occurs only positively in $\mu A \cdot \Phi$ (resp. $\nu A \cdot \Phi$) also occurs only positively in $\neg \nu A \cdot \Phi'$ (resp. $\neg \mu A \cdot \Phi'$).

Conclude that we could have defined either μ or ν as an abbreviation.

3.2   CTL

Another useful modal logic for representing properties of execution paths inside automata is *CTL*, or *computation tree logic*. It has been created to describe specifications of hardware circuits, but shares lots of properties with, notably, the modal μ-calculus.

Its syntax is that of propositional logic augmented with unary modalities AX, EX and binary modalities $A(\_U\_)$ and $E(\_U\_)$, that is, if Φ and Φ' are CTL formulas, then $AX\Phi$, $EX\Phi$, $A(\Phi U \Phi')$, $E(\Phi U \Phi')$ are formulas.

These notations come from the fact that CTL is a subset of the more powerful path logic CTL*, where the modalities are A (all), E (exists), X (next time) and $\_U\_$ (until), which we won't describe here. (See (Emerson, 1990).)

The meaning of CTL formulas is the following. First, the semantics of CTL (or CTL*) is defined on state-transition diagrams such that each state has at least one successor. Then $AX\Phi$ holds at state s of a state-transition diagram if and only if Φ holds at every successor s' of s; $EX\Phi$ holds at s if and only if Φ holds at some successor s' of s. Finally, $A(\Phi U\Phi')$ (resp. $E(\Phi U\Phi')$) holds at s if and only if on all paths (resp. on some path) $s_1 \longrightarrow s_2 \longrightarrow \ldots$ starting from $s_1 = s$, eventually Φ' holds, (say, the first time at state s_n) and moreover Φ holds on all states $s_1, \ldots, s_{n-1}$.

CTL and CTL* can be embedded in the so-called (weak) *modal μ-calculus*, which is the strong modal μ-calculus we described in Section 3.1, with Axiom (D) ($\Box\Phi \Rightarrow \Diamond\Phi$, recast here as $AX\Phi \Rightarrow EX\Phi$) added. This is because we assumed that each state in the automata on which CTL formulas express properties had at least one successor, i.e. CTL, CTL* and the modal μ-calculus are serial logics.

The embedding inside the modal μ-calculus works in the following way. (We shall consider this as a definition of CTL.) We assume that $\mathcal{A}$ consists of exactly one letter, call it X. Then, $AX\Phi$ translates to $[X]\Phi$, $EX\Phi$ translates to $\langle X\rangle\Phi$, $A(\Phi U\Phi')$ translates to $\mu B \cdot \Phi' \vee (\Phi \wedge [X]B)$, and $E(\Phi U\Phi')$ translates to $\mu B \cdot \Phi' \vee (\Phi \wedge \langle X\rangle B)$, where we assume B to be a new variable, which does not occur free in either Φ or Φ'.

3.3 Propositional Dynamic Logic

Propositional Dynamic Logic (PDL for short) is another modal logic in the style of K or of HML, but where modalities are labelled with full programs. If P is a (maybe non-deterministic) program, and Φ is a formula, then $[P]\Phi$ means that after every possible (terminating) execution of program P, Φ holds. We define again $\langle P\rangle\Phi$ as $\neg[P]\neg\Phi$, so that $\langle P\rangle\Phi$ means that there is an execution of P which terminates and such that, afterwards, Φ holds.

Formulas are built as usual with $\wedge$, $\vee$, $\neg$, $\Rightarrow$ and the modalities $[P]$, where P is a program. *Programs* are built with a minimal language for describing computations: a program may be an *action constant* a; or a *test* $\Phi?$, where Φ is a formula; or a *sequence* $P_1; P_2$ where P_1 and P_2 are two programs; or the *non-deterministic choice* $P_1 \cup P_2$ between two programs P_1 and P_2; or the *repetition* P_1^* of a program P_1.

Kripke frames (W, R) for PDL are such that R is a binary relation between worlds (or *states*) in W, indexed by the action constants. That is, for every action constant a, there is a binary relation R_a between worlds, and R is the direct sum of all R_a's. Then, given an assignment ρ from propositional variables to subsets of W, the meaning of a formula Φ is, as usual, the set of states w such that $w, \rho \models \Phi$, and the meaning of a program P is a binary relation $[\![P]\!]\rho$ between states, defined as follows:

- $[\![a]\!]\rho = R_a$, if a is an action constant;

- is Φ is a formula, then $[\![\Phi?]\!]\rho$ is the relation r such that $w\ r\ w'$ if and only if $w = w'$ and $w, \rho \models \Phi$; (Φ? just proceeds to execute the rest of the program if Φ holds, otherwise the program aborts, i.e., does not terminate.)

- $[\![P;Q]\!]\rho = [\![Q]\!]\rho \circ [\![P]\!]\rho$, i.e. $w\ [\![P;Q]\!]\rho\ w'$ if and only if there is an intermediate state w'' such that $w\ [\![P]\!]\rho\ w''$ and $w''\ [\![Q]\!]\rho\ w'$;

- $[\![P \cup Q]\!]\rho = [\![P]\!]\rho \cup [\![Q]\!]\rho$;

- $[\![P^*]\!]\rho$ is the reflexive-transitive closure of $[\![P]\!]\rho$;

Then the meaning of a formula is as usual in modal logics, and the meaning of a modal formula $[P]\Phi$ is described by:

- $w, \rho \models [P]\Phi$ if and only if for every state w' such that $w\ ([\![P]\!]\rho)\ w'$, $w', \rho \models \Phi$.

Other forms of programs (which define a variant of Edsger W. Dijkstra's *guarded command language*) may be derived from these primitives:

- skip, the program that does nothing, is **T**?, where **T** is any valid formula;

- fail, the program that never terminates, is **F**?, where **F** is any unsatisfiable formula;

- if $\Phi_1 \rightarrow P_1 \mid \ldots \mid \Phi_n \rightarrow P_n$ fi, the alternative guarded command (a form of case statement), abbreviates $(\Phi_1?; P_1) \cup \ldots \cup (\Phi_n?; P_n)$;

- do $\Phi_1 \rightarrow P_1 \mid \ldots \mid \Phi_n \rightarrow P_n$ od, the iterative guarded command, abbreviates (if $\Phi_1 \rightarrow P_1 \mid \ldots \mid \Phi_n \rightarrow P_n$ fi)*; $(\neg\Phi_1 \wedge \ldots \wedge \neg\Phi_n)$;

- if Φ then P else Q, the if-then-else statement, abbreviates if $\Phi \rightarrow P \mid \neg\Phi \rightarrow Q$ fi;

- while Φ do P od, the while loop, abbreviates do $\Phi \rightarrow P$ od;

- repeat P until Φ, the repeat loop, abbreviates P; while $\neg\Phi$ do P od;

- finally, Hoare's partial correctness assertion $\{\Phi\}P\{\Psi\}$ is defined as $\Phi \Rightarrow [P]\Psi$.

A Hilbert-style system (PDL) for PDL has (MP) and (Nec) as deduction rules, and all propositional tautologies, the Axiom (K), and the following axioms:

- $[P \cup Q]\Phi \Leftrightarrow [P]\Phi \wedge [Q]\Phi$;

- $[P; Q]\Phi \Leftrightarrow [P][Q]\Phi$;

- $[\Psi?]\Phi \Leftrightarrow (\Psi \Rightarrow \Phi)$;

- $[P^*]\Phi \Rightarrow \Phi \wedge [P][P^*]\Phi$;

- $\Phi \Rightarrow [P^*](\Phi \Rightarrow [P]\Phi) \Rightarrow [P^*]\Phi$.

The last axiom is called the *PDL induction axiom*: indeed, is says that, to prove that $[P^*]\Phi$ holds, we can prove Φ at the initial time (base case), and prove the induction hypothesis $[P^*](\Phi \Rightarrow [P]\Phi)$.

Because of the induction axiom, PDL is not compact:

Theorem 5.10 (Non-Compactness) *There is an infinite set Γ of PDL formulas, and a PDL formula Φ, such that $\Gamma \models \Phi$, but for every finite subset Δ of Γ, $\Delta \not\models \Phi$.*

Proof: Let Γ be the set containing $\langle a^* \rangle A$, and $\neg A$, $\neg \langle a \rangle A$, $\neg \langle a \rangle \langle a \rangle A$, $\ldots$, and let Φ be any unsatisfiable formula.

We do have $\Gamma \models \Phi$, because Γ is satisfied in no model.

However, for every finite subset Δ of Γ, we can build a model of Δ in the following way: let k be any integer such that for every $j \geq k$, $\neg \underbrace{\langle a \rangle \ldots \langle a \rangle}_{j \text{ times}} A$ is not in Δ, build the oriented graph consisting of exactly one cycle of $k+1$ states, numbered from 0 to k; define R_a as the relation mapping any vertex on the cycle to its successor (i.e., mapping i to $i+1$ if $0 \leq i < k$, and k to 0), and let ρ be an assignment making A true in state k only. At state 0, $\langle a^* \rangle A$ holds, and for every $j < k$, $\neg \underbrace{\langle a \rangle \ldots \langle a \rangle}_{j \text{ times}} A$ holds, too, since A does not hold at state j. So this is a model of Δ, but it is not one of Φ (since Φ has no model), hence $\Delta \not\models \Phi$.   □

Despite this apparent ugliness, PDL remains rather nice:

Theorem 5.11 (Soundness, Completeness) *The Hilbert-style system (PDL) is sound and complete for PDL.*

Proof: See (Kozen and Tiuryn, 1990).   □

and:

Theorem 5.12 (Decidability) *The satisfiability problem for a PDL formula is decidable.*

Proof: See (Kozen and Tiuryn, 1990). This again comes from a small model property, obtained through filtration. However, the proof is more complicated than for S4, as we cannot take filtrations defined by subformulas only. (We use what is called the *Fischer-Ladner closure* of a formula, which is basically, the set of all subformulas modulo arbitrary expansions of program modalities, like expanding $[P^*]$ into $[P][P^*]$.)   □

PDL is very expressive, and the satisfiability problem in PDL is complete for deterministic exponential time (DEXPTIME-complete). That is, we can express and characterize all exponential-time computations as the satisfiability (or validity, for that respect) of a PDL formula. The strict deterministic variant SDPDL of PDL, where ∪, ? and * are replaced by the deterministic skip, fail, if _ then _ else and while _ do _ od constructions, and where the formulas in the latter two constructions are required not to use any modalities, is only PSPACE-complete. (Semantically, we also need to constrain the R_a to be functions, not arbitrary relations, to give meaning to SDPDL formulas.)

Notice also that in SDPDL, $\Phi \Rightarrow \langle P \rangle \Psi$ is a total correctness assertion ("if Φ holds, then P terminates in a state where Ψ holds"), while this is not so in PDL ("if Φ holds, then *there exists a computation of P* which terminates in a state where Ψ holds). To add total correctness assertions to PDL, we can add a least fixpoint operator μ as in the modal μ-calculus. The calculus then remains decidable, and actually DEXPTIME-complete and not more complicated, but this is very difficult to prove: filtrations indeed do not work any longer, (although the small model property still holds) and no complete finite axiomatisation is known for the enhanced logic.

▶ **EXERCISE 5.5**

(Heiki Tuominen) Define a translation from HML formulas to PDL formulas that preserves validity and satisfiability.

▶ **EXERCISE 5.6**

(Heiki Tuominen) Define a translation from S4 to PDL preserving validity and satisfiability.

 ▶ **EXERCISE 5.7**

Define a translation from PDL to the strong modal μ-calculus preserving validity and satisfiability.

4 MODEL-CHECKING

In all logics of the kind we described in Section 3, the satisfiability problem, or the validity problem are not so important as in other logics. A formula Φ in these logics L is a *specification* of an intended property of a program. Implementations are systems of transitions (CCS programs for Milner, state-transition diagrams of hardware systems for example in the CTL community, etc.), i.e. *models* (Kripke frames and assignments). Satisfiability checking means checking that it is possible to implement the specification. Although this may be valuable, this is usually unrealistic, because it requires the satisfiability checker to invent an implementation of the specification, a task best left to man.

The real problem we are faced with, usually, is the following: given a specification of a system (as a modal formula Φ), and an implementation of this system (a model (W, R, ρ)), does the implementation satisfy the implementation? In logical terms:

$$\text{Is } (W, R, \rho) \text{ a model for } \Phi?$$

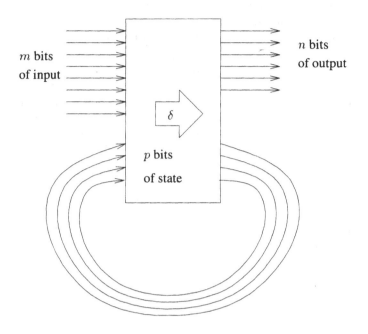

Figure 5.1. A sequential machine

This problem is called *model-checking*.

4.1   Sequential Machines and Symbolic Model-Checking

The prototypical example of model-checking comes from the domain of hardware
verification. Implementations, or circuits, are described as sequential machines with
m bits of input, n bits of output, and p bits of internal state. Sequential machines
are driven by an external clock, and at each clock tick, the transition relation δ of
the sequential machine is computed from the current input and the current state, to
yield the next output and the next internal state, as depicted in Figure 5.1. The transition relation δ may map input/current state combinations to several output/next
state combinations, thus expressing non-determinism.

   Formally, the transition relation is a relation δ from $\mathbb{B}^{m+p}$ to $\mathbb{B}^{n+p}$. A sequential machine defines a Kripke frame whose worlds are are $(m + n + p)$-tuples of
Booleans, or bits, representing the current input, the current output and the current
state,   and   whose   accessibility   relation R is   given   by $(i_1, \ldots, i_m,$
$s_1, \ldots, s_p, o_1, \ldots, o_n)$ R $(i_1, \ldots, i_m, s_1', \ldots, s_p', o_1, \ldots, o_n)$ if and only if

$(i_1, \ldots, i_m, s_1, \ldots, s_p) \; \delta \; (s'_1, \ldots, s'_p, o_1, \ldots, o_n)$.

We now describe what the technique of *symbolic model-checking* is all about. This technique, in combination with BDDs (see Chapter 2, Section 4.4), has been widely successful in hardware verification.   For more information, see (McMillan, 1993). We give the example of symbolic model-checking of formulas of the modal μ-calculus with one modality (which we shall write $\Box$, as in S4, but meaning "at each successor state") with respect to sequential machines.

Notice how large the Kripke frame is: it has 2^{m+n+p} states, and so it is impossible to represent it as a graph in memory for even small values of m, n, and p. But we can represent it as a propositional formula, which we'll denote by δ again, built on $m + n + 2p$ variables $i_1, \ldots, i_m, s_1, \ldots, s_p, s'_1, \ldots, s'_p, o_1, \ldots, o_n$. The size of this formula is roughly proportional to the number of gates in the circuit implementing the sequential machine, and is therefore much more palatable.

The variables $s_1, \ldots, s_p$ are called *state variables*. A state of the sequential machine is identified with an assignment to the state variables, of the input variable and of the output variables. Then, every propositional formula, and even every quantified propositional formula (see Chapter 2, Section 5.3) over the set of input, state and output variables denotes a set of input/state/output configurations (or worlds): the set of assignments that satisfy the formula. For example, if A and B are two state variables, then $A \lor B$ represents the set of all states in which A is true or in which B is true.

The basic idea of symbolic model-checking is, given a Kripke frame represented as a propositional formula δ on input, current state, next state and output variables, and given a modal formula Φ to be checked on the Kripke frame, to compute a propositional (non-modal) formula Ψ on the input, state and output variables representing the set of worlds where the modal formula Φ holds. Then, Φ holds on the given Kripke frame if and only if Ψ is valid.

We need to recapitulate and introduce a few notions on propositional (non-modal) formulas. If Φ is a propositional formula, and A is a propositional variable, recall that $\forall A \cdot \Phi$ is the formula $\Phi[\mathbf{T}/A] \wedge \Phi[\mathbf{F}/A]$, and that $\exists A \cdot \Phi$ is the formula $\Phi[\mathbf{T}/A] \lor \Phi[\mathbf{F}/A]$. We also make the following observation:

Lemma 5.13 *Let Φ be a propositional formula, and A a propositional variable, such that $A \Rightarrow A'$ implies $\Phi \Rightarrow \Phi[A'/A]$. (We say that Φ is* monotonic *in A.)*

Then there is a propositional formula $\mu A \cdot \Phi$ (resp. $\nu A \cdot \Phi$) such that:

- *$\mu A \cdot \Phi$ is logically equivalent to $\Phi[(\mu A \cdot \Phi)/A]$ (resp. $\nu A \cdot \Phi$ is logically equivalent to $\Phi[(\nu A \cdot \Phi)/A]$),*

- *and for every Φ' such that Φ' is logically equivalent to $\Phi[\Phi'/A]$, then $\mu A \cdot \Phi$ implies Φ' (resp. Φ' implies $\nu A \cdot \Phi$).*

Proof: Let $\mathcal{A}$ be the set of assignments on the free variables of Φ.

Let f be the following function from $\mathcal{P}(\mathcal{A})$ to $\mathcal{P}(\mathcal{A})$: if S is the set of assignments satisfying some formula Φ', then $f(S)$ is the set of assignments satisfying

$\Phi[\Phi'/A]$. This is a total function, because every S can be written as the set of assignments satisfying some formula Φ' built on $\mathcal{A}$: let Φ' be the disjunction of all $A_1^\rho \wedge A_2^\rho \wedge \ldots \wedge A_n^\rho$, where ρ ranges over S, $\mathcal{A} = \{A_1, \ldots, A_n\}$ and A^ρ is defined as A if $\rho(A)$ is true and as $\neg A$ otherwise.

Because Φ is monotonic with respect to A, f is a non-decreasing function from $\mathcal{P}(\mathcal{A})$ to $\mathcal{P}(\mathcal{A})$. By Knaster-Tarski's Fixed Point Theorem, f has a least fixed point $\underline{S}$ and a greatest fixed point $\overline{S}$. (This Theorem states that the set of fixed points of a monotonic function from a complete lattice, say $\mathcal{P}(\mathcal{A})$, to itself is itself a non-empty sub-lattice.) We let $\mu A \cdot \Phi$ be any formula satisfied by exactly those assignments in $\underline{S}$, and $\nu A \cdot \Phi$ be any formula satisfied by exactly those assignments in $\overline{S}$. $\square$

The proof of the Lemma gives a way of computing least and greatest fixed points of monotonic propositional functions. In the case of finite lattices (and $\mathcal{P}(\mathcal{A})$ is indeed finite), the proof of Knaster-Tarski's Theorem is constructive. The least fixed point of f is obtained as the union of $f(\varnothing)$, $f(f(\varnothing))$, $\ldots$, $\underbrace{f(\ldots f(\varnothing)\ldots)}_{k \text{ times}}$, $\ldots$, this sequence being stationary at the latest when $k \geq 2^n$, where n is the number of free variables of Φ. Symmetrically, the greatest fixed point of f is obtained as the intersection of $f(\mathcal{A})$, $f(f(\mathcal{A}))$, $\ldots$, $\underbrace{f(\ldots f(\mathcal{A})\ldots)}_{k \text{ times}}$, $\ldots$, this sequence being stationary at the latest when $k \geq 2^n$ again. Translating this construction to the level of formulas, we can compute $\mu A \cdot \Phi$ in the following way:

1. Initialise Φ' to **F**;

2. compute $\Phi'' = \Phi' \vee \Phi[\Phi'/A]$;

3. if Φ'' is logically equivalent to Φ', then stop and return Φ';

4. otherwise, set Φ' to Φ'', and go back to step 2.

and symmetrically, we can compute $\nu A \cdot \Phi$ in the following way:

1. Initialise Φ' to **T**;

2. compute $\Phi'' = \Phi' \wedge \Phi[\Phi'/A]$;

3. if Φ'' is logically equivalent to Φ', then stop and return Φ';

4. otherwise, set Φ' to Φ'', and go back to step 2.

Then, the essence of symbolic model-checking is, as McMillan notices:

Theorem 5.14 *Let* (W, δ) *be a Kripke frame, where worlds are assignments on input variables* i_k, $1 \leq k \leq m$, *state variables* s_k, $1 \leq k \leq p$, *and output variables* o_k, $1 \leq k \leq n$; *and the transition relation is encoded as a propositional formula* δ *on the* i_k, s_k, o_k *and next state variables* s'_k, $1 \leq k \leq p$.

Let $\overline{\Phi}$ *be the set of assignments satisfying a quantified propositional formula* Φ. *Then, for every propositional formula* Φ *built on the* i_k, s_k *and* o_k *only, and for every world* w, $w \models \Phi$ *is characterised as follows:*

- $w \models A$ *if and only if* $w \in \overline{A}$, *with* A *one of the* i_k, s_k *or* o_k;

- $w \models \Phi \vee \Phi'$ *if and only if* $w \in \overline{\Phi \vee \Phi'}$ *(and similarly for* $\wedge$, $\neg$, $\Rightarrow$);

- $w \models \Box\Phi$ *if and only if* $w \in \overline{\forall s'_1 \cdot \ldots \forall s'_p \cdot \delta \Rightarrow \Phi[s'_1/s_1, \ldots, s'_p/s_p]}$;

- $w \models \mu A \cdot \Phi$ *if and only if* $w \in \overline{\mu A \cdot \Phi}$;

- $w \models \nu A \cdot \Phi$ *if and only if* $w \in \overline{\nu A \cdot \Phi}$;

where in the last two cases, we have assumed that Φ *was monotonic in* A.

Proof: Derives directly from the definitions and from Lemma 5.13.   □

The Theorem provides the intended translation of modal formulas Φ to non-modal propositional formulas Ψ such that the configurations of (W, δ) where Φ holds are represented exactly by the assignments satisfying Ψ. The translation algorithm works as follows:

- if Φ is a variable A, return the formula A itself;

- if $\Phi = \Phi' \vee \Phi''$, let Ψ' be the translation of Φ' and Ψ'' be the translation of Φ'', then return $\Psi' \vee \Psi''$; (similarly for $\wedge$, $\neg$, $\Rightarrow$;)

- if $\Phi = \Box\Phi'$, let Ψ' be the translation of Φ', and return $\forall s'_1 \cdot \ldots \forall s'_p \cdot \delta \Rightarrow \Psi'[s'_1/s_1, \ldots, s'_p/s_p]$;

- if $\Phi = \mu A \cdot \Phi'$, let Ψ' be the translation of Φ', and return $\mu A \cdot \Psi'$;

- if $\Phi = \nu A \cdot \Phi'$, let Ψ' be the translation of Φ', and return $\nu A \cdot \Psi'$.

That the algorithm is correct is easy, except for two things, both related to μ and ν formulas. First, the variable A is not among the input, state or output variables, which, strictly speaking, prevents us from using Theorem 5.14. We fix this by adding A to the set of, say, input variables, and let the reader check that δ is left unchanged. Second, the translations of the μ and ν operations are not obviously well-defined, because we must first ensure that the formula Ψ' is monotonic in A. The reason why it works is the following:

Lemma 5.15 *Let* Φ' *be a modal formula, where* A *only occurs positively. Let* Ψ' *be the translation of* Φ'. *Then* Ψ' *is monotonic in* A.

Proof: In fact, A only occurs positively in Ψ', as an easy structural induction shows.   □

4.2 Binary Decision Diagrams

To check whether the modal formula Φ holds in the given Kripke frame, it is necessary and sufficient to check that the translated formula Ψ is valid. All methods of Chapter 2 can then be used.

However, the computation of fixed points $\mu A \cdot \Psi$ and $\nu A \cdot \Psi$ is arduous, because we must test for logical equivalence at each pass through the loop. Notice that checking for logical equivalence of two formulas Ψ_1 and Ψ_2 means checking the validity of $\Psi_1 \Leftrightarrow \Psi_2$, and this may take exponential time in the sizes of Ψ_1 and Ψ_2.

We can offset a great part of this cost in practical applications if we represent propositional formulas in canonical form on which propositional operations are fast. Binary Decision Diagrams, or BDDs, (see Chapter 2, Section 4.4) have all the desired properties. In particular, negation operates in constant time (by BDDs we mean TDGs, as is customary), and all binary operations operate in time proportional to smn, where m and n are the sizes of the BDDs to combine, and s is the cost of sharing.

To use BDDs efficiently, however, we still need to define substitutions on them. On formulas built on $\vee, \wedge, \neg, \Rightarrow$, substitution meant textual replacement. But if Φ is a BDD, what we want $\Phi[\Phi'/A]$ to represent is the BDD of the formula $\Psi[\Phi'/A]$, where Ψ is any formula represented by the BDD Φ. This is not a mere textual replacement, which would not yield a BDD in general.

We first define $\Phi[\mathbf{T}/A]$ and $\Phi[\mathbf{F}/A]$ by structural induction on Φ:

- $\mathbf{T}[\mathbf{T}/A] = \mathbf{T}, \mathbf{F}[\mathbf{T}/A] = \mathbf{F}, \mathbf{T}[\mathbf{F}/A] = \mathbf{T}, \mathbf{F}[\mathbf{F}/A] = \mathbf{F}$;

- if $\Phi = A \longrightarrow \Phi_+ ; \Phi_-$, then $\Phi[\mathbf{T}/A] = \Phi_+, \Phi[\mathbf{F}/A] = \Phi_-$;

- if Φ is of the form $A' \longrightarrow \Phi_+ : \Phi_-$, with $A' \neq A$,
  then $\Phi[\mathbf{T}/A] = \mathtt{BDDmake}(A', \Phi_+[\mathbf{T}/A], \Phi_-[\mathbf{T}/A])$,
  and $\Phi[\mathbf{F}/A] = \mathtt{BDDmake}(A', \Phi_+[\mathbf{F}/A], \Phi_-[\mathbf{F}/A])$.

Then we may compute $\Phi[\Phi'/A]$ as $(\Phi' \Rightarrow \Phi[\mathbf{T}/A]) \wedge (\neg\Phi' \Rightarrow \Phi[\mathbf{F}/A])$, or as $(\Phi' \wedge \Phi[\mathbf{T}/A]) \vee (\neg\Phi' \wedge \Phi[\mathbf{F}/A])$.

This is all that we need to compute propositional quantifications and fixed points: we have all the tools that we need to do symbolic model-checking with BDDs.

The fact that the basic operations take polynomial time does not imply that any polynomial-length sequence of basic operations can still be done in polynomial time: computing the conjunction of two BDDs of size n takes time $O(n^2)$, but yield a BDD whose size may be of the order of n^2; so, computing the conjunction of the latter with a new BDD of size n may take $O(n^3)$ time, yielding a BDD of size $O(n^3)$, ..., and computing the conjunction of n BDDs may produce a BDD of size $O(n^n)$ in exponential time.

BDDs can be too big to be built (either because memory runs out, or because it takes a very long time). They are practical in hardware verification because it seems that for many practical problems, they can be maintained rather small.

▶ **EXERCISE 5.8**

Give a direct definition of a recursive procedure for computing $\forall A \cdot \Phi$ and $\exists A \cdot \Phi$ (i.e., without using substitution as an auxiliary operation.)

▶ **EXERCISE 5.9**

Define substitution $\Phi[\Phi'/A]$ in purely logical terms, by using propositional quantifications.

4.3 Developments

The encoding of the transition relation δ as a propositional formula between m input variables, $2p$ current and next state variables, and n output variables is not economical. In particular, if the circuit is *deterministic*, δ is actually a function from $\mathbb{B}^{m+p}$ (input and current state) to $\mathbb{B}^{p+n}$ (next state and output). Then δ can be coded as a $(p + n)$-tuple of functions $\delta_1, \ldots, \delta_{p+n}$ from $\mathbb{B}^{m+p}$ to $\mathbb{B}$, i.e. as a $(p + n)$ tuple of formulas on $m + p$ variables. This is very important, as the size of a BDD on k variables may be of the order of 2^k. (More precisely, the least upper bound is $2^k/k$.) To give an example, in the case $m = n = 0$, the BDD for δ on the $2p$ current and next state variables may have a size proportional to the square of the greatest size of the BDDs for $\delta_i, 1 \le i \le p+n$: in practice, the tuple representation allows us to check circuits in a few megabytes of memory, which would need several gigabytes or even terabytes in the previous representation. (Not talking about the time we would need to build such structures in the first place.)

If the circuit is non-deterministic, we can still use a variant of the tuple representation. We can usually describe the degree of non-determinism in a few bits only: that is, we know that there will never be more than, say, 2^k worlds in relation with any given world, where usually k is small (at least compared to p). Then, we introduce k auxiliary propositional variables $x_1, \ldots, x_k$, and encode δ as the union of 2^k parameterized functions δ_ρ, where ρ is an assignment of truth-values to the x_i, $1 \le i \le k$.

Model-checking goes through as before, except that the third case in Theorem 5.14 must now be changed to:

- if $\Phi = \Box\Phi'$, let Ψ' be the translation of Φ', and return $\forall x_1 \cdot \ldots \forall x_k \cdot \Psi'[\delta_1(i_1, \ldots, i_m, s_1, \ldots, s_p)/s_1, \ldots, \delta_p(i_1, \ldots, i_m, s_1, \ldots, s_p)/s_p]$;

Notice that in the deterministic case, we don't even need any quantification.

Another variation on the basic method concerns the way we represent sets of configurations. We have used formulas Φ to represent sets of configurations by their satisfying assignments. That is, considering Φ as a function from $\mathbb{B}^{m+p+n}$ to $\mathbb{B}$, we represent sets of configurations by the inverse image $\Phi^{-1}(\{\mathbf{T}\})$. A dual way is to take a $(m+p+n)$-tuple of Boolean functions Φ_i, $1 \leq i \leq m+p+n$, from $\mathbb{B}^{m+p}$ to $\mathbb{B}$, and to represent sets of configurations as the direct image of $\mathbb{B}^{m+p}$ by the tuple of functions.

Finally, we should mention that checking that a modal formula Φ always holds (i.e., at all worlds) is usually not what we want. A sequential machine is indeed started in a given configuration (say, all inputs, and all state bits set to $\mathbf{F}$), then it only goes through the space of configurations that are *reachable* from this initial configuration. For example, a 2-bit counter intended to count 0, 1, 2 and repeat (corresponding to the configurations $(\mathbf{F}, \mathbf{F})$, $(\mathbf{F}, \mathbf{T})$, $(\mathbf{T}, \mathbf{F})$ respectively) will never reach a configuration where both bits are $\mathbf{T}$. What we want, then, is to check that the modal formula Φ holds in all reachable worlds, not necessarily in all worlds.

A naïve way of achieving this goal is to compute the non-modal translation Ψ translation of Φ with respect to the given Kripke frame as before, to compute a formula Ψ' whose satisfying assignments are exactly the reachable configurations, then to check that $\Psi' \Rightarrow \Psi$ is valid. This, however, may be too costly in practice, and it is interesting to prune BDDs Ψ by the set of reachable configurations while building it.

More discussions on model-checking with BDDs and applications can be found in (McMillan, 1993). (Coudert and Madre, 1995) is a good source of references. A more in-depth treatment of modal and temporal logics in computer science can be found in (Stirling, 1992; Emerson, 1990). Modal logics are the topic of (Gabbay and Guenthner, 1984).

CHAPTER 6

FIRST-ORDER CLASSICAL LOGIC

In this chapter, we introduce a definitely more expressive logic than most of the propositional logics we saw in the previous chapters. This is done by refining the language of logic, and replacing propositional variables by more elaborate formulas describing properties of values in a domain of interest. The result is called *first-order logic*, or *predicate calculus*. Most propositional logics can be extended to first-order logics (and even to higher-order logics). We present here *classical* first-order logic.

1 DEFINITIONS

1.1 Syntax

The syntax of first-order logics is two-tiered: there is one level for terms, which represent the objects of interest, and one for formulas, i.e. properties of and between these objects.

Let $\mathcal{V}$ be a countably infinite set of so-called *individual variables* x, y, $\ldots$, $\mathcal{F}$ an at most countable set of so-called *function symbols* f, g, $\ldots$, and $\mathcal{P}$ an at most countable set, disjoint from the latter, of *predicate symbols* P, Q, $\ldots$ Let also $\mathbf{m}$: $\mathcal{F} \cup \mathcal{P} \to \mathbb{N}$ be a function, called *arity function*; the arity m of f is defined by $m = \mathbf{m}(f)$.

Definition 6.1 (Terms, Formulas) *We call* terms s, t, $\ldots$ *the elements of the smallest set $\mathcal{T}$ such that:*

- *every variable is a term;*

- *for every $f \in \mathcal{F}$ of arity m, for all terms $t_1, \ldots, t_m$ in $\mathcal{T}$, the $(m+1)$-tuple $(f, t_1, \ldots, t_m)$ is also in $\mathcal{T}$.*

To make reading easier, this tuple, called application *of f to $t_1, \ldots, t_m$, is written $f(t_1, \ldots, t_m)$.*

We call atomic formula *or* atom A, B, $\ldots$ *every application $P(t_1, \ldots, t_m)$ of a predicate symbol P of arity m to m terms.*

The formulas Φ, Ψ, $\ldots$ *are the atomic formulas, the negations $\neg\Phi$ of formulas, the conjunctions $\Phi \wedge \Phi'$, the disjunctions $\Phi \vee \Phi'$, the implications $\Phi \Rightarrow \Phi'$, the universal quantifications $\forall x \cdot \Phi$ and the existential quantifications $\exists x \cdot \Phi$.*

185

We shall also write $\forall x_1 x_2 \ldots x_n \cdot \Phi$ instead of $\forall x_1 \cdot \forall x_2 \cdot \ldots \forall x_n \cdot \Phi$, and $\exists x_1 x_2 \ldots x_n \cdot \Phi$ instead of $\exists x_1 \cdot \exists x_2 \cdot \ldots \exists x_n \cdot \Phi$.

Intuitively, terms represent values in a given domain of objects. Constants are coded as function symbols of arity 0, and other function symbols represent basic operations on values (addition, multiplication, for example). Atomic formulas represent statements of basic properties on values (say, being less than or equal to, being odd, and so on), and are combined together through the use of the usual propositional connectives, supplemented by quantifications: $\forall x \cdot \Phi$ ("for all x, Φ") means that, interpreting Φ as a function $x \mapsto f(x)$ mapping a value denoted by x to the truth-value of Φ, then $f(v)$ is true of all values v; symmetrically, $\exists x \cdot \Phi$ ("there exists x such that Φ") holds whenever $f(v)$ is true of some value v.

Notice that, exactly like propositional formulas in Chapter 2, terms are directed acyclic graphs, and have associated structural induction and structural recursion principles. The set of propositional formulas, the set of terms, the set of first-order formulas can all be described as particular *free algebras*:

Definition 6.2 (Algebra) *Let T be a non-empty set, and F_n be a set of functions from T^n to T, for each $n \in \mathbb{N}$. Let F be the union of the F_n's, $n \in \mathbb{N}$. Then (T, F) is called an* algebra.

Let X be a non-empty set, F_n be a set of function symbols of arity n, $n \in \mathbb{N}$, and F be the union of the F_n's. The free algebra $T(X, F)$ *over X and F is the smallest set such that:*

- $X \subseteq T(X, F)$;

- *and whenever $f \in F_n$, for any n, and $t_1, \ldots, t_n$ are in $T(X, F)$, then the $(n+1)$-tuple $(f, t_1, \ldots, t_n)$ is in $T(X, F)$.*

We write $f(t_1, \ldots, t_n)$ for $(f, t_1, \ldots, t_n)$.

Therefore, propositional formulas as defined in Chapter 2, Section 1 are the elements of the free algebra $T(\mathcal{X}, \{\wedge, \vee, \neg, \Rightarrow\})$, where $\mathcal{X}$ is the set of propositional variables. The set $\mathcal{T}$ of first-order terms is the free algebra $T(\mathcal{V}, \mathcal{F})$. We can think of first-order formulas as the elements of the free algebra $T(\mathcal{T}, \{\wedge, \vee, \neg, \Rightarrow\} \cup \{\forall x \cdot \mid x \in \mathcal{V}\} \cup \{\exists x \cdot \mid x \in \mathcal{V}\})$, where the operators $\forall x \cdot$ and $\exists x \cdot$ have arity 1. (We say they are *unary*.)

Definition 6.3 (Free, Bound Variables) *If t is a term or a formula, we define its set of* free variables $\mathrm{fv}(t)$ *and its set of* bound variables $\mathrm{bv}(t)$ *by structural induction:*

- $\mathrm{fv}(x) = \{x\}$, $\mathrm{bv}(x) = \varnothing$ *for every variable $x \in \mathcal{V}$;*

- $\mathrm{fv}(f(t_1, \ldots, t_m)) = \mathrm{fv}(t_1) \cup \ldots \cup \mathrm{fv}(t_m)$, $\mathrm{bv}(f(t_1, \ldots, t_m)) = \varnothing$;

- $\mathrm{fv}(P(t_1, \ldots, t_m)) = \mathrm{fv}(t_1) \cup \ldots \cup \mathrm{fv}(t_m)$, $\mathrm{bv}(P(t_1, \ldots, t_m)) = \varnothing$;

- $\text{fv}(\neg\Phi) = \text{fv}(\Phi)$, $\text{bv}(\neg\Phi) = \text{bv}(\Phi)$;

- $\text{fv}(\Phi \vee \Phi') = \text{fv}(\Phi \wedge \Phi') = \text{fv}(\Phi \Rightarrow \Phi') = \text{fv}(\Phi) \cup \text{fv}(\Phi')$, $\text{bv}(\Phi \vee \Phi') = \text{bv}(\Phi \wedge \Phi') = \text{bv}(\Phi \Rightarrow \Phi') = \text{bv}(\Phi) \cup \text{bv}(\Phi')$;

- $\text{fv}(\forall x \cdot \Phi) = \text{fv}(\exists x \cdot \Phi) = \text{fv}(\Phi) \setminus \{x\}$, $\text{bv}(\forall x \cdot \Phi) = \text{bv}(\exists x \cdot \Phi) = \text{bv}(\Phi) \cup \{x\}$.

A term or a formula is said to be closed, *or* ground, *if its set of free variables is empty. A closed formula is called a* sentence.

A theory T is a set of sentences.

Contrarily to the propositional case, a variable x may occur in a formula without occurring free: for example, $\forall x \cdot P(x)$ has no free variables, and x only occurs bound (here, by the universal quantification).

As in the propositional case, we can define substitutions, except that we don't substitute atoms but first-order variables. The definitions are the same, because these are actually definitions on objects in an arbitrary free algebra:

Definition 6.4 (Substitution) *A substitution σ is a function from V to T, such that the set $\text{dom } \sigma = \{x \in V \mid x \neq \sigma(x)\}$, called the* domain *of σ, is finite.*

The range *of σ is by definition $\text{rng } \sigma = \{\sigma(x) \mid x \in \text{dom } \sigma\}$, and we let yield $\sigma = \bigcup\{\text{fv}(t) \mid t \in \text{rng } \sigma\}$.*

We also write σ as $[\sigma(x_1)/x_1, \ldots, \sigma(x_n)/x_n]$, where $x_1, \ldots, x_n$ contain all the variables in $\text{dom } \sigma$ and are pairwise distinct.

In particular, $[]$ is the empty (or identity) substitution.

As in the propositional case, the function σ extends to a unique morphism $t \mapsto t\sigma$ from T to T by:

- $x\sigma = \sigma(x)$

- $f(t_1, \ldots, t_m)\sigma = f(t_1\sigma, \ldots, t_m\sigma)$

Definition 6.5 (Instances) *Let t be a term: its* instances *are all terms of the form $t\sigma$, for σ a substitution.*

The composition *$\sigma\sigma'$ of two substitutions is defined by $t(\sigma\sigma') = (t\sigma)\sigma'$.*

A substitution σ' is said to be less general *than σ, and we write $\sigma' \preceq \sigma$, if and only there is a substitution σ'' such that $\sigma\sigma'' = \sigma'$.*

Substitution composition is well-defined: the proof is as in Theorem 2.5, page 15; this is no surprise, as this is actually a theorem for all free algebras and the associated notion of substitution. Then, substitution composition is again associative and has $[]$ has unit element. Moreover, $\preceq$ is a preorder. Intuitively, σ' is less general than σ if and only if all the instances of $t\sigma'$ are also instances of $t\sigma$, for any term t; more succinctly, a substitution is more general than another if it has at least all the instances of the second substitution.

We can again extend substitution to formulas by structural induction:

- $P(t_1, \ldots, t_m)\sigma = P(t_1\sigma, \ldots, t_m\sigma)$

- $(\neg\Phi)\sigma = \neg(\Phi\sigma)$

- $(\Phi \oplus \Phi')\sigma = \Phi\sigma \oplus \Phi'\sigma$, where $\oplus \in \{\vee, \wedge, \Rightarrow\}$

- for every $Q \in \{\forall, \exists\}$, $(Qx \cdot \Phi)\sigma = (Qx' \cdot \Phi[x'/x]\sigma)$, where x' is a variable outside yield $\sigma \cup (\mathrm{fv}(\Phi) \setminus \{x\})$.

As the last rule is not deterministic, there are several possible instances of a quantified formulas by σ. However, all these formulas will be equivalent semantically and computationally. Transforming $Qx \cdot \Phi$ into $Qx' \cdot \Phi[x'/x]$ with $x' \notin \mathrm{fv}(\Phi) \setminus \{x\}$ will then be harmless: this is exactly the same α-*renaming* process as in Chapter 4, where we have applied it to λ-terms. (Exercise: check that the semantic rules of Section 2 and the deduction systems of Section 3 are indeed insensitive to α-renaming.)

2   SEMANTICS

As in the propositional case, the semantics of first-order logic is given by means of an interpretation. However, the language of first-order logic is richer, and we have to account for the added complexity by introducing new definitions:

Definition 6.6 (Interpretation) *An interpretation I is a non-empty set D_I, called the* domain *of the interpretation, together with a function $I(f)$ from D_I^m to D_I for each symbol function f of arity m, and a function $I(P)$ from D_I^m to $\mathbb{B}$ for each predicate symbol P of arity m.*

An assignment *ρ is a function from $\mathcal{V}$ to D_I. We let $\mathcal{A} = \mathcal{V} \rightarrow D_I$ be the set of all assignments.*

Intuitively, an interpretation provides a set of values D_I which we describe by terms in the language of first-order logic, a set of operations $I(f)$ on these values, and a set of basic predicates $I(P)$ on tuples of values.

Contrarily to the propositional case, interpretations and assignments now have different meanings: an assignment just maps variables to values, while an interpretation describes the domain of values and the semantics of the function and predicate symbols.

Interpretations allow us to define the semantics of first-order terms (as values in D_I) and formulas (as truth-values in $\mathbb{B}$):

Definition 6.7 (Semantics) *For any assignment ρ, let $\rho[v/x]$ be the assignment mapping every variable y other than x to $\rho(y)$, and mapping x to v.*

In an interpretation I, and modulo an assignment ρ, the semantics of terms and formulas is defined by:

- $\llbracket x \rrbracket I \rho = \rho(x)$;

- $\llbracket f(t_1, \ldots, t_m) \rrbracket I \rho = I(f)(\llbracket t_1 \rrbracket I \rho, \ldots, \llbracket t_m \rrbracket I \rho)$;

- $\llbracket P(t_1, \ldots, t_m) \rrbracket I \rho = I(P)(\llbracket t_1 \rrbracket I \rho, \ldots, \llbracket t_m \rrbracket I \rho)$;

- $\llbracket \neg \Phi \rrbracket I \rho = \overline{\neg} \llbracket \Phi \rrbracket I \rho$;

- $\llbracket \Phi \vee \Phi' \rrbracket I \rho = \llbracket \Phi \rrbracket I \rho \overline{\vee} \llbracket \Phi' \rrbracket I \rho$;

- $\llbracket \Phi \wedge \Phi' \rrbracket I \rho = \llbracket \Phi \rrbracket I \rho \overline{\wedge} \llbracket \Phi' \rrbracket I \rho$;

- $\llbracket \Phi \Rightarrow \Phi' \rrbracket I \rho = \llbracket \Phi \rrbracket I \rho \overline{\Rightarrow} \llbracket \Phi' \rrbracket I \rho$;

- $\llbracket \forall x \cdot \Phi \rrbracket I \rho = \bigwedge_{v \in D_I} \llbracket \Phi \rrbracket I(\rho[v/x])$;

- $\llbracket \exists x \cdot \Phi \rrbracket I \rho = \bigvee_{v \in D_I} \llbracket \Phi \rrbracket I(\rho[v/x])$;

where $\bigwedge$ denotes distributed conjunction and $\bigvee$ denotes distributed disjunction, and $\overline{\neg}$, $\overline{\wedge}$, $\overline{\vee}$ and $\overline{\Rightarrow}$ denote the usual Boolean functions.

A formula is valid *if it is true in every interpretation under every assignment; otherwise, it is* invalid. *A formula is* unsatisfiable *if it is false in every interpretation modulo every assignment; otherwise, it is* satisfiable.

An interpretation in which a formula Φ is satisfied is called a model of Φ. *A model of a theory is a model of all the formulas that it contains. If F is a formula or a theory, we write $I \models F$ the relation "I is a model of F."*

The notion of semantic consequence, *also written $\models$, relates a theory (resp. a formula) F to another theory (resp. formula) F': $F \models F'$ if every model of F is also a model of F'.*

Then we have the analogue of Theorem 2.10, page 17, which says that assignments are the semantic counterpart of substitutions:

Theorem 6.8 *For every substitution σ and every assignment ρ, let $\sigma\rho$ be the assignment mapping each variable x to $\llbracket \sigma(x) \rrbracket \rho$.*

Then, for every term or formula F, for every substitution σ, for every interpretation I and every assignment ρ, $\llbracket F\sigma \rrbracket I \rho = \llbracket F \rrbracket I(\sigma\rho)$.

Proof: On terms, the proof is the same as for Theorem 2.10. On formulas, we prove it by structural induction again, but we have to examine the case of quantifications in more detail. If F is $\forall x \cdot \Phi'$, then $\llbracket F\sigma \rrbracket I \rho = \bigwedge_{v \in D_I} \llbracket \Phi'\sigma \rrbracket I(\rho[v/x])$, where we assume that x was chosen outside dom $\sigma \cup$ yield σ (apply α-renaming). By induction hypothesis, this is $\bigwedge_{v \in D_I} \llbracket \Phi' \rrbracket I(\sigma(\rho[v/x]))$. But $\sigma(\rho[v/x])$ maps every variable y to $\llbracket \sigma(y) \rrbracket I(\rho[v/x])$, so because $x \notin$ dom $\sigma \cup$ yield σ, it maps every y other than x to $\llbracket \sigma(y) \rrbracket I \rho$, and x to v. So $\sigma(\rho[v/x]) = (\sigma\rho)[v/x]$, and $\llbracket F\sigma \rrbracket I \rho = \bigwedge_{v \in D_I} \llbracket \Phi' \rrbracket I(\sigma\rho)[v/x] = \llbracket F \rrbracket I(\sigma\rho)$. The case of $\exists x \cdot \Phi'$ is similar. □

As in Chapter 2, Section 5.2, we could have defined the semantics of first-order formulas by using any *complete* Boolean algebra instead of $\mathbb{B}$. A Boolean algebra is complete if and only if every family of elements has a least upper bound and a greatest lower bound (not just the finite families). The need for complete Boolean algebras stems from the distributed conjunctions and disjunctions, which are defined in terms of least upper and greatest lower bounds respectively. Then, first-order logic is still sound and complete with this Boolean-valued semantics. (See (Manin, 1977).) Observe also that this meaning of "complete" is not related at all with other uses of the same word in other parts of this book.

3   DEDUCTION SYSTEMS

The deduction systems for first-order logic are all extensions of those for propositional logic, where the mention "propositional variable" must be replaced by "atomic formula", and where new rules to take quantifications into account have to be introduced.

For example, classical first-order logic has Hilbert-style systems. An example of such a system is Andrews' System $\mathcal{F}$ (Andrews, 1986), which is the following extension of System $\mathcal{P}$ (see Chapter 2, Section 3.1) defined as follows:

- Axioms:

    (1) $\Phi \vee \Phi \Rightarrow \Phi$ for every formula Φ,

    (2) $\Phi \Rightarrow \Phi' \vee \Phi$ for all formulas Φ, Φ',

    (3) $(\Phi \Rightarrow \Phi') \Rightarrow (\Phi'' \vee \Phi \Rightarrow \Phi' \vee \Phi'')$ for all Φ, Φ', Φ'',

    (4) $(\forall x \cdot \Phi) \Rightarrow \Phi[t/x]$ for every formula Φ, every variable x and every term t,

    (5) $(\forall x \cdot \Phi \vee \Phi') \Rightarrow \Phi \vee \forall x \cdot \Phi'$, for all formulas Φ, Φ' and for every variable x such that x is not free in Φ.

- Rules:

    (MP) from Φ and $\Phi \Rightarrow \Phi'$, deduce Φ', for all Φ, Φ',

    (Gen) from Φ, deduce $\forall x \cdot \Phi$, for any formula Φ and any variable x.

where only Axioms (4), (5) and Rule (Gen) are new. As in System $\mathcal{P}$, only $\neg$ and $\vee$ are primitive connectives in this system, so that $\Phi \Rightarrow \Phi'$ is really an abbreviation for $\neg \Phi \vee \Phi'$, $\Phi \wedge \Phi'$ is an abbreviation for $\neg(\neg \Phi \vee \neg \Phi')$. Moreover, $\exists$ is also an abbreviation: $\exists x \cdot \Phi$ is taken to abbreviate $\neg \forall x \cdot \neg \Phi$.

The (Gen) rule is called the *generalisation rule*. If we have managed to prove a formula Φ, then Φ is valid (provided the logic is sound), so it holds for every possible assignment, hence for any value of x, so that $\forall x \cdot \Phi$ must be valid, hence provable (if we want the deduction system to be complete).

$$\frac{\Gamma \longrightarrow \Phi[y/x]}{\Gamma \longrightarrow \forall x \cdot \Phi} \, (\forall I) \qquad\qquad \frac{\Gamma \longrightarrow \forall x \cdot \Phi}{\Gamma \longrightarrow \Phi[t/x]} \, (\forall E)$$

(y not free in Γ)

$$\frac{\Gamma \longrightarrow \Phi[t/x]}{\Gamma \longrightarrow \exists x \cdot \Phi} \, (\exists I) \qquad \frac{\Gamma \longrightarrow \exists x \cdot \Phi \qquad \Gamma, \Phi[y/x] \longrightarrow \Psi}{\Gamma \longrightarrow \Psi} \, (\exists E)$$

(y not free in Γ)

Figure 6.1. Natural Deduction in sequent form: quantifier rules

Notice the similarity between the (Gen) rule and the (Nec) rule of modal logics in Chapter 5. This should not be a surprise, as the Kripke semantics of modal logics dictates that □ is a kind of universal quantifier anyway.

Classical first-order logic also has natural deduction systems. To get one, we add the rules of Figure 6.1 to the system $\mathcal{ND}$ of Chapter 2, Section 3.2. (See Figure 2.2, page 28.)

For the purpose of analysing the structure of proofs, we shall be more interested in the Gentzen system **LK** (See Figure 6.2). This is also an extension of the system **LK**$_0$ that we presented in Chapter 2, Section 3.3, to which we have added the quantifier rules ∀L, ∀R, ∃L, ∃R.

All these systems are sound and complete for first-order logic. For Andrews' System $\mathcal{F}$, see (Andrews, 1986, Chapter 2). We shall only prove it for **LK** in Section 5.3 after we have introduced all relevant notions.

4   EXPRESSIVE POWER

First-order logics are much more expressive than the corresponding propositional logics, and in particular classical first-order logic is much more expressive than classical propositional logic.

First, first-order logic can express properties on infinite domains. In propositional logic, to express properties on values, we could only encode values on a finite number of propositional variables. This only yields finitely many values, as there are only finitely many assignments to the free propositional variables of a propositional formula. In first-order logic, we can build formulas like:

$$\forall x, y, z \cdot x < y \wedge y < z \Rightarrow x < z$$
$$\wedge \quad \forall x, y \cdot x < y \Rightarrow \exists z \cdot P(x, z) \wedge \neg P(y, z)$$
$$\wedge \quad \forall x \cdot \exists y \cdot x < y$$

where P and $<$ are two binary predicate symbols, $<$ being written in infix notation. This formula is satisfiable: take $<$ to be the "less than" relation on integers, and P

$$\frac{}{\Gamma, \Phi \longrightarrow \Delta, \Phi} \text{ Ax}$$

$$\frac{\Gamma, \Phi, \Phi' \longrightarrow \Delta}{\Gamma, \Phi \wedge \Phi' \longrightarrow \Delta} \wedge\text{L} \qquad\qquad \frac{\Gamma \longrightarrow \Delta, \Phi \quad \Gamma \longrightarrow \Delta, \Phi'}{\Gamma \longrightarrow \Delta, \Phi \wedge \Phi'} \wedge\text{R}$$

$$\frac{\Gamma, \Phi \longrightarrow \Delta \quad \Gamma, \Phi' \longrightarrow \Delta}{\Gamma, \Phi \vee \Phi' \longrightarrow \Delta} \vee\text{L} \qquad\qquad \frac{\Gamma \longrightarrow \Delta, \Phi, \Phi'}{\Gamma \longrightarrow \Delta, \Phi \vee \Phi'} \vee\text{R}$$

$$\frac{\Gamma \longrightarrow \Phi, \Delta \quad \Gamma, \Phi' \longrightarrow \Delta}{\Gamma, \Phi \Rightarrow \Phi' \longrightarrow \Delta} \Rightarrow\text{L} \qquad\qquad \frac{\Gamma, \Phi \longrightarrow \Delta, \Phi'}{\Gamma \longrightarrow \Delta, \Phi \Rightarrow \Phi'} \Rightarrow\text{R}$$

$$\frac{\Gamma \longrightarrow \Delta, \Phi}{\Gamma, \neg\Phi \longrightarrow \Delta} \neg\text{L} \qquad\qquad \frac{\Gamma, \Phi \longrightarrow \Delta}{\Gamma \longrightarrow \Delta, \neg\Phi} \neg\text{R}$$

$$\frac{\Gamma, \Phi[t/x] \longrightarrow \Delta}{\Gamma, \forall x \cdot \Phi \longrightarrow \Delta} \forall\text{L} \qquad\qquad \frac{\Gamma \longrightarrow \Phi[y/x], \Delta}{\Gamma \longrightarrow \forall x \cdot \Phi, \Delta} \forall\text{R}$$
$$\text{(}y \text{ not free in } \Gamma, \Delta\text{)}$$

$$\frac{\Gamma, \Phi[y/x] \longrightarrow \Delta}{\Gamma, \exists x \cdot \Phi \longrightarrow \Delta} \exists\text{L} \qquad\qquad \frac{\Gamma \longrightarrow \Phi[t/x], \Delta}{\Gamma \longrightarrow \exists x \cdot \Phi, \Delta} \exists\text{R}$$
$$\text{(}y \text{ not free in } \Gamma, \Delta\text{)}$$

$$\frac{\Gamma \longrightarrow \Delta, \Phi \quad \Gamma', \Phi \longrightarrow \Delta'}{\Gamma, \Gamma' \longrightarrow \Delta, \Delta'} \text{ Cut}$$

Figure 6.2. Gentzen's System **LK**

be the "less than or equal to" relation. But its models are all infinite: $<$ must denote a strict partial order by the first two conjuncts, with no upper bound by the third.

We can build more useful theories. For example, as above, we can take a binary (infix) predicate symbol $<$, and express that it is a strict order, that is, an irreflexive:

$$\forall x \cdot \neg x < x$$

and transitive:

$$\forall x \cdot \forall y \cdot \forall z \cdot x < y \wedge y < z \Rightarrow x < z$$

relation.

Equality $\doteq$ is a bit harder to axiomatize, as we want to be able to replace equals by equals, i.e. if $a \doteq b$ holds, then $s[a/x] \doteq t[a/x]$ must hold for any terms s and t; also $\Phi[a/x]$ must be logically equivalent to $\Phi[b/x]$. The trick to axiomatize equality is to first fix the language we are to work on, i.e. the set of function and predicate symbols we shall need, and issue one axiom per symbol to express that each symbol preserves equality. This yields the following theory:

- Reflexivity: $\forall x \cdot x \doteq x$;

- Symmetry: $\forall x, y \cdot x \doteq y \Rightarrow y \doteq x$;

- Functional congruence: for every function symbol f, of any arity n:

$$\forall x_1 \cdot \ldots \forall x_n \cdot \forall y_1 \cdot \ldots \forall y_n \cdot$$
$$x_1 \doteq y_1 \wedge \ldots \wedge x_n \doteq y_n \Rightarrow f(x_1, \ldots, x_n) \doteq f(y_1, \ldots, y_n)$$

- Predicate congruence: for every predicate symbol P, of arity n:

$$\forall x_1 \cdot \ldots \forall x_n \cdot \forall y_1 \cdot \ldots \forall y_n \cdot$$
$$x_1 \doteq y_1 \wedge \ldots \wedge x_n \doteq y_n \wedge P(x_1, \ldots, x_n) \Rightarrow P(y_1, \ldots, y_n)$$

Assume that we wish to prove a single formula. Its language is finite, therefore we need only finitely many axioms for $\doteq$. If we are to extend the language, for example by introducing new function symbols (like Skolem function symbols, see Section 5.1), we shall then need to add the corresponding congruence axioms.

With these constructions, we can recast most modal logics, as well as intuitionistic logic in terms of classical first-order logic, by expressing their Kripke semantics. (See Exercise 6.1.) It is nonetheless usually a bad idea to use these translations, as the specialized proof-search procedures for the modal or intuitionistic logics are in general more efficient than the general procedures for first-order logic.

A lot of mathematics can be built with first-order logic with equality. For example, the theory of groups extends that of equality by adding the following axioms on the language 0 (constant), $+$ (binary function):

$$\forall x \cdot 0 + x \doteq x \wedge x \doteq x + 0 \qquad \text{(0 is a neutral element)}$$
$$\forall x \cdot \forall y \cdot \forall z \cdot x + (y + z) \doteq (x + y) + z \quad \text{(associativity)}$$
$$\forall x \cdot \exists y \cdot x + y \doteq 0 \wedge y + x \doteq 0 \qquad \text{(inverse)}$$

The theory of rings adds new function symbols 1 (constant), $*$ (binary function) and the following axioms:

$$\forall x \cdot \forall y \cdot x + y \doteq y + x \qquad\qquad \text{(commutativity of } +)$$
$$\forall x \cdot 1 * x \doteq x \wedge x \doteq x * 1 \qquad\quad \text{(1 is neutral for } *)$$
$$\forall x \cdot \forall y \cdot \forall z \cdot x * (y * z) \doteq (x * y) * z \qquad \text{(associativity)}$$
$$\forall x \cdot \forall y \cdot \forall z \cdot x * (y + z) \doteq x * y + x * z$$
$$\forall x \cdot \forall y \cdot \forall z \cdot (y + z) * x \doteq y * x + z * x \qquad \text{(distributivity)}$$

Division rings add the axiom:

$$\forall x \cdot x \neq 0 \Rightarrow \exists y \cdot x * y \doteq 1 \wedge y * x \doteq 1$$

We can in the same sort develop theories for fields, modules, vector spaces, and so on.

These theories model the respective mathematical notions quite adequately. However, there are notions that cannot be captured accurately by any recursively enumerable first-order theory. We say that such notions are not *first-order axiomatizable*. (Trivially, any mathematical notion can be captured by some first-order theory, namely the one that lists all valid sentences about the notion; but they may not be recursively enumerable.) For example, there is no axiomatizable first-order theory whose models are all finite groups, and only finite groups. (Although we can describe groups of cardinal p, for any fixed integer p.) We cannot translate PDL or the modal μ-calculus in first-order logic faithfully either, that is, we cannot axiomatize the constructs of the logic in first-order logic. This is because the induction axiom of PDL, or the induction rules for the modal μ-calculus cannot be translated faithfully to first-order logic. For the same reason, there is no first-order axiomatization whose models are the set IN of natural integers with the usual arithmetical operations.

We can nevertheless build some form of arithmetic, called *first-order Peano arithmetic* **PA$_1$**, but to do this we shall need infinitely many axioms, and still it won't describe arithmetic completely—although it will describe a good part of it. The language of **PA$_1$** consists of the constant 0, the unary function s (successor), the binary functions $+$ and $*$ (addition and multiplication, which we write as infix operations), the binary relations $\doteq$ and $\leq$ (equality and the "less than or equal to" relation, written as infix operations; we use dots to distinguish them from equality and inequality statements). The axioms of **PA$_1$**—the elements of the theory **PA$_1$**—are the following:

- Equality axioms:

    - Reflexivity: $\forall x \cdot x \doteq x$;
    - Symmetry: $\forall x, y \cdot x \doteq y \Rightarrow y \doteq x$;
    - Transitivity: $\forall x, y, z \cdot x \doteq y \wedge y \doteq z \Rightarrow x \doteq z$;

– Congruence:

$$\forall x, y \cdot x \doteq y \Rightarrow s(x) \doteq s(y)$$
$$\forall x, y, z, t \cdot x \doteq z \wedge y \doteq t \Rightarrow x + y \doteq z + t$$
$$\forall x, y, z, t \cdot x \doteq z \wedge y \doteq t \Rightarrow x * y \doteq z * t$$
$$\forall x, y, z, t \cdot x \doteq z \wedge y \doteq t \wedge x \stackrel{.}{\leq} y \Rightarrow z \stackrel{.}{\leq} t$$

- Basic Peano axioms:

  – $\forall x \cdot x \doteq 0 \vee \exists y \cdot x \doteq s(y)$ (every integer is either 0 or a successor);

  – $\forall x \cdot s(x) \neq 0$ (0 is the successor of no integer);

  – $\forall x, y \cdot s(x) \doteq s(y) \Rightarrow x \doteq y$ (s is injective; in particular, every non-zero integer has a unique predecessor);

  – $\forall x \cdot x + 0 \doteq x, \forall x, y \cdot x + s(y) \doteq s(x + y)$ (definition of addition);

  – $\forall x \cdot x * 0 \doteq 0, \forall x, y \cdot x * s(y) \doteq (x * y) + x$ (definition of multiplication);

  – $\forall x, y \cdot x \stackrel{.}{\leq} y \Leftrightarrow \exists z \cdot x + z \doteq y$ (definition of $\stackrel{.}{\leq}$);

- Induction scheme: for every first-order formula Φ in the language of $\mathbf{PA}_1$, and for every variable x:

$$\Phi[0/x] \wedge (\forall x \cdot \Phi \Rightarrow \Phi[s(x)/x]) \Rightarrow \forall x \cdot \Phi$$

The induction scheme is an infinite collection of axioms. We would like to write instead something like:

$$\forall P \cdot P(0) \wedge (\forall x \cdot P(x) \Rightarrow P(s(x))) \Rightarrow \forall x \cdot \Phi$$

where P is a predicate variable. (This is called the *second-order induction axiom*.) We cannot do this in first-order logic, since indeed we can quantify only on variables, and variables denote values, not predicates over values. However, the harm done is limited: although $\mathbf{PA}_1$ has infinitely many axioms, the set of axioms is recursive, that is, we can decide by machine whether a given formula is an axiom of $\mathbf{PA}_1$.

One model of $\mathbf{PA}_1$ is the intended model: the set $\mathbb{N}$ of natural integers, where the constant 0 is interpreted as zero, s is interpreted as the function mapping n to $n + 1$, $+$ is interpreted as addition, $*$ as multiplication, $\doteq$ as equality and $\stackrel{.}{\leq}$ as the natural ordering $\leq$ verifies all the axioms. This is called the *standard model* of $\mathbf{PA}_1$.

But $\mathbf{PA}_1$ has other models as well, the so-called *non-standard models* of arithmetic, as should be expected from our statement that Peano arithmetic is not axiomatisable in first-order logic. (See Exercise 6.2 for an example.) This really is an inherent limitation of first-order logic, and can be explained in this case by the Löwenheim-Skolem Theorem, which we shall present in Section 5.2.

We could repair this in second-order logic, by using the second-order induction axiom. This would yield *second-order Peano arithmetic* $\mathbf{PA}_2$. $\mathbb{N}$ is the only model of $\mathbf{PA}_2$, up to isomorphism. The problem is that there is no effective, sound and complete deduction system for second-order logic or for $\mathbf{PA}_2$, because of Gödel's incompleteness theorem (see Theorem 6.36, page 221).

Leon Henkin managed in 1950 to repair this problem of higher-order logics by defining *general models* of higher-order formulas, which generalise the intended standard models. But then, there are non-standard general models of $\mathbf{PA}_2$ again, so we are back to where we started. There is, in fact, no way out of this dilemma. (See (Andrews, 1986) for higher-order logic and general models.)

First-order logic can not only express this weak form of arithmetic, but in fact it is expressive enough to express *set theory*, for example by Zermelo and Frænkel's axioms, on which all mathematics (except for a few facts, of interest to logicians only) can be founded. See (Johnstone, 1992) for a discussion of Zermelo-Frænkel set theory and its basic concepts and results.

▶ **EXERCISE 6.1**

Using the Kripke semantics for S4 as a guide, describe a translation of S4 into first-order classical logic preserving validity.

▶ **EXERCISE 6.2**

Let D be the set of all complements of all finite subsets of $\mathbb{N}$. Show that D is a filter (see Appendix B). Conclude that there is an ultrafilter $\mathcal{F}$ containing all complements of all finite subsets of $\mathbb{N}$.

Let N denote the set of all functions from $\mathbb{N}$ to $\mathbb{N}$, and define the following interpretation I of the symbols of arithmetic in N:

- $I(0)$ is the constant function $\lambda n \cdot 0$;

- $I(s)$ maps every f in N to $\lambda n \cdot f(n) + 1$;

- $I(+)$ maps every f, g in N to $\lambda n \cdot f(n) + g(n)$;

- $I(*)$ maps every f, g in N to $\lambda n \cdot f(n)g(n)$;

- $I(\dot{\leq})(f, g)$ holds whenever $\{n \mid f(n) \leq g(n)\} \in \mathcal{F}$;

- $I(\dot{=})(f, g)$ holds whenever $\{n \mid f(n) = g(n)\} \in \mathcal{F}$.

Show that N is a model of $\mathbf{PA}_1$. (Hint: for every assignment ρ on N, for each $n \in \mathbb{N}$, let ρ_n denote the assignment on $\mathbb{N}$ mapping each variable x to $\rho(x)(n)$; show by structural induction on any first-order formula Φ that $\llbracket \Phi \rrbracket I \rho$ holds if and only if $\{n \mid \llbracket \Phi \rrbracket I_0 \rho_n\}$ is in $\mathcal{F}$, where I_0 is the standard interpretation mapping 0 to zero, + to the usual addition function over integers, and so on.)

Show that the constant functions define an embedding on $\mathbb{N}$ in N. Show that there is an element of N greater (with respect to $I(\dot{\leq})$) than all elements of $\mathbb{N}$ (viewed as constant functions). Therefore, N is not isomorphic to $\mathbb{N}$.

5 META-MATHEMATICAL PROPERTIES

First-order logic has lots of interesting logical properties, and most of them can be approached from either the semantic viewpoint, using interpretations (this is called *model theory*), or from the syntactic viewpoint, using arguments on proofs, say in **LK** (this is called *proof theory*).

5.1 Prenex and Skolem Forms

Our goal in automated deduction is to find whether formulas are valid (resp. provable). However, before we look for proofs, it is practical to simplify the first-order logic formulas we want to prove. What this means is the following: given a first-order formula Φ, we wish to compute a simpler (in some sense to be made precise) formula Φ', whose validity status (resp. provability status) is the same as that of Φ.

First, for any first-order formula Φ, we can push its quantifiers outwards without changing the meaning of the formula, by the following rules (we assume $\Rightarrow$ has been expanded in terms of $\neg$ and $\vee$):

- $\neg \forall x \cdot \Psi \longrightarrow \exists x \cdot \neg \Psi$,

- $\neg \exists x \cdot \Psi \longrightarrow \forall x \cdot \neg \Psi$,

- $(\forall x \cdot \Psi) \wedge \Psi' \longrightarrow \forall x' \cdot \Psi[x'/x] \wedge \Psi'$, where $x' \notin \mathrm{fv}(\Psi')$,

- $\Psi \wedge (\forall x \cdot \Psi') \longrightarrow \forall x' \cdot \Psi \wedge \Psi'[x'/x]$, where $x' \notin \mathrm{fv}(\Psi)$,

- $(\exists x \cdot \Psi) \wedge \Psi' \longrightarrow \exists x' \cdot \Psi[x'/x] \wedge \Psi'$, where $x' \notin \mathrm{fv}(\Psi')$,

- $\Psi \wedge (\exists x \cdot \Psi') \longrightarrow \exists x' \cdot \Psi \wedge \Psi'[x'/x]$, where $x' \notin \mathrm{fv}(\Psi)$,

- $(\forall x \cdot \Psi) \vee \Psi' \longrightarrow \forall x' \cdot \Psi[x'/x] \vee \Psi'$, where $x' \notin \mathrm{fv}(\Psi')$,

- $\Psi \vee (\forall x \cdot \Psi') \longrightarrow \forall x' \cdot \Psi \vee \Psi'[x'/x]$, where $x' \notin \mathrm{fv}(\Psi)$,

- $(\exists x \cdot \Psi) \vee \Psi' \longrightarrow \exists x' \cdot \Psi[x'/x] \vee \Psi'$, where $x' \notin \mathrm{fv}(\Psi')$,

- $\Psi \vee (\exists x \cdot \Psi') \longrightarrow \exists x' \cdot \Psi \vee \Psi'[x'/x]$, where $x' \notin \mathrm{fv}(\Psi)$.

This rewriting system terminates. Indeed, for each occurrence of a quantified subformula in Φ, let its *rank* be the number of $\neg$, $\wedge$ and $\vee$ wrappers around it, and let the rank of Φ be the sum of the ranks of occurrences of its quantified subformulas. (Formally, we define the rank $r(\Phi)$ and the number of occurrences of quantified subformulas $n(\Phi)$ of Φ by structural recursion in the following way: $n(A) = r(A) = 0$ for all atoms A; $n(\Phi \wedge \Phi') = n(\Phi) + n(\Phi')$, $r(\Phi \wedge \Phi') = r(\Phi) + r(\Phi') + n(\Phi \wedge \Phi')$, and similarly for $\vee$; $n(\neg \Phi) = n(\Phi)$, $r(\neg \Phi) = r(\Phi) + n(\Phi)$; $n(\forall x \cdot \Phi) = n(\Phi) + 1$, $r(\forall x \cdot \Phi) = r(\Phi)$ and similarly for $\exists$.) Then, the rank of Φ is a non-negative integer that decreases at each rewriting step, so the rewriting must terminate.

The normal form that we get after we have finished rewriting may not be unique, and depends on the way we rewrote the formula. Moreover, we can make the rewriting more efficient by adding rules like:

- $(\forall x \cdot \Psi) \wedge (\forall x' \cdot \Psi') \longrightarrow \forall x \cdot \Psi \wedge \Psi'[x/x']$,

- $(\exists x \cdot \Psi) \vee (\exists x' \cdot \Psi') \longrightarrow \exists x \cdot \Psi \vee \Psi'[x/x']$,

where $x \notin \mathrm{fv}(\Psi')$, which don't hinder termination, but may help generating less quantifications in the end result.

The normal form that we get after this rewriting is called a *prenex form*:

Definition 6.9 (Prenex) *A formula is called* prenex *if and only if it is of the form* $Q_1 x_1 \cdot \ldots Q_n x_n \cdot \Phi$, *where* $Q_1, \ldots, Q_n$ *are quantifiers and* Φ *is quantifier-free.*

That Φ is quantifier-free means that it is built without using any quantifications. Alternatively, it is actually a *propositional formula* where atoms play the rôle of propositional variables.

Theorem 6.10 *For every first-order formula* Φ, *there exists a prenex formula* Φ', *computable from* Φ, *such that* Φ *and* Φ' *are (semantically, resp. provably) equivalent.*

Proof: We apply the rewrite rules above. The normal form Φ' we obtain in the end must have rank 0, since otherwise one of the rules applies; but a formula of rank 0 is a prenex formula. The fact that this derivation terminates proves that the rewriting rules define an algorithm for computing such a prenex form. It remains to show that Φ and Φ' are (semantically, resp. provably) equivalent, which we can do by induction on the length of the rewriting derivation, by showing that each rule has (semantically, resp. provably) equivalent left-hand and right-hand sides, and that (semantic, resp. provable) equivalence is preserved by applications of logical operators.

Semantic equivalence is then clear. Provable equivalence must be checked by finding corresponding proofs in **LK**, i.e. for each rule $\Phi \longrightarrow \Phi'$, we must find proofs for $\Phi \Leftrightarrow \Phi'$, or equivalently we must find proofs of the sequents $\Phi \longrightarrow \Phi'$ and $\Phi' \longrightarrow \Phi$. We leave it to the reader. (See Exercise 6.3.) □

Theorem 6.10 shows that instead of working with an arbitrary formula, we can work with prenex formulas, which are much better structured. We can still improve on it, because of the following idea.

Take a formula of the form, say, $\forall x \cdot \Phi$: it is valid if and only if for every x, Φ holds. Let us replace x by a new constant c in Φ. Then $\forall x \cdot \Phi$ is valid if and only if, for every possible interpretation of c, $\Phi[c/x]$ holds, that is, if and only if $\Phi[c/x]$ is valid. In short, we can encode the universal quantification on x by using a constant c and the implicit universal quantification over models in the statement of validity.

Now, take a formula of the form $\exists x \cdot \forall y \cdot \Phi$: it is valid if and only if there is a value for x such that for every y, Φ holds. A counterexample to the validity would then be, for each given value of x, a value for y such that Φ does not hold; in other terms, a counterexample for y is a function $f(x)$, where f is a new function symbol, not just a constant c. Without resorting to the notion of counterexamples, we may say that $\exists x \cdot \forall y \cdot \Phi$ is valid if and only if $\exists x \cdot \Phi[f(x)/y]$ is valid, where the universal quantification on y is expressed implicitly, in the statement of what it means to be valid.

Formally, this leads to the following:

Definition 6.11 *An existential formula* is a formula of the form $\exists x_1 \cdot \ldots \exists x_n \cdot \Phi$, *where Φ is quantifier-free.*

A universal formula *is a formula of the form* $\forall x_1 \cdot \ldots \forall x_n \cdot \Phi$, *where Φ is quantifier-free.*

The following is due mostly to Thoralf Skolem for universal formulas, but Jacques Herbrand made extensive use of it for existential formulas:

Theorem 6.12 (Herbrand-Skolem) *Let Φ be a first-order formula.*

There exists an existential formula Φ', computable from Φ, such that Φ' is valid if and only if Φ is valid. We say that Φ' is obtained by herbrandizing Φ, *and that Φ' is a* Herbrand form *of Φ.*

Dually, there is a universal formula Φ'', computable from Φ, such that Φ'' is unsatisfiable if and only if Φ is unsatisfiable. We say that Φ'' is obtained by skolemizing Φ, *and that Φ'' is a* Skolem form *of Φ.*

Proof: We may assume Φ to be in prenex form, by Theorem 6.10. We shall deal with skolemization, as herbrandizing means taking the negation of a skolemization of a negation. (And conversely.)

So Φ is $Q_1 x_1 \cdot \ldots Q_n x_n \cdot \Psi$, where Ψ is quantifier-free. For each $1 \le i \le n$ such that Q_i is the existential quantifier $\exists$, let $x_{i_1}, \ldots, x_{i_{m_i}}$ be the universally quantified variables of index less than i, let f_i be a new function symbol of arity m_i and t_i be the term $f_i(x_{i_1}, \ldots, x_{i_{m_i}})$. Then, let Ψ' be Ψ where every existentially quantified variable x_i has been replaced by t_i, and Φ' be $\forall x_{i_1} \cdot \ldots \forall x_{i_m} \cdot \Psi'$, where $x_{i_1}, \ldots, x_{i_m}$ are the universally quantified variables. (I.e., we replace existential variables by fresh functions of all universal variables before them. In herbrandization, we replace universal variables by fresh functions of all existential variables before them.)

We claim that Φ is satisfiable if and only if Φ' is satisfiable. To prove this, we show that any sentence $\forall y_1 \cdot \ldots \forall y_k \cdot \exists z \cdot \Phi_1$ is satisfiable if and only if $\forall y_1 \cdot \ldots \forall y_k \cdot \Phi_1[f(y_1, \ldots, y_k)/z]$ is satisfiable, where f is a function symbol not appearing in Φ_1. The claim then follows by induction on the number of existential quantifiers in the prefix of Φ.

If $\forall y_1 \cdot \ldots \forall y_k \cdot \Phi_1[f(y_1, \ldots, y_k)/z]$ is satisfiable, then there is a model I, and an interpretation of f in I, such that $\Phi_1[f(y_1, \ldots, y_k)/z]$ is true for all assignments of the variables $y_1, \ldots, y_k$ to values $v_1, \ldots, v_k$ respectively. Then $I(f)(v_1, \ldots, v_k)$ is a value for z that makes Φ_1 true in I, for any assignment of y_i to v_i, $1 \leq i \leq k$. Therefore I satisfies $\forall y_1 \cdot \ldots \forall y_k \cdot \exists z \cdot \Phi_1$.

Conversely, assume that there is an interpretation I satisfying $\forall y_1 \cdot \ldots \forall y_k \cdot \exists z \cdot \Phi_1$. For every assignment of y_1 to some value $v_1, \ldots$, of y_k to v_k, let $G(v_1, \ldots, v_k)$ be the set of values for z that make Φ_1 true. By assumption, $G(v_1, \ldots, v_k)$ is nonempty for every k-tuple of values $(v_1, \ldots, v_k)$. We now use the Axiom of Choice to deduce that there must be a function g mapping each $(v_1, \ldots, v_k)$ to some element of $G(v_1, \ldots, v_k)$. Extending the interpretation I to a new interpretation I' such that $I'(s) = I(s)$ for all symbols in Φ_1, and such that $I'(f) = g$, we then see that I' satisfies $\forall y_1 \cdot \ldots \forall y_k \cdot \Phi_1[f(y_1, \ldots, y_k)/z]$.   □

Whereas conversion to prenex form yields a logically equivalent formula, herbrandization or skolemization does not. Herbrandization only conserves the validity status, and skolemization only conserves the satisfiability status.

In fact, herbrandization also conserves the provability status, and skolemization the consistency status of the formula, too, so this process is not only justified semantically, but also proof-theoretically. We shall demonstrate this in Section 5.3.

Semantically, the essence of herbrandization or skolemization lies in the use of the Axiom of Choice. Consider the following informal argument: extend the language of first-order logic by adding a new variable-binding operator ϵ, called *Hilbert's symbol*: $\epsilon x \cdot \Phi$, for any variable x and formula Φ, is a term whose set of free variables is $\mathrm{fv}(\Phi) \setminus \{x\}$; $\epsilon x \cdot \Phi$ is meant to describe some value for x such that Φ holds, if such a value exists. The semantics of ϵ is defined in the following way: given an interpretation I of domain D_I, there is a choice function g such that for every non-empty subset S of D_I, $g(S) \in S$ (this is the Axiom of Choice); extend g so that $g(\varnothing)$ is an arbitrary element of D_I, and let $[\![\epsilon x \cdot \Phi]\!]I\rho$ be $g(\{v \in D_I \mid [\![\Phi]\!]I\rho[v/x]\})$. If $\exists x \cdot \Phi$ holds, then $\epsilon x \cdot \Phi$ is a term t such that $\Phi[t/x]$ holds, i.e. $\Phi[\epsilon x \cdot \Phi/x]$ holds; conversely, if $\Phi[\epsilon x \cdot \Phi/x]$ holds, then $\exists x \cdot \Phi$ also holds. To sum up, $\exists x \cdot \Phi$ and $\Phi[\epsilon x \cdot \Phi/x]$ are equivalent formulas in the extended language.

In particular, we can use ϵ to justify skolemization. Let $x_1, \ldots, x_n$ be the free variables of $\epsilon x \cdot \Phi$. Then, we can encode $\epsilon x \cdot \Phi$ as $(\lambda x_1, \ldots, x_n \cdot \epsilon x \cdot \Phi)(x_1, \ldots, x_n)$, where $\lambda x_1, \ldots, x_n \cdot \epsilon x \cdot \Phi$ is a new symbol, whose meaning is intended to be the function that takes $x_1, \ldots, x_n$ as arguments, and returns $\epsilon x \cdot \Phi$. Considering $\lambda x_1, \ldots, x_n \cdot \epsilon x \cdot \Phi$ as a symbol f means translating $\exists x \cdot \Phi$ into $\Phi[f(x_1, \ldots, x_n)/x]$, and this is precisely what skolemization is all about. For a formalisation of this argument, see Exercise 6.4.

▶ **EXERCISE 6.3**

For each rule $\Phi \longrightarrow \Phi'$ of conversion to prenex form, show that the sequents $\Phi \longrightarrow \Phi'$ and $\Phi' \longrightarrow \Phi$ are derivable in **LK**. (Try not to use Cut, as well.)

▶ **EXERCISE 6.4**

Define the following two functions for herbrandization, h, and for skolemization, s:

- if Φ is atomic, then $h(\Phi) = s(\Phi) = \Phi$;

- $h(\Phi' \wedge \Phi'') = h(\Phi') \wedge h(\Phi'')$, $s(\Phi' \wedge \Phi'') = s(\Phi') \wedge s(\Phi'')$;

- $h(\Phi' \vee \Phi'') = h(\Phi') \vee h(\Phi'')$, $s(\Phi' \vee \Phi'') = s(\Phi') \vee s(\Phi'')$;

- $h(\neg\Phi') = \neg s(\Phi')$, $s(\neg\Phi') = \neg h(\Phi')$,

- $h(\Phi' \Rightarrow \Phi'') = s(\Phi') \Rightarrow h(\Phi'')$, $s(\Phi' \Rightarrow \Phi'') = h(\Phi') \Rightarrow s(\Phi'')$;

- $h(\exists x \cdot \Phi') = h(\Phi')$, $s(\exists x \cdot \Phi') = s(\Phi')[f(x_1, \ldots, x_n)/x]$, where $x_1, \ldots,$ x_n are the free variables of $\exists x \cdot \Phi'$ and f is a new n-ary function symbol;

- $h(\forall x \cdot \Phi') = h(\Phi')[f(x_1, \ldots, x_n)/x]$, where $x_1, \ldots, x_n$ are the free variables of $\exists x \cdot \Phi'$ and f is a new n-ary function symbol, and $s(\forall x \cdot \Phi') = s(\Phi')$.

Let $\overline{h}(\Phi)$ be $\exists x_1 \cdot \ldots \cdot \exists x_n \cdot h(\Phi)$, and $\overline{s}(\Phi)$ be $\forall x_1 \cdot \ldots \cdot \forall x_n \cdot s(\Phi)$, where $x_1, \ldots, x_n$ are the free variables of $h(\Phi)$.

Show that, for every formula Φ, $\overline{h}(\Phi)$ is valid if and only if Φ is valid, and that $\overline{s}(\Phi)$ is satisfiable if and only if Φ is satisfiable. (Show that we can even choose the same Skolem function symbols for two logically equivalent occurrences of quantified subformulas.)

The purpose of this exercise is to show that there are herbrandization/skolemization procedures that do not need to put the formula in prenex form first. (They are to preferred to the naïve procedures at all costs, in automated deduction, as they generate less Skolem functions, and with reduced arities: this makes skolemisation faster, but also considerably facilitates subsequent proof search.)

5.2   Semantic Herbrand Theory

Having narrowed the class of formulas we need to consider for validity to existential formulas, we introduce a few fundamental notions, introduced by Jacques Herbrand in the 1930s:

Definition 6.13 (Herbrand Universe) *Let B be a set of so-called* base constants, *such that $B \neq \varnothing$ if there is no zero-ary function symbol. We let B_0 be the smallest of these B, i.e. $\varnothing$ if there are zero-ary function symbols, otherwise $\{a\}$ for some new constant symbol.*

The set $H(B)$ of terms in $T(B, \mathcal{F})$ of closed terms on the language expanded with the set B of constants is called the Herbrand universe on B. *The* Herbrand universe *is defined as $H(B_0)$.*

A Herbrand interpretation *I is a set of closed atoms in the language expanded with the set B of constants. It defines an interpretation on H as follows:*

- $I(f)(t_1, \ldots, t_n)$, *where the t_i's are closed terms, is the closed term* $f(t_1, \ldots, t_n)$;

- $I(P)(t_1, \ldots, t_n)$, where the t_i's are closed terms, is $\top$ if and only if the closed atom $P(t_1, \ldots, t_n)$ is in the set I, and $\perp$ otherwise.

Herbrand universes are the most simple-minded interpretations we can think of, given that we know the syntax of first-order logic. Indeed, they consist in interpreting terms as closed terms, as the following Lemma shows. The strange condition on the emptiness of B or of B_0 ensures that $H(B)$ is the domain of some interpretation, that is, that $H(B)$ is not empty.

Lemma 6.14 *Let ρ be an assignment on $H(B)$. Then ρ is also a closed substitution, and for any Herbrand interpretation I, for any term t, $[\![t]\!]I\rho = t\rho$.*

Proof: Notice that $t\rho$ in the right-hand side means t on which the substitution ρ has been applied. The proof is then an immediate structural induction on t.   □

This is a kind of converse to Theorem 6.8: on Herbrand universes, closed substitutions *are* assignments.

But the main interest of Herbrand universes and interpretations lie in the following theorems. The following Lemma asserts that Herbrand interpretations are no more particular than general interpretations. That is, Definition 6.13 asserts that all Herbrand interpretations define interpretations on $H(B)$. The following Lemma states the converse:

Lemma 6.15 *Let I be an interpretation over a Herbrand universe $H(B)$. Then there is a Herbrand interpretation I' over $H(B)$ such that for all formulas Φ and assignments ρ, $[\![\Phi]\!]I\rho = [\![\Phi]\!]I'\rho$.*

Proof: We let I' be the set of closed atoms A such that $[\![A]\!]I\rho = \top$. So, by definition, for every atom A—not necessarily closed—$[\![A]\!]I'\rho = [\![A\rho]\!]I'[]$ by Lemma 6.14, i.e. $[\![A]\!]I'\rho = [\![A\rho]\!]I[]$ (by definition of I') $= [\![A]\!]I\rho$ (by Theorem 6.8). The Lemma then follows by structural induction on the formula Φ, of which we have just proved the base case; the other cases are straightforward.   □

From now on, we shall therefore just confuse the two notions of Herbrand interpretation and of interpretation over $H(B)$.

The following shows that we don't need more than syntax (Herbrand universes and interpretations) to decide validity:

Theorem 6.16 *Let Φ be an existential first-order formula. (A herbrandized formula.)*

Then the following are equivalent:

(i) Φ *is valid;*

(ii) *for every Herbrand universe $H(B)$, Φ holds in every Herbrand interpretation over B;*

(iii) Φ *holds in every Herbrand interpretation over* B_0.

Proof: (i) implies (ii), which implies (iii). It remains to show that (iii) implies (i). As Φ is existential, it has the form $\exists x_1 \cdot \ldots \exists x_n \cdot \Psi$, where Ψ is quantifier-free. By (iii), whatever the Herbrand interpretation I, there are closed terms $t_1, \ldots, t_n$ such that $\Psi[t_1/x_1, \ldots, t_n/x_n]$ holds in I, i.e. there is a closed instance $\Psi\sigma$ of Ψ that holds in I.

Now, take any interpretation I' (not necessarily on $H(B_0)$) on an arbitrary non-empty domain D. This interpretation induces a Herbrand interpretation I as the set of closed atoms A such that A is true under I'. Then by (iii), there is a closed instance $\Psi\sigma$ of Ψ that holds in I, hence in I'. (The argument is as for Lemma 6.15.) But then, the assignment mapping each x_i to the interpretation of the closed term $x_i\sigma$ in D satisfies Ψ. So, I' satisfies $\exists x_1 \cdot \ldots \exists x_n \cdot \Psi$, that is, Φ. As I' is arbitrary, (i) is proved. $\qquad\square$

> The theorem does not hold if Φ is not herbrandized first. We indeed need to show the existence of closed terms $t_1, \ldots, t_n$ denoting values in D to prove the Theorem. Another viewpoint is the following: if Φ is not existential, then we might first herbrandize it to an existential formula Φ', apply the Theorem above on Φ', and try to translate the result to Φ. The problem is that Φ' is built on a strictly larger alphabet of function symbols, so its Herbrand universe is not the same as that of Φ, and we cannot translate the result to Φ.

This Theorem has the following important consequence:

Theorem 6.17 (Herbrand, Semantic Version) *Let* Φ *be an existential formula* $\exists x_1 \cdot \ldots \exists x_n \cdot \Psi$, *where* Ψ *is quantifier-free. Assume moreover that there is at least one constant in the language.*

Then Φ *is valid if and only if there is an integer k, and k closed instances* $\Psi\sigma_1$, *..., $\Psi\sigma_k$ of Ψ such that* $\Psi\sigma_1 \vee \ldots \vee \Psi\sigma_k$ *is propositionally valid.*

Proof: If $\Psi\sigma_1 \vee \ldots \vee \Psi\sigma_k$ is propositionally valid, then Φ is clearly valid, since in every possible interpretation, one of the $\Psi\sigma_i$, $1 \leq i \leq k$ must hold.

Conversely, if Φ is valid, by Theorem 6.16, for every Herbrand interpretation I over $H(B_0)$, there is a closed instance $\Psi\sigma_I$ satisfied by I. (Notice that, since there is at least one constant in the language, B_0 is empty, and $H(B_0)$ is exactly the set of all closed terms.) In particular, the set of all $\neg\Psi\sigma_I$ is propositionally unsatisfiable, so by the Propositional Compactness Theorem (Theorem 2.17, page 22), there is a finite propositionally unsatisfiable subset $\neg\Psi\sigma_1, \ldots, \neg\Psi\sigma_k$. Therefore, $\Psi\sigma_1 \vee \ldots \vee \Psi\sigma_k$ is propositionally valid. $\qquad\square$

Herbrand's Theorem is important because it provides a procedure for testing whether a given formula is valid: herbrandize the formula to get an existential formula Φ as above, then enumerate all sequences of closed instances $\Psi\sigma_1, \ldots, \Psi\sigma_k$ and test for the propositional validity of their disjunction. This may never terminate, as we don't know any a priori bound on k, but at least if the formula is valid, we will eventually find it out by this process. (We say that this process is *complete*. It is obviously sound, i.e. it cannot declare valid any invalid formula.) This method was explored until the early 1960s, where more efficient methods for using Herbrand's Theorem were devised. These methods will be the subject of the chapters to come.

Herbrand's Theorem is usually presented in the following dual form:

Theorem 6.18 (Herbrand, Dually) *Let Φ be a universal formula $\forall x_1 \cdot \ldots \forall x_n \cdot \Psi$, where Ψ is quantifier-free. Assume moreover that there is at least one constant in the language.*

Then Φ is unsatisfiable if and only if there is an integer k, and k closed instances $\Psi\sigma_1, \ldots, \Psi\sigma_k$ of Ψ such that $\Psi\sigma_1 \wedge \ldots \wedge \Psi\sigma_k$ is propositionally unsatisfiable.

This is of course the same as the previous theorem, modulo negation.

An intriguing component of this theorem is that it needs an integer k, where we might have guessed $k = 1$. That is, we might expect that $\exists x \cdot \Psi$ would be valid if and only if there was a term t such that $\Psi[t/x]$ is valid. Such a term t would be called a *witness* of the existential formula $\exists x \cdot \Psi$, and would be a term representation of a value v such that $\Psi[v/x]$ holds. This is true in intuitionistic logic, but in classical logic not all values can be represented as terms. For example, $\exists x \cdot P(a) \wedge P(b) \Rightarrow P(x)$ is classically valid, and a value v for x would be a if $P(a)$ holds, and b otherwise, i.e. we might choose "if $P(a)$ then a else b" as a witness ... but it is not a first-order term! We've just gained new insight into Herbrand's Theorem: it says that $\exists x \cdot \Psi$ is valid if and only if we can find witnesses for x in the form of finite if-then-else programs that return terms.

Herbrand's Theorem has the following direct consequence:

Theorem 6.19 (Recursive Enumerability) *The set of all first-order valid sentences is recursively enumerable. That is, there is a procedure that enumerates all first-order valid sentences, and only them.*

Proof: Notice that if A is a recursively enumerable set, then the set A^n of n-tuples over A is recursively enumerable: let f be a computable function from $\mathbb{N}$ to A, then enumerate all integers k in ascending order, and for each k enumerate the set of n-tuples of integers of sum k, and for each such n-tuple $(i_1, \ldots, i_n)$, produce $(f(i_1), \ldots, f(i_n))$. Enumerating all tuples in this way is called *dovetailing*.

Similarly, the set A^* of sequences of elements of A is recursively enumerable, and all elements of free algebras over A is recursively enumerable.

Therefore, we can recursively enumerate all couples $(\Phi, \overline{\sigma})$, where Φ is a first-order formula, and $\overline{\sigma}$ is a sequence of substitutions $\sigma_1, \ldots, \sigma_k$. For each of these couples, we herbrandize Φ as an existential formula $\exists x_1 \cdot \ldots \exists x_n \cdot \Psi$, check whether $\Psi\sigma_1 \vee \ldots \Psi\sigma_k$ is a propositional tautology, and output Φ precisely when this is so. This enumerates exactly the set of all first-order theorems, by machine. □

All recursive properties P of formulas (i.e., decidable properties, those that can be checked by a terminating algorithm) are recursively enumerable: enumerate all formulas, and check whether P holds for each. However, there are recursively enumerable properties that are not recursive, and unfortunately first-order validity is indeed not decidable:

Theorem 6.20 (Undecidability) *There is no terminating algorithm that, given a first-order formula Φ as input, returns true whenever Φ is classically valid.*

Proof: Since we have not given any precise definition of *computable* or *recursive* (see (Johnstone, 1992; Davis and Weyuker, 1985)), and since our purpose is not to do so, we won't be able to give any formal proof of it. The usual argument goes as follows: the halting problem (roughly, given the text for a procedure C in some programming language and an input x to this procedure, decide whether this procedure halts on this input) is undecidable; then, encode this problem as a first-order existential formula.

The theory behind this would take up more space than we wish to use, so instead we shall give a reduction from a more manageable undecidable problem, *Post's correspondence problem* (from the logician Emil Post). The undecidability of the latter is deduced from the undecidability of the halting problem through the use of a clever encoding of Turing machines.

Let A be a finite alphabet, containing at least two distinct letters. A^* is the set of (possibly empty) words over A, i.e. finite sequences of letters. Let E be a finite collection of couples (u_i, v_i), $1 \leq i \leq n$, which we customarily write as $u_i \doteq v_i$. Given a finite sequence $i_1, i_2, \ldots, i_k$ of indices between 1 and n, we may form the concatenated word $u_{i_1} u_{i_2} \ldots u_{i_k}$, and also $v_{i_1} v_{i_2} \ldots v_{i_k}$. Post's correspondence problem is: given E, is there a non-empty sequence as above such that $u_{i_1} u_{i_2} \ldots u_{i_k} = v_{i_1} v_{i_2} \ldots v_{i_k}$?

See Figure 6.3 for an example: we have taken four equations (on the left of the Figure), and indicated one solution to Post's problem on the right.

Post's problem is undecidable; it in fact expresses the halting problem for some strange computational paradigm that is powerful enough to simulate every computation of a Turing machine. See (Davis and Weyuker, 1985). We can moreover assume that E does not contain the equation (ϵ, ϵ), where ϵ is the empty word, and that the alphabet contains only two distinct letters a and b.

To encode Post's problem in first-order logic, we create two unary function symbols to represent the letters, and name them a and b again; we also create a constant ϵ representing the empty word, and a binary predicate symbol P representing conversion by Post's rules. We encode words w, say $a_{j_1} a_{j_2} \ldots a_{j_m}$, as terms of the form $a_{j_m}(\ldots a_{j_2}(a_{j_1}(x)) \ldots)$, where $a_{j_1}, \ldots, a_{j_m}$ are letters and x is a variable; let this term be denoted as $w(x)$. We also write w for $w(\epsilon)$. We can then axiomatize Post's problem as follows:

- Axioms for E: $\forall x, y \cdot P(x, y) \Rightarrow P(u_i(x), v_i(x))$, for every $1 \leq i \leq n$;

- Success condition: $P(\epsilon, \epsilon)$.

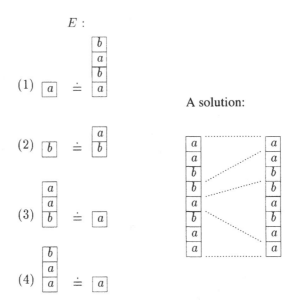

Figure 6.3. An example of Post's Correspondence Problem

Now, we take the conjunction C of these formulas, and let Φ be the formula $C \Rightarrow \exists x \cdot P(a(x), a(x)) \vee P(b(x), b(x))$. Intuitively, C states that $P(w, w')$ holds either when $w = w' = \epsilon$ or when $w = w_1 u_i$ and $w' = w'_1 v_i$ for some i and some w_1, w'_1 such that $P(w_1, w'_1)$. The formula Φ states that we can generate some non-empty word (i.e. some word of the form wa or wb for some possibly empty w) from these rules. Because E does not contain (ϵ, ϵ), this is equivalent to Post's problem.

For instance (see Figure 6.3), we have $P(aababbaa, aababbaa)$ because $P(\epsilon, \epsilon)$, hence $P(aab, a)$ by equation (4), hence $P(aaba, aabab)$ by equation (1), hence $P(aabab, aababba)$ by equation (2), and hence $P(aababbaa, aababbaa)$ by equation (3).

We now prove that Φ is valid if and only if Post's problem has a solution. First, putting Φ in prenex form yields an existential formula. So, if Φ is valid, then Herbrand's Theorem applies, and we can prove Φ by showing that the disjunction of $\neg P(\epsilon, \epsilon)$, of finitely many ground instances of negated axioms:

$$P(w_j, w'_j) \wedge \neg P(u_{i_j}(w_j), v_{i_j}(w'_j))$$

with $1 \leq j \leq m$ and for every j, $1 \leq i_j \leq n$; and of finitely many ground instances of the conclusion:

$$P(a(v_k), a(v_k))$$
$$P(b(v_k), b(v_k))$$

where $1 \leq k \leq p$, is propositionally valid. Equivalently, the set of clauses:

$$P(\epsilon, \epsilon)$$
$$\neg P(w_j, w'_j) \lor P(u_{i_j}(w_j), v_{i_j}(w'_j))$$
$$\neg P(a(v_k), a(v_k))$$
$$\neg P(b(v_k), b(v_k))$$

where $1 \leq j \leq m$ and $1 \leq k \leq p$, is propositionally unsatisfiable. By the completeness of propositional resolution, there is a derivation of the empty clause from the above. Say that (w, w') is a *Post pair* if and only if $w = u_{j_0} u_{j_1} \ldots u_{j_l}$ and $w' = v_{j_0} v_{j_1} \ldots v_{j_l}$ for some $l \geq 0$ and some indexes $j_0, \ldots, j_l$ between 1 and n. An easy induction on the length of the resolution derivation shows that the only non-empty clauses that we can derive are:

- of the form $\neg P(w_{j_0}, w'_{j_0}) \lor P(w(w_{j_0}), w'(w'_{j_0}))$, where (w, w') is a Post pair;

- or of the form $\neg P(w_{j_0}, w'_{j_0})$, where there is a Post pair (w, w') such that $w(w_{j_0}) = w'(w'_{j_0})$ is not the empty word;

- or of the form $P(w(\epsilon), w'(\epsilon))$ with (w, w') a Post pair.

The empty clause can only be derived by resolving on clauses of the last two kinds: this entails that there are two Post pairs (w_1, w'_1) and (w_2, w'_2) such that $w_2(w_1(\epsilon)) = w'_2(w'_1(\epsilon))$, i.e. such that $w_1 w_2 = w'_1 w'_2$, and the latter is not the empty word. So if Φ is valid, then Post's problem has a solution.

Conversely, if Post's problem has a solution, it is clear that Φ is valid.

It follows that the kind of first-order problems that encode Post problems is undecidable, hence that validity first-order logic in general is undecidable. (Of course, this holds also for satisfiability.)   □

This has not deterred researchers from trying to find efficient proof search procedures. Theorem 6.20 means that any sound and complete proof search procedure must fail to terminate on some input (some propositions whose validity is sought); we can still use such a procedure, and stop it when it takes too long, in which case we do not know whether the input proposition was valid or not. (To give a finer analysis, we can think of this failure as an indication that the input proposition, if valid, is all the more difficult to prove as it resists for a longer time.) Alternatively, any sound terminating proof procedure must fail to find proofs for some valid propositions. All these compromises are in fact acceptable, so long as we strive to decide the largest possible class of propositions in reasonable time. Clearly, the enumeration procedure of the proof of Theorem 6.19 is one of the most inefficient we can conceive, and we shall examine other proof search methods in subsequent chapters.

On the logical, or meta-mathematical level, the Herbrand constructions also yield the following results:

Theorem 6.21 (Compactness) *If* $\Gamma \models \Phi$, *then there is a finite subset* Δ *of* Γ *such that* $\Delta \models \Phi$.

Proof: It is enough to show that if Γ has no model, then there is a finite subset Δ of Γ that has no model. We assume that all formulas in Γ have been skolemized, without loss of generality. If Γ has no model, it has no Herbrand model over B_0, which means that the set of closed instances of quantifier-free formulas Ψ where $\forall x_1 \cdots \ldots \forall x_n \cdot \Psi \in \Gamma$ has no model, by Theorem 6.16, item (iii). The latter set is a propositionally unsatisfiable set of propositional formulas: by the Propositional Compactness Theorem, there is a propositionally unsatisfiable finite subset E of these formulas. We can now pick any finite subset Δ of Γ such that whenever Ψ' is in E, it is an instance of some Ψ such that $\forall x_1 \cdots \ldots \forall x_n \cdot \Psi \in \Delta$. $\qquad \square$

It follows:

Theorem 6.22 (Löwenheim-Skolem, Weak Version) *If* Φ *has a model, then it has models of all infinite cardinalities.*

Proof: If Φ has a model, then it has a Herbrand model over $H(B)$ for any B, by Theorem 6.16. If all function symbols in $\mathcal{F}$ have arity 0, and choosing B of cardinal $\alpha \geq \aleph_0$, the cardinality of $H(B)$ is that of $\mathcal{F}$ plus α, i.e. α since the cardinal of $\mathcal{F}$ is less than or equal to $\aleph_0$. Otherwise, if there is a non-nullary function symbol in $\mathcal{F}$, then the cardinal of $H(B)$ is $\alpha.\aleph_0$, that is, α since α was assumed greater than or equal to $\aleph_0$. As α is arbitrary among infinite cardinals, the Theorem is proved. $\square$

But we have cheated: in the above model, we have introduced many values that may actually be provably equal (if we have an equality predicate). We can in fact prove the following rather surprising result:

Theorem 6.23 (Löwenheim-Skolem) *Assume that* $\doteq$ *is a binary predicate symbol. We say that a model is* equational *if and only if* $\doteq$ *is interpreted as equality in this model.*

Then, if Φ *has an infinite equational model, then it has equational models of all infinite cardinalities.*

Proof: Assume that Φ has an infinite equational model. Alternatively, there is an interpretation I on a domain D that satisfies Φ and all axioms of equality $\doteq$, such that the quotient of D by the equivalence relation $I(\doteq)$ is infinite.

Let β be the cardinality of this quotient, and let α be any infinite cardinal such that $\alpha \leq \beta$. Create α fresh constant symbols c_i, $0 \leq i < \alpha$ and extend the interpretation I so that each c_i is mapped to a distinct equivalence class modulo $I(\doteq)$. We can do this, as there are $\beta \geq \alpha$ such equivalence classes. Let S be the set of formulas containing Φ and all the inequations $\neg c_i \doteq c_j$, $i \neq j$ (viewed as a conjunction); S clearly admits I as an equational model. But then, I translates to a Herbrand interpretation I' on $H(B)$, where $B = B_0 \cup \{c_i \mid 0 \leq i < \alpha\}$, so that I' satisfies S. In particular, I' satisfies Φ, and because I' satisfies all the $\neg c_i \doteq c_j$,

$i \neq j$, its quotient by $I'(\doteq)$ has cardinality at least α. Then, B has cardinality α, so $H(B)$ has cardinality at most $\alpha.\aleph_0$, that is, exactly α. In particular, $H(B)/I'(\doteq)$ is an equational model of Φ of cardinality α. This proves the so-called *descending* version of the Theorem, namely that if Φ has an infinite equational model, then it has equational models of all lower infinite cardinalities.

To prove the corresponding *ascending* version, we build an equational model of cardinality $\alpha > \beta$ by first building one of cardinality β^α, then by using the descending version to get down to α. Indeed, because $\beta \geq 2$, we have $\alpha \leq \beta^\alpha$. Define D^α as being the product of α distinct copies of D (its cardinal if β^α), and define an interpretation I^α over D^α by $I^\alpha(f)(v_1, \ldots, v_n) = (I(f)(v_{1i}, \ldots, v_{ni}))_{0 \leq i < \alpha}$ (pointwise application) and letting $I^\alpha(P)(v_1, \ldots, v_n)$ be the conjunction of all Booleans $I(P)(v_{1i}, \ldots, v_{ni})$, $0 \leq i < \alpha$. Notice that $I^\alpha(\doteq)$ is equality over D^α.

For every assignment ρ of variables to D^α, and for every $0 \leq i < \alpha$, let ρ_i be the assignment mapping x to the ith component of $\rho(x)$. By a straightforward structural induction on formulas, for every formula Ψ, we have $\bigwedge_{0 \leq i < \alpha} [\![\Psi]\!] I\rho_i \leq [\![\Psi]\!] I^\alpha \rho \leq \bigvee_{0 \leq i < \alpha} [\![\Psi]\!] I\rho_i$, where $\bigwedge$ denotes distributed conjunction, $\bigvee$ denotes distributed disjunction and $\leq$ is the ordering defined on Booleans by $\bot \leq \top$, $\bot \neq \top$. But for $\Psi = \Phi$, both the conjunction and the disjunction evaluate to $\top$, hence $[\![\Phi]\!] I^\alpha \rho = \top$, i.e. I^α is indeed an (equational) model of Φ.

(A simpler, but less constructive proof of the ascending version is the following: given any cardinal α, add α new constants c_i, $0 \leq i < \alpha$, and consider the set of formulas $S = \{\Phi\} \cup \{\neg c_i \doteq c_j \mid i \neq j\}$. If S had no equational model, by compactness some finite subset of S would be inconsistent; but any finite subset of S can be interpreted by mapping each c_i appearing in the finite subset to pairwise distinct elements of the infinite model I. Therefore, S has an equational model, which is of cardinality at least α, and we use the descending version to get one of cardinality exactly α.)   □

In particular, we cannot hope to characterize the integers or the reals in first-order logic, even with infinitely many axioms (because of compactness, we get back to the finitely-many-axioms case, hence to the one-formula case). This is because, whatever first-order axiomatization we choose, there will be non-countable models of arithmetic. The problem is similar with the real numbers: any first-order axiomatisation of the reals must have countable models, although the reals are not countable. (The theory of real-closed fields, also called the first-order theory of the reals, in particular has as smallest model the countable set of all algebraic numbers.)

The Löwenheim-Skolem Theorem is all the more surprising as Zermelo-Frænkel set theory, or ZF for short, is a first-order theory. So, if it has a model, in particular it has a countable model. But we can develop arithmetic and real analysis in any model of ZF, and the reals are not countable. This paradox is called Löwenheim's Paradox, and seems to imply that ZF is inconsistent. This is not the case, and the error lies in a sloppy use of language. Take a countable model Z of ZF. In it, we can find an object N that represents the set of natural numbers $\mathbb{N}$, and an object R that represents the set of real numbers $\mathbb{R}$. Saying that R is not countable *in Z* means that there is no object f in Z that represents a bijection between N and R. Of course, this bijection exists between the set of objects in Z that belong (in the Z sense) to N and those that belong (in the Z sense) to R. What the Paradox means is that the model Z is too small to be able to contain such a bijection.

In other terms, the small model Z consists of objects that have limited observational power. If we live inside Z, we can deduce that we shall never observe any bijection between N and R. However, if we look at Z from the outside, we may observe such a bijection, but it won't be in Z.

▶ **EXERCISE 6.5**

Let T the theory **PA**$_1$ plus all the axioms $\hat{m} \leq c$, for every $m \in \mathbb{N}$, where c is a new constant and $\hat{m}$ denotes $\underbrace{s(\ldots s(\,0)\ldots)}_{m \text{ times}}$. Show that T is consistent. Conclude that **PA**$_1$ has non-standard *countable* models. Compare with Exercise 6.2.

▶ **EXERCISE 6.6**

The *Bernays-Schönfinkel class* (also named the $\exists^*\forall^*$ class) is the class of formulas of the form $\exists x_1 \cdot \ldots \exists x_n \cdot \forall y_1 \cdot \ldots \forall y_m \cdot \Psi$, where Ψ is a quantifier-free formula *without any function symbols*. Show that deciding the satisfiability of formulas of this class is decidable. (That is, the set of satisfiable $\exists^*\forall^*$ formulas is recursive.)

5.3   Cut-Elimination and Syntactic Herbrand Theory

Everything that we have developed in the last sections can also be developed in a purely syntactic framework, replacing notions of validity and satisfiability by the corresponding proof-theoretic notions of provability and consistency. Doing this is harder than doing it semantically, but we shall gain more insight into what really happens.

The basic tool for this is the Cut Elimination Theorem (Theorem 6.26), that is, the fact that we can transform any proof of **LK** into a proof without Cut. To prove it, we first need a few definitions and lemmas. We shall consider that sequents are sets of formulas, so that contractions, weakenings and exchanges are dealt with implicitly; the reader is invited to check that this does not invalidate our arguments.

Definition 6.24 *In an* **LK** *deduction rule*

$$\frac{\Gamma_i \longrightarrow \Delta_i, 1 \leq i \leq n}{\Gamma \longrightarrow \Delta}$$

the sequents $\Gamma_i \longrightarrow \Delta_i$ *are called the* premises, *and* $\Gamma \longrightarrow \Delta$ *is called the* conclusion *of the rule.*

In a Cut rule

$$\frac{\Gamma \longrightarrow \Phi, \Delta \qquad \Gamma', \Phi \longrightarrow \Delta'}{\Gamma, \Gamma' \longrightarrow \Delta, \Delta'}$$

Φ *is called the* cut *formula.*

In all rules, the active formulas *(See Figure 6.2) in the premises are* Φ, Φ' *in the* $\wedge L$, $\wedge R$, $\vee L$, $\vee R$, $\Rightarrow L$, $\Rightarrow R$ *cases,* Φ *in the* $\neg L$ *and* $\neg R$ *cases,* $\Phi[t/x]$ *in the* $\forall L$ *and* $\exists R$ *cases,* $\Phi[y/x]$ *in the* $\forall R$ *and* $\exists L$ *cases, and* Φ *(the cut formula) in the case of* Cut.

In all rules, the principal formula *in the conclusion is* Φ *in the Ax case,* $\Phi \wedge \Phi'$ *in the* $\wedge L$ *and* $\wedge R$ *cases,* $\Phi \vee \Phi'$ *in the* $\vee L$ *and* $\vee R$ *cases,* $\Phi \Rightarrow \Phi'$ *in the* $\Rightarrow L$ *and* $\Rightarrow R$ *cases,* $\neg\Phi$ *in the* $\neg L$ *and* $\neg R$ *cases,* $\forall x \cdot \Phi$ *in the* $\forall L$ *and* $\forall R$ *cases,* $\exists x \cdot \Phi$ *in the* $\exists L$ *and* $\exists R$ *cases,*

Let Φ *be a formula. The* depth $d(\Phi)$ *of* Φ *is defined by:* $d(A) = 0$ *if* A *is an atom,* $d(\Phi \vee \Phi') = 1 + \max(d(\Phi), d(\Phi'))$ *(and similarly for* $\wedge$, $\neg$, $\Rightarrow$*),* $d(\forall x \cdot \Phi') = 1 + d(\Phi')$ *(and similarly for* $\exists$*).*

Let π *be an* **LK** *proof, which we see as an inverted tree of sequents.*

The depth $d(\pi)$ *of* π *is the depth of the tree representing it, i.e., the depth of an axiom* (Ax) *is 0, and the depth of a proof ending in a rule with premises the conclusions of proofs* π_i, $1 \le i \le n$, *is* $1 + \max(d(\pi_1), \ldots, d(\pi_n))$.

Let's say that a cut is topmost *if it is a cut between two cut-free proofs. The* cut-rank $r(\pi)$ *of* π *is defined as 0 if* π *is cut-free, otherwise as the maximum of* $d(\pi') + d(\Phi)$, *where* π' *ranges over all subproofs of* π *ending in a topmost cut, with cut formula* Φ.

The cut-rank is a measure of the complexity of the proof π. The following Lemma shows that we can always decrease this complexity, at the expense of a possible increase in size of the proof:

Lemma 6.25 (Cut Elimination, One Step) *Let* π *be an* **LK** *proof of non-zero cut-rank* r. *Then there is an* **LK** *proof of cut-rank at most* $r - 1$ *of the same sequent.*

Proof: Let n be the number of cut-free subproofs π' of π ending in a topmost cut on Φ such that $d(\pi') + d(\Phi) = r$, i.e. of maximal rank. By assumption, $n \ne 0$. π' has the form:

$$
\begin{array}{cc}
\pi_1 & \pi_2 \\
\vdots & \vdots \\
\end{array}
$$

$$
\dfrac{\Gamma \longrightarrow \Delta, \Phi \qquad \Gamma', \Phi \longrightarrow \Delta'}{\Gamma, \Gamma' \longrightarrow \Delta, \Delta'} \ \text{Cut}
$$

where π_1 and π_2 are cut-free. Then, the idea is to bubble up the Cut above through π_1 and π_2, so that the Cut above of maximal rank is eliminated, leaving only Cuts of lesser rank. More formally, we show that we can transform the proof π into a proof of the same sequent, with at most $n - 1$ subproofs of rank r and none of higher rank.

To do this, we show that we can transform π' into a proof of rank at most $r - 1$. The proof is by induction on $d(\pi_1) + d(\pi_2)$. There are two cases:

 – if Φ is not the principal formula in the last rule of π_1 (or symmetrically, if Φ is not the principal formula in the last rule of π_2), then it is a rule

R of the form

$$\frac{\Gamma_i \longrightarrow \Delta_i, \Phi \quad (1 \le i \le k)}{\Gamma \longrightarrow \Delta, \Phi}$$

so we transform π' into:

$$
\begin{array}{cc}
\pi'_1 \qquad \pi_2 & \pi'_k \qquad \pi_2 \\
\vdots \qquad \vdots & \vdots \qquad \vdots \\
\cfrac{\cfrac{\Gamma_1 \longrightarrow \Delta_1, \Phi \quad \Gamma', \Phi \longrightarrow \Delta'}{\Gamma_1, \Gamma' \longrightarrow \Delta_1, \Delta'} \text{ Cut} \quad \cdots \quad \cfrac{\Gamma_k \longrightarrow \Delta_k, \Phi \quad \Gamma', \Phi \longrightarrow \Delta'}{\Gamma_k, \Gamma' \longrightarrow \Delta_k, \Delta'} \text{ Cut}}{\Gamma, \Gamma' \longrightarrow \Delta, \Delta'} R
\end{array}
$$

where $\pi'_1, \ldots, \pi'_k$ are the subproofs whose conclusions are $\Gamma_i \longrightarrow \Delta_i, \Phi$, $1 \le i \le k$, and where the last rule is indeed an instance of R, as can be checked for all rules R. (Remember that Φ was not the principal formula.)

By induction hypothesis, (notice that $d(\pi'_i)+d(\pi_2)$ is strictly less than $d(\pi_1)+d(\pi_2)$ for every i) the proofs that end in the k Cuts above can be transformed into proofs of rank at most $r - 1$. This induces a new proof of rank at most $r - 1$.

Notice that if π' was an Ax rule, then $k = 0$ and the Cut simply disappears, leaving only the original Ax instance.

- if Φ is the principal formula in both last rules of π_1 and π_2, then there are again two cases:

  - if π_1 (symmetrically, π_2) is an instance of Ax, then π' has the following form:

$$
\begin{array}{cc}
& \pi_2 \\
& \vdots \\
\cfrac{\cfrac{}{\Gamma, \Phi \longrightarrow \Delta, \Phi} \text{ Ax} \quad \Gamma', \Phi \longrightarrow \Delta'}{\Gamma, \Gamma', \Phi \longrightarrow \Delta, \Delta'} \text{ Cut}
\end{array}
$$

    but we can prove the same end sequent by *weakening* the proof π_2 to a new proof π'_2, i.e. consistently adding Γ on the left and Δ on the right of every sequent appearing in π_2. (This is a straightforward proof by structural induction on π_2.) π'_2 then has no Cut of rank greater than or equal to r: in fact, since π_2 is cut-free, so is π'_2.

  - otherwise, both proofs end in rules whose principal formula are the cut formula Φ, and these rules are not Ax or Cut. Now, if Φ is a conjunction, the end rules must be $\wedge$L and $\wedge$R; if Φ is a disjunction, they must be $\vee$L and $\vee$R; and so on, similarly for negation, implication, universal and existential quantifications.

We deal with the conjunction case; the other propositional cases are similar or simpler. Then π' has the form:

$$
\begin{array}{ccc}
\pi'_1 & \pi'_2 & \pi''_2 \\
\vdots & \vdots & \vdots
\end{array}
$$

$$
\cfrac{\cfrac{\Gamma, \Phi', \Phi'' \longrightarrow \Delta}{\Gamma, \Phi' \wedge \Phi'' \longrightarrow \Delta}\ \wedge L \qquad \cfrac{\Gamma' \longrightarrow \Delta', \Phi' \qquad \Gamma' \longrightarrow \Delta', \Phi''}{\Gamma' \longrightarrow \Delta', \Phi' \wedge \Phi''}\ \wedge R}{\Gamma, \Gamma' \longrightarrow \Delta, \Delta'}\ \text{Cut}
$$

where $\Phi = \Phi' \wedge \Phi''$. We transform it into:

$$
\begin{array}{cc}
\pi'_1 & \pi'_2 \\
\vdots & \vdots
\end{array}
$$

$$
\cfrac{\cfrac{\Gamma, \Phi', \Phi'' \longrightarrow \Delta \qquad \Gamma' \longrightarrow \Delta', \Phi'}{\Gamma, \Gamma', \Phi'' \longrightarrow \Delta, \Delta'}\ \text{Cut} \qquad \begin{array}{c} \pi'_3 \\ \vdots \end{array} \ \ \Gamma' \longrightarrow \Delta', \Phi''}{\Gamma, \Gamma' \longrightarrow \Delta, \Delta'}\ \text{Cut}
$$

(Notice that we might have chosen to cut Φ'' first and then on Φ', instead of the other way around as above; this is largely irrelevant here.) The two cuts above have rank strictly smaller than r. Indeed, the upper Cut has rank $\max(d(\pi'_1) + 1, d(\pi'_2) + 1) + d(\Phi') < d(\pi') + d(\Phi') < d(\pi') + d(\Phi) = r$; and the lower Cut has rank $\max(d(\pi'_1) + 2, d(\pi'_2) + 2, d(\pi'_3) + 1) + d(\Phi'') \le d(\pi') + d(\Phi'') < d(\pi') + d(\Phi) = r$.

We now come to the case of quantifications; we deal with universal quantifications only, as the existential case is similar. π has the form:

$$
\begin{array}{cc}
\pi'_1 & \pi'_2 \\
\vdots & \vdots
\end{array}
$$

$$
\cfrac{\cfrac{\Gamma, \Phi[t/x] \longrightarrow \Delta}{\Gamma, \forall x \cdot \Phi \longrightarrow \Delta}\ \forall L \qquad \cfrac{\Gamma' \longrightarrow \Phi[y/x], \Delta'}{\Gamma' \longrightarrow \forall x \cdot \Phi}\ \forall R}{\Gamma, \Gamma' \longrightarrow \Delta, \Delta'}\ \text{Cut}
$$

where y is not free in Γ', Δ'.

Let π''_2 be the proof obtained by consistently replacing y by t in every sequent of π'_2. By a straightforward structural induction, and because π'_2 is cut-free, π''_2 is indeed a valid proof. (The only difficulty would be that replacing y by t would break the side-conditions on $\forall R$ and $\exists L$ rules inside π'_2, but we check that it cannot be the case when π'_2 contains no Cut.) Then π''_2 proves $\Gamma'[t/x] \longrightarrow \Phi[t/x]$,

$\Delta'[t/x]$, i.e. $\Gamma' \longrightarrow \Phi[t/x], \Delta'$. We therefore get the new proof:

$$
\begin{array}{cc}
\pi'_1 & \pi''_2 \\
\vdots & \vdots \\
\Gamma, \Phi[t/x] \longrightarrow \Delta \qquad & \Gamma' \longrightarrow \Phi[t/x], \Delta'
\end{array}
$$
$$
\frac{\Gamma, \Phi[t/x] \longrightarrow \Delta \qquad \qquad \Gamma' \longrightarrow \Phi[t/x], \Delta'}{\Gamma, \Gamma' \longrightarrow \Delta, \Delta'} \text{Cut}
$$

which is one level shallower, so that the resulting Cut has rank at most $r - 1$.

We then conclude that the Lemma holds by induction on the number n of Cuts of maximal rank between cut-free proofs.   □

Theorem 6.26 (Cut Elimination) *A sequent* $\Gamma \longrightarrow \Delta$ *is provable in* **LK** *if and only if it is provable in* **LK** *without the Cut rule.*

Proof: By induction on the cut-rank of a proof of $\Gamma \longrightarrow \Delta$, and using Lemma 6.25.   □

Observe also that the proof says more: from every proof of $\Gamma \longrightarrow \Delta$, we can *compute* a cut-free proof of the same sequent.

The next properties follow straightforwardly:

Definition 6.27 (Subformula) *Let* Φ *be a formula of first-order logic. The set of subformulas of* Φ *is the smallest set of formulas such that:*

- Φ *is a subformula of* Φ;

- *if* $\Phi' \wedge \Phi''$ *is a subformula of* Φ, *then* Φ' *and* Φ'' *are subformulas of* Φ *(similarly with* $\vee, \neg, \Rightarrow$*).*

- *if* $\forall x \cdot \Phi'$ *is a subformula of* Φ, *then all formulas of the form* $\Phi'[t/x]$, *where t is any term, are subformulas of* Φ. *(Similarly with existential quantifications.)*

The set of subformulas is in general infinite, and is in general much bigger than the set of nodes in the graph representing Φ: this is because any instance of Φ' is considered a subformula of $\forall x \cdot \Phi'$ or of $\exists x \cdot \Phi'$. The rationale behind this is that we can recognize subformulas of a given formula, by pattern-matching.

Corollary 6.28 (Subformula Property) *In a cut-free proof of a sequent* $\Gamma \longrightarrow \Delta$, *all sequents consist only of subformulas of formulas in* Γ *or* Δ.

Proof: By a straightforward structural induction on the proof, and noticing that the set of subformulas of formulas in the premise of any rule other than Cut is included in the set of subformulas of formulas in the conclusion.   □

Corollary 6.29 $\longrightarrow \exists x \cdot \Phi$ *is provable in* **LK** *if and only if there is a finite number* k *of terms* $t_1, \ldots, t_k$ *such that* $\longrightarrow \Phi[t_1/x], \ldots, \Phi[t_k/x]$ *is provable in* **LK**.

Proof: Let π be a cut-free proof of $\longrightarrow x \cdot \Phi$. Let $t_1, \ldots, t_k$ be subterms such that $\Phi[t_i/x]$ appears in π. There can be only finitely many such formulas, as the proof is finite.

We transform the proof π into a proof of $\longrightarrow \Phi[t_1/x], \ldots, \Phi[t_k/x]$ in the following way: for each sequent in the proof, erase any occurrence of $\exists x \cdot \Phi$ from the right-hand side of sequents, and add all formulas $\Phi[t_1/x], \ldots, \Phi[t_k/x]$ to these same right-hand sides. Notice that $\exists x \cdot \Phi$ can only occur on the right-hand sides of sequents. Therefore, our transformation preserves axioms Ax. It is also left invariant by all rules other than Cut and quantifier rules. But the only quantifier rule we may use is $\exists$R, which we simply erase, since the translated premise equals the translated conclusion. We finally get a proof of $\longrightarrow \Phi[t_1/x], \ldots, \Phi[t_k/x]$. $\square$

Corollary 6.30 (Herbrand, Syntactic Version) *Let* Φ *be an existential formula* $\exists x_1 \cdot \ldots \exists x_n \cdot \Psi$, *where* Ψ *is quantifier-free.*

If $\longrightarrow \Phi$ *is provable in* **LK**, *then there is a finite number* k, *and* k *instances* $\Psi\sigma_1,$ $\ldots, \Psi\sigma_k$ *of* Ψ *such that* $\longrightarrow \Psi\sigma_1, \ldots \Psi\sigma_k$ *is provable in* **LK**.

Proof: Direct induction on n, using Corollary 6.29. $\square$

Another direct application of Cut Elimination is the following:

Corollary 6.31 (Consistency) *System* **LK** *is consistent, i.e. there is no proof of the absurd sequent* $\longrightarrow$ *in* **LK**.

Proof: If there was one, there would be a cut-free one. But the set of subformulas of $\longrightarrow$ is empty, so we can write no cut-free proof of it. $\square$

Our goal is however to show that **LK** is sound and complete with respect to the semantics of classical first-order logic. The semantic version of Herbrand's Theorem entails that it is indeed the case for existential formulas, and we also know how to relate general formulas to existential formulas having the same validity status, by herbrandization; it remains to show that we can again relate general formulas to existential formulas so as to preserve their provability status.

This is done by noticing that **LK** rules *permute* in cut-free proofs, as shown by Stephen C. Kleene (1967). For instance, we can prove $A \wedge B \longrightarrow B \wedge A$ either by:

$$\cfrac{\cfrac{\quad}{A, B \longrightarrow B}\text{Ax} \qquad \cfrac{\quad}{A, B \longrightarrow A}\text{Ax}}{\cfrac{A, B \longrightarrow B \wedge A}{A \wedge B \longrightarrow B \wedge A}\wedge\text{L}}\wedge\text{R}$$

or by:

$$\cfrac{\cfrac{}{A, B \longrightarrow B} \text{Ax}}{A \wedge B \longrightarrow B} \wedge \text{L} \qquad \cfrac{\cfrac{}{A, B \longrightarrow A} \text{Ax}}{A \wedge B \longrightarrow A} \wedge \text{L}}{A \wedge B \longrightarrow B \wedge A} \wedge \text{R}$$

where the rules $\wedge$R and $\wedge$L have been permuted. In general, we have the following result:

Theorem 6.32 (Permutability) *For any two non-Cut rules R_1 and R_2, we say that R_1 permutes under R_2, or that the pair R_1/R_2 permutes, if and only if for every proof of the form:*

$$\cfrac{\cfrac{\overset{\pi_i^1}{\vdots}}{\underbrace{\Gamma_i^1 \longrightarrow \Delta_i^1}_{1 \le i \le m}} R_1 \quad \cdots \quad \cfrac{\cfrac{\overset{\pi_i^n}{\vdots}}{\underbrace{\Gamma_i^n \longrightarrow \Delta_i^n}_{1 \le i \le m}}}{\Gamma^n \longrightarrow \Delta^n} R_1}{\Gamma \longrightarrow \Delta} R_2$$

where the principal formula in the conclusion of the R_1 rules is not active in the premise of the R_2 rule, we also have a proof with rules R_1 and R_2 permuted, i.e. of the form:

$$\cfrac{\cfrac{\cfrac{\overset{\pi_1^j}{\vdots}}{\underbrace{\Gamma_1^j \longrightarrow \Delta_1^j}_{1 \le j \le n}}}{\Gamma_1 \longrightarrow \Delta_1} R_2 \quad \cdots \quad \cfrac{\cfrac{\overset{\pi_m^j}{\vdots}}{\underbrace{\Gamma_m^j \longrightarrow \Delta_m^j}_{1 \le j \le n}}}{\Gamma_m \longrightarrow \Delta_m} R_2}{\Gamma \longrightarrow \Delta} R_1$$

Then the only pairs of rules that do not permute in **LK** *are* $\forall L/\forall R,\ \forall L/\exists L,\ \exists R/\forall R.$

Proof: By inspection of all possible pairs of rules. We omit it here, as it is fastidious and is not particularly instructive. The idea, as the $\wedge$R/$\wedge$L example demonstrates, is that if R_1 and R_2 act on different parts of the sequents (in the example, $\wedge$R was used to build $B \wedge A$ on the right, while $\wedge$L was used to build $A \wedge B$ on the left), we can apply them in any order we wish.

It fails on $\forall L/\forall R$, because to prove $\forall x \cdot \Phi \longrightarrow \forall x \cdot \Phi \vee \Phi'$, we have to use $\forall R$ last, otherwise we would be forced to write:

$$\cfrac{\cfrac{\vdots}{\Phi[t/x] \longrightarrow \Phi[y/x] \vee \Phi'[y/x]} \forall R}{\cfrac{\Phi[t/x] \longrightarrow \forall x \cdot \Phi \vee \Phi'}{\forall x \cdot \Phi \longrightarrow \forall x \cdot \Phi \vee \Phi'} \forall L}$$

where we need $y \notin \mathrm{fv}(t)$ to apply $\forall R$, but in general we cannot prove $\Phi[t/x] \longrightarrow$ $\Phi[y/x] \vee \Phi'[y/x]$ unless $t = y$. (Take Φ atomic, and Φ' such that $x \notin \mathrm{fv}(\Phi')$, then it can be derived only by Ax then $\vee R$.) The other impermutabilities are of a similar nature.   □

Therefore, to find a proof of a given sequent $\Gamma \longrightarrow \Delta$, we may choose to expand any formula from Γ or from Δ, unless dependencies among quantifications prohibit this. This leaves us a considerable degree of freedom as to which expansion strategy to choose to look for a proof of a given sequent.

> Things are a bit more complicated in the first-order case than in the propositional case, for two reasons: first, the quantifier rules $\forall L$ and $\exists R$ force us to guess a term t from possibly infinitely many (this will be solved by *unification*, see Chapter 7); second, quantifier rules may yield infinitely long expansions, like (read it from the bottom-up, to follow the process of proof search from the goal):
>
> $$
> \cfrac{
> \cfrac{
> \cfrac{
> \begin{array}{c} \vdots \\ \longrightarrow \Phi[t_1/x], \ldots, \Phi[t_{k-1}/x], \Phi[t_k/x], \exists x \cdot \Phi \end{array}
> }{\longrightarrow \Phi[t_1/x], \ldots, \Phi[t_{k-1}/x], \exists x \cdot \Phi} \; \exists R
> }{
> \begin{array}{c} \vdots \\ \longrightarrow \Phi[t_1/x], \Phi[t_2/x], \exists x \cdot \Phi \end{array}
> }
> }{
> \cfrac{\longrightarrow \Phi[t_1/x], \exists x \cdot \Phi}{\longrightarrow \exists x \cdot \Phi} \; \exists R
> } \; \exists R
> $$
>
> To avoid such infinite expansions, we shall need to change our goals from time to time, i.e. stop trying to expand $\exists x \cdot \Phi$ to give a chance to other formulas in the sequent. In other terms, we shall need *fair* expansion strategies.

We can then use permutabilities to generalise Corollary 6.30 to the case where Ψ is not necessarily quantifier-free:

Theorem 6.33 (Herbrand, Augmented Syntactic Version) *Let* Φ *be of the form* $\exists x_1 \cdot \ldots \exists x_n \cdot \Psi$.

If $\longrightarrow \Phi$ *is provable in* **LK***, then there is a finite number* k, *and* k *instances* $\Psi \sigma_1$, $\ldots$, $\Psi \sigma_k$ *of* Ψ *such that* $\longrightarrow \Psi \sigma_1, \ldots \Psi \sigma_k$ *is provable in* **LK**.

Proof: Take a cut-free proof of $\longrightarrow \Phi$. We can permute $\exists R$ rules that introduce quantifications $\exists x_i$, $1 \leq i \leq n$ with any other rules, pushing the existential rules down the proof to get a proof of the form:

$$
\begin{array}{ccc}
\pi_1 & \cdots & \pi_m \\
\vdots & & \vdots \\
\Gamma_1 \longrightarrow \Delta_1 & \cdots & \Gamma_n \longrightarrow \Delta_m \\
& (\pi) & \\
& \vdots & \\
& \longrightarrow \Phi &
\end{array}
$$

where (π) is the maximal set of $\exists$R rule instances introducing quantifications $\exists x_i$, $1 \leq i \leq n$. The process of pushing these rules downwards must terminate, as each exchange strictly decreases the sum of the depths of $\exists$R instances, where the depth of an instance of a rule is the number of inferences between its conclusion and the conclusion of the whole proof.

By structural induction on (π), we must have $m = 1$ ($\exists$R has only one premise); then, the sequent $\Gamma_1 \longrightarrow \Delta_1$ must be of the form:

$$\longrightarrow \exists x_{i_1} \cdot \ldots \exists x_n \cdot \Psi \sigma_1, \ldots, \exists x_{i_k} \cdot \ldots \exists x_n \cdot \Psi \sigma_k$$

and by maximality of (π), the existential prefixes in the latter formulas must be empty, i.e. the conclusion of the sub-proof π_1 must be $\longrightarrow \Psi \sigma_1, \ldots, \Psi \sigma_k$. $\qquad \square$

Theorem 6.34 (Syntactic Herbrand-Skolem) *A formula Φ is provable in* **LK** *if and only if its herbrandization is provable in* **LK**.

Dually, Φ is consistent in **LK** *if and only if its skolemization is consistent in* **LK**.

Proof: We leave as an exercise to the reader to check that Φ is provable (resp. consistent) if and only if any of its prenex forms is provable (resp. consistent): see Exercise 6.3.

Assume therefore Φ to be in prenex form. If it has at least one universal quantifier, write Φ as $\exists x_1 \cdot \ldots \exists x_n \cdot \forall y \cdot \Psi$. Then $\longrightarrow \Phi$ is provable if and only if there are k substitutions σ_i, $1 \leq i \leq k$, of domain inside $\{x_1, \ldots, x_n\}$, such that $\longrightarrow (\forall y \cdot \Psi)\sigma_1, \ldots, (\forall y \cdot \Psi)\sigma_k$ is provable, by Theorem 6.33.

Because of permutabilities (Theorem 6.32), and as in the proof of Theorem 6.33, we may assume that the latter is provable if and only if it is provable with the $\forall$R rule applies as late as possible, i.e. with a proof of the form:

$$
\begin{array}{c}
\pi \\
\vdots \\
\hline
\dfrac{\longrightarrow \Psi[y_1/y]\sigma_1, \Psi[y_2/y]\sigma_2, \ldots, \Psi[y_k/y]\sigma_k}{\longrightarrow (\forall y \cdot \Psi)\sigma_1, \Psi[y_2/y]\sigma_2, \ldots, \Psi[y_k/y]\sigma_k} \ \forall\text{R} \\
\vdots \\
\dfrac{\longrightarrow (\forall y \cdot \Psi)\sigma_1, \ldots, (\forall y \cdot \Psi)\sigma_{k-1}, \Psi[y_k/y]\sigma_k}{\longrightarrow (\forall y \cdot \Psi)\sigma_1, \ldots, (\forall y \cdot \Psi)\sigma_k} \ \forall\text{R}
\end{array}
$$

where $y_1, \ldots, y_k$ are new, distinct variables. We shall also assume, without loss of generality, that the formulas $(\forall y \cdot \Psi)\sigma_i$, $1 \leq i \leq n$, are pairwise distinct.

On the one hand, if Φ is provable, then we may replace in the proof above y_i by any arbitrary term t_i, provided that the free variables of t_i do not break any side-condition on the instances of $\forall$R or $\exists$L in the proof. (Recall that these side-conditions say that we can introduce $\forall x$ on the right or $\exists x$ on the left if and only if x is not free in the rest of the sequent.) To this purpose, it is enough to check that no universally quantified variable (or no new variables as the y_i's above) occur free in t_i. Choosing $f(x_1, \ldots, x_n)\sigma_i$ for t_i is safe, for each $1 \leq i \leq n$; substituting t_i for y_i above, and adding as many $\exists$R rules as needed at the end, we get a proof of $\exists x_1 \cdot \ldots \exists x_n \cdot \Psi[f(x_1, \ldots, x_n)/y]$.

On the other hand, if the latter is provable, then there is a proof of $\longrightarrow \Psi[f(x_1, \ldots, x_n)/y]\sigma_1, \ldots, \Psi[f(x_1, \ldots, x_n)/y]\sigma_k$. We can also assume without loss of generality that all formulas $\Psi[f(x_1, \ldots, x_n)/y]\sigma_i$, $1 \leq i \leq k$, are pairwise distinct. Because f is assumed to be a new function symbol, the only subterms in these formulas built with f are $f(x_1, \ldots, x_n)\sigma_i$, $1 \leq i \leq n$. Because we can assume the proof cut-free, the subformula property holds. So, as the formulas above are quantifier-free, the only subterms built with f in the whole proof are the above. We now replace $f(x_1, \ldots, x_n)\sigma_i$ by as many new variables y_i, $1 \leq i \leq n$ throughout the proof: notice that we can only do this if, whenever $i \neq j$, $f(x_1, \ldots, x_n)\sigma_i$ and $f(x_1, \ldots, x_n)\sigma_j$ are distinct terms. But we assumed that $\Psi[f(x_1, \ldots, x_n)/y]\sigma_i$ and $\Psi[f(x_1, \ldots, x_n)/y]\sigma_j$ were distinct, hence there is a free variable in $\Psi[f(x_1, \ldots, x_n)/y]$ that is bound to two different terms by σ_i and by σ_j. And all of these free variables are among $x_1, \ldots, x_n$; so $f(x_1, \ldots, x_n)\sigma_i$ and $f(x_1, \ldots, x_n)\sigma_j$ are distinct.

Having replaced $f(x_1, \ldots, x_n)\sigma_i$ by y_i for each i throughout the proof, we get a proof of $\longrightarrow \Psi[y_1/y]\sigma_1, \Psi[y_2/y]\sigma_2, \ldots, \Psi[y_k/y]\sigma_k$. We now use $\forall$R on these k formulas in turn (we can, because the y_i's are pairwise distinct), to get a proof of $\longrightarrow (\forall y \cdot \Psi)\sigma_1, \ldots, (\forall y \cdot \Psi)\sigma_k$, to which we add all necessary instances of $\exists$R to get a proof of Φ.

This construction shows how we can eliminate one universal quantifier in favour of a Herbrand symbol. The result that herbrandization preserves provability then follows by induction on the number of universal quantifications (or Herbrand symbols). The result than skolemization preserves satisfiability comes by duality via negation.   □

We conclude by the completeness theorem for classical first-order logic, which was proved for the first time by Kurt Gödel in 1930, although in a different form:

Theorem 6.35 (Soundness, Completeness) **LK** *is sound and complete for the semantics of classical first-order logic, i.e.* $\models \Phi$ *if and only if* $\vdash^{\text{LK}} \Phi$.

Proof: That it is sound is clear.

To prove completeness, assume that $\models \Phi$. Let $\Phi' = \exists x_1 \cdot \ldots \exists x_n \cdot \Psi$ be a herbrandized form of Φ. Then $\models \Phi'$. By the semantic version of Herbrand's Theorem, there are k closed instances $\Psi\sigma_1, \ldots, \Psi\sigma_k$ of Ψ such that $\models \Psi\sigma_1 \vee \ldots \vee \Psi\sigma_k$. By the Completeness Theorem for **LK**$_0$, $\vdash^{\text{LK}_0} \Psi\sigma_1, \ldots, \Psi\sigma_k$, hence $\vdash^{\text{LK}} \Psi\sigma_1, \ldots, \Psi\sigma_k$ since **LK**$_0$ is a subset of **LK**. By using $\exists$R k times, we infer $\vdash^{\text{LK}} \Phi'$, and by Theorem 6.34, $\vdash^{\text{LK}} \Phi$.   □

A more direct proof of the completeness of a deduction system for classical first-order logic, not using any prenex, herbrandized or skolemized forms, makes use of the so-called *Henkin constants*, introduced by Leon Henkin in 1949. The proof, in our opinion, is at least as complicated, and offers less insight as to how proof search can be conducted practically. (See (Manin, 1977).)

▶**EXERCISE 6.7**

Let **LJ** be the same as System **LK**, but restricted so that the right-hand side of sequents consists of at most one formula, where ∨R is replaced by the following rules:

$$\frac{\Gamma \longrightarrow \Phi}{\Gamma \longrightarrow \Phi \vee \Phi'} \vee R_1 \qquad \frac{\Gamma \longrightarrow \Phi'}{\Gamma \longrightarrow \Phi \vee \Phi'} \vee R_2$$

where ⇒L is replaced by:

$$\frac{\Gamma \longrightarrow \Phi \qquad \Gamma, \Phi' \longrightarrow \Delta}{\Gamma, \Phi \Rightarrow \Phi' \longrightarrow \Delta} \Rightarrow L$$

and ¬L is replaced by:

$$\frac{\Gamma \longrightarrow \Phi}{\Gamma, \neg \Phi \longrightarrow \Delta} \neg L$$

(This describes first-order intuitionistic logic.)

Show that $\exists x_1 \cdot \ldots \exists x_n \cdot \Psi$ is provable if and only if some instance $\Psi\sigma$ of Ψ is provable. (Hint: assume that Cut-elimination holds in **LJ**; the curious reader might want to replay the proof of Theorem 6.26.)

The purpose of this exercise is to show that intuitionistic logic has *constructive existence proofs*.

▶**EXERCISE 6.8**

What permutabilities fail in **LJ**? Can we put every intuitionistic formula in clausal, resp. prenex, resp. herbrandized, resp. skolemized form?

6   DIGRESSIONS

Permutabilities, and syntactic herbrandization and skolemization seem to be overkill for what we are interested in. Indeed, we are mostly interested in truth, i.e. validity of formulas. The semantic notion of herbrandization or skolemization reduces the validity problem to that of the validity of existential formulas, and we could declare ourselves content with Herbrand's Theorem (Theorem 6.30), from which completeness of **LK** restricted to existential formulas follows immediately.

Why then did we bother with re-deriving herbrandization and skolemization by syntactic methods? The answer is: to gain insight in what it really means, and because sometimes we cannot do otherwise.

6.1 Non-classical logics

Consider for example the case with intuitionistic logic or linear logic. These logics, as well as many others, are better understood by their associated deduction systems. Although Kripke models for intuitionistic logic provide a fast way of checking that certain formulas are not intuitionistically valid, the most fruitful semantics for intuitionistic logic is the Brouwer-Heyting-Kolmogorov interpretation, or the Curry-Howard interpretation (extended to the first-order, here): but this semantics is just the proof theory of the logic, recast in functional notation.

The case of linear logic is more striking. Although several semantics for first-order linear logic are known, no concrete such semantics is known to be complete. Most results on linear logic (consistency, notably) could only be deduced from cut elimination and the properties of cut-free sequent proof.

From the automated deduction perspective, a cut-free sequent system also yields directly a tableaux method, and therefore a natural proof-search method. (Although finding *efficient* proof search methods, avoiding redundancies in the search, is far less obvious.)

6.2 Arithmetic

Some other deduction systems are incomplete: looking for a proof of Φ in such systems is not the same as checking whether Φ is valid. We might be tempted to add new deduction rules, then, to get a complete deduction system, but there are cases where this is impossible. This is a consequence of Gödel's Incompleteness Theorem, which we cite at last:

Theorem 6.36 (Gödel) *Let D be a deduction system, such that we can interpret first-order Peano arithmetic $\mathbf{PA}_1$ in it, i.e. there are formulas $Z(x)$ ("x is zero"), $S(x,y)$ ("y is the successor of x"), $P(x,y,z)$ ("z = x+y"), $T(x,y,z)$ ($z = xy$"), $E(x,y)$ ("x equals y"), $L(x,y)$ ("x is less than or equal to y") in the language of D provably satisfying satisfying all axioms of $\mathbf{PA}_1$.*

Then one of the following must hold:

- *D is incomplete, i.e. there is a formula Φ built with the logical connectives, and the quantifiers on the formulas Z, S, P, T, E, L above, such that Φ holds in the standard model $\mathbb{N}$ of arithmetic, but is not provable in D;*

- *or D is inconsistent, i.e. every formula is provable in D;*

- *or D is non-effective, i.e. there is no terminating algorithm for deciding whether a formula is (literally) an axiom of D or whether a deduction rule is a rule of D;*

which leaves a rather narrow choice. It should be clear that we usually prefer D to be incomplete, but effective and (hopefully) consistent. (For a proof of this theorem, see (Johnstone, 1992) or (Manin, 1977).)

We give a proof sketch of this remarkable theorem, using Gödel's original arguments. Assume that D is effective and consistent, then we derive the existence of a sentence whose truth is equivalent to its own unprovability.

The first idea is that $\mathbf{PA}_1$, and therefore D, is strong enough to build not only integers, but also couples, lists and trees of integers. (It suffices to build couples, then the rest follows; Gödel does this by coding the couple (m, n) as $p^m q^n$, where p and q are two distinct primes; observe that this encoding is injective.) Then, we can encode any formula Φ of the language as an integer "Φ" (the double quotes are there to remind the reader that this is an *encoding* of Φ as an integer), and similarly any proof π (as a tree of formulas; we have to check that rules are properly chained up, but we can define this in $\mathbf{PA}_1$). We can in fact build a predicate $Form_D(n)$ that is true exactly when n is the code of some formula, and a predicate $Pr_D(p, n)$, such that $Pr_D(p, n)$ holds if and only if p is the code "π" of some proof π in D, n is the code "Φ" of some formula Φ, and Φ is the conclusion of the proof π. (We say that proof-checking is definable in D. We can do this easily if D is PA_1; in the general case, we need to prove that every recursive function, in particular every effective method for recognizing the axioms can be coded up as an arithmetical formula.) The predicate $\exists p \cdot Pr_D(p, n)$ then just says that n is the code of some provable sentence. (We say that provability in D is definable in D.) And, given a fixed variable x, we can also define a predicate $Sub(m, n, k)$ that holds whenever n is "Φ" for some Φ and m equals "$\Phi[k/x]$". (We say that substitution is definable in D.)

If $\Phi(x)$ is a formula with at most one free variable x, and n is "$\Phi(x)$", then we write $\Phi(x)$ as $\Phi_n(x)$. Gödel's paradoxical formula $C(n)$ states that $\Phi_n(n)$ is not provable: using the concepts above, we may define $C(x)$ as $\neg \exists p, m \cdot Sub(m, x, x) \wedge Pr_D(p, m)$. But clearly $C(x)$ is a formula with at most x as free variable, so, letting n be "$C(x)$", we have $C(x) = \Phi_n(x)$. In particular, $C(n)$ holds if and only if $\Phi_n(n)$ holds. On the other hand, by the very definition of $C(x)$, for any m $C(m)$ holds if and only if $\Phi_m(m)$ is not provable. In particular, if we take $m = n$, $C(n)$ holds if and only if $\Phi_n(n)$ holds, if and only $\Phi_n(n)$ is unprovable.

$C(n)$ is in fact a clever encoding of the liar's paradox, which says "I am lying", where n takes the place of "I" and lying means "you cannot show that I'm telling the truth".

More technically, assume that $\Phi_n(n)$ is provable: then it is true, since D is consistent; then it is unprovable by construction. So $\Phi_n(n)$ is in fact unprovable. Similarly, if $\neg\Phi_n(n)$ was provable, then it would be true, hence $\Phi_n(n)$ would be false and $\Phi_n(n)$ would be provable as well, which contradicts the consistency of D. So $\neg\Phi_n(n)$ is unprovable as well. We call such a sentence F, such that neither F nor $\neg F$ is provable, an undecidable sentence, or better a sentence *independent* of the deduction system D. There is no hope of filling such gaps in a theory D: if F is independent of D, we may just decide to add F or $\neg F$ as a new axiom to D, giving D'. But D' again has an undecidable sentence F', so we consider a richer system D'', and so on. This infinite regress can only stop when we have got a deduction system that is so huge that it is not effective any longer...

Another mind-boggling perspective is that, since $\Phi_n(n)$ is unprovable, it is actually true by construction. Moreover, we have *proved* it. The point is that we cannot prove it *in D*, rather we have used a richer system, which is basically D plus the axiom $\exists n \cdot Form_D(n) \wedge \neg\exists p \cdot Pr_D(p, n)$. The latter axiom says that there is an unprovable formula in D, i.e. that D is consistent, and it is usually written $Con(D)$. We can encode the whole proof above, using the same tricks, so as to get an encoded proof (an integer) of Gödel's incompleteness theorem in the enriched theory $D + Con(D)$. In particular, saying that $\Phi_n(n)$ is true translates to the fact that $D + Con(D)$ proves $\Phi_n(n)$. Since D cannot prove it, it follows that $Con(D)$ cannot be proved in D; and since D is consistent, $D + Con(D)$ is also consistent, so $Con(D)$ is in fact independent of D as well! This is called Gödel's second incompleteness theorem. Observe that it has the following consequence: no effective deduction system D that can encode $\mathbf{PA}_1$ can prove its own consistency. Or, in a more startling manner: every effective deduction system D that can encode PA_1 and which can prove its own consistency is inconsistent...

Because of Gödel's incompleteness theorem, there are deduction systems, like those describing first-order Peano arithmetic or Zermelo-Frænkel set theory (in which we can build the natural integers), where we cannot in fact reason semantically! Some valid sentences indeed have no proof in the corresponding system. We are then forced to reason syntactically, i.e. on the deduction system directly.

Gerhard Gentzen invented sequent systems so as to be able to do just this, and derive consistency proofs for arithmetic, which he managed to do in 1934. The idea is that the theorems of $\mathbf{PA}_1$ are exactly the theorems of the sequent system Z, which is $\mathbf{LK}$ augmented with infinitely many axioms (all closed atomic sequents that are propositionally derivable from instances of axioms of $\mathbf{PA}_1$ except the induction axioms), and the following ω-rule:

$$\frac{(\Gamma \longrightarrow \Delta)[0/x] \quad (\Gamma \longrightarrow \Delta)[s(0)/x] \cdots (\Gamma \longrightarrow \Delta)[\underbrace{s(\cdots s(0)\cdots)}_{n \text{ times}}/x] \cdots}{\Gamma \longrightarrow \Delta}$$

stating that if a sequent is derivable when the integer variable x is replaced by an arbitrary integer, then it is derivable. (This expresses induction in a different form.)

Cut-elimination works for Z exactly as it worked for $\mathbf{LK}$, and we can therefore derive the usual properties of cut-elimination, consistency, and the subformula property. (Compactness fails because the ω-rule is infinitary. Moreover, cut-elimination *fails* for $\mathbf{LK}$ plus the induction axiom: it only works for Z.)

But cut-elimination yields more. The proof of the cut-elimination theorem is a *termination proof*, showing that the process of moving Cuts upwards terminates. Using the Curry-Howard correspondence between proofs and programs, this means that a certain language having first-order formulas as types is normalising. In fact, this language is strongly normalising, i.e. *all* strategies for eliminating cuts terminate. This language is Gödel's system T (Girard *et al.*, 1989), and is basically a typed λ-calculus extended with constants 0 (zero), s (successor of an integer), and recursors R obeying the computation rules:

$$Rfz0 \rightarrow z$$
$$Rfz(sn) \rightarrow fzn(Rfzn)$$

with the appropriate types. More clearly, for every definable function f, Rfz is a function g defined by primitive recursion (on any type, even higher-order) by $g(0) = z$ and $g(n + 1) = f(z, n, g(n))$.

Extensions of this correspondence between deduction systems and programs have led Jean-Yves Girard to prove that $\mathbf{PA}_2$ (second-order Peano arithmetic, with the full induction axiom), and that even $\mathbf{PA}_\omega$ (higher-order Peano arithmetic, with the full induction axiom and full higher-order reasoning) are consistent, by interpreting Cut-elimination as the fundamental reduction rule of a programming language, Church's λ-calculus restricted to those terms typable in Girard's System F (resp. F_ω), and proving that all terms of the language are strongly normalising. System F is basically quantified propositional (intuitionistic) logic, but it captures all peculiarities of $\mathbf{PA}_2$ from the viewpoint of typability.

Because of Cut-elimination, $\mathbf{PA}_1$ also enjoys some form of Herbrand's Theorem. This is Georg Kreisel's *no-counterexample interpretation* for $\mathbf{PA}_1$. Whereas Herbrand's Theorem essentially says that $\exists x \cdot \Psi$ is provable in $\mathbf{LK}$ if and only if we can find witnesses for x in the form of finite if-then-else programs that return terms, Kreisel's result says that $\exists x \cdot \Psi$ is provable in $\mathbf{PA}_1$ (or Z, for that matter) if and only if we can find witnesses for x in the form of finite programs built with primitive recursion of any finite type (i.e., of the kind allowed by Gödel recursors). That is, $\mathbf{PA}_1$ enriches the set of programs we need for witnesses by allowing a form of controlled, terminating recursion called primitive recursion. (See (Shoenfield, 1967), especially Chapter 8. See also (Schwichtenberg, 1977).)

6.3 Higher-order logics

$\mathbf{PA}_2$ and $\mathbf{PA}_\omega$ are axiomatisable in first-order logic, but as rather complicated theories. If we wish to reason about arithmetic, it is better to abandon the realm of first-order logic altogether, either by setting up a new deduction system designed exclusively for arithmetic, or by using an even more expressive framework.

One such framework is *higher-order logic*, where we are able to quantify not only on individuals, but also on predicates, on functions, on predicates of predicates, and so on. Higher-order logic is very expressive, and we can indeed formalise $\mathbf{PA}_\omega$, as well as second-order and higher-order analysis (the theory of the real line $\mathbb{R}$) and most other mathematical theories in it rather easily. The price to pay is that almost all the nice properties we had before fail in higher-order logic: compactness and completeness fail, the subformula property fails also (unless we accept a fairly useless notion of subformula), although cut-elimination still works. Even worse, we cannot even perform herbrandization or skolemization in advance, and proof search is considerably more complicated than in the first-order case, as almost every search subproblem becomes undecidable. (See (Huet, 1973; Huet, 1975) for problems, explanations, examples and details.)

Although we won't study higher-order logics, it is interesting to give a short description of the syntax, the proof rules, and the semantics of higher-order logics, so that problems become more apparent.

Syntax

The easiest formulation of higher-order logics is due to Church, who based it on the simply-typed λ-calculus, and added constants to represent the basic quantifiers. Syntactically indeed, we can use the variable-binding λ operator to represent all binding operations, and consider $\forall x \cdot \Phi$ as an abbreviation for $\forall(\lambda x \cdot \Phi)$, where $\forall$ is a constant that takes a Boolean-valued function f, and returns $\top$ if and only if f is the $\top$ constant function. The need for using a *typed* λ-calculus instead of the simpler untyped λ-calculus stems from the fact that the system without types is inconsistent. Such an untyped system was invented by Gottlob Frege in the 1890's to formalize mathematics; but we can state a self-contradictory formula in it, Russell's paradox, which is (again) the essence of the liar's paradox. Define indeed F as the function:

$$\lambda x \cdot \neg x(x)$$

and consider $F(F)$. By β-reduction, $F(F)$ is just the same as $\neg F(F)$, from which we deduce every formula.

We first define the algebra of simple types. Let the base types be $I\!B$ (the type of formulas) and $\mathbb{T}$ (the type of terms). We might as well refine $\mathbb{T}$ into several different types, and this is basically what sorted logics do: see Chapter 9, Section 4.1 for a glimpse of sorted first-order logics. The *simple types* are the elements of the smallest set containing the base types, and such that if τ and τ' are simple types, then $\tau \rightarrow \tau'$ is a simple type. As usual, we assume that arrows associate to the left, so $\tau \rightarrow \tau' \rightarrow \tau''$ means $\tau \rightarrow (\tau' \rightarrow \tau'')$.

To define expressions in higher-order logics, we fix a *signature* Σ, which is a set of constants of all types. We usually assume that there is at least one constant of each type, as this is not a limitation. Similarly, we assume that there are infinitely many variables x_τ of each type τ. We shall omit type subscripts when they are not necessary. The *syntax* of higher-order expressions e is then defined together with the typing judgments $e : \tau$ as follows:

- x_τ is an expression of type τ (we say $x_\tau : \tau$);

- if c is a constant of type τ, then c is an expression of type τ;

- if $e : \tau \rightarrow \tau'$ and $e' : \tau$, then ee', the *application* of e to e' is an expression of type τ';

- if $e : \tau'$, then $\lambda x_\tau \cdot e$, the *abstraction* of e over x, is an expression of type $\tau \rightarrow \tau'$.

Finally, we assume that Σ contains the following logical constants: $\wedge : I\!B \rightarrow I\!B \rightarrow I\!B$ (conjunction), $\vee : I\!B \rightarrow I\!B \rightarrow I\!B$ (disjunction), $\Rightarrow : I\!B \rightarrow I\!B \rightarrow I\!B$ (implication), $\neg : I\!B \rightarrow I\!B$ (negation), $\forall_\tau : (\tau \rightarrow I\!B) \rightarrow I\!B$ (universal quantification on objects of type τ, for each τ), and $\exists_\tau : (\tau \rightarrow I\!B) \rightarrow I\!B$ (existential quantification on objects of type τ, for each τ). Free variables, substitutions are defined as in first-order logic, with the proviso that bound variables should be renamed to avoid spurious capture of variable names.

Notice that we can encode first-order terms and formulas in this setting. We translate n-ary functions f into constants f of type $\underbrace{\mathbb{T} \rightarrow \ldots \rightarrow \mathbb{T}}_{n \text{ times}} \rightarrow \mathbb{T}$, and n-ary predicate symbols P into constants P of type $\underbrace{\mathbb{T} \rightarrow \ldots \rightarrow \mathbb{T}}_{n \text{ times}} \rightarrow I\!B$.

Therefore, an application term $f(t_1, \ldots, t_n)$ is reencoded as the application $(\ldots(ft_1)t_1 \ldots t_n)$. If we use the convention that application associates to the left, we can also write this as $ft_1 \ldots t_n$.

Standard Semantics

The *standard semantics* is essentially what you can expect from the definition, knowing what the semantics of first-order logic is. An *standard interpretation* is a non-empty set D_I, together with interpretations for all constants that respect their types. More precisely, we define the interpretation of types as follows:

- $[\![\mathbb{T}]\!]I = D_I$, $[\![\mathbb{B}]\!]I = \mathbb{B}$;
- $[\![\tau \to \tau']\!]I$ is the space of all (total) functions from $[\![\tau]\!]I$ to $[\![\tau']\!]I$.

Then, for the interpretation $I(c)$ to respect the type τ of c, we must enforce the constraint: $I(c) \in [\![\tau]\!]I$, i.e. constants of type $\mathbb{T}$ should denote values in D_I, constants of type $\mathbb{T} \to \mathbb{T}$ should denote unary functions from D_I to D_I, and so on.

Naturally, we also enforce the constraint that all the logical constants $\wedge$, $\vee$, $\Rightarrow$, $\neg$, $\forall$, $\exists$ be given their intended meanings, and:

- $[\![ee']\!]I\rho = [\![e]\!]I\rho([\![e']\!]I\rho)$;
- $[\![\lambda x_\tau \cdot e]\!]I\rho$ is the function mapping every $v \in [\![\tau]\!]I$ to $[\![e]\!]I(\rho[v/x])$.

In the standard semantics, we can define arithmetic by all the basic Peano axioms, plus the induction axiom (which can now be written as only one axiom, thanks to higher-order quantification):

$$\forall P_{\mathbb{T} \to \mathbb{B}} \cdot P(0) \wedge (\forall x_{\mathbb{T}} \cdot P(x) \Rightarrow P(s(x))) \Rightarrow \forall x_{\mathbb{T}} \cdot P(x)$$

and we would get *higher-order arithmetic*. We can then prove that $\mathbb{N}$ is a model of higher-order arithmetic, and that it is the only one, up to isomorphism.

The main problem with higher-order arithmetic is that, because of Gödel's theorem, and because we do have a model ($\mathbb{N}$), every sound axiomatic system for it will be incomplete. Moreover, we can encode arithmetic in higher-order logic itself, without needing extra symbols like s or extra axioms like the above, by encoding integers n as Church's integers $\lambda f_{\mathbb{T} \to \mathbb{T}} \cdot \lambda x_{\mathbb{T}} \cdot \underbrace{f(f(\ldots f(x)\ldots)}_{n \text{ times}}$. Again, because of Gödel's theorem, every sound axiomatic system for higher-order logic is incomplete.

Deduction Systems

We now describe how the first-order sequent system **LK** can be extended to higher orders. The corresponding system is consistent, as Girard managed to prove, so it must be incomplete.

We define reduction rules as the usual reduction rules of the λ-calculus, which we have already seen in Chapter 4:

(β) $(\lambda x \cdot e)e' \rightarrow e[e'/x]$

(η) $\lambda x \cdot ex \rightarrow e$ if $x \notin \mathrm{fv}(e)$

This defines two binary relations (denoted as $\rightarrow$ above), which we shall write β and η respectively. Let λ stand for β or for $\beta \cup \eta$. We call $\longrightarrow_\lambda$ the smallest relation containing λ and stable by context application (that is, such that if $e \longrightarrow_\lambda e'$, then $eu \longrightarrow_\lambda eu'$, $ue \longrightarrow_\lambda u'e$, and $\lambda x \cdot e \longrightarrow_\lambda \lambda x \cdot e'$). Call $\longrightarrow_\lambda^*$ the reflexive transitive closure of $\longrightarrow_\lambda$, and $=_\lambda$ its reflexive symmetric transitive closure. Notice that, if $e =_\lambda e'$, then trivially e and e' must have the same standard interpretations.

The simply-typed λ-calculus has the remarkable property that we can decide whether $e =_\lambda e'$ by first normalizing e and e' (reducing them until we cannot any more, getting *normal forms*), and comparing the normal forms syntactically. This is because this calculus is confluent and terminating (see Chapter 4); for more information on the λ-calculus, see (Barendregt, 1984).

We can then extend the **LK** deduction system by just replacing the quantifier deduction rules by the following, typed versions:

$$\frac{\Gamma, \Phi[t/x] \longrightarrow \Delta}{\Gamma, \forall x_\tau \cdot \Phi \longrightarrow \Delta} \forall L \qquad \frac{\Gamma \longrightarrow \Phi[y_\tau/x], \Delta}{\Gamma \longrightarrow \forall x \cdot \Phi, \Delta} \forall R \text{ (y not free in Γ, Δ)}$$

$$\frac{\Gamma, \Phi[y_\tau/x] \longrightarrow \Delta}{\Gamma, \exists x_\tau \cdot \Phi \longrightarrow \Delta} \exists L \text{ (y not free in Γ, Δ)} \qquad \frac{\Gamma \longrightarrow \Phi[t/x], \Delta}{\Gamma \longrightarrow \exists x \cdot \Phi, \Delta} \exists R$$

where t is of type τ, and the following rule:

$$\frac{(\Gamma \longrightarrow \Delta)[u/x] \qquad u =_\lambda v}{(\Gamma \longrightarrow \Delta)[v/x]} (\lambda)$$

meaning that we can always replace any expression by a λ-equivalent one. In practice, we maintain all formulas in normal form with respect to $\longrightarrow_\lambda$, and this takes care of the (λ) rule.

But this only takes care of the (λ) rule if all instances are known in advance. Consider indeed the $\exists R$ rule: to prove $\Gamma \longrightarrow \exists x_\tau \cdot \Phi, \Delta$, we have to guess an expression t of type τ such that $\Gamma \longrightarrow \Phi[t/x], \Delta$ is provable. We shall see that, in the first order case, we don't have to guess this term, because a special *unification procedure* will find it for us. First-order unification means finding a substitution σ such that $t\sigma = t'\sigma$, given two terms t and t'. At higher orders, *higher-order unification* is enough for tableaux systems also; this is the problem of finding substitutions σ mapping variables to expressions of the same type, such that $t\sigma =_\lambda t'\sigma$, given two terms t and t' of the same type. Unfortunately, this problem is undecidable as soon as function or predicate variables are allowed (and provided we have at least two constants, too).

To see that there are many other difficulties with the higher-order logic system above, consider the following example:

$$\neg \exists F_{\pmb{T} \to \pmb{T} \to \pmb{B}} \cdot \forall P_{\pmb{T} \to \pmb{B}} \cdot \exists x_{\pmb{T}} \cdot \forall y_{\pmb{T}} \cdot F x y \Leftrightarrow P y$$

which expresses a weak form of Cantor's Theorem (there is no surjective function F from D to $\mathcal{P}(D)$, where D is represented as type $\pmb{T}$, and the powerset $\mathcal{P}(D)$ is represented as the set of characteristic functions of type $\pmb{T} \to \pmb{B}$; the formula expresses surjectivity by saying that for every subset P of D, there is an x such that $F x = P$).

We might be tempted to skolemize, and try to refute:

$$F(cX)y \Leftrightarrow Xy$$

where $X_{\pmb{T} \to \pmb{B}}$ and $y_{\pmb{T}}$ are the two variables (X stands for P), F is a Skolem constant of type $\pmb{T} \to \pmb{T} \to \pmb{B}$, and c is a unary Skolem function, of type $(\pmb{T} \to \pmb{B}) \to \pmb{T}$. We can then conclude by using the substitution $[\lambda x_{\pmb{T}} \cdot \neg(F x x)/ X, c(\lambda x_{\pmb{T}} \cdot \neg(F x x))/y]$.

But the problem is that, although skolemization is sound with respect to the standard semantics, it is *unsound* for the deduction system. Skolemization indeed depends on the axiom of choice:

$$\forall P_{\tau \to \tau' \to \pmb{B}} \cdot (\forall x_{\tau} \cdot \exists y_{\tau'} \cdot P x y) \Rightarrow (\exists f_{\tau \to \tau'} \cdot \forall x_{\tau} \cdot P x (f x))$$

which is unprovable in this system, although it holds in the standard semantics.

The usual solution is to enrich the deduction system with some rule expressing this axiom of choice. We are then faced with the converse problem: proving a formula by first skolemizing it, then trying to find a proof of the resulting quantifier-free formula is incomplete. In fact, we sometimes have to use the axiom of choice (or skolemization of some quantifiers only) in the middle of the search for a proof. This is due to the fact that substituting higher-order variables may completely change the shape of the formula. Consider:

$$\forall P_{\pmb{B} \to \pmb{B}} \cdot P(\exists x_{\pmb{T}} \cdot \Phi(x))$$

By the substitution $[\lambda y \cdot y/P]$, we can infer:

$$\exists x_{\pmb{T}} \cdot \Phi(x)$$

where the existential quantification appears positively, but by $[\neg/P]$, we get:

$$\neg \exists x_{\pmb{T}} \cdot \Phi(x)$$

where the existential quantification now appears negatively. There is no way to decide in advance how to skolemize the existential.

Substituting P for more complicated Boolean expressions may in fact instantiate the formula $P(\exists x_{\mathbb{T}} \cdot \Phi(x))$ to any arbitrary formula. This is why we say that the subformula property fails for cut-free proofs. (Unless we are ready to admit that any formula is a subformula of formulas of the form above.) Cut-elimination works, but is of little help here for automating proof search.

General Semantics

We have seen that there were at least four different variant of higher-order logics, depending on whether we admit β-equalities or β and η equalities, and depending on whether we admit the axiom of choice or not. These logics really have different sets of theorems, although we cannot see this from the standard semantics.

To reflect more precisely what happens in these various systems, Leon Henkin proposed a more general semantics for higher-order logics in 1950 (see (Andrews, 1986)). The trick is that the denotations of functional types may become too big: for example, if $\mathbb{T}$ is interpreted as $\mathbb{N}$, $\mathbb{T} \to \mathbb{T}$ must be interpreted as the set of all functions from $\mathbb{N}$ to $\mathbb{N}$ in the standard semantics. But only countably many such functions can be described by formulas, and we can think of this as a reasonable cause of incompleteness.

Henkin therefore devised a more abstract notion of interpretation of types, of application and abstraction. A *general interpretation* replaces the non-empty set D_I by a family D of sets D_τ indexed by the types τ, together with an *application operator* $@_{\tau,\tau'}$, or $@$ for short, from $D_{\tau \to \tau'} \times D_\tau$ to $D_{\tau'}$. The couple $(D, @)$ is then called an *applicative structure* if and only if all expressions have interpretations, where interpretation is defined by modifying the notion of standard interpretation:

- $[\![ee']\!]I\rho = @([\![e]\!]I\rho, [\![e']\!]I\rho)$;

- $[\![\lambda x_\tau \cdot e]\!]I\rho$ is some element f in $D_{\tau \to \tau'}$, where $e : \tau'$, such that for each $v \in [\![\tau]\!]I$, $@(f, v) = [\![e]\!]I(\rho[v/x])$.

Being an applicative structure therefore means containing enough elements that can be interpreted as functions from D_τ to $D_{\tau'}$.

We can now define the concept of a *general structure*. A general structure is an applicative structure, together with a *valuation*, that is, a function v from D_B to $\mathbb{B}$ that interprets elements of D_B as Booleans, and such that $v(@([\![\neg]\!]I\rho, x))$ is the negation of $v(x)$, $v(@(@([\![\wedge]\!]I\rho, x), y))$ is the conjunction of $v(x)$ and $v(y)$, and so on. Indeed, applicative structures do not impose that the interpretation of D_B be $\mathbb{B}$.

Notice also that we did not require $f \in D_{\tau \to \tau'}$ to be unique among all elements representing the function $\lambda x_\tau \cdot e$ in the definition of general interpretations. And indeed, there are non-extensional applicative structures, i.e. applicative structures in which $\forall x \cdot fx = gx$ does not imply $f = g$, for some terms f and g where x is not free. In fact, η-equality is the special case where g is $\lambda y \cdot fy$ (by (β), for every x, $fx = (\lambda y \cdot fy)x$, so extensionality implies $f = \lambda y \cdot fy$ in this case), and there are applicative structures where (η) is not valid; in particular, (η) is not provable from β-equality alone.

Notice finally that applicative structures may not be rich enough to satisfy the axiom of choice. They only need to contain functions for representing λ-abstractions, but not for representing choice functions. Hence the axiom of choice is not provable from the higher-order sequent system of the last section.

General structures play much the same rôle for the higher-order sequent system than first-order interpretations play for **LK**. We can then replay some form of Herbrand theory. In particular, a particularly acceptable general structure is that of all equivalence classes of typed λ-terms modulo the congruence relation generated by (β) and (η), where @ is the ordinary application of λ-calculus and $D_{\mathbf{B}}$ is the set of all Church Booleans for instance. This syntactic interpretation is quite similar to Herbrand's interpretation over the universe of ground terms in first-order logic.

It is not very surprising that the higher-order sequent system of last section is sound and complete for the general structure semantics. Of course, we then lose the possibility of axiomatising $\mathbb{N}$ or $\mathbb{R}$ by higher-order logics with general structure semantics, because of Gödel's theorem.

In particular, there are still unprovable valid formulas. A particularly frustrating one is due to Michael Kohlhase (1995):

$$\neg cb \lor c(\neg\neg b)$$

is unprovable in our sequent system, or Kohlhase's first intensional (i.e., non-extensional) tableau system. This is because the only way of proving this is to show that $b = \neg\neg b$. Unfortunately, we can prove $b \Leftrightarrow \neg\neg b$, and although we might think that logical equivalence and equality should be the same on objects of type $\mathbb{B}$, general structures allow for the contrary. (Take $D_{\mathbf{B}}$ to contain more than two elements.)

This can be repaired by considering *general models*, which are general structures in which $D_{\mathbf{B}}$ is required to be exactly $\mathbb{B}$, and v to be the identity function. In the deduction system, we add the following Boolean extensionality axiom (which was therefore unprovable from the others):

$$\forall F_{\mathbf{B}} \cdot \forall G_{\mathbf{B}} \cdot (F \Leftrightarrow G) \Leftrightarrow F \doteq G$$

where $\doteq$ is the definable equality of higher-order logic, namely $F \doteq G$ equals $\forall P \cdot P(F) \Rightarrow P(G)$. (This is known as *Leibniz's equality*, saying that two objects are equal if and only if they have the same properties.) The resulting deduction system is then sound and complete for general models (but of course incomplete again for the standard semantics).

We end here our study of the basic notions and pitfalls of higher-order logic. For more information, we refer the reader to (Andrews, 1986), a nice book on logic and proofs with a large part on a minimal system for higher-order logic (with the η rule, but without the axiom of choice). There are automated proof methods for higher-order logics: Huet's higher-order resolution (Huet, 1973) and Kohlhase's higher-order tableaux (Kohlhase, 1995) are two. All these methods need some form of unification modulo the theory of β (resp. $\beta\eta$) conversion of typed λ-terms (we shall see in Chapter 7, Section 2 what unification is in first-order logic, and in Chapter 9, Section 3.2 what unification modulo a theory means). Huet's higher-order unification algorithm is the reference (Huet, 1975), and is usually efficient in practice. Some form of Herbrand theory can also be extended to higher orders, as shown by Miller (1987).

CHAPTER 7

RESOLUTION

We have already introduced resolution as an automated proof method for classical propositional logic in Chapter 2. Although it was not one of the most efficient methods in this context, it can be optimised in a number of ways. Moreover, it is useful to introduce the basic tools we need to look for proofs in classical first-order logic without bothering with too many complications stemming from purely propositional problems. Finally, resolution is the computational core of logic programming languages, of which Prolog was the first, and which we shall examine in Chapter 10.

1 FUNDAMENTAL IDEAS

The main idea is due to J. Alan Robinson in 1965. It improves on all previous proof procedures, like the one we hinted at in the proof of Theorem 6.19 (page 204). There are two ingredients in Robinson's method: propositional resolution, which we have already examined in Chapter 2, Section 4.2, and unification, which makes it usable when lifting the method to the first-order case.

Consider a first-order formula Φ, which we want to test for validity. As in propositional resolution, we first negate it, and try to derive a contradiction from it. To do, this, we skolemize $\neg\Phi$ (recall that the skolemized form is unsatisfiable if and only if $\neg\Phi$ is), getting a formula of the form $\forall x_1 \cdot \ldots \forall x_n \cdot \Psi$. We can now write Ψ in clausal form, as in the propositional case: Ψ is the conjunction of clauses $C_1, \ldots, C_m$. Then, by Herbrand's construction (see in particular Theorem 6.16, page 202), $\neg\Phi$ is unsatisfiable if and only if the conjunction of finitely many closed instances of the clauses C_i, $1 \leq i \leq m$, is unsatisfiable, i.e. *propositionally unsatisfiable*, since no quantifiers remain.

Then, we can just use propositional resolution, as in Chapter 2, Section 4.2. In particular, Φ is valid if and only if there is a finite set S of closed instances of the clauses C_i, such that we can deduce the empty clause $\longrightarrow$ from S by the Cut rule only.

Robinson's real invention was to design unification as a device to factor away all commonalities between applications of Cut on different closed instances of the same clauses. To understand it, look at Figure 7.1, where we have represented two clauses $C = \Gamma \longrightarrow \Delta, A_1, \ldots, A_m$ and $C' = A'_1, \ldots, A'_n, \Gamma' \longrightarrow \Delta'$, of respective closed instances $\Gamma\sigma \longrightarrow \Delta\sigma, A$ and $A, \Gamma'\sigma' \longrightarrow \Delta\sigma'$, where $A = A_1\sigma =$

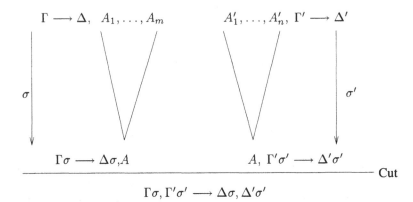

(where $A = A_1\sigma = \ldots = A_m\sigma = A'_1\sigma' = \ldots = A'_n\sigma'$)

Figure 7.1. Lifting

$\ldots = A_m\sigma = A'_1\sigma' = \ldots = A'_n\sigma'$. By the propositional Cut rule, the latter two yield the closed clause $\Gamma\sigma, \Gamma'\sigma' \longrightarrow \Delta\sigma, \Delta'\sigma'$.

Now, assume that $A_1, \ldots, A_m$ are the only atoms on the right-hand side of C such that $A_i\sigma = A$, and that $A'_1, \ldots, A'_n$ are the only atoms on the left-hand side such that $A'_i\sigma' = A$. Any propositional Cut between the closed clauses $C\sigma$ and $C'\sigma$ has the same structure as above, whatever the values of σ and σ', but provided that they all identify exactly $A_1\sigma, \ldots, A_m\sigma$ on the right-hand side of C and $A'_1\sigma', \ldots, A'_n\sigma'$ on the left-hand side of C'. We can then factor all these deductions in one single rule, the resolution rule:

- from the two clauses $C = \Gamma \longrightarrow \Delta, A_1, \ldots, A_m$ and $C' = A'_1, \ldots, A'_n, \Gamma' \longrightarrow \Delta'$, infer the clauses $\Gamma\sigma, \Gamma'\sigma' \longrightarrow \Delta\sigma, \Delta'\sigma'$, for every substitutions σ and σ' that identify precisely $A_1\sigma, \ldots, A_m\sigma$ on the right-hand side of C and $A'_1\sigma', \ldots, A'_n\sigma'$ on the left-hand side of C'.

This does not look better than enumerating all closed instances of clauses and cutting. It is much better in fact, because the unification process, which we shall detail in Section 2, gives a concise representation of the set of all possible substitutions σ and σ', as the set of instances of a single substitution, called the *most general unifier* of $A_1, \ldots, A_m$, and renamed versions of $A'_1, \ldots, A'_n$.

2 UNIFICATION

Unification is the particular case where, given two terms or atoms s and t, we want to find the set of all substitutions σ such that $s\sigma = t\sigma$. (From now on, we shall just consider the case of terms, as atoms have the same structure as terms.)

Definition 7.1 (Unifier) *Let s and t be two terms. A substitution σ is called a unifier of s and t if and only if $s\sigma = t\sigma$.*

This particular case is enough to deal with the problem at hand. We prove it in two lemmas, for which we need the following definition first:

Definition 7.2 (Renaming) *A substitution ρ is called a* renaming *if and only if it maps variables to variables, and is one-to-one.*

Then:

Lemma 7.3 (Unification with Renaming) *Let s and t be two terms, and let ρ be a renaming mapping the free variables of t to variables not free in s.*
  Then the couples (σ, σ') of substitutions such that $s\sigma = t\sigma'$ are exactly the couples $(\sigma, \rho\sigma)$, where σ unifies s and $t\rho$.

Proof: If $s\sigma = t\rho\sigma$, then letting $\sigma' = \rho\sigma$, we trivially have $s\sigma = t\sigma'$. Conversely, if $s\sigma = t\sigma'$, then $s\sigma = (t\rho)(\rho^{-1}\sigma')$. Without loss of generality, we may assume that $\mathrm{dom}\,\sigma \subseteq \mathrm{fv}(s)$, and $\mathrm{dom}\,(\rho^{-1}\sigma') \subseteq \mathrm{fv}(t\rho)$. By definition of ρ, $\mathrm{fv}(s) \cap \mathrm{fv}(t\rho) = \varnothing$, so σ and $\rho^{-1}\sigma'$ have distinct domains: letting σ'' be the disjoint union of σ and $\rho^{-1}\sigma'$, we therefore have $s\sigma'' = t\rho\sigma''$. □

Lemma 7.4 (Multiple Unification) *Let $t_1, \ldots, t_n$ be n terms, $n \geq 2$. Then there is a substitution σ such that $t_1\sigma = \ldots = t_n\sigma$ if and only if there is σ_1 such that $t_1\sigma_1 = t_2\sigma_1$, and σ_2 such that $t_2\sigma_1\sigma_2 = t_3\sigma_1\sigma_2$, and $\ldots$, and σ_{n-1} such that $t_{n-1}\sigma_1 \ldots \sigma_{n-1} = t_n\sigma_1 \ldots \sigma_{n-1}$.*

Proof: If $t_1\sigma = \ldots = t_n\sigma$, let σ' be the substitution mapping all variables in yield σ to a variable outside $\mathrm{dom}\,\sigma$. Let θ be $\sigma\sigma'$: then $t_1\theta = \ldots = t_n\theta$, and moreover $\theta\theta = \theta$ by construction of σ'. We therefore let $\sigma_1 = \ldots = \sigma_{n-1} = \theta$.
  Conversely, if we have $\sigma_1, \ldots, \sigma_{n-1}$ as in the lemma, then we let σ be $\sigma_1\sigma_2 \ldots \sigma_{n-1}$, so clearly $t_1\sigma = \ldots = t_n\sigma$. □

To find two substitutions σ and σ' such that $A_1\sigma = \ldots = A_m\sigma = A_1'\sigma' = \ldots = A_n'\sigma'$, it suffices to rename the A_i''s, $1 \leq i \leq n$, into $A_i'\rho$, where ρ maps the variables in $\mathrm{fv}(A_1') \cup \ldots \cup \mathrm{fv}(A_n')$ to $\mathcal{V} \setminus (\mathrm{fv}(A_1) \cup \ldots \cup \mathrm{fv}(A_m))$. Then, by Lemma 7.3, it remains to unify $A_1, \ldots, A_m, A_1'\rho, \ldots, A_n'\rho$. By Lemma 7.4, we can do this by unifying them pairwise, composing substitutions along the way.
  We now turn to solve the unification problem:

Given s and t, how can we compute the set of unifiers of s and t?

As example of unification problems, consider two constants a and b. If $a = b$, every substitution is a unifier of a and b, i.e. for every σ, $a\sigma = b\sigma$. On the other hand, if a and b are distinct constants, then they have no unifier. In fact, two application terms $f(s_1, \ldots, s_m)$ and $g(t_1, \ldots, t_n)$ are unifiable only if $f = g$. The condition is not sufficient, as for example, $f(a)$ is not unifiable with $f(b)$ when $a \neq b$.

If x is a variable, x is unifiable with, say, $f(a)$, and the unifiers are all substitutions mapping x to $f(a)$. But x is not unifiable with, for example, $f(x)$: indeed, any unifier of x and $f(x)$ should map x to a term t such that $t = f(t)$, but the depth of the term $f(t)$ is one plus that of t, so there is no such term.

The fundamental result about unification of first-order terms is the following:

Theorem 7.5 (Unification) *Let s and t be two terms. Then, either there is no unifier of s and t, and we say that s and t are not unifiable; or there exists a* most *general unifier (for short,* mgu*), i.e. a substitution σ such that:*

- σ *is a unifier of s and t, that is, $s\sigma = t\sigma$;*

- *and every unifier σ' of s and t is an instance of σ, i.e. there is a substitution σ'' such that $\sigma' = \sigma\sigma''$.*

Or shortly, using the notations of Chapter 6, Section 1.1, σ is a mgu if and only if it is a unifier that is more general than every other unifier σ' (i.e., $\sigma \succeq \sigma'$). **Proof:** We give a constructive proof of it, which can be understood as a naïve program to compute an mgu or fail if there is no unifier. The proof is by induction on the sum of the sizes of the terms s and t (as number of nodes in the directed acyclic graph representation of terms):

- (first, a useful special case) if $s = t$, then the empty substitution $[]$ is clearly an mgu of s and t;

- otherwise, if s and t are both variables, we claim that $[t/s]$ is an mgu of s and t; indeed, the unifiers of s and t are those substitutions that map s and t to the same term, and these are exactly the instances of $[t/s]$.

- otherwise, if exactly one of s and t is a variable, we may assume without loss of generality that s is a variable, and t is not. We then have two cases.

  If $s \in \mathrm{fv}(t)$ (we say that the *occurs-check fails*), then s and t are not unifiable, since any unifier should map s to some term u of depth n, such that $u = t[u/s]$, where $t[u/s]$ has depth at least $1 + n$, and this is impossible: in this case, s and t are not unifiable (the unification algorithm fails).

  Otherwise, if $s \notin \mathrm{fv}(t)$, then every unifier σ must map s to a term equal to $t\sigma$, and so must be an instance of $[t/s]$. Moreover, $[t/s]$ is indeed a unifier of s and t, since $t[t/s] = t$ by $s \notin \mathrm{fv}(t)$, and clearly $s[t/s] = t$.

- otherwise, s and t are both applications: $s = f(s_1, \ldots, s_m)$, $t = g(t_1, \ldots, t_n)$. If $f \neq g$, then s and t are not unifiable. Otherwise, we have $f = g$ and $n = m$, and we claim that the set of simultaneous unifiers of s_1 with t_1, $\ldots$, s_n with t_n is either empty or has an mgu by induction on n.

  If $n = 0$, the claim is trivially proved, the mgu being $[]$.

  Otherwise, every simultaneous unifier of the s_i's with the t_i's must unify s_n with t_n. If s_n and t_n are not unifiable, then there is no simultaneous unifier. Otherwise, let σ be the mgu of s_n and t_n. Because σ is most general, the set of simultaneous unifiers of the s_i's with the t_i's, $1 \leq i \leq n$, is the set of $\sigma\sigma'$'s, where σ' ranges over all unifiers of the $s_i\sigma$'s with the $t_i\sigma$'s, $1 \leq i \leq n - 1$.

Notice that, in each case, the sum of the sizes of the terms to unify is strictly less than the sum of the sizes of s and t, even though we may apply arbitrary unifiers in the last case: this is because unifiers always map variables to already existing subgraphs of the previous terms (informally, each new variable binding collapses one variable node with one subterm node). □

The mgu that the procedure above finds has the property that it is idempotent:

Definition 7.6 (Idempotence) *A substitution σ is said to be* idempotent *if and only if $\sigma\sigma = \sigma$.*

A more operational characterisation of idempotent substitutions is the following:

Lemma 7.7 *A substitution σ is idempotent if and only if* $\mathrm{dom}\ \sigma \cap \mathrm{yield}\ \sigma = \varnothing$.

Proof: If $\mathrm{dom}\ \sigma \cap \mathrm{yield}\ \sigma = \varnothing$, then for every term t in the range of σ, $t\sigma = t$ since no variable free in t is in $\mathrm{dom}\ \sigma$.

Conversely, if $\sigma\sigma = \sigma$, then for every $t \in \mathrm{rng}\ \sigma$, $t\sigma = t$, which can only happen if $\mathrm{fv}(t) \cap \mathrm{dom}\ \sigma = \varnothing$. Then, $\mathrm{yield}\ \sigma \cap \mathrm{dom}\ \sigma = (\bigcup_{t \in \mathrm{rng}\ \sigma} \mathrm{fv}(t)) \cap \mathrm{dom}\ \sigma = \bigcup_{t \in \mathrm{rng}\ \sigma}(\mathrm{fv}(t) \cap \mathrm{dom}\ \sigma) = \varnothing$. □

From the characterisation, it should be clear that the procedure used in the proof of Theorem 7.5 returns only idempotent mgus, if it does not fail. There may be non-idempotent mgus: for example, $[z/x, z/y, x/z]$ is a non-idempotent mgu of the variables x and y.

In particular, the mgu may not be unique, as for example $[x/y]$ and $[y/x]$ are both mgus of the two variables x and y. But some form of uniqueness can be recovered, by using renamings:

Lemma 7.8 (Uniqueness) *Let s and t be two terms. Then the mgus of s and t are unique modulo renaming, that is, for any two mgus σ and σ' of s and t, there is a renaming such that $\sigma = \sigma'\rho$ (and $\sigma' = \sigma\rho^{-1}$).*

We then write $mgu(s, t)$ the unique (modulo renaming) mgu of s and t.

Proof: σ and σ' are instances of each other, i.e. there are substitutions θ and θ' such that $\sigma = \sigma'\theta'$ and $\sigma' = \sigma\theta$. Let $R(\sigma)$ be $\{x \mid x \in \mathrm{fv}(y\sigma), y \in \mathcal{V}\}$, i.e. yield $\sigma \cup (\mathcal{V} \setminus \mathrm{dom}\,\sigma)$, and similarly for σ'. Without loss of generality, we may assume $\mathrm{dom}\,\theta \subseteq R(\sigma)$, $\mathrm{dom}\,\theta' \subseteq R(\sigma')$.

Then, $\sigma = \sigma\theta\theta'$. Therefore, $\theta\theta'$ is the identity on $R(\sigma)$, so in particular θ maps the variables of $R(\sigma)$ to variables. And, if $x \in \mathrm{dom}\,\theta$, then $x\theta \neq x$, so $x\theta \in \mathrm{dom}\,\theta' \subseteq R(\sigma')$. So θ maps the variables of $\mathrm{dom}\,\theta \subseteq R(\sigma)$ to variables of $\mathrm{dom}\,\theta'$. Symmetrically, θ' maps the variables of $\mathrm{dom}\,\theta'$ to variables of $\mathrm{dom}\,\theta$. Recall then that $\theta\theta'$ is the identity on $\mathrm{dom}\,\theta$ (since $\mathrm{dom}\,\theta \subseteq R(\sigma)$), and symmetrically that $\theta'\theta$ on $\mathrm{dom}\,\theta'$. Therefore θ and θ' are mutually inverse renamings. $\square$

The procedure in the proof of Theorem 7.5 yields a variant of Robinson's unification algorithm. It is potentially rather inefficient: when unifying $f(s_1, s_2)$ and $f(t_1, t_2)$, it computes the mgu σ of s_1 and t_1, and then explicitly applies σ to s_1 and s_2 before continuing to look for unifiers of $s_1\sigma$ and $t_1\sigma$. In fact, there are couples of terms on which Robinson's unification algorithm runs in exponential time.

Indeed, consider the terms $f(x_1, x_2, \ldots, x_n)$ of size $n + 1$ and $f(g(x_2, x_2), g(x_3, x_3), \ldots, g(x_{n+1}, x_{n+1}))$ (of size $2n + 1$ as a graph, of size $3n + 1$ as a tree). Then, the algorithm unifies x_n with $g(x_{n+1}, x_{n+1})$, yielding $\sigma_n = [g(x_{n+1}, x_{n+1})/x_n]$. It then applies σ_n to the remaining terms, and tries to unify x_{n-1} with $g(g(x_{n+1}, x_{n+1}), g(x_{n+1}, x_{n+1}))$, which succeeds. At the $n - i$th step, it tries to unify x_i with the term t_i defined by $t_i = g(t_{i+1}, t_{i+1})$ and $t_{n+1} = x_{n+1}$. Notice that the size of t_i is $2^{n+2-i} - 1$ as a tree (although it has only $n + 2 - i$ different subterms), so that a simple recursive descent along these terms to apply the substitutions or to do the occurs-check test takes exponential time.

Memoizing the couples of terms to unify helps, as the example above shows. The problem, then, is that most of what we have memoized becomes useless as soon as we apply the substitutions we have found, because applying substitutions in general creates new terms. The trick is to avoid applying these substitutions:

Definition 7.9 (Triangular Form) *A triangular form is a finite sequence $[t_1/x_1; t_2/x_2; \ldots; t_n/x_n]$ of bindings t_i/x_i. We then say that the triangular form represents* the substitution σ *if and only if* $\sigma = [t_{p(1)}/x_{p(1)}][t_{p(2)}/x_{p(2)}] \ldots [t_{p(n)}/x_{p(n)}]$ *for some permutation p of $\{1, 2, \ldots, n\}$.*

For instance, a triangular form for the substitution in the example above would be $[g(x_2, x_2)/x_1; g(x_3, x_3)/x_2; \ldots; g(x_{n+1}, x_{n+1})/x_n]$, which has linear size.

Another improvement we can bring to Robinson's unification procedure is to change strategies: when unifying $f(s_1, \ldots, s_n)$ and $f(t_1, \ldots, t_n)$, instead of first unifying s_n with t_n, we may unify pairs of terms (s_i, t_i) in any order we wish.

The resulting unification algorithm can then be described by the rules of Figure 7.2, in the style of Martelli and Montanari (1982). The algorithm works by transforming couples (F, σ), where F is a set, or a multiset of pairs of terms (represented as unoriented formal equations between terms) that are to be unified simul-

(Delete)	$(F \cup \{t \doteq t'\}, \sigma) \quad \rightarrow \quad (F, \sigma)$ if $t = t'$
(Decomp)	$(F \cup \{t \doteq t'\}, \sigma) \quad \rightarrow \quad (F \cup \{t_1 \doteq t'_1, \ldots, t_m \doteq t'_m\}, \sigma)$ if $t \neq t'$ and $t = f(t_1, \ldots, t_m), t' = f'(t'_1, \ldots, t'_{m'})$, with $f = f'$ and $m = m'$
(Bind)	$(F \cup \{x \doteq t\}, [t_1/x_1; \ldots; t_n/x_n])$ $\rightarrow (F, [t/x; t_1/x_1; \ldots; t_n/x_n])$ if $x \neq t$ and $x \notin \{x_1, \ldots, x_n\} \cup \mathrm{fv}(t[t_1/x_1] \ldots [t_n/x_n])$
(Merge)	$(F \cup \{x_i \doteq t\}, [t_1/x_1; \ldots; t_n/x_n])$ $\rightarrow (F \cup \{t_i \doteq t\}, [t_1/x_1; \ldots; t_n/x_n])$ if $x_i \neq t, 1 \leq i \leq n$

Figure 7.2. Rules for unification

taneously, and σ is the triangular form of the mgu under construction. A *multiset* is a map from a set A to $\mathbb{N}$, that is, it is a "set where multiplicities of elements are recorded". Multiset union is then the pointwise sum of functions, i.e. we put together elements by adding the number of times they occur in each multiset. Computationally speaking, multisets are implemented as lists, where we consider the order of elements as unimportant.

Initially, to unify s and t, we launch the algorithm on $(\{s \doteq t\}, [])$. Then, we choose a rule to apply according to a given strategy until no rule applies. The strategy does the following: if F is empty, then σ is the desired mgu; otherwise, we choose an equation $t \doteq t'$ to solve from F: if t equals t' (rule **(Delete)**), it is already solved, and we just remove it from F; if both t and t' are applications of the same function symbol, then only rule **(Decomp)** applies, and we replace $t \doteq t'$ by the indicated equations; if t or t' is a variable, say t, either t is yet unbound, i.e. it is not in the domain of σ, and we must apply rule **(Bind)** to bind t to t'; or it is bound, and we must replace the equation $t \doteq t'$ by a new, substituted equation by rule **(Merge)**. If no rule applies and F is non-empty, then unification fails.

This presentation of the algorithm clearly decouples the unification task from its *control*, i.e. the strategy we use. This is beneficial, as we are not tied to traversing the terms to unify in Robinson's way, and we can use more efficient traversals. One possible strategy is the following: use **(Delete)** whenever possible; otherwise, use **(Decomp)** if possible; otherwise, use **(Merge)**; finally, use **(Bind)** if this is our only choice. (This is because testing the occurs-check is possibly costly.)

To represent F, we simply use a list of values, whose order is unimportant (hence the abstraction of F as a multiset). To make the algorithm efficient, we can maintain F as the following compound data-structure: first, a list F_a of equations between non-variable terms, i.e. on which we can apply only rules **(Delete)** and **(Decomp)**.

Then, a sorted list F_v of equations between variables and terms where duplicate equations have been removed, so as to have the least possible work to do, binding variables. To achieve this, we can for instance design a canonical representation of equations: choose a total ordering $<$ on variables, extend it to an ordering on terms such that $x < t$ whenever x is a variable and t is an application term, then normalise equations $s \doteq t$ between variables and terms as couples (s, t), such that the first component is the least one ($s < t$; $s = t$ is impossible, as we have chosen to apply (**Delete**) as soon as we can.) Then, represent F_v as the list of all these couples, sorted lexicographically, for instance, and with duplicates removed.

Rule (**Bind**) then eliminates one equation from the list F_v, whatever the number of occurrences in which this equation appeared in the original terms to unify. Rule (**Merge**) instantiates a variable side of an equation in F_v, and moves it to F_a, provided the instantiated equation becomes an equation between non-variable terms.

There still remains one possible source of inefficiency in the algorithm: the occurs-check test $x \notin \{x_1, \ldots, x_n\} \cup \mathrm{fv}(t[t_1/x_1] \ldots [t_n/x_n])$ of the (**Bind**) rule. One way of implementing it is the following. Represent the triangular form σ as a sorted list of equations in normal form, i.e. exactly like F_v above. Testing whether a variable x is bound by σ can be done by dichotomy search in the list, in $O(\log n)$ time, where n is the number of bound variables. And testing whether $x \in \mathrm{fv}(t[t_1/x_1] \ldots [t_n/x_n])$ can be done by computing $occ(x, t)$, where $occ(x, t)$ is defined as follows: if $t = x$, then return "yes" (the occurs-check fails); if t is one of the variables x_j, $1 \le j \le n$, then return $occ(x, t_j)$; if t is an application $f(t_1, \ldots, t_n)$, then return "yes" if some $occ(x, t_k, i)$ returns "yes", $1 \le k \le n$, otherwise return "no".

The resulting algorithm always takes polynomial time in the sizes of the original terms to unify ($O(n^2 \log n)$, where n is the maximum size of the input terms). We can do better by using and maintaining an *occurs-check ordering* to test for the occurs-check incrementally, and get almost linear-time unification algorithms (see Exercise 7.1), or even (destructive) linear-time algorithms. In practice, however, these algorithms are not more efficient than the above one. See (Jouannaud and Kirchner, 1991) for a survey on unification, and more.

▶ **EXERCISE 7.1**

Let σ be an idempotent substitution, $[t_1/x_1; \ldots; t_n/x_n]$ a triangular form for σ. Define the relation $\prec$ between variables as follows: $x_i \prec x$ if and only if $x \in \mathrm{fv}(t_i)$, for each $1 \le i \le n$. Let $\prec^+$ be the transitive closure of $\prec$ (notice that $x \prec^+ y$ means "x definitely depends on y in the triangular form"). Show that $\prec^+$ is a strict ordering, i.e. that it is irreflexive and transitive. (This is called the *occurs-check ordering* of the triangular form.)

◈   ▶ **EXERCISE 7.2**

Let σ be an idempotent substitution, and $[t_1/x_1; \ldots; t_n/x_n]$ a triangular form for σ, sorted according to the occurs-check ordering of Exercise 7.1. (I.e., if

$x_i \prec^+ x_j$, then $i < j$.) Show that $\sigma = [t_1/x_1]\ldots[t_n/x_n]$. How can we get back an idempotent substitution efficiently from one of its triangular forms?

▶ **EXERCISE 7.3**

Let σ and σ' be two substitutions. A *unifier* of σ and σ' is a substitution σ'' such that $\sigma\sigma'' = \sigma'\sigma''$. Show that either σ and σ' have no unifier, or they have a most general unifier. (This extends Theorem 7.5 to the case of substitutions.)

3   RESOLUTION

Having examined unification, we now return to the resolution rule. Recall that we stated it as:

- from the two clauses $C = \Gamma \longrightarrow \Delta, A_1, \ldots, A_m$ and $C' = A'_1, \ldots, A'_n$, $\Gamma' \longrightarrow \Delta'$, infer the clauses $\Gamma\sigma, \Gamma'\sigma' \longrightarrow \Delta\sigma, \Delta'\sigma'$, for every substitutions σ and σ' that identify precisely $A_1\sigma, \ldots, A_m\sigma$ on the right-hand side of C and $A'_1\sigma', \ldots, A'_n\sigma'$ on the left-hand side of C'.

Because of the uniqueness of the mgu, we know that all result clauses $\Gamma\sigma$, $\Gamma'\sigma' \longrightarrow \Delta\sigma, \Delta'\sigma'$ are actually the instances of a unique clause, namely $\Gamma\sigma$, $\Gamma'\rho\sigma \longrightarrow \Delta\sigma, \Delta'\rho\sigma$, where ρ renames the free variables of the second clause to variables not free in the first clause, and σ is the mgu of $A_1, \ldots, A_m, A'_1\rho, \ldots, A'_n\rho$.

This leads us to the resolution principle, as it is really used in resolution theorem proving:

Definition 7.10 (Resolution) *The* binary resolution rule *is the following:*

$$\frac{\Gamma \longrightarrow \Delta, A \qquad A', \Gamma' \longrightarrow \Delta'}{\Gamma\sigma, \Gamma'\sigma \longrightarrow \Delta\sigma, \Delta'\sigma} \; \sigma = mgu(A, A')$$

where we assume the two clauses $\Gamma \longrightarrow \Delta, A$ *and* $A', \Gamma' \longrightarrow \Delta'$ *to be renamed so as to have no free variables in common. The conclusion of the rule is called the* binary resolvent *of the two clauses, the premises are called the* parent clauses, *and* A *(resp. $\neg A'$) is the literal* resolved upon *in the first clause (resp. second clause).*

We say that the literals A and $\neg A'$ are complementary *whenever A unifies with* A'.

The factoring rule *is the following twin rule:*

$$\frac{\Gamma \longrightarrow \Delta, A, B}{\Gamma\sigma \longrightarrow \Delta\sigma, A\sigma} \; \sigma = mgu(A, B) \qquad \frac{A, B, \Gamma \longrightarrow \Delta}{A\sigma, \Gamma\sigma \longrightarrow \Delta\sigma} \; \sigma = mgu(A, B)$$

The conclusion of the rule is called a binary factor *of the clause in premise.*

A factor *of a clause is either the clause itself, or a binary factor of one of its factors. A* proper factor *of a clause is a factor of one of its binary factors.*

A resolvent *of two clauses is a binary resolvent of factors of each clause, not necessarily proper.*

A resolution derivation *is an inverted tree, whose nodes are decorated with clauses, leaves are decorated with clauses in the initial set of clauses, and inner nodes have their parent clauses as predecessors.*

A resolution refutation *is a derivation ending in the empty clause, i.e. a derivation whose root node is decorated with* $\longrightarrow$.

Notice how the use of a separate factoring rule encodes multiple unifications in resolution. We have the following Theorem:

Theorem 7.11 (Soundness, Completeness) *Let S_0 be a set of clauses, and S_n the sequence of sets of clauses defined by letting S_{n+1} be the set of all resolvents of clauses in S_n (S_n is called the* level set *at level n).*

Then resolution is a sound *and* complete *refutation search procedure, i.e. the set S_0, viewed as a universally quantified conjunction of disjunctive clauses, is unsatisfiable if and only if there some level set S_n, $n \in \mathbb{N}$, contains the empty clause.*

Proof: Soundness, i.e. the fact that $\longrightarrow \in S_n$ for some n implies that S_0 is unsatisfiable, comes as in the propositional case from the fact that any conjunction of two clauses (seen here as universally quantified) implies their resolvents. Indeed, every clause, seen as a formula, implies any of its instances; and resolvents are obtained by taking instances and applying Cut, which we have already seen to be sound (see Theorem 2.39, page 49).

Completeness is the fact that if S_0 is unsatisfiable, then some level set contains the empty clause. But, by the observations of Section 1, if S_0 is unsatisfiable, then the set of closed instances of clauses in $\bigcup_{n \in \mathbb{N}} S_n$ contains the empty clause. (We derived this from the completeness of propositional resolution and Herbrand's Theorem.) Now, the only clause which has the empty clause as instance is the empty clause itself. So the empty clause is in $\bigcup_{n \in \mathbb{N}} S_n$, i.e. in some S_n. □

We might think that factoring is useless, and that factors would be generated as needed by the unifiers produced by the binary resolution rule. This would indeed simplify the implementation of resolution. Unfortunately, factors may be indispensable. For example, consider the set of two clauses:

$$\longrightarrow P(x, a), P(a, x)$$
$$P(y, a), P(a, y) \longrightarrow$$

This is unsatisfiable, because $\longrightarrow P(a, a)$ is a factor of the first clause, $P(a, a) \longrightarrow$ is a factor of the second clause, and they resolve to the empty clause. However, the only clauses we can get by binary resolution alone are the ones above,

$P(a, x) \longrightarrow P(a, x)$, $P(y, a) \longrightarrow P(y, a)$ and $P(a, a) \longrightarrow P(a, a)$, but not the empty clause: binary resolution alone is incomplete.

Look at (Chang and Lee, 1973) for another proof of the completeness of resolution. Chang and Lee proceed by purely semantic arguments, by using so-called *Herbrand trees*. The construction is as follows: enumerate all closed atoms A_1, ..., A_n, ..., and build the (in general infinite) decision tree on these atoms, thus mimicking the completeness proof we gave for propositional logic in Chapter 2, Section 4.2. Each branch of the tree defines a unique Herbrand interpretation, just like propositional decision trees defined unique interpretations of propositional variables. This tree is called a *Herbrand tree*.

As in the propositional case (See Figure 2.6, page 50), we define the *failure nodes* for each clause C in S as the highest nodes such that any full Herbrand interpretation going through this node fails to satisfy C (where we see C as a universally quantified formula). Then if S is unsatisfiable, there is at least one failure node on each branch (i.e., Herbrand interpretation). We define the *closed Herbrand tree* as the Herbrand tree, chopped off at failure nodes (which become leaves of the closed Herbrand tree). Then, if S is unsatisfiable, all branches of the closed Herbrand tree are of finite length. We then use the following Lemma to conclude that the closed Herbrand tree is finite:

Lemma 7.12 (König's Lemma) *Any finitely-branching tree whose branches are all finite is finite (i.e., has finitely many nodes, hence has finite depth and finite size).*

We can then deduce Herbrand's Theorem, in the following form: if S is unsatisfiable, then there is a finite unsatisfiable set of closed instances of clauses of S (indeed, the closed Herbrand tree has finite depth, so it used only finitely many closed atoms, which can only be combined as finitely many clauses). Moreover, resolution has exactly the same effect as in the propositional case, namely to push failure nodes upwards by adding resolvents, until the closed Herbrand tree is reduced to its sole root (provided it was finite to begin with). And, as in the propositional case, when this happens, we must have produced the empty clause.

König's Lemma, although seemingly obvious, is actually equivalent to a weak form of the Axiom of Choice, called the Axiom of Dependent Choices. The idea of the proof is the following: assume T to be a finitely-branching infinite tree, then its root r_0 has a successor that is the root r_1 of another finitely-branching infinite tree T_1, which in turn has a finitely-branching infinite immediate subtree T_2, and so on; then $r_0, r_1, \ldots$, describes an infinite branch in T. (More formally, we can define a function F that maps each root r of an infinite subtree of T to the set of the successors of r that are roots of infinite subtrees; for each r, $F(r)$ is not empty because the subtree rooted at r is finite, but r has only finitely many successors; therefore, by the Axiom of Choice, there is a function f mapping each r to some element of $F(r)$; defining the relation $<$ by the smallest transitive relation such that $f(r) < r$, we then see that $<$ is not well-founded, i.e. there is an infinite branch in T.)

The most important thing to notice is that using König's Lemma to prove that the closed Herbrand tree is finite, hence of finite depth, parallels our use of Tychonoff's Theorem to prove that propositional logic is compact, hence that Herbrand's Theorem holds.

4   OPTIMISATIONS

As we have already noticed in Chapter 2, Section 4.2, resolution is rather inefficient at the propositional level. If we separate first-order resolution in its two components, propositional resolution and lifting by unification, then only its first component is weak. Fortunately, we can refine resolution in a number of ways, so as to
make it efficient. Some ideas come from the Davis-Putnam procedure (see Chapter 2, Section 4.3), some others try to restrict resolution derivations to almost canonical derivations, to avoid redundancy.

4.1   Clause Selection Strategies

The first thing we have to look at is the clause selection process. The only such process that we have presented, level-saturation search, is complete, but may generate
lots of irrelevant clauses. In fact, if the set at level 0 contains n clauses of at most
m literals, the set at level 1 may contain up to $mn(n+1)/2$ additional clauses of at
most $2m$ literals, and in general the set at level k may contain up to $2^k mn^{2^k}$ clauses
of up to $2^k m$ literals. So, if the least k we need to generate the empty clause is too
high, and the level sets grow too quickly, then we shall never be able to derive the
empty clause in practice. (Exponentials become huge very fast! For $k = 50$, 2^{50} is
about 10^{15}; no computer has yet 10^{15} bytes of memory, and even if we could save
enough space, and assuming we could perform one operation every microsecond,
doing 10^{15} operations would then need a bit less than 32 years.)

We shall strongly decrease the chances of stumbling on such worst cases by using various optimisations, as presented in Sections 4.2 or 4.4. We still have to complement this by examining cleverer strategies for exploring a space of resolution
derivations.

Level-saturation search generates all possible derivations in parallel. First, all
derivations of depth 1, yielding as conclusions all clauses in S_1, then all derivations
of depth 2, yielding all clause in S_2, and so on. We are not tied to such a strategy.
In fact, all that we want is to explore the space of derivations so that no resolution
step is dismissed for indefinitely long. This is a fairness condition:

Definition 7.13 (Fairness) *A clause selection strategy is said to be* fair *if and only
if no pair of clauses is ignored by the strategy for indefinitely long. (Equivalently,
if the set of resolvents generated using the strategy is exactly the union of all level
sets.)*

It is then clear that not only level-saturation search, but any other fair strategy preserves the completeness of resolution. As we noticed, level-saturation search is expensive both in terms of space and time usage. Depth-first search uses memory
more sparingly: choose a pair of clauses, resolve, add the resolvent to the set of
clauses, and repeat; then, backtrack and choose another pair of clauses to resolve

if we realize later on that this pair does not lead to a refutation. Then the number of generated clauses we need to keep in memory at level k is $O(k)$, much less that the huge amount described above.

The problem is that depth-first search is not a fair strategy. This can be repaired by the following trick, called *iterative deepening*: fix a limit $k \in \mathbb{N}$, then use depth-first search as above, but backtrack if the empty clause has not been generated in less than k resolution steps; if no refutation can be produced in less than k resolution steps, we shall find it in finite time; then, increase k and start over. By doing this, we shall redo all the computations that we have already done, generating resolvents at levels less than the old value of k, but this usually takes no more time than exploring the new resolvents at level $k + 1$. (For instance, assume that we need $O(2^k)$ time to explore the clauses in S_k, then to explore those in S_{k+1}, we shall need $O(2^{k+1})$ time, of which only half is devoted to re-exploring old clauses.)

We now turn to reducing the set of clauses worth generating. This will reduce the space of interesting resolution derivations, but in general not so drastically that it becomes finite. In every case, we shall need fair clause selection strategies.

4.2   Deletion Strategies

The *deletion strategies* eliminate inessential clauses that, if left in, would pollute the proof process by artificially increasing the size of the set of clauses.

Elimination of Subsumed Clauses

Definition 7.14 (Subsumption) *Let C be $\Gamma \longrightarrow \Delta$ and C' be another clause. We say that C subsumes C', and we write $C \geq C'$, if and only if C' has the form $\Gamma', \Gamma\sigma \longrightarrow \Delta\sigma, \Delta'$ for some substitution σ.*

That is, C subsumes all the clauses obtained by weakening any of its instances. Look at C and C' as universally quantified disjunctions of literals: semantically, if C subsumes C', then C logically implies C'.

Subsumed clauses are normally useless, because of the following Lemma:

Lemma 7.15 *Let C_1 and C_1' be two clauses, such that C_1 subsumes C_1'. Then, for any resolvent C' between C_1' and C_2, C' is subsumed by C_1 or by some resolvent of C_1 and C_2.*

Moreover, the only clause that subsumes the empty clause is the empty clause.

Proof: Write C_1 as $\Gamma_1 \longrightarrow \Delta_1$, and C_1' as $\Gamma_1', \Gamma_1\sigma_1 \longrightarrow \Delta_1\sigma_1, \Delta_1'$. If C' is a factor of C_1', it is clearly subsumed by C_1. We now check the result on binary resolvents. Assume we get C' by binary resolution on some atom A_1' on the right-hand side of C_1' and on some atom A_2 on the left-hand side of C_2. (The symmetric case is similar.)

As a first case, we may have $A_1' \in \Delta_1'$. Let $\Delta_1'' = \Delta_1' \setminus \{A_1'\}$, $C_2 = A_2, \Gamma_2 \longrightarrow \Delta_2$, and $\sigma' = mgu(A_1', A_2)$. Then $C_1' = \Gamma_1', \Gamma_1 \sigma_1 \longrightarrow \Delta_1 \sigma_1, \Delta_1''$, A_1', $C' = \Gamma_2 \sigma', \Gamma_1' \sigma', \Gamma_1 \sigma_1 \sigma' \longrightarrow \Delta_1 \sigma_1 \sigma', \Delta_1'' \sigma', \Delta_2 \sigma'$. So, $C' \leq C_1$.

The second case is $A_1' \in \Delta_1 \sigma_1$. Let A_1' be $A_1'' \sigma_1$, $\Delta_1'' = \Delta_1 \setminus \{A_1''\}$, $C_2 = A_2, \Gamma_2 \longrightarrow \Delta_2$, and $\sigma' = mgu(A_1'' \sigma_1, A_2)$. Since $A_1'' \sigma_1 \sigma' = A_2 \sigma'$, and since $\mathrm{fv}(A_1'') \cap \mathrm{fv}(A_2) = \varnothing$, the substitution $\sigma_1 \sigma' \cup \sigma'$ is a unifier of A_1'' and A_2, say $\sigma \theta$, where $\sigma = mgu(A_1'', A_2)$. Then we can resolve C_1 with C_2, on A_1'' and A_2 respectively, getting $C = \Gamma_1 \sigma, \Gamma_2 \sigma \longrightarrow \Delta_1'' \sigma, \Delta_2 \sigma$. As $C' = \Gamma_2 \sigma \theta, \Gamma_1' \sigma \theta, \Gamma_1 \sigma \theta \longrightarrow \Delta_1'' \sigma \theta, \Delta_1' \sigma \theta, \Delta_2 \sigma \theta$, C subsumes C'.

Finally, if $\Gamma \longrightarrow \Delta$ subsumes the empty clause, then Γ and Δ must be empty, so only the empty clause subsumes the empty clause. $\square$

This lemma has the following easy corollary:

Corollary 7.16 *Let S be a set of clauses, and S' be a subset of S. Assume that every clause C in S is in S' or is subsumed by a clause in S'.*

If the empty clause is derivable from S in k resolution steps, then it is derivable from S' in at most k resolution steps.

Proof: By induction on k. If $k = 0$, then the empty clause is in S, so by assumption it is in S' or it is subsumed by some clause in S'; in any case, the empty clause must already be in S'. Otherwise, assume the result to hold for $k - 1$ steps, and assume that we can derive the empty clause from S in k steps, $k > 0$. Examine the first resolution step in the derivation: we resolve two clauses C_1 and C_2 to yield the resolvent C.

By assumption, C_1 is subsumed by some clause C_1' in S' (possibly C_1 itself), and C_2 is subsumed by some clause C_2' in S'. By Lemma 7.15, C is subsumed by some clause C', which may be C_1', C_2' or a resolvent of C_1' and C_2'. Then, $S \cup \{C\}$ and $S' \cup \{C'\}$ are two sets of clauses verifying the assumptions of the Theorem, and we can derive the empty clause in $k - 1$ steps from $S \cup \{C\}$. By induction hypothesis, we can derive the empty clause in at most $k - 1$ steps from $S' \cup \{C'\}$, i.e. in at most k steps from S'. $\square$

Thus, we can delete subsumed clauses in the following way. If S is the current set of clauses, we delete from S any clause that is subsumed by some other clause in S until we cannot progress any further. This yields a minimal set S' as in the Corollary. Then, we produce one new resolvent, and delete the subsumed clauses in the resulting set of clauses, and so on. This is complete because if we are k resolvents away from the empty clause, deleting subsumed clauses leaves us at most k resolvents away from the empty clause, then computing the right resolvent puts us at most $k - 1$ resolvents from it, and so on.

A wrong way of deleting subsumed clauses is to identify all subsumed clauses in S and deleting them *en masse*. Indeed, the subsumption preordering may have cycles when several different renamed versions of the same clause are present in S. Consider the following example: $S = \{\longrightarrow P(x), \longrightarrow P(x'), P(y) \longrightarrow\}$.

S is unsatisfiable, but the first two clauses are subsumed by each other ; while $\{ \longrightarrow P(x'), P(y) \longrightarrow \}$ and $\{ \longrightarrow P(x), P(y) \longrightarrow \}$ are both unsatisfiable subsets of S (subsets S' as in Corollary 7.16), the result of eliminating all subsumed clauses from S yields the only clause $P(y) \longrightarrow$, which is satisfiable.

Subsumption can be used in mainly two ways. *Forward subsumption* is the following: when we generate resolvents from clauses in the current clause set S, we only keep those resolvents that are not subsumed by any clause in S. On the other hand, *backward subsumption* consists in the following: when we generate resolvents C from clauses in the current set S, we erase from S all clauses subsumed by C. A combination of both can be used, say forward subsumption, then backward subsumption if the new resolvent survives.

Note also that forward subsumption should only be used on resolvents, not all clauses, and in particular not on factors. Indeed, a factor of a clause C is always subsumed by C: if forward subsumption on factors were complete, then we would not need factoring for completeness.

To compute whether a clause C is subsumed by some other clause C' is hard. In fact, this is an NP-complete problem. Indeed, although determining whether an atom B is of the form $A\sigma$ is easy (this is called *matching*, and can be tested in polynomial time, by running a modified form of the unification algorithm, where the occurs-check test is no longer needed), to decide subsumption, we first have to guess, or find by backtracking, a mapping from literals A' of C' to literals A of C, such that $A = A'\sigma$ for some common substitution σ. This can be improved by noticing that A' can match A only when they have the same top predicate symbol, so that we explore only atoms A with the same predicate symbol as A', for each A'. This can even be improved by considering the function symbols appearing at given positions in each atom, but this is more complicated, and we won't describe it here.

The problem is made harder in practice by the fact that we are not interested in knowing whether C is subsumed by a given clause C', but by some (unknown) clause in the possibly large set S of clauses. There are efficient methods to solve this problem in ways that are usually very fast. Let us merely hint at a possible solution, similar to Larry Wos' *discrimination nets* (as used in the prover Otter), and used by Andrei Voronkov in his Vampire prover. We compile any atom, then any clause, as a sequence of instructions of an abstract machine (with instructions to push and pop terms to and from a stack, to test equality of terms, to bind variables to terms, and so on), whose rôle is to test whether the given atom or clause subsumes an atom or a clause given as argument. Because these programs are strings of op-codes, we can share them by their common prefixes, creating choice nodes, and making a tree of the set of programs. Sharing prefixes of these programs has the effect that the resulting tree program makes subsumption tests on usually many clauses in parallel.

Elimination of Tautologies

The simplest deletion strategy is *elimination of tautologies*. A tautological clause is a clause of the form $\Gamma, A \longrightarrow A, \Delta$ for some atom A. (If we see clauses as disjunctions, having both A and $\neg A$ is indeed the only way to make it a tautology.) Eliminating tautologies means the following: each time we generate a resolvent, we add it to the current set of clauses only if it is not a tautology (and it is not subsumed by previous clauses).

Intuitively, tautologies contribute nothing to the task of discovering a proof, because they are always true. This is no justification, however. But eliminating tautologies is justified by the following Lemma:

Lemma 7.17 *Let C be a tautological clause. Then any resolvent of C with any clause C' is a tautology or is subsumed by C'.*

Proof: Any factor of C' is subsumed by C. We check the binary resolvents of C and C'. Let C be $A, \Gamma \longrightarrow \Delta, A$. If we do not resolve on A, then the result is again a tautology. Otherwise, let C' be $A, \Gamma' \longrightarrow \Delta'$ (the symmetric case is similar), then the resolvent is $A, \Gamma, \Gamma' \longrightarrow \Delta', \Delta$, which is subsumed by C. $\square$

Moreover, no tautology can ever be the empty clause. Therefore, generating new tautologies is useless, as well as subsumed clauses.

Comparing with the Davis-Putnam method in the propositional case (see Chapter 2, Section 4.3), eliminating tautologies in first-order resolution is analogous to the elimination of tautologies in the Davis-Putnam method.

Elimination of Pure Clauses

Another simplification that we can lift from the Davis-Putnam method to the first-order resolution case is the elimination of pure clauses.

Definition 7.18 *Let S be a set of clauses.*
A clause C is said to be pure *in S if and only if it contains a pure literal, i.e. it is of the form $\Gamma \longrightarrow \Delta, A$, (resp. $A, \Gamma \longrightarrow \Delta$) where A unifies with no atom of the left-hand side (resp. right-hand side) of any clause in S. If this is the case, we say that A (resp. $\neg A$) is the pure literal in the clause C.*

We can always eliminate pure clauses, irrespectively of whether they are newly generated clauses or clauses already present in the current set of clauses S, because of the following Lemma:

Lemma 7.19 *Any resolvent of a pure clause with any other clause is still pure.*

Proof: Any factor of a pure clause is clearly still pure. Then, any binary resolution between a clause where A (resp. $\neg A$) is pure and some other clause must operate on some other literal B. The resulting binary resolvent must therefore feature A (resp. $\neg A$) as a literal, which is then pure. $\square$

Noticing that no pure clause can be empty, we conclude that pure clauses are dead-ends, as far as finding empty clauses is concerned. Therefore, we can eliminate all pure clauses without sacrificing completeness.

In fact, we can combine all the deletion strategies we have mentioned (subsumption, elimination of tautologies, elimination of pure clauses) while keeping resolution complete. Look at what we have now got on the propositional level (where factoring is not needed): binary resolution is similar to one step in Davis and Putnam's splitting rule (doing all possible resolutions on some given atom A indeed generates all clauses produced by splitting—this will be made clearer in Section 4.3); elimination of tautologies, resp. of pure literals, directly correspond to the operation of the same names in the Davis-Putnam method. The only differences are: in propositional resolution, forward subsumption amounts to not generating weakenings of previous clauses, thus generalising the memoization we needed to make it terminate; generating resolvents is usually less efficient than doing splitting as in the Davis-Putnam
method; and we don't have the equivalent of the unit rule in resolution.

Unit Resolution

To mirror the unit rule of the Davis-Putnam procedure, we define a special case of the resolution rule, called *unit resolution*:

Definition 7.20 *A* unit clause *is a clause of the form* $\longrightarrow A$ *or* $A \longrightarrow$, *where* A *is an atom.*

Unit resolution *is the variant of resolution where binary resolution is restricted to have at least one unit parent clause.*

In other words, in unit resolution, to apply binary resolution on non-unit clauses, we first have to factor at least one of the clauses to a unit clause.

Unfortunately, unit resolution is *not* complete. For example, the set of propositional clauses:

$$\longrightarrow A, B$$
$$A \longrightarrow B$$
$$A, B \longrightarrow$$
$$B \longrightarrow A$$

is unsatisfiable, but cannot be shown so by unit resolution, because it contains no unit clause.

Unit resolution asks much more than applying the unit rule in the Davis-Putnam method: the unit rule was applied as soon as we could, but not exclusively. Therefore, we might want to relax the constraint on unit resolution in the following way: use unit resolution whenever possible, otherwise use unconstrained resolution. But this is also incomplete, i.e., this strategy is unfair: consider the set of clauses above, plus the clauses:

$$\longrightarrow P(a)$$
$$P(x) \longrightarrow P(f(x))$$

The resulting set is still unsatisfiable, but applying unit resolution is always possible, and we can only get clauses of the form $\longrightarrow P(f^k(a))$, $k \geq 1$, this way, never the empty clause.

In important special case is the following:

Definition 7.21 (Horn) A Horn clause *is a clause with at most one atom on the right-hand side (alternatively, an intuitionistic atomic sequent).*

A Horn set *of clauses is a set of Horn clauses.*

We have:

Lemma 7.22 *Unit resolution is complete for Horn sets.*

This is an easy and interesting exercise:

▶ **EXERCISE 7.4**

Using the completeness of resolution, show that any set Horn clauses with non-empty left-hand sides is satisfiable. Conclude that unit resolution is complete for Horn clauses. (Show that, in fact, we can constrain unit resolution even a bit more.)

4.3   Ordering Strategies

In the Davis-Putnam method, splitting eliminates one atom A by rewriting the set S of clauses into two sets $S[\mathbf{T}/A]$ and $S[\mathbf{F}/A]$, where A does not occur any longer. This process eliminates atoms one by one, until none remains. The resolution rule also eliminates one atom, but only between two clauses. One major problem, then, is the highly non-deterministic nature of this elimination process.

Consider for example the four clauses:

$$\longrightarrow A, B$$
$$A \longrightarrow B$$
$$B \longrightarrow A$$
$$A, B \longrightarrow$$

We might choose to resolve most clause pairs on A, or on B. If we choose to resolve on A, the only choices that do not yield tautological clauses are between the first and second clauses, yielding $\longrightarrow B$, or between the third and fourth clauses, yielding $B \longrightarrow$. Then, choosing to resolve on B on the last two clauses, we get $\longrightarrow$. We might have chosen to resolve on B first, then on A, and we would have got the same result: this is because, to get the empty clause, we must eliminate all atoms in the clauses by resolution, and we can do it in any order we wish.

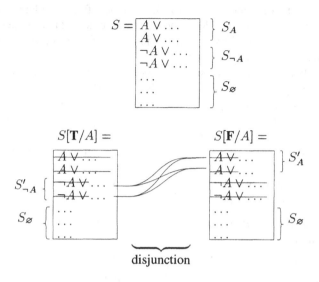

Figure 7.3. Splitting simulated by resolution

To sum up, there is no real non-determinism in the order in which we eliminate atoms by the resolution rule, as all orders must yield the same result —derivability or underivability of the empty clause. (This is a so-called *confluence* argument.) To eliminate this redundancy, a solution is to fix an ordering on atoms in advance, and to only allow to resolve on the highest possible atoms in the ordering.

> We are here justifying an *irrevocable* choice of atoms by a *confluence* argument. Confluence just says: whether you decide to eliminate the atoms $A_1, A_2, \ldots, A_m$ in this order, or to eliminate $B_1, B_2, \ldots, B_n$, then there is always a possibility of completing each series of eliminations so that, at some point, we get to the same situation in each case. But if the atoms A_i, $1 \leq i \leq m$, were chosen by some strategy $\mathcal{A}$, and the atoms B_j, $1 \leq j \leq n$, were chosen by some other strategy $\mathcal{B}$, confluence does not say that we can get to the same situation in each case *by continuing to apply strategy $\mathcal{A}$ in the first case and strategy $\mathcal{B}$ in the second case*. The latter condition (irrevocability) is a stronger condition than confluence, and is in fact the only one that interests us: an irrevocable strategy does not have to backtrack. Confluence is easier to justify, but we shall prove that the current situation is in fact irrevocable by showing that ordering strategies preserve completeness. Tableaux expansion strategies (see Chapter 8), even when they can be made confluent by some clever implementation mechanism (cf. model elimination, in Chapter 8, Section 3.4), are usually not irrevocable, i.e. these methods still have to backtrack.

Why it works is due to the fact that on the propositional level, splitting on A (the highest atom) means computing all resolvents of clauses on A. Look at Figure 7.3: write S as the union of three sets: the set S_A of clauses of the form $A \vee C$, alternatively sequents with A on the right-hand side; the set $S_{\neg A}$ of clauses of the form $\neg A \vee C$, alternatively sequents with A on the left-hand side; and the set $S_\varnothing$ of all clauses where A and $\neg A$ do not occur. We assume that tautologies have been removed, therefore these three sets are disjoint. Let S'_A be S_A where A has been erased, and similarly for $S'_{\neg A}$ and $S_{\neg A}$.

Observe that $S[\mathbf{T}/A]$ is precisely the set of clauses in $S'_{\neg A}$ and $S_\varnothing$. Intuitively, $S[\mathbf{T}/A] = S'_{\neg A} \wedge S_\varnothing$; and similarly, $S[\mathbf{F}/A] = S'_A \wedge S_\varnothing$. The Davis-Putnam method looks for a satisfying interpretation of $S[\mathbf{T}/A]$ or of $S[\mathbf{F}/A]$. This is one way of looking for a satisfying interpretation of $S[\mathbf{T}/A] \vee S[\mathbf{F}/A]$. The latter is, intuitively, $S_\varnothing \wedge (S_A \vee S_{\neg A})$. And $S_A \vee S_{\neg A}$ is just, by the distributivity rules, the same as the set of all resolvents between clauses on S on A and $\neg A$.

On the propositional level, we may eliminate all atoms in any order. Alternatively, we may fix an ordering on atoms, and always eliminate the highest atom. Not all orderings work in the first order case, though, so we first define:

Definition 7.23 (Stability) *A strict ordering $>$ on atoms is said to be* stable *if and only if for all atoms A, B, and for every substitution σ, $A > B$ implies $A\sigma > B\sigma$.*

Recall that a strict ordering is an irreflexive and transitive relation. Strict stable orderings (i.e., orderings $>$ such that $A > B$ implies $A\sigma > B\sigma$) are also called *A-orderings* in the literature.

A strict stable ordering cannot be total in general. Indeed, if $P(x) > P(y)$, for instance, then we must also have $P(y) > P(x)$, by using the substitution $[x/y, y/x]$. Since we want to restrict the atoms we resolve upon to maximal atoms with respect to $>$, it is interesting to have as little choice as possible, i.e. to have as few incomparable atoms as possible. To solve the incomparability problem is again impossible in general, but we can improve $>$ by forcing it to have the following property:

Definition 7.24 (Ground Totality) *A strict ordering $>$ is said to be a* ground total *if and only if for any two ground atoms A and B, either $A < B$, or $A = B$, or $A > B$.*

Ground total orderings exist. One of the simplest family of such orders are the *lexicographic orderings*, defined on atoms and terms alike as follows. Fix a strict total ordering $\succ$ on the set of function symbols, then two terms (or atoms) s and t are such that $s > t$ if and only if:

* $s = f(s_1, \ldots, s_m)$, $t = g(t_1, \ldots, t_n)$, and either $f \succ g$, or $f = g$ and $s_1 = t_1, \ldots, s_{k-1} = t_{k-1}, s_k > t_k$ for some k, $1 \leq k \leq n$.

In this definition, variables are comparable with no other term. This is called a *left-to-right lexicographic ordering*, as arguments of applications are compared from left to right. There are clearly many other possibilities.

We then define the following refinement of resolution:

Definition 7.25 (Ordering Strategy) *Let $>$ be a stable strict ordering on atoms (possibly, and preferably, total on ground atoms). Let $>^{\#}$ be the relation defined* *by $s >^{\#} t$ if and only if $s > t$ or s and t are incomparable with respect to $>$, or equivalently if $s \not< t$. We say that an atom A is maximal in a set S of atoms if and only if $A >^{\#} B$ for every $B \in S \setminus \{A\}$, i.e. if and only if A is greater than every comparable atom in S, i.e. no other atom is greater than A in S.*

The ordering strategy *is the refinement of resolution where factors of clauses C are computed by unifying maximal atoms in C only, and where binary resolvents of clauses C and C' are computed by unifying complementary literals that are maximal in their respective clauses.*

As expected, we have:

Theorem 7.26 (Completeness) *The ordering strategy is sound and complete.*

Proof: It is clearly sound, as a restriction of resolution, which is sound.

It is complete at the ground level. Indeed, we use the argument above to simulate Davis-Putnam by resolution. There is still a snag: when A or $\neg A$ is pure, i.e. when $S_{\neg A}$ or S_A is empty, there is no resolvent on A. For instance, A is pure in the set of three clauses $A \longrightarrow B$, $\longrightarrow B$ and $B \longrightarrow$, so there is no resolvent on A; but this set of clauses is still unsatisfiable: resolve on B instead. We therefore have to ignore pure literals in the proof.

Let S be a finite unsatisfiable set of ground clauses. We simulate Davis and Putnam's splitting on the (non-pure) atom A by computing the set of all resolvents of clauses in S upon A; let $S(A)$ be the latter, union all clauses in S where A and $\neg A$ do not occur. Then $S(A)$ is a cnf for the disjunction of $S[\mathbf{T}/A]$ and $S[\mathbf{F}/A]$, conjoined with $S_{\varnothing}$, as illustrated on Figure 7.3.

Since S is unsatisfiable, S must either contain the empty clause, and we are done; or there must be at least one non-pure atom A in S, otherwise S would be satisfiable. Choose A maximal among all non-pure atoms in S, and compute $S(A)$: it is a set of ordered resolvents from S on A, union the subset $S_{\varnothing}$ of S. Furthermore, $S(A)$ contains strictly less pure literals than S, since no new literals were created, and the pure literals in S remain pure in $S(A)$; and $S(A)$ is clearly unsatisfiable.

This argument leads to an easy argument of completeness by induction on the number of non-pure atoms in S.

To lift the result to the first-order case, we apply Herbrand's Theorem: a set S of first-order clauses is unsatisfiable if and only if there is a finite propositionally unsatisfiable set S' of instances of clauses of S. Then, we can use ordered resolution at the ground level to derive the empty clause from S', as we have just shown. Finally, this ground ordered resolution can be simulated by first-order resolution steps as in Theorem 7.11, and these resolution steps are ordered resolution steps, because $>$ is stable. □

Ordered resolution is also compatible with the deletion of tautologies, of subsumed clauses and of pure clauses. In other words, combining these techniques still yield a complete refutation procedure. In fact, the proof above already shows that eliminating pure clauses is compatible with the ordering strategy. As far as subsumption and tautology elimination are concerned, the key point is that Lemma 7.15 continues to work in the ordered case, because by stability of the ordering every maximal literal in some clause $\Gamma', \Gamma\sigma \longrightarrow \Delta\sigma, \Delta'$ subsumed by $\Gamma \longrightarrow \Delta$ is either in Γ', Δ' or an instance of some maximal literal in $\Gamma \longrightarrow \Delta$.

A subtlety of ordered resolution is the following. Assume that we eliminate pure clauses as soon as they come up. As the proof of Theorem 7.26 shows, at the ground level, we are allowed to choose maximal atoms not only in some given clauses, but from the whole set of (non-pure) clauses S. We might think that we can refine ordered resolution again in the first-order case by first choosing a maximal atom A in the whole set S, then computing the resolvents upon A only. This is wrong, because this clause selection process is not fair any longer. Consider for instance the set of clauses:

$$\longrightarrow A, B$$
$$A \longrightarrow B$$
$$A, B \longrightarrow$$
$$B \longrightarrow A$$
$$\longrightarrow P(a)$$
$$P(x) \longrightarrow P(f(x))$$

which we have already used to prove that unit resolution was incomplete; and use the lexicographic ordering based on $P \succ A \succ B$. This would force us into using only clauses based on P, generating clauses of the form $\longrightarrow P(f^k(a))$, but not the empty clause.

▸ **EXERCISE 7.5**

Let $>$ be a total ordering on ground terms, defined by an enumeration $(A_i)_{i \in \mathbb{N}}$ of the ground atoms; in other words, $A_i > A_j$ if and only if $i > j$, and for any two (non-ground) atoms A and B, $A > B$ implies that $A\sigma > B\sigma$ for every ground substitution σ. Show that the proof of Theorem 2.39 already shows the completeness of the ordering strategy. In what respect is this proof weaker than the one we gave of Theorem 7.26?

▸ **EXERCISE 7.6**

Show that the following *extended ordering strategy* is complete. We produce the resolvent of $\Gamma \longrightarrow \Delta, A_1, \ldots, A_m$ and $A'_1, \ldots, A'_{m'}, \Gamma' \longrightarrow \Delta'$, where σ is the mgu of $A_1, \ldots, A_m, A'_1, \ldots, A'_{m'}$, only when $A_1\sigma$ is maximal among $\Gamma\sigma, \Delta\sigma, \Gamma'\sigma$, $\Delta'\sigma$. What is the difference with the ordering strategy?

4.4   Semantic Resolution and Set-of-Support Strategies

The *semantic resolutions* guide the resolution rules by using an interpretation I. Consider the Davis-Putnam procedure again. At the ground level, the ordering strategy amounts to fixing an ordering of the ground atoms $A_1, A_2, \ldots, A_n$; we then

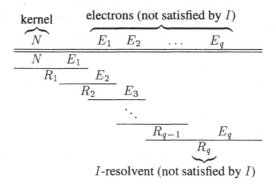

Figure 7.4. I-clash in semantic resolution

split on A_1, then on A_2, and so on. The semantic resolutions control which branch we take at each splitting. Whenever we wish to split on A_i, we may try to explore the case A_i true first, then the case A_i false; or the converse. We may encode this choice by an interpretation I: if I makes A_i true, then we choose to explore the case A_i true first, otherwise we explore the case A_i false first.

> Interpretations then only encode *uniform* such strategies. For instance, if I makes A true, then we must always test the case A true first. We cannot change our mind and decide to test the case A true first on some clause sets and the case A false first on other clause sets that we encounter while running the Davis-Putnam procedure.

> Similarly, the ordering strategy used a uniform enumeration of the atoms, whereas the Davis-Putnam procedure allows one to enumerate atoms in a more versatile way; for example, on the A_1 true branch, to test A_2 then A_3, and on the A_1 false branch, to test A_3 then A_2.

Let us say that I *satisfies* a clause C if and only if it satisfies C for all possible assignments of its free variables. Alternatively, letting $x_1, \ldots, x_n$ be the free variables of C, if it satisfies the universal closure $\forall x_1 \cdot \ldots \cdot \forall x_n \cdot C$ of C. An alternative view of semantic resolution is as follows.

For example, assume that our initial set of clauses S consists in, first, a set S' of clauses expressing enough axioms of arithmetic, and second, a set S'' of clauses representing the skolemized version of the negation of a proposition that we wish to prove in system S'. Because arithmetic is consistent, it has a model I, namely the usual set $\mathbb{N}$ of natural numbers with the corresponding operations. We can then use this model I to guide us into a proof of our proposition. The idea of semantic resolution is to try to recognise which clauses are already satisfied by I, and to disallow any resolution step between two such clauses. Intuitively, this would not help in progressing towards the derivation of the empty clause, since we cannot derive it from a satisfiable set of clauses.

Any auxiliary interpretation works for semantic resolution. One of the simplest such interpretations is the *positive Herbrand interpretation*, in which all closed atoms are assumed to be true. Then I satisfies a clause C if and only if its right-hand side is not empty. We call a clause $\Gamma \longrightarrow$, which is not satisfied by the positive interpretation, a *negative clause*.

Symmetrically, in the *negative Herbrand interpretation*, all closed atoms are assumed to be false. In this interpretation, a clause is satisfied if and only its left-hand side is not empty, i.e. it is not a *positive clause* $\longrightarrow \Delta$.

All this yields the following refinement of resolution:

Definition 7.27 (Semantic Resolution) *Let I be an interpretation. Let A_I be a terminating algorithm such that, if A_I returns "yes" on input C, then I satisfies C.*

An I-clash is a finite set of clauses $\{N, E_1, \ldots, E_q\}$, $q \le 1$, such that:

- *A_I returns "no" on any of $E_1, \ldots, E_q$;*

- *there is a resolvent R_1 of N and E_1, a resolvent R_2 of R_1 and E_2, ..., a resolvent R_q of R_{q-1} and E_q;*

- *A_I returns "yes" on $R_1, \ldots, R_{q-1}$, and "no" on R_q.*

The E_i's are called the electrons, *N is called the* nucleus *of the I-clash, and R_q is called an I-resolvent of the I-clash.*

The algorithm A_I is the strange part of the definition. In the positive (resp. negative) interpretation, A_I can just be taken to return "yes" on negative (resp. positive) clauses, and "no" on all other clauses, and this characterises exactly those clauses that are satisfied by the interpretation. But in more sophisticated uses of I-resolution, satisfaction by I may not be decidable: in this case, the constraint we put on A_I is one of *safeness*. In the conservative case where A_I always returns "no" —a very safe case indeed— semantic resolution is just ordinary, unrestricted resolution, and we have not gained much. We shall therefore want to have A_I return "yes" on as many satisfied clauses as it can detect.

See Figure 7.4 for a pictorial representation of an I-clash. Notice that the nucleus may or may not have been determined to be satisfied by I. In a more informal way, semantic resolution works as follows. Take a set S of clauses, and split it into the set S' of clauses on which A_I returns "yes" (the set of clauses definitely satisfied by I), and the set S'' of clauses on which A_I returns "no" (the set of clauses possibly not satisfied by I). Then, choose a clause N from $S' \cup S''$, and a clause E_1 from S'', and resolve to get a new clause R_1. (If we chose N and E_1 both from S', we would know that I satisfied R_1, so it would be a useless lemma.) If A_I returns "no" on R_1, then R_1 is possibly not satisfied by I, and this brings us some new information, so we add R_1 to S'', and this terminates the I-resolution step. Otherwise, we resolve between R_1 and some clause E_2 from S'', getting R_2, and so on until we come to a clause R_q on which A_I returns "no", and then we add R_q to S'', and this finishes the I-resolution step.

In general, however, we cannot proceed in this manner, as we do not know any a priori bound on q, and we might be taken into an infinite loop, searching for a I-resolvent on the wrong nucleus. We therefore have to interleave the searches on all possible nuclei. In practice, very simple instances of semantic resolution are used, like hyper-resolution or the set-of-support strategy, where there is no such problem. In fact, semantic resolution is mostly a theoretical device for exploring more practical refinements of resolution.

We prove the expected completeness result before describing the most important aforementioned examples of semantic resolution.

Theorem 7.28 (Completeness) *Given any arbitrary interpretation I, I-resolution is sound and complete.*

Proof: I-resolution is sound, as a restriction of resolution. Now, we claim that it is complete. We prove it at the ground level first, then lift the result to the first-order case by Herbrand's Theorem.

Let S be a finite unsatisfiable set of ground clauses. We prove by induction on the number of atoms appearing in S that S can be proven unsatisfiable by an I-resolution refutation. If this number is 0, then S must be $\{\longrightarrow\}$, and the claim is proved. Otherwise, let A be some atom occurring in S, and assume without loss of generality that A is true in I. (It makes sense to say that A is true or false in I, since A is a ground atom.)

Since S is unsatisfiable, $S[\mathbf{T}/A]$ is also unsatisfiable. By induction hypothesis, there is an I-resolution derivation D_1 of $\longrightarrow$ from $S[\mathbf{T}/A]$. We translate it to an ordered I-resolution derivation D_1' of $A \longrightarrow$ as follows.

First, D_1 can be seen as an inverted tree, where nodes are clauses, leaves are clauses in $S[\mathbf{T}/A]$, and the root is the empty clause $\longrightarrow$. Build D_1' by adding back A to the left of the clauses $\Gamma \longrightarrow \Delta$ in $S[\mathbf{T}/A]$ at the leaves which came from clauses $A, \Gamma \longrightarrow \Delta$ in S, and by recursively adding A to the left of clauses at inner nodes if we have added A to the left of one of their parents. (This transformation may not be unique.) D_1' is still an I-resolution derivation, because the electrons augmented with A on the left remain false in I. The conclusion of D_1' may be $\longrightarrow$ again, in which case we are done; or $A \longrightarrow$, which is false in I. Assume the latter.

Now, since S is unsatisfiable, so is $S[\mathbf{F}/A]$, so by induction hypothesis, there is an I-resolution refutation D_2 of $S[\mathbf{F}/A]$. We transform D_2 into an I-resolution refutation D_2' of $S \cup \{A \longrightarrow\}$ as follows. The idea is to put back A on the right of clauses in D_2, but this does not work: some electrons and some I-resolvents might become true in I.

But we can get back an I-resolution by resolving with the electron $A \longrightarrow$. Let $N, E_1, \ldots, E_q$ be an arbitrary I-clash in D_2. By adding A to the right of some of the latter clauses, we get a sequence $N', E_1', \ldots, E_q'$ of clauses, where N' is N or

N plus A on the right, and for every i, $1 \leq i \leq q$, E_i' is either E_i or E_i plus A on the right. For every i, if $E_i' = E_i$, then E_i' is an electron, and we don't modify it; otherwise, we replace E_i' by the I-clash E_i' (nucleus), $A \longrightarrow$ (electron) to deduce E_i (which is indeed false in I). This allows us to get a sequence of resolution steps between N', E_1, ..., E_q. If $N' = N$, then we have rebuilt the original I-clash in D_2. Otherwise, we produce the I-clash N', $A \longrightarrow$, E_1, ..., E_q: since resolving between N' and $A \longrightarrow$ produces N, the latter has exactly the same end clause as N, E_1, ..., E_q. By induction on the length of D_2, we therefore produce an I-resolution refutation from $S \cup \{A \longrightarrow\}$.

To lift the result to the first-order case, we apply Herbrand's Theorem: a set S of first-order unsatisfiable if and only if there is a finite propositionally unsatisfiable set S' of instances of clauses of S. Then, we can use I-resolution at the ground level to derive the empty clause from S', as we have just shown. Finally, this ground resolution can be simulated by first-order ordered resolution steps as in Theorem 7.26, and these resolution steps are in fact I-resolution steps. Indeed, whenever a ground instance E' of a clause E is false in I, then A_I must return "no" on E: by the safety assumption, if A_I returns "yes", then E is definitely satisfied by I. $\square$

Not only is semantic resolution complete, for any interpretation I, but the deletion strategies (tautologies, subsumption, pure clauses) do not break completeness either. Replay indeed the proof of Lemma 7.15: the key point is that if an electron is subsumed by some clause, then the latter is an electron too, i.e. it is falsified under I.

However, *ordered* semantic resolution is in general not complete: see Exercise 7.7 in the following section. We can however use the following refinement:

Definition 7.29 Semi-ordered *semantic resolution is the refinement of semantic resolution where we require that the literal on which we resolve at each step is maximal in its electron.*

That is, we do require that it is maximal in each E_i, but not in R_i, $1 \leq i \leq q$. Then:

Theorem 7.30 *Given any arbitrary interpretation I, and any strict stable ordering $>$, semi-ordered I-resolution is sound and complete.*

Proof: Similar as Theorem 7.28, except that we choose A to be a minimal atom. When we add A to an electron E, the literal resolved upon in E remains maximal in E augmented with A. Moreover, in $A \longrightarrow$, A is necessary maximal. So all the derivations that are produced in the proof are still semi-ordered. $\square$

Clearly, semi-ordered resolution with the deletion strategies continues to be complete as well.

Hyper-resolution

The first instance of semantic resolution is *hyper-resolution*:

Definition 7.31 (Hyper-resolution) Positive hyper-resolution *is ordered I-resolution, where electrons and resolvents are constrained to be positive clauses. (Alternatively, the interpretation I is the negative Herbrand interpretation.)*

Negative hyper-resolution *is ordered I-resolution, where electrons and resolvents are constrained to be negative clauses. (Alternatively, the interpretation I is the positive Herbrand interpretation.)*

By Theorem 7.30, positive and negative hyper-resolution, and even their semi-ordered variants, are complete. To implement, say, positive hyper-resolution, we can relax the nucleus-electron framework: just pick a positive clause, resolve it with some other clause, and add the resolvent to the current set of clauses. Semi-ordered positive hyper-resolution then means that we only resolve on maximal literals in the chosen positive clause.

As Chang and Lee point out, positive hyper-resolution is similar to thinking forward, i.e. to use disjunctive facts (positive clauses) as electrons, so as to deduce more facts, until we get the empty clause, that is, the contradictory set of facts. If we want to prove $A \Rightarrow B$, we generate a fact clause $\longrightarrow A$, and a goal clause $B \longrightarrow$: positive hyper-resolution will then deduce facts from $\longrightarrow A$, and cancel them using the negative information on B to get the empty clause.

Conversely, negative hyper-resolution is similar to thinking backward, i.e. to reduce the goal to prove to the empty set of subgoals. In the example above, negative hyper-resolution will generate negative clauses from $B \longrightarrow$, until all subgoals (atoms on the left-hand side of negative clauses) are all canceled by facts derived from A.

▶ **EXERCISE 7.7**

Consider the following set of clauses:

$$\begin{aligned} &\longrightarrow A, B \\ B &\longrightarrow C \\ A &\longrightarrow \\ C &\longrightarrow \end{aligned}$$

Show that it is unsatisfiable, but that ordered positive hyperresolution cannot derive $\longrightarrow$ in the order $A < B < C$. Conclude that ordered semantic resolution is in general incomplete.

Set-of-Support

The other major instance of semantic resolution is the *set-of-support strategy*. The idea is the following: we don't usually want to prove single formulas Φ, but rather we want to prove Φ, given a set of axioms $\mathcal{A}$. That is, assuming $\mathcal{A}$ is finite, we want to prove $\mathcal{A} \Rightarrow \Phi$, where we have used $\mathcal{A}$ to denote the conjunction of all axioms in the set $\mathcal{A}$. We can usually prove or at least assume that $\mathcal{A}$ is consistent: axioms are not temporary assumptions, they are meant to describe a model. The problem with proving $\mathcal{A} \Rightarrow \Phi$, then, is that proof-search will devote a considerable portion of its time to refute $\mathcal{A}$, which is known—or believed—to be useless in advance. This is particularly clear in resolution, where to prove $\mathcal{A} \Rightarrow \Phi$, we would generate clauses for $\mathcal{A}$, and a few clauses for $\neg\Phi$, and all resolutions between clauses for $\mathcal{A}$ are then useless!

This can be solved by using semantic resolution, where the guiding interpretation I is any model of the axioms. As we know nothing about I in advance, we choose for A_I one of the simplest safe algorithms: answer "yes" on input C if and only if C is one of the axiom clauses.

Because A_I is so simple, the set-of-support strategy can be implemented in the following way: divide the initial set of clauses S in two disjoint subsets S' (the set of clauses coming from the axioms in $\mathcal{A}$) and S'' (the other clauses, corresponding to the negation of the goal Φ to prove). The set S'' is then the *set of support* for the resolution process. Then, we apply resolution repetitively, selecting parent clauses so that at least one of the two parent clauses comes from the set of support S'', and we add the resolvents to the set of support S''. (Note that we can also implement hyperresolution this way.)

As a special case of semantic resolution, the set-of-support strategy is then complete, provided that the axioms have at least one model. We may also safely implement the semi-ordered restriction without losing completeness, with all deletion strategies. In practice, although semantic resolution looks promising, it is considerably less efficient than what could be expected. One possible source of inefficiency is that resolvents are always put in the set of support S'', and are never recognised to be satisfied by I, even if they are: the algorithm A_I above is too simple-minded, we would really need some more specialised algorithm.

4.5   The Linear Strategy

The level of non-determinism in the choice of the clauses to resolve, even in ordered resolution or semi-ordered I-resolution, can be appalling. One refinement of resolution where this level of non-determinism is reduced is *linear resolution*, where we choose some clause C_0 from the initial clause set S, resolve it with some clause to get a resolvent C_1, which we add to S; we then proceed to resolve C_1 with some other clause to get C_2, which we add to S, and so on, until we get the empty clause.

Non-determinism is reduced, because one of the clauses to resolve is always fixed: it is the last clause we generated by resolution. The term "linear resolution", which has nothing to do with linear logic, comes from the special shape of resolution derivations we get in this way: see Figure 7.5. We define:

Definition 7.32 (Linear Resolution) *A linear resolution derivation is a derivation from a set S of clauses of clauses $C_0, C_1, \ldots, C_i$, such that $C_0 \in S$, and for every*

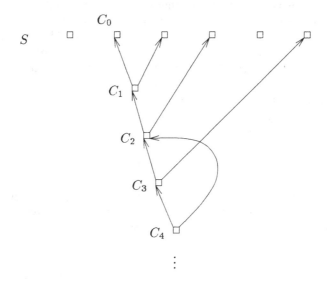

$$S$$

Figure 7.5. Linear resolution

$i \geq 0$, C_{i+1} *is a resolvent of* C_i *with some clause in* $S \cup \{C_1, \ldots, C_i\}$.

If C_{i+1} *is a resolvent of* C_i *with some clause* C, *we say that* C_i *is the* centre clause *of the resolution step,* C *is the* side clause, *and* C_{i+1} *is the new* top clause. *The clauses* $C_0, \ldots, C_i$ *are called the* ancestor clauses. *The clauses in* S *are called the* input clauses.

A linear resolution derivation is said to be semi-ordered *with respect to a strict stable ordering* $>$ *if and only if the literal resolved upon in the centre clause is maximal under* $>$.

Then we have:

Theorem 7.33 (Completeness) *Linear resolution, and semi-ordered linear resolution are sound and complete.*

Proof: It is sound, as it is a refinement of resolution.

The main idea is to replay the proof of Theorem 7.28, using the following fact: given an unsatisfiable set S of clauses, there is a *minimally unsatisfiable* set S' of clauses included in S, that is one such that every proper subset is satisfiable. This minimal subset may not be unique, but it does not matter. Without loss of generality, assume that S itself is minimally unsatisfiable.

Observe that a minimally unsatisfiable set S of clauses cannot contain any pure clause or any tautology. It follows that for every atom A occurring in S, A occurs on the left of some clause and on the right of some other clause in S; and that A cannot occur both on the left and on the right of the same clause.

We show that there is a linear resolution refutation from S by induction on the number of atoms appearing in S. If this number is 0, then S must be $\{\longrightarrow\}$, and the claim is proved. Otherwise, the result follows from the following more precise fact, which we prove by induction: if S' is a satisfiable set of clauses, and $S' \cup \{C_0\}$ is unsatisfiable, then there is a linear resolution refutation of $S' \cup \{C_0\}$.

If C_0 is the empty clause, the claim is proved. Otherwise, let A be any atom occurring in C_0. Without loss of generality, let A occur on the left, that is to say, C_0 is of the form $A, \Gamma_0 \longrightarrow \Delta_0$.

Since S is unsatisfiable, $S[\mathbf{T}/A]$ is also unsatisfiable, so by induction hypothesis there is a linear refutation D_1 of $\longrightarrow$ from $S[\mathbf{T}/A]$. As in the proof of Theorem 7.28, we transform D_1 into a derivation D_1' by adding back A to the left of all relevant clauses in D_1. D_1' is a linear derivation, and it derives either $\longrightarrow$ or $A \longrightarrow$. In the first case, we are done, so assume the latter.

Now $S' \cup \{A \longrightarrow\}$ is also unsatisfiable, because any interpretation satisfying both S' and $A \longrightarrow$ also satisfies S' and $A, \Gamma_0 \longrightarrow \Delta_0$, i.e., S. In particular, $(S' \cup \{A \longrightarrow\})[\mathbf{F}/A]$ is unsatisfiable, that is, $S'[\mathbf{F}/A]$ is unsatisfiable. Let S'' be a minimally unsatisfiable subset of $S'[\mathbf{F}/A]$. Since S'' is unsatisfiable, S'' is not a subset of S', which is satisfiable. So there is a clause $C_1 = \Gamma_1 \longrightarrow \Delta_1$ in S'' that is not in S'. This clause must come from some clause in S' which is not C_1 itself, so from $\Gamma_1 \longrightarrow \Delta_1, A$.

By induction hypothesis, there is a linear refutation D_2 of $S'[\mathbf{F}/A]$ with initial top clause C_1. Transform D_2 into a linear refutation D_2' of $S' \cup \{A \longrightarrow\}$ with initial top clause $A \longrightarrow$ as follows. First, D_2' resolves the top clause $A \longrightarrow$ with $\Gamma_1 \longrightarrow \Delta_1, A$ (which is in S') to get the new top clause C_1. If C_1 is the empty clause, then we are done. Otherwise, the first step in D_2 was a resolution between the top clause C_1 and some other clause C_1' in $S'[\mathbf{F}/A]$ to get a new top clause C_2. If C_1' is in S', then D_2' just resolves between the same two clauses to get the same new top clause C_2. Otherwise, there is a clause $\Gamma_1 \longrightarrow \Delta_1, A$ in S' such that $\Gamma_1 \longrightarrow \Delta_1 = C_1'$. Then D_2' first resolves between C_1 and $\Gamma_1 \longrightarrow \Delta_1, A$ to get C_2 with an added A on the right as new top clause, and then resolves the latter with the ancestor clause $A \longrightarrow$ to get C_2 again. By an induction that we do not formalize here over the length of D_2, we obtain a linear refutation D_2' of $S' \cup \{A \longrightarrow\}$ with initial top clause $A \longrightarrow$.

Concatenating D_1' with D_2' then yields a linear resolution refutation of S with initial top clause C_0, as claimed. We then lift the result to the first-order case as usual.

Semi-ordered linear resolution is complete by a similar argument: just choose A to be a minimal atom, and C_0 a clause containing it. This is as in Theorem 7.30. $\square$

Linear resolution is not complete for an arbitrary choice of C_0; instead, there is a choice of the initial clause C_0 which will eventually lead to the empty clause, if S

is unsatisfiable. We can combine this with the set-of-support strategy, and choose C_0 outside of the set of axiom clauses. In particular, if the goal is a universally quantified conjunction of atomic formulas, there is only one non-axiom clause, and the choice is then fixed.

The top clause C_i may also have to be resolved not only with the input clauses, i.e. those in S, but also with ancestor clauses C_j, $1 \leq j < i$. The restriction of linear resolution where side clauses are input clauses is called *input resolution*. Unfortunately, input resolution is not complete. For instance, there is no input resolution refutation of $\longrightarrow A, B, A \longrightarrow B, B \longrightarrow A, A, B \longrightarrow$. In fact, there is an input resolution of a set S of clauses if and only if there is a unit resolution of S (See (Chang and Lee, 1973), pages 134–135), and we have already seen that there was no unit resolution refutation of the latter set of clauses.

Finally, note that *no* deletion strategy is compatible with the linear strategy: erasing the current clause is just meaningless.

▸ **EXERCISE 7.8**

A Horn clause with a non-empty right-hand side is called a *definite clause*.

Show that any set of definite clauses is satisfiable, and conclude that input resolution is complete for Horn clauses directly, i.e. without using the fact that input resolution is equivalent to unit resolution.

4.6 *Other Refinements of Resolution*

There are a lot of other refinements of resolution. See (Chang and Lee, 1973) for a survey, and for the details on how and why they work.

Ordered resolution itself may be defined by different notions of ordering. Robert Boyer's *lock resolution*, for example, assigns arbitrary ranks to occurrences of literals in clauses; factoring and binary resolution are then used only on literals of lowest rank in each clause; the literals L in resolvents are assigned ranks by *merging low*, that is, by inheriting the lowest rank of a literal L' that got instantiated to L in the process. This strategy is still complete, and reported to be efficient, although using it in conjunction with most other refinements of resolution (elimination of tautologies, of pure clauses, and so on) breaks completeness.

An interesting variant of linear resolution, called *OL-resolution* (for ordered linear resolution, but this is not exactly what we have called ordered linear resolution in Section 4.5), is linear resolution where clauses are thought of as *lists* of literals without repetition; the order in which literals appears plays the rôle of the ordering on atoms that we have used to constrain resolution derivations. The way that clauses and their resolvents are kept sorted looks a lot like Boyer's lock resolution technique, but the method remains complete even by using deletion strategies. Moreover, clever implementation techniques can be used to represent only the top clause C_i, and encode the centre clauses $C_1, \ldots, C_{i-1}$ inside C_i. (See (Chang and

Lee, 1973), Section 7.4.) We shall in fact see in Chapter 8, Section 3.4 how this technique can be thought of as a technique of tableaux.

Robert Kowalski's *connection graph resolution* is more involved, as it is both an specific implementation and a method for preventing many spurious clauses from being generated. In this method, unifiable pairs of complementary literals are represented as *links*, or edges, in a graph of clauses; resolvents inherit some of the links of their parent clauses, so that the number of unifications to perform is minimised. A clause with a literal without any link is pure, and is eliminated right away. This might in turn eliminate other links, thus creating new pure clauses, and so on. To avoid redundancies, once a link has been used to produce a resolvent, it is cut out, so as not to be used again. This is a very efficient implementation of resolution. However, cutting out used links may break completeness with certain strategies; and when connection graph resolution is complete, the proof of completeness is usually complex.

These problems are alleviated in Jean-Paul Billon's *instance subtraction*, but instance subtraction does not work with resolution. Instead, Billon implements instance subtraction on top of Lee and Plaisted's calculus (Lee and Plaisted, 1992), where instances of clauses, not resolvents, are generated using the following rule:

$$\frac{C \wedge A \qquad \neg A' \wedge C'}{(C \wedge A)\sigma \quad (\neg A' \wedge C')\sigma}$$

where the clauses below the line are added to the current set of clauses, and σ is the most general unifier of A and A'. This idea by Lee and Plaisted allows the prover to decouple the task of finding unifiers (by unifying complementary literals) from the task of showing the propositional unsatisfiability of an underlying set of ground clauses. This second task can be left to any propositional method, including resolution, but is not limited to resolution.

On top of this, Billon's idea is to attach to each clause a set of substitutions that have been used to generate instances, and which must not be reused again. This avoids generating the same clause over and over again in most cases. Instead of cutting links, it forbids making unifications along previous links by comparing substitutions. This is both more general as Kowalski's technique, as it is not limited to cutting the link on which we resolved, and easier to prove complete. Billon has, since then, streamlined the idea to the point that we don't even need this recording of substitutions, by a clever usage of a notion of paths through the set of clauses (Billon, 1996).

5   RESOLUTION AS CUT-ONLY PROOFS

We end this chapter with an analysis of what we have done from a proof-theoretical point of view. We have indeed developed most of resolution, as is customary, from a mostly semantic point of view. However, the resolution rule looks so much like the Cut rule in Gentzen's system **LK** that there must be a connection between them. The connection, as can be expected, comes from a variant of the Cut-Elimination Theorem (Theorem 6.26, page 214). This will also answer the riddle we gave in Chapter 2, Section 4.2: since we can eliminate all cuts, how does this justify resolution, where all rules *but* Cuts are eliminated?

First, extend slightly the Cut Elimination Theorem for first-order logic:

Theorem 7.34 (Extended Cut Elimination) *Let S be a set of atomic sequents. Let* **LK** $+ S$ *be the sequent system obtained by adding to the rules of* **LK** *all rules of the form:*

$$\frac{}{\Gamma \longrightarrow \Delta}$$

where $\Gamma \longrightarrow \Delta$ is a sequent in S. Such rules are called non-logical axioms.

Let S_∞ be the smallest set of sequents containing S and stable by the Cut rule. (That is, if $\Gamma \longrightarrow \Delta, \Phi$ and $\Phi, \Gamma' \longrightarrow \Delta'$ are in S_∞, then so is $\Gamma, \Gamma' \longrightarrow \Delta, \Delta'$.) We say that a Cut is non-logical *if and only if its premises are both in S_∞. (Then, its conclusion is in S_∞, too.)*

Then, an atomic sequent is provable in **LK** $+ S$ *if and only if it is a weakened sequent of a sequent provable in* **LK** $+ S$ *by a proof where all Cuts are non-logical.*

Proof: Notice that weakening steps can be encoded as Cuts with Ax, provided that the weakened sequent is non-empty. Moreover, any sequence of weakenings is equivalent to a single weakening, so we implicitly assume that the conclusion of a weakened sequent is never cut with Ax.

We show that a sequent is provable in **LK** $+ S$ if and only if it is provable by a proof where the Cuts are either non-logical, or are Cuts between a weakened version of a sequent in S_∞ and Ax (yielding again a weakened version of the same sequent in conclusion).

The proof is the same as for Theorem 6.26, and rests on a variant of Lemma 6.25. We say that a Cut is *essential* if it is non-logical and it is not a Cut with Ax, and we shall eliminate all inessential Cuts.

We redefine the cut-rank $r(\pi)$ of a proof π in **LK** $+ S$ as 0 if π only contains essential Cuts (instead of π being cut-free in the **LK** case), and otherwise as the maximum of $d(\pi') + d(\Phi)$, where π' ranges over all subproofs of π ending in a logical Cut between two proofs with only essential Cuts, and with cut formula Φ. Then, we have to show that, whenever π is an **LK** $+ S$ proof of non-zero cut-rank r, we can transform it into an **LK** $+ S$ proof of rank at most $r - 1$ of the same sequent, where no non-logical Cut lies below a Cut with Ax.

To ensure that no non-logical Cut lies below a Cut with Ax, we can always push Cuts with Ax (weakenings) down the derivation: If $\Gamma \longrightarrow \Delta$ produces $\Gamma, \Gamma' \longrightarrow \Delta, \Delta'$ by weakening, and if we infer a sequent $\Gamma'' \longrightarrow \Delta''$ from the latter by some rule R, then either R acts on some formula already in $\Gamma \cup \Delta$, and then the rules permute, or R acts on some other formula, and then $\Gamma'' \longrightarrow \Delta''$ must be a weakened

version of $\Gamma \longrightarrow \Delta$, as can be readily checked on all **LK** rules. We shall tacitly
assume that this normalisation is always performed.

As in the Lemma, choose a topmost subproof π' of π ending on an inessential
Cut on Φ, with $d(\pi') + d(\Phi)$ maximal. Then π' has the form:

$$
\begin{array}{cc}
\pi_1 & \pi_2 \\
\vdots & \vdots
\end{array}
$$

$$
\frac{\Gamma \longrightarrow \Delta, \Phi \qquad \Gamma', \Phi \longrightarrow \Delta'}{\Gamma, \Gamma' \longrightarrow \Delta, \Delta'} \text{ Cut}
$$

where π_1 and π_2 do not contain any inessential Cuts. Because we have assumed
that we would always push down weakenings below any other Cuts, π_1 and π_2 do
not contain any logical Cut at all (whether inessential or weakenings).

By construction, the last rule of π_1 is not a non-logical Cut, or the last rule of
π_2 is not a non-logical Cut. Indeed, the contrary would mean that the Cut above
is non-logical, hence essential. Assume without loss of generality that π_1 does not
end in a non-logical Cut. Then, it does not end with Cut at all, since π_1 does not
contain any logical Cut. If π_2 does not end with Cut, then we proceed as in the
proof of Lemma 6.25.

Otherwise, π_2 ends in a non-logical Cut, and π_1 does not end with Cut. Because
S_∞ contains only atomic sequents, the cut formula Φ must be atomic. Then, either
Φ is not the principal formula in the last rule of π_1, so the first case in the proof of
Lemma 6.25 applies, and we move the Cut along π_1; or Φ is the principal formula,
but since Φ is atomic, π_1 must end in an instance of Ax, so that $\Gamma, \Gamma' \longrightarrow \Delta, \Delta'$
must be a weakening of $\Gamma', \Phi \longrightarrow \Delta'$; but this contradicts our assumption that the
Cut above was inessential.

Finally, the process terminates as in Lemma 6.25, yielding a proof where the
only Cuts are non-logical Cuts and Cuts with Ax, with the Cuts with Ax pushed all
the way down to the conclusion of the proof. The latter weakenings can be contracted to at most one Cut with Ax, yielding the desired result.   □

We infer the corresponding notion of subformula property:

Corollary 7.35 (Extended Subformula Property) *Let S be a set of atomic sequents.
If $\Gamma \longrightarrow \Delta$ is provable by a proof in* **LK** $+ S$ *where all Cuts are non-logical, then
all formulas occurring in the proof are subformulas of formulas in Γ, Δ, or appear
in S.*

Proof: By structural induction on the proof. The only extension to Corollary 6.28
is when we come to the conclusion of a non-logical Cut rule; but then all formulas,
including the cut formula, appear in S_∞. We claim that if they appear in S_∞, then
they appear in S. Indeed, let S_0 be S, S_1 be the set of all conclusions of Cut rules
on sequents in S_0, and in general S_n be the level set at level n; we have $S_\infty = \bigcup_{n \in \mathbb{N}} S_n$, and an easy induction on n proves that the set of formulas appearing in
S_n is exactly the set of formulas appearing in S.   □

In particular, if the sequent to be proved is the empty sequent, we get back the proof-theoretic version of the Completeness Theorem for propositional resolution (Theorem 2.39, page 49), in a purely syntactical way:

Theorem 7.36 (Resolution) *Let Φ be a propositional formula in cnf. We view Φ as a set S of atomic sequents. Then Φ is inconsistent if and only if the empty sequent $\longrightarrow$ can be deduced from S and the Cut rule only.*

Proof: By Theorem 7.34, if Φ is inconsistent, then we can derive the empty sequent from S, using non-logical Cuts and the rules of **LK** other than Cut, then possibly doing a weakening step. But the only sequent that weakens to the empty sequent is the empty sequent itself. By Corollary 7.35, this proof only involves formulas appearing in S. But this means that no rule except Cut has been used, hence the result. $\square$

Then, to lift this result to first-order clauses, we use the syntactic version of Herbrand's Theorem (Theorem 6.30) to conclude that a set of first-order clauses S is inconsistent if and only if some finite set of instances of clauses of S is inconsistent.

Unification is needed because we do not know which instances we shall need in advance. That is, we replace the usual Cut rule:

$$\frac{\Gamma \longrightarrow \Delta, \Phi \qquad \Gamma', \Phi \longrightarrow \Delta'}{\Gamma, \Gamma' \longrightarrow \Delta, \Delta'}$$

by the following "lazy Cut" rule:

$$\frac{\Gamma \longrightarrow \Delta, \Phi_1, \ldots, \Phi_n \qquad \Gamma', \Phi'_1, \ldots, \Phi'_{n'} \longrightarrow \Delta'}{\Gamma\sigma, \Gamma'\sigma' \longrightarrow \Delta\sigma, \Delta'\sigma'}$$

where n and n' are non-zero, and $\Phi_1\sigma = \ldots = \Phi_n\sigma = \Phi'_1\sigma' = \ldots = \Phi'_{n'}\sigma'$. That is, instead of generating all instances of the premises, and finding some that we can cut, we embed the process of computing these instances in the rule itself. We say that this is a "lazy" rule, because we don't have to know all bindings of variables to terms to be able to use it.

The developments on unification of Section 2 carry through, and we can again encode this rule as the combination of a factoring rule and a binary resolution rule.

The reason why we may need factoring, which is a costly rule in terms of non-determinism in proof-search, takes a new meaning here: it is the expression of the implicit contraction rule of classical logic. Indeed, when we apply the substitution σ to some sequent, it σ unifies several formulas in the sequent, then an automatic contraction step occurs, which we are forced to guess by factoring. Such surprising behaviours in a logical system are precisely what linear logic aims at controlling. This does not mean that finding proofs in linear logic is simple, however.

In fact, it is rather complicated to find proofs in linear logic. It is true that in the exponential-free fragment, also called MALL1 (Multiplicative Additive Linear Logic, at the first order), provability is decidable. This justifies that the rules on exponentials are the core of the undecidable part of first-order logic (look at Herbrand's Theorem: without contraction, the number k of instances we need without exponentials is necessary exactly one). But provability in MALL1 is complete for NEXPTIME, the class of problems that we can solve in non-deterministic exponential time. This is a very high complexity class indeed.

Other techniques also have a proof-theoretical justification. For example, deleting tautologies amounts to ignoring instances of Ax: as Theorem 7.34 shows, we cannot ignore Ax in general, as we might need to cut with axioms Ax as a final step in the proof of a sequent $\Gamma \longrightarrow \Delta$ in $\mathbf{LK} + S$; but if Γ and Δ are empty, it is all right to do so, since the only sequent that weakens to the empty sequent is the empty sequent itself. The reason why this works, then, is that we are able to push all Cuts with Ax down the proof, as shown in the proof of Theorem 7.34.

Eliminating subsumed clauses is then partly justified by the same argument. Recall that a clause $C = \Gamma \longrightarrow \Delta$ subsumes C' if and only if C' has the form $\Gamma', \Gamma\sigma \longrightarrow \Delta\sigma, \Delta'$ for some substitution σ, i.e. if C' is a weakened version of an instance of C. Weakening is just cutting with Ax, and can therefore be pushed down to the bottom of the proof. To explain why using instantiated clauses is useless (except as factors), we have to resort to Lemma 7.15, which was already proved completely syntactically. (I.e., contrarily to the completeness of ordered resolution, for example, we did not have to resort to semantic considerations, we merely pushed substitutions around.)

Finally, it can be observed that the various ordering strategies, semantic resolution, linear resolution, and so on, are expressions of the fact that Cut on some formula permutes with Cut on any other formula, extending Theorem 6.32, page 216. These restrictions of resolution can then be thought of as further restrictions on the shape of Cut-only proofs of the empty sequent, so as to make them more and more canonical. However, the proofs that we can permute Cuts is complicated by the fact that we have to do this modulo substitutions. This is why the completeness proofs for semantic resolution and the ordering strategy proceeded by reasoning in the fully instantiated (non-lazy, or ground) case, then by lifting the result to the first-order (lazy) case.

CHAPTER 8

TABLEAUX, CONNECTIONS AND MATINGS

A more direct use of Gentzen's system **LK** is, as in the propositional case, the method of Tableaux. Although it is not as well-developed, in terms of refinements, as resolution, it is grounded in a cleaner formalism, namely **LK** without the Cut rule. Moreover, as cut-free proofs are already (almost) canonical without any need for refinements of the method, we can hope for a considerably smaller space of possible proofs than in resolution. As we shall see, this is not quite true.

We present the basic method in Section 1. We refine it to get the method of *free-variable tableaux* in Section 2, and describe in Section 3 the quite similar method of connections, a.k.a. matings. We finally relate these methods to some variants of resolution in Section 3.2, showing that the methods are not that far apart.

1   FIRST-ORDER TABLEAUX

Because of the Cut Elimination Theorem for first-order logic (Theorem 6.26, we can restrict our search space for a proof of a given sequent $\Gamma \longrightarrow \Delta$ to the space of cut-free proofs. This is exactly what we did in the propositional case in Chapter 2, Section 4.1.

In this section, we examine how we may look for a proof of an existential formula Φ. To prove an arbitrary first-order formula, we can always first herbrandize it to get an existential formula, so we don't lose any generality in doing this. We shall refine this to handle general first-order formulas directly in Section 2.

Looking for an **LK** proof of $\Phi = \exists x_1 \ldots \exists x_n \cdot \Psi$, where Ψ is quantifier-free, can be done in much the same way as with propositional tableaux. The only difference is that we have to handle the additional $\exists$R rule. This poses a new problem: to prove $\Gamma \longrightarrow \Delta, \exists x \cdot \Phi$ using $\exists x \cdot \Phi$ as the principal formula, we have to guess an instance $\Phi[t/x]$ of Φ, and produce $\Gamma \longrightarrow \Delta, \exists x \cdot \Phi, \Phi[t/x]$, or $\Gamma \longrightarrow \Delta, \Phi[t/x]$.

Because weakening on the right-hand side of sequents is admissible in the classical system **LK**, if $\Gamma \longrightarrow \Delta, \Phi[t/x]$ is provable, then so is $\Gamma \longrightarrow \Delta, \exists x \cdot \Phi, \Phi[t/x]$. So it is enough to search for proofs of the latter. However, since $\exists x \cdot \Phi$ remains on the right-hand side, we can repeat the process ad infinitum: proof search is not guaranteed to terminate. But we are forced to introduce such duplications to remain complete. What we have done, logically speaking, is applying $\exists$R coupled with an instance of the contraction rule. Linear logic (Chapter 3) extends naturally to the first-order case, and solves this problem. Intuitionistic logic (see Exercise 6.7) also solves the problem, by entirely disallowing weakenings on the right. But in both these logics, proof search is made considerably more difficult than in classical logic; notably, we cannot even put formulas in prenex forms to begin with.

1.1   Using Herbrand's Theorem

The simplest solution is to use Herbrand's Theorem (Theorem 6.30): Φ is provable if and only if there is an integer k, and k substitutions $\sigma_1, \ldots, \sigma_k$ such that $\longrightarrow \Psi\sigma_1, \ldots, \Psi\sigma_k$ is derivable in **LK**. Because Cuts can be eliminated from **LK** proofs, and because cut-free formulas obey the subformula property, the latter condition is equivalent to: $\longrightarrow \Psi\sigma_1, \ldots, \Psi\sigma_k$ is derivable in $\mathbf{LK}_0$ without Cut.

We can therefore extend the propositional tableaux method to the first-order case in the following way:

- Guess an integer k;

- guess k substitutions $\sigma_1, \ldots, \sigma_k$;

- apply the propositional tableaux method to determine whether there is a cut-free proof of $\longrightarrow \Psi\sigma_1, \ldots, \Psi\sigma_k$.

This naïve procedure rests on a lot of guesswork. The work of guessing the substitutions σ_i can indeed be replaced by computing unifiers, as was already the case in resolution.

First, we replace the problem of guessing k substitutions by that of guessing just one. Let $\Psi_1, \Psi_2, \ldots,$ be renamings of the formula Ψ with pairwise distinct sets of free variables. Then, we have to find k substitutions such that $\longrightarrow \Psi_1\sigma_1, \ldots, \Psi_k\sigma_k$ is provable, or equivalently just one substitution σ such that $(\longrightarrow \Psi_1, \ldots, \Psi_k)\sigma$ is provable: let σ be $\sigma_1 \cup \ldots \cup \sigma_k$, and conversely, σ_i be σ restricted to $\mathrm{fv}(\Psi_i)$, $1 \leq i \leq k$.

So, without loss of generality, we have to find a substitution σ that instantiates a given sequent $\Gamma \longrightarrow \Delta$ to one that we can prove in $\mathbf{LK}_0$ without Cut. Instead of guessing σ, we compute it by letting the tableaux expansion guide us. Notice that the α-rules and the β-rules of Figure 2.5 are unaffected by our knowing or not knowing the substitution to apply to the formulas in the current path. The only change is in the way we close paths, i.e. in the way we use Ax: instead of closing a path $\Gamma \longrightarrow \Delta$ by finding a formula that is in both Γ and Δ, we have to find a formula Φ_1 in Γ and a formula Φ_2 in Δ such that $\Phi_1\sigma = \Phi_2\sigma$ for some substitution σ. When Φ_1 and Φ_2 are atomic formulas, this is just unification. The substitution σ that we find must then be applied to the rest of the proof under construction before we proceed to close other paths. We make this clearer by defining the method, extending that of Chapter 2, Section 4.1.

Keeping up with the tradition of using little memory, here is the usual *depth-first* way of expanding tableaux, defined as a non-deterministic function `prove`. `prove` now takes two arguments, a list l of signed formulas (the current path) and the current substitution σ. `prove`(l, σ) either returns a new substitution σ' that closes all paths that expand from $l\sigma$, or a special token fail, meaning that no such substitution can be found:

- Non-deterministically choose between closing the current branch (finding a positive atomic formula $+\Phi$ and a negative atomic formula $-\Phi'$ in l such that $\Phi\sigma$ and $\Phi'\sigma$ are unifiable with mgu σ', and returning $\sigma\sigma'$), or the following:

- if l contains only signed atoms, return fail;

- otherwise, apply the expansion strategy to choose a non-atomic signed formula Φ in l, and let l' be l with Φ removed.

    - if Φ is an α-type formula (yielding formulas $\Phi_1, \ldots, \Phi_n$), then: if $\texttt{prove}(\Phi_1 :: l', \sigma)$ returns fail, then return fail; otherwise, let σ_2 be the returned substitution, then if $\texttt{prove}(\Phi_2 :: l', \sigma_2)$ returns fail, then return fail; otherwise, ..., let σ_n be $\texttt{prove}(\Phi_{n-1} :: l', \sigma_{n-1})$, and return $\texttt{prove}(\Phi_n :: l', \sigma_n)$.

    - if Φ is a β-type formula (yielding formulas $\Phi_1, \ldots, \Phi_n$), then return $\texttt{prove}(\Phi_1 :: \ldots :: \Phi_n :: l', \sigma)$.

This is almost the implementation we described for propositional tableaux in Chapter 2, Section 4.1, except for the puzzling non-deterministic choice in the first step. Let us illustrate what it means on an example. Assume we wish to prove:

$$\exists x \cdot \exists y \cdot \exists z \cdot$$
$$(P(f(y), z) \wedge P(y, f(z))) \vee P(f(a), f(a))$$
$$\Rightarrow P(x, f(y)) \vee P(f(x), y)$$

Using Herbrand's Theorem, and guessing the needed number of instances k to be 1, we have to expand and close the tableau with initial path:

$$+(P(f(y), z) \wedge P(y, f(z))) \vee P(f(a), f(a))$$
$$\Rightarrow P(x, f(y)) \vee P(f(x), y)$$

As there are no complementary pairs of atomic formulas, we have to expand the formula above. After a few expansion steps, we get:

$$+P(x, f(y))$$
$$+P(f(x), y)$$
$$-P(f(y), z) \mid -P(f(a), f(a))$$
$$-P(y, f(z)) \mid$$

Now, we can decide to close the leftmost path first for example, but we have four possibilities of complementary pairs, yielding three different mgus: $[z/x, z/y]$, $[f(y)/x, f(y)/z]$ and $[z/x, f(z)/y]$. If we choose $[z/x, z/y]$, then the current substitution σ changes from $[]$ to $[z/x, z/y]$, and the rightmost path instantiates by the latter to $\{+P(z, f(z)), +P(f(z), z), -P(f(a), f(a))\}$, which cannot be closed. We therefore have to backtrack and choose another mgu, say, $[f(y)/x, f(y)/z]$. The current substitution σ is now $[f(y)/x, f(y)/z]$, and the rightmost path instantiates to $\{+P(f(y), f(y)), +P(f(f(y)), y), -P(f(a), f(a))\}$, which can be closed by unifying $P(f(y), f(y))$ with $P(f(a), f(a))$, yielding the final substitution $[f(a)/x, f(a)/z]$.

Because of this need for backtracking, Prolog is usually considered a language of choice for implementing proof-search in tableaux methods. An efficient and small tableaux prover (for small problems at least) is described in (Beckert and Posegga, 1994). This is however a prover based on the *free variable tableaux* variant; see Section 2.

Theorem 8.1 (Soundness, Completeness) *Let Φ be an existential formula $\exists x_1 \cdot \ldots \exists x_n \cdot \Psi$.*

*Then Φ is provable in **LK** if and only if there is an integer k, a substitution σ, and a tableau expansion from the path $\{\Psi_1, \ldots, \Psi_k\}$ that is closed by σ.*

More precisely, given any expansion strategy f, then if Φ is provable, the tableaux procedure above terminates and returns such a substitution σ as a result for some high enough k.

Proof: If σ closes some tableau expansion from $\{\Psi_1, \ldots, \Psi_k\}$, then $\longrightarrow \Psi_1\sigma$, $\ldots, \Psi_k\sigma$ is derivable in **LK**, so Φ is provable.

Conversely, assume Φ provable in **LK**. By Herbrand's Theorem, there is a substitution σ such that $\longrightarrow \Psi_1\sigma, \ldots, \Psi_k\sigma$ is derivable in **LK$_0$** without Cut. Therefore, the fully expanded (propositional) tableau T stemming from the path $\{+\Psi_1\sigma, \ldots, +\Psi_k\sigma\}$ has all its paths closed. We lift this to the first-order case by considering the fully expanded tableau T' stemming from the path $\{+\Psi_1, \ldots, +\Psi_k\}$: let C be any path in this fully expanded tableau, then $C\sigma$ must also be a fully expanded path in T', so it must be closed by σ. Now, let $C_1, \ldots, C_m$ be the paths in T': for every i, $1 \leq i \leq m$, σ unifies some pair of atomic formulas A_i and B_i of opposite signs in C_i. Therefore, there is a simultaneous mgu of all the pairs A_i, B_i, $1 \leq i \leq m$, which will eventually be found by choosing non-deterministically the right pairs and computing the simultaneous mgu incrementally, as in the proof of Lemma 7.4. $\qquad\qquad\qquad\qquad\qquad\qquad\qquad\qquad\qquad\qquad\qquad\quad$ $\square$

The proof shows that we don't actually need to apply the non-deterministic choice step between closing and expanding systematically: the procedure is still complete if we force the expansion of any path that contains some non-atomic formula, and try and close only atomic paths. Doing so may reduce non-determinism, but in general forces us to wait until all paths have been fully expanded before closing them, that is, to wait until we have possibly exponentially many paths to close. Therefore, expanding before closing is not realistic in general, and we prefer to try and close each path as soon as possible, although it may add some non-determinism.

This is done as follows. Given a current path, we first try to unify atomic formulas of opposite sign on the path, and return the corresponding unifier. If the search subsequently fails, then the prover backtracks to the next choice of complementary pairs to unify. If there are no complementary pairs left, then the prover tries to expand some α or β formula. If this is not possible, the prover backtracks to the last choice point, or fails if there is not any. This can be improved by having the prover avoid unifying complementary pairs that have already turned out to lead to no proof: maintain the path as a list *old* of old atomic literals, a list *new* of new atomic literals, and a list *exp* of formulas to expand. When expanding a formula in *exp*, we first add all literals in *new* to *old*, and reset *new* to the empty list; then we expand the formula, putting the generated subformulas in *exp* again if they are not atomic, or in *new*. We then only need to try and close a path by unifying a literal in *new* with some literal of opposite sign in *old*.

The problem, given k, of finding whether $\longrightarrow \Psi_1, \ldots, \Psi_k$ has a provable instance is decidable. What is its complexity, then? It turns out that the problem is complete for some new complexity class, named Σ_2^P: this class is that of problems that we can solve by using a non-deterministic Turing machine, which has in addition access to an oracle in NP. That is, not only can the machine guess solutions, but it can also have the answers to NP-complete problems in unit time, by asking an oracle. Logically speaking, whereas NP is the class of problems of the form "does there exist an x (guessable in polynomial time) such that $P(x)$ holds (checkable in polynomial time)?", Σ_2^P is the class of problems of the form "does there exist an x (guessable in polynomial time), such that for all y (again, guessable in polynomial time), such that $P(x,y)$ holds (checkable in polynomial time)?". All problems in Σ_2^P are in PSPACE, hence can be solved in deterministic exponential time, but the fact that Σ_2^P is larger, and probably strictly larger, than NP suggests that the problem is harder to solve efficiently than NP-complete problems, like propositional satisfiability.

1.2 Extension Steps

Given k, we are now able to decide whether $\longrightarrow \Psi_1, \ldots, \Psi_k$ has an instance that is provable in $\mathbf{LK}_0$. How do we guess k?

The answer is: we cannot. At least, we cannot compute the least k we need, or even any conservative upper bound for this least k. The reason is the undecidability of first-order logic (Theorem 6.20, page 205). Indeed, assume that we can compute a value of k from the existential formula Φ, such that if Φ is provable, then $\longrightarrow$ $\Psi_1, \ldots, \Psi_k$ has a provable instance. Then we decide whether Φ is provable in the following way: compute such a k, then apply the terminating algorithm described in the previous section. Therefore, this would provide us with a means of deciding the provability of Φ by machine, which would contradict Theorem 6.20.

So, we really have to guess k. One way is to use *iterative deepening*. Initialise k to 1, then apply the tableaux procedure to decide whether $\longrightarrow \Psi_1$ has a provable instance. If it does, then Φ is proved. Otherwise, increment k and apply the tableaux procedure again to decide whether $\longrightarrow \Psi_1, \Psi_2$ has a provable instance. If it does, then Φ is proved. Otherwise, increment k, and so on.

Iterative deepening, combined with depth-first search as in the previous section, is a popular way of implementing tableaux. As we have already discussed in Chapter 7, Section 4.1, iterative deepening depth-first search is a reasonable alternative to breadth-first search, in so far as it saves lots of memory. Unfortunately, the time requirements for proving formulas in general grows exponentially with k and the size of the formula, so that only small and easy theorems can be proved with this strategy.

In general, to represent the incrementing of k, we add a new *extension step* in the tableau expansion procedure: at any time in the proof, we are free to add a new instance $\Psi\rho$ of Ψ to the current path, where ρ renames all free variables in Ψ to new variables. We can then use heuristics to favour opportunistic extension steps, in the hope of closing the current path faster than with the current formulas on the path. The only thing we have to monitor is the *fairness* of the process: we must not give precedence to extension steps for indefinitely long, otherwise the procedure would become incomplete. Then, finding the right heuristics is much of a black art, and we won't delve into it.

▶ **EXERCISE 8.1**

We can already improve the tableaux proof-search procedure by exploiting symmetries. In particular, $\longrightarrow \Psi_1, \ldots, \Psi_k$ is the same sequent as $\longrightarrow \Psi_{s(1)}, \ldots, \Psi_{s(k)}$ for any permutation s. How can you use this observation to restrict the tableaux proof search?

▶ **EXERCISE 8.2**

Propositional tableaux are scalable in the sense that, under reasonable assumptions, for any propositional formula Φ, if Φ is provable by tableaux, then any substitution instance of Φ is provable in much the same time by tableaux. (See Exercise 2.19.)

In the same vein, let Φ be a first-order formula, and generate a formula Φ' by replacing predicate symbols by full predicates (formulas). More precisely, by replacing a n-ary predicate symbol P by a formula F with n free variables $x_1, \ldots, x_n$, we mean consistently replacing in Φ all atoms of the form $P(t_1, \ldots, t_n)$ by formulas $F[t_1/x_1, \ldots, t_n/x_n]$.

Are first-order tableaux scalable? Discuss palliatives.

2 FREE VARIABLE TABLEAUX

2.1 Presentation

An evolution of the basic tableaux method that we presented in Section 1 is the *free variable tableaux* method, as Melvin Fitting calls it (Fitting, 1983).

In free variable tableaux, extension steps are integrated in the proof-search process itself, by using lazy versions of the ∃R and ∀L rules of **LK**, in much the same

γ-rules

γ	γ_1
$+\exists x \cdot \Phi$	$+\Phi[x'/x]$
$-\forall x \cdot \Phi$	$-\Phi[x'/x]$

(with x' a new variable)

Figure 8.1. γ-rules in free-variable tableaux

way as resolution could be explained in terms of a lazy Cut rule (see Chapter 7, Section 5). These rules are:

$$\frac{\Gamma \longrightarrow \Delta, \exists x \cdot \Phi, \Phi[x'/x]}{\Gamma \longrightarrow \Delta, \exists x \cdot \Phi} \exists R' \qquad \frac{\Gamma, \forall x \cdot \Phi, \Phi[x'/x] \longrightarrow \Delta}{\Gamma, \forall x \cdot \Phi \longrightarrow \Delta} \forall L'$$

where x' is a new variable, standing for the term t in the corresponding **LK** rule. We have seen that, instead of guessing t, the tableaux method computed it by binding x' to t when computing a most general unifier to close a path (equivalently, to make a sequent an instance of Ax). This yields the so-called γ-rules of free-variable tableaux: see Figure 8.1.

The $\forall$R and $\exists$L rules can also be adapted, so that we won't need to herbrandize the formula to prove as a preprocessing step. One way of doing this is to introduce replacements for $\forall$R and $\exists$L as well, in the style of the above rules, but with the proviso that the newly introduced variable x' *must not be bound during unification*. We can achieve this by creating a new constant instead of a new variable. This yields the following rules:

$$\frac{\Gamma \longrightarrow \Delta, \Phi[c/x]}{\Gamma \longrightarrow \Delta, \forall x \cdot \Phi} \forall R' \qquad \frac{\Gamma, \Phi[c/x] \longrightarrow \Delta}{\Gamma, \exists x \cdot \Phi \longrightarrow \Delta} \exists L'$$

where c is a new constant.

This is not enough, however, because we must be sure, not only that c does not appear in Γ or Δ, but that it will never appear in any instance of Γ and Δ that we get by unification. Consider indeed the sequent:

$$\longrightarrow \exists x \cdot (P(x) \Rightarrow \forall y \cdot P(y))$$

which is not valid. Applying $\exists$R', $\Rightarrow$R and $\forall$R' without any further care yields the sequent:

$$P(x') \longrightarrow P(c)$$

which can be closed by binding x' to c.

To forbid this, a database of *dependencies* is maintained. Whenever we create a new constant c by the $\forall$R' or $\exists$L' rules, we record the *dependency constraints*

δ-rules

δ	δ_1
$+\forall x \cdot \Phi$	$+\Phi[c/x]$
$-\exists x \cdot \Phi$	$-\Phi[c/x]$

(with c a new constant, for which we add the constraint $c < y$ for every free variable y in the current path.)

Figure 8.2. δ-rules in free-variable tableaux

$c < y$ for every free variable y in the current sequent: this states the side condition on universal variables c that c should not depend on y.

Then, the occurs-check test in unification is extended so as to cope with these extra dependency constraints: we say that a term t depends on x if and only if $t = x$, or t is a constant c with the constraint $c < x$ in the database, or $t = f(t_1, \ldots, t_n)$ and some t_i, $1 \leq i \leq n$, depends on x. We then require that most general unifiers be found as substitutions σ such that for no $x \in \operatorname{dom} \sigma$, $x\sigma$ depends on any variable in $\operatorname{dom} \sigma$; this is an extension of the idempotence condition on most general unifiers.

In tableaux jargon, the $\forall R'$ and $\exists L'$ rules give rise to the δ-rules shown in Figure 8.2.

We can now extend the notion of paths in free-variable tableaux so that: if C is a path containing a γ-formula Φ, then $C \cup \{\gamma_1\}$ is a path again, where γ_1 is defined on Figure 8.1; and if C is a path containing a δ-formula Φ, then $C \setminus \{\Phi\} \cup \{\delta_1\}$ is a path again, where δ_1 is defined on Figure 8.2. Then, a *fair expansion strategy* is one which does not do any infinite consecutive sequence of γ-expansions (or extension steps), and, just as in Section 1, any fair expansion strategy gives rise to a sound and complete proof search procedure.

2.2   Herbrandizing On-the-fly

The dependency database of free-variable tableaux can be coded more straightforwardly by generating, not constants c with constraints $c < y_1, \ldots c < y_n$, but a unique term $f(y_1, \ldots, y_n)$, where we encode the dependency constraint directly inside the term $f(y_1, \ldots, y_n)$. This is just another way of doing herbrandization, on the fly.

> Introducing new constants or new function symbols is safe in first-order logic. In higher-order logic, it is not, because instantiating higher-order variables might propagate the introduction of constants or variables into other formulas. In fact, we cannot herbrandize: there, we have to stick to free-variable tableaux, and even consider c not as constants, but as variables that cannot be instantiated to any other term (Kohlhase, 1995).

There is another reason why we would want to avoid generating constants like c. Creating a new constant each time we want to expand a positive universal quantifi-

$$\delta^{++}\text{-rules}$$

δ	δ_1
$+\forall x \cdot \Phi$	$+\Phi[f(x_1,\dots,x_n)/x]$
$-\exists x \cdot \Phi$	$-\Phi[f(x_1,\dots,x_n)/x]$

$$(\mathrm{fv}(\Phi) \setminus \{x\} = \{x_1,\dots,x_n\}, f = \lambda x_1 \cdot \dots, \lambda x_n \cdot \epsilon x \cdot \Phi)$$

Figure 8.3. δ^{++}-rules in free-variable tableaux

cation or a negative existential quantification may indeed be inefficient. The reason is that two constants c and c' created at different times in the expansion process never unify. This may prevent the tableau procedure from closing paths quickly, although it might have been possible to do so by instantiating the formula Φ above more smartly.

Using the idea that what the δ-rules accomplish is just herbrandization/ skolemization, we examine the proof of Theorem 6.34 more deeply: to expand a δ-formula, say $+\forall x \cdot \Phi$, we have to replace x in Φ by some term t, such that different instances of $\forall x \cdot \Phi$ produce different instances of t. In the rules above, we produced a new term t (in the form of a new constant) each time we instantiated Φ, so the condition is verified. But we can do better, and in particular use the δ^{++}-rule of Figure 8.3, due to Peter Schmitt, Reiner Hähnle and Bernhard Beckert (Beckert *et al.*, 1993). The rule is shown on Figure 8.3 and corresponds to the following sequent rules:

$$\frac{\Gamma \longrightarrow \Delta, \Phi[f(x_1,\dots,x_n)/x]}{\Gamma \longrightarrow \Delta, \forall x \cdot \Phi} \forall\mathrm{R}^+ \qquad \frac{\Gamma, \Phi[f(x_1,\dots,x_n)/x] \longrightarrow \Delta}{\Gamma, \exists x \cdot \Phi \longrightarrow \Delta} \exists\mathrm{L}^+$$

where $\mathrm{fv}(\Phi) \setminus \{x\} = \{x_1,\dots,x_n\}$, and f is the symbol $\lambda x_1 \cdot \dots, \lambda x_n \cdot \epsilon x \cdot \Phi$, instead of $\forall\mathrm{R}$ and $\exists\mathrm{L}$. (For ϵ and λ, compare with our remarks on Hilbert's ϵ symbol at the end of Chapter 6, Section 5.1.)

To build the Skolem function symbol f, we do the following: first, we rename all variables in Φ so that they take on canonical names. Then, we produce the expression $\lambda x_1 \cdot \dots, \lambda x_n \cdot \epsilon x \cdot \Phi$, where $x_1, \dots, x_n$ are the canonical names of variables, and Φ is assumed to be renamed. Such an expression, where λ is a binding operator, just as $\forall$ or $\exists$, can be represented in the same way as formulas, namely as a tree-like object. We now consider the whole expression, which is neither a term nor a formula of first-order logic, as a new symbol in itself: this is the desired Skolem function symbol.

It might seem surprising that we decide that a whole expression, and in fact a λ-expression, is to be considered as a symbol, since we are accustomed to thinking of symbols as mere character strings; but symbols may be anything we like, the only constraint being that there are infinitely many of them.

We might also wonder why we produce such seemingly complicated symbols, when generating a fresh constant sufficed. As we have said before, generating new constants may considerably increase the size of a proof to be found for the goal formula. The δ^{++}-rule is one of the most favourable ones in terms of proof length. Moreover, generating new function symbols in this way is quite simple to implement, as we already have all the machinery we need for building and comparing formulas.

This technique is in particular important when two quantified subformulas of the formula to prove are the same, or when it contains logical equivalences. For example, assume we wish to prove the formula:

$$\left(P(a) \Leftrightarrow \left((\exists x \cdot Q(x)) \Leftrightarrow P(a) \right) \right) \Rightarrow \exists x \cdot Q(x)$$

If we expand the corresponding tableau, we get:

$$(1) -P(a) \Leftrightarrow ((\exists x \cdot Q(x)) \Leftrightarrow P(a)) \ (*)$$
$$(2) + \exists x \cdot Q(x)$$
$$(3) + Q(x_1) \quad (\gamma \text{ from } (2))$$

$$
\begin{array}{c|c}
\begin{array}{l}
(6) + P(a) \\
(8) +(\exists x \cdot Q(x)) \Leftrightarrow P(a) \ (*) \\
(10)-\exists x \cdot Q(x) \ (*) \quad (11) + \exists x \cdot Q(x) \\
(12) + P(a) \qquad\qquad (13) - P(a) \\
(14) - Q(c) \\
(\delta^{++} \text{ from } (10))
\end{array}
&
\begin{array}{l}
(7) - P(a) \\
(9) -(\exists x \cdot Q(x)) \Leftrightarrow P(a) \ (*) \\
(15)-\exists x \cdot Q(x) \ (*) \quad (16) + \exists x \cdot Q(x) \\
(17) - P(a) \qquad\qquad (18) + P(a) \\
(19) - Q(c) \\
(\delta^{++} \text{ from } (15))
\end{array}
\end{array}
$$

where (6), (7), (8), (9) were obtained by α and β-rules from (1); (10), (11), (12), (13) were obtained by α and β-rules from (8), and (15), (16), (17), (18) were obtained by α and β-rules from (9). Notice how the same quantification $-\exists x \cdot Q(x)$ in (10) and (15) gives rise to the same Herbrand-Skolem symbol c in (14) and (19): this tableaux can then be closed by the substitution $[c/x_1]$. However, if we had introduced new constants to Herbrandize (10) and (15), (14) would be $-Q(c)$ and (19) would be $-Q(c')$, for two distinct constants c and c', and we cannot close the tableaux before we re-apply an extension step (a γ-rule) on (2) to get enough variables to bind to the constants we created.

Then, we have:

Theorem 8.2 (Soundness, Completeness) *Let Φ be a closed formula. Then Φ is provable in* **LK** *if and only if there is a substitution σ that closes some expansion of the initial tableau $+\Phi$.*

Proof: Let **LK$^+$** be the sequent system formed by adding the rules $\forall R^+$, $\exists L^+$, $\forall L^+$ and $\exists R^+$ to **LK$_0$** without Cut. Expanding tableaux means developing **LK$^+$** proofs from the goal up.

We therefore have to prove that we can transform any cut-free **LK** proof π into a couple (π', σ) where π' is an **LK**$^+$ derivation and σ is a substitution that instantiates the sequents without premises in π' to instances of Ax. This is done by induction on the depth of π. The interesting cases are:

- when we translate a derivation ending in an application of $\exists$R (resp. $\forall$L): then we derived $\Gamma \longrightarrow \Delta, \exists x \cdot \Phi$ from $\Gamma \longrightarrow \Delta, \Phi[t/x]$, which is proved by a shallower proof. Weaken the latter to $\Gamma \longrightarrow \Delta, \exists x \cdot \Phi, \Phi[t/x]$, and translate it by induction hypothesis into an **LK**$^+$ derivation of $\Gamma \longrightarrow \Delta, \exists x \cdot \Phi, \Phi[t/x]$ and a substitution σ'; this is also an **LK**$^+$ derivation of $\Gamma \longrightarrow \Delta, \exists x \cdot \Phi, \Phi[x'/x]$, where x' is a new variable, with substitution $\sigma'[t/x']$; then apply $\exists$R' to get the desired **LK**$^+$ derivation and substitution $\sigma = \sigma'[t/x']$. (Similarly with $\forall$L.)

- when we translate a derivation ending in an application of $\forall$R (resp. $\exists$L): then we derived $\Gamma \longrightarrow \Delta, \forall x \cdot \Phi$ from $\Gamma \longrightarrow \Delta, \Phi[x'/x]$, with x' new. The latter is proved by a shallower proof. Replace x' throughout by the Herbrand-Skolem term $f(x_1, \ldots, x_n)$, where $x_1, \ldots, x_n$ are the free variables of Φ minus x, then we get a proof of the same depth of $\Gamma \longrightarrow \Delta, \Phi[f(x_1, \ldots, x_n)/x]$, which we translate by induction hypothesis into an **LK**$^+$ derivation of the same sequent with substitution σ. Apply $\forall$R$^+$, then, to get the desired **LK**$^+$ derivation and substitution σ. (Similarly with $\exists$L.)

And conversely, we have to prove that we can transform any couple (π', σ), where π' is an **LK**$^+$ derivation and σ turns all sequents without premises into instances of Ax, into an **LK** proof. This follows the same pattern as before, except that in the $\forall$R and $\exists$L cases, we replay the arguments of Theorem 6.34 to replace the Herbrand-Skolem terms back by variables and get a proof in **LK**. □

The Theorem can then be strengthened by saying that, if such a substitution σ exists, then we can find one by trying all possible simultaneous unifiers of pairs of complementary formulas in the paths of the expanded tableau for $+\Phi$, as in Section 1. This terminates our justification of free-variable tableaux.

2.3 Discussion

The first advantage of free-variable tableaux over the tableaux of Section 1 is that we don't need to do any preprocessing step on the formula we want to prove. This is not much of an advantage in classical logic, since we shall in fact redo the herbrandization/skolemization work during the proof instead of before the proof. However, this is crucial to adapt the method to non-classical logics, like intuitionistic logic. Observe in particular that free-variable tableaux operate in mostly unchanged form for intuitionistic logic: we merely have to check that no path has ever more than one positive formula in it. Recall also that converting to prenex form or skolemizing in advance does not preserve provability in intuitionistic logic, so free-variable tableaux are basically the only reasonable method. The argument would be similar for linear, relevant or modal logics.

The second advantage of free-variable tableaux is that the search for proofs is confined to a search for cut-free proofs, which is a much narrower space than the space of all general proofs. This does not compete yet with the size of the search-space for the most efficient refinements of resolution in classical logic, but we shall see in Section 3.4 that we do not need to constrain tableaux much to get one of the most powerful refinements of resolution, namely model elimination.

The third advantage of free-variable tableaux, which they share with the first tableaux method we presented in this chapter, are that even if we insist on skolemizing in advance, we don't need to convert the goal formula to a clausal form. This is a benefit over resolution, since computing a clausal form may take exponential time in the size of the input formula —although this rarely happens in practice.

Tableaux have major drawbacks, however. They try to find cut-free proofs by developing **LK** proofs, or **LK**-like proofs, as trees, where no sharing between paths is performed, except for prefixes of paths. In particular, much work is redone from one path to another. This is one defect that the refinements of resolution do not exhibit in general, as all attempts of finding a refutation are always recorded in the same set of clauses, where they are always available. Optimising tableaux in this respect would need to keep some information on what has been done and what has failed from path to path. The only positive point of that absence of sharing between computations on different paths is that it eases the parallelisation of tableaux method. However, parallelising is not enough to make tableaux much better than sequential algorithms, and the most sophisticated parallel implementations of tableaux already use additional techniques to avoid redundancies between paths.

Another defect of tableaux is that, although we count on unification to guide the search for a proof, we might have to expand paths fully or almost fully to reach the first point where we can unify atomic formulas. By the time we have expanded paths enough to compute non-trivial unifiers, we may be stuck with too many paths to be able to close them all in reasonable time. There might indeed be an exponential number of them in terms of the length of the path.

A final defect of tableaux is that there is still some redundancy in the space of all cut-free proofs due to permutabilities (Theorem 6.32, page 216). In classical logic, this is not serious: most rules permute, so we may use any expansion strategies that we want, at least on propositional connectives. The only non-permutabilities arise from the quantifier rules, and we have encoded them by the use of Skolem terms (or dependency constraints): the essence of Theorem 8.2 is indeed that we have replaced **LK**, a system with three impermutabilities, by $\mathbf{LK}^+$, a system without impermutabilities but with a more sophisticated unifiability condition. This idea of encoding impermutabilities as Skolem terms can be extended to handle the impermutabilities of sequent systems for non-classical logics as well (Shankar, 1992).

3   CONNECTIONS, MATINGS AND MODEL ELIMINATION

3.1   *Connections and Matings*

The *connection method* was invented in late 1979 by Wolfgang Bibel, and about at the same time by Peter Andrews, under the name *method of matings*. Interestingly enough, they are mostly alternative implementations of the basic tableau method of Section 1.

The main change is one in vocabulary and in presentation. As far as presentation is concerned, a two-dimensional one is usually chosen. By convention, assume that we want to prove existential formulas $\exists x_1 \cdot \ldots \exists x_n \cdot \Psi$, where Ψ is quantifier-free, and that Ψ is in negation normal form, or nnf (see Chapter 2, Section 4.2). That is, Ψ is built with the connectives $\wedge$, $\vee$, $\neg$ only, and negations can only apply to atomic subformulas.

Then, represent disjunction subformulas in Ψ as columns, and conjunction subformulas in Ψ as rows. This yields the so-called *matrix representation* of Ψ. For example, $A \vee B$ is:

$$\boxed{\begin{array}{c} A \\ \hline B \end{array}}$$

$\neg A \wedge C$ is:

$$\boxed{\neg A \quad C}$$

and $(\neg A \wedge C) \vee A \vee (B \wedge (\neg C \vee D))$ is represented as:

This representation may seem unimportant at first, but it has several virtues. First, we implicitly merge any two columns on top of each other, and any two rows that are side by side: this takes care of the associativity of $\wedge$ and $\vee$. Second, it makes it easier to see the duality between $\wedge$ and $\vee$, as the following definition shows:

Definition 8.3 (Paths) *Let Ψ be a quantifier-free formula in nnf, represented as a matrix.*

The set of vertical paths *in Ψ is the smallest set such that:*

- *$\{\Psi\}$ is a vertical path;*

- *if C is a vertical path containing a formula Ψ' of the form:*

then $C \setminus \{\Psi'\} \cup \{\Psi_1, \ldots, \Psi_m\}$ *is a vertical path;*

- *if* C *is a vertical path containing a formula* Ψ' *of the form:*

$$\boxed{\Psi_1 \quad \ldots \quad \Psi_m}$$

then $C \setminus \{\Psi'\} \cup \{\Psi_i\}$, $1 \leq i \leq m$, *are all vertical paths.*

In the example of the formula $(\neg A \wedge C) \vee A \vee (B \wedge (\neg C \vee D))$ above, the fully expanded vertical paths (those which contain only literals) are $\{\neg A, A, B\}$, $\{\neg A, A, \neg C, D\}$, $\{C, A, B\}$, $\{C, A, \neg C, D\}$.

The notion of vertical paths is quite similar to that of paths in tableaux. In fact, we can turn any atomic tableau path into a fully expanded vertical path by converting formulas $+A$ into A and $-A$ into $\neg A$. And conversely, any fully expanded vertical path can be turned into an atomic tableau path by translating A to $+A$ and $\neg A$ to $-A$.

The connection method then proceeds as the tableaux method of Section 1: we expand vertical paths and try to close all of them by unification; and to represent extension steps, we add new generic renamings of Ψ at the bottom of the whole current matrix, thus extending all vertical paths automatically.

A complementary pair of literals in a vertical path is then called a *connection* by Bibel; the analogue of tableaux, viewed as sets of paths, is called a *pre-mating* by Andrews, and a *spanning set of paths* by Bibel. A *spanning set of connections* is a set of connections such that each path in a pre-mating has one connection in the set. In the tableaux method, we looked forward to close all paths; here we search for a unifiable spanning set of connections. Apart from the vocabulary, few things change when compared with the tableaux method, so we won't describe the method further. Its main advantage is computational, since we may represent connections as links in a so-called *connection graph*, on which various efficient combinatorial operations can be performed. See (Bibel, 1987) for a full technical description.

3.2 Relation with Resolution

There are some parallels between tableaux and resolution, as we are going to demonstrate. The first thing we have to observe, though, is that resolution works mostly in classical logic only (at least not without some modifications), so that the comparison we shall operate will be limited to classical logic, and in fact to first-order classical logic.

The second difficulty we face is that resolution and tableaux do not solve the same problem, as we have defined these methods. Because we have tried to present both methods by deriving them from Gentzen-style sequent systems, we have described resolution as a method looking for refutations (trying to show a formula unsatisfiable, semantically), while tableaux look for proofs (they try to show a formula valid, semantically). To compare them, we therefore need either to modify resolution so that it becomes a proof procedure, or to modify the tableaux method so that it becomes a refutation procedure.

Each task is equally easy, because Φ is provable (resp. valid) if and only if $\neg\Phi$ is inconsistent (resp. unsatisfiable). But the process needs some mental agility. For historical reasons, both resolution and the tableaux method are usually presented as refutation procedures. We feel that this is unnatural, at least in the case of tableaux: refutation-oriented tableaux usually force us to reason in terms of models, i.e. semantically, while tableaux can be used at no extra cost for logics whose semantics are cumbersome, like intuitionistic or linear logic.

For this reason, and to show that there is no way in which refutation procedures may be superior to direct proof procedures (a common mistake), we first give again the definition of resolution, this time in validity form (compare with Definition 7.10, page 241):

Definition 8.4 (Positive Resolution) *Assume we want to prove an existential formula* $\Phi = \exists x_1 \cdot \ldots \exists x_n \cdot \Psi$, *where* Ψ *is expressed in disjunctive normal form (dnf), i.e.* Ψ *is a disjunction of conjunctive clauses.*

The positive binary resolution rule *is the following:*

$$\frac{C \wedge A \qquad \neg A' \wedge C'}{C\sigma \wedge C'\sigma} \; \sigma = mgu(A, A')$$

where we assume the two clauses $C \wedge A$ *and* $\neg A' \wedge C'$ *to be renamed so as to have no free variables in common. The conclusion of the rule is called the* binary resolvent *of the two clauses, the premises are called the* parent clauses, *and* A *(resp.* $\neg A'$*) is the* literal resolved upon *in the first clause (resp. second clause).*

We say that the literals A *and* $\neg A'$ *are* complementary *whenever* A *unifies with* A'.

The positive factoring rule *is the following twin rule:*

$$\frac{C \wedge A \wedge B}{C\sigma \wedge A\sigma} \; \sigma = mgu(A, B) \qquad \frac{\neg A \wedge \neg B \wedge C}{\neg A\sigma \wedge C\sigma} \; \sigma = mgu(A, B)$$

The conclusion of the rule is called a binary factor *of the clause in premise.*

A factor *of a clause is either the clause itself, or a binary factor of one of its factors. A* proper factor *of a clause is a factor of one of its binary factors.*

A resolvent *of two clauses is a binary resolvent of factors of each clause, not necessarily proper.*

A positive resolution derivation *is an inverted tree, whose nodes are decorated with conjunctive clauses, leaves are decorated with clauses in the initial set of clauses, and inner nodes have their parent clauses as predecessors.*

A positive resolution proof *is a derivation ending in the empty conjunctive clause, i.e. a derivation whose root node is decorated with the empty conjunction* **T***. (Standing for true.)*

All of Chapter 7 can then be translated to apply to positive resolution. Here, resolvents are not implied by their parent clauses, rather resolvents imply the disjunction of their parent (conjunctive) clauses; whenever we derive $\mathbf{T}$, the disjunction of the ancestor clauses of $\mathbf{T}$ that are in the original set of clauses is valid, hence proving the original set of clauses. Also, to prove a sentence Φ in a set of axioms $\mathcal{A}$, putting $\mathcal{A} \Rightarrow \Phi$ in dnf means putting $\neg\mathcal{A}$ and Φ in dnf; this is the exact counterpart of putting $\mathcal{A}$ and $\neg\Phi$ in cnf, in refutation resolution. Finally, all refinements of resolution go through almost without modification.

The counterpart of conjunctive clauses of positive resolution in tableaux (or the connection method) is the notion of paths orthogonal to the one we used in Section 3.1:

Definition 8.5 (Paths) *Let Ψ be a quantifier-free formula in nnf, represented as a matrix.*

The set of horizontal paths *in Ψ is the smallest set such that:*

- *$\{\Psi\}$ is a horizontal path;*

- *if C is a horizontal path containing a formula Ψ' of the form:*

$$\boxed{\;\Psi_1 \quad \ldots \quad \Psi_m\;}$$

then $C \setminus \{\Psi'\} \cup \{\Psi_1, \ldots, \Psi_m\}$ is a horizontal path;

- *if C is a horizontal path containing a formula Ψ' of the form:*

$$\boxed{\begin{array}{c} \Psi_1 \\ \vdots \\ \Psi_m \end{array}}$$

then $C \setminus \{\Psi'\} \cup \{\Psi_i\}$, $1 \leq i \leq m$, are all horizontal paths.

For example, the fully expanded horizontal paths of our example $(\neg A \wedge C) \vee A \vee (B \wedge (\neg C \vee D))$ are $\{\neg A, C\}$, $\{A\}$, $\{B, \neg C\}$, $\{B, D\}$. Alternatively, a dnf for the formula is the following set of conjunctive clauses:

$$\neg A \wedge C$$
$$A$$
$$B \wedge \neg C$$
$$B \wedge D$$

To sum up, while tableaux try to close vertical paths, (positive) resolution combines horizontal paths together. Moreover, resolution automatically integrates the extension steps of tableaux by systematically adding resolvents as new, renamed

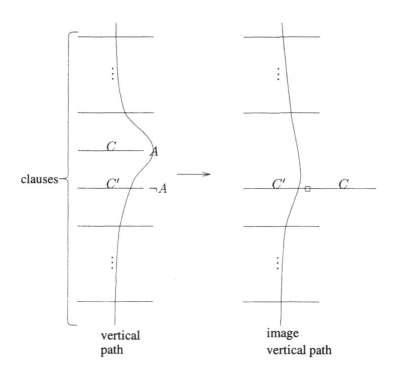

Figure 8.4. Effect of computing resolvents on vertical paths

clauses to the set of current clauses. On the other hand, resolution works by representing explicitly all clauses, whereas they are left implicit in tableaux. (But it is easy to adapt resolution to work on the set of horizontal paths of a formula without converting them to an explicit set of clauses: this is called *non-clausal resolution*.)

In the following two sections, we shall see two ways of relating resolution and tableaux. The first one adapts resolution to work like tableaux, by dissociating extension steps from closing steps. The second one adapts tableaux techniques to conjunctive clause form, and yields the so-called *model elimination resolution* technique.

3.3   Chang's V-Resolution

When we create new resolvents by the resolution rule, we automatically rename all of its free variables, so that no two clauses have any free variables in common. Implicitly, clauses C in positive resolution are quantified existentially (they mean

$\exists x_1 \cdot \ldots \exists x_n \cdot C$), so that generating an instance of the clause with new variables is exactly the same as using a γ-rule.

Chin-Lian Chang's idea (Chang and Lee, 1973) is then to limit this effect by not renaming the free variables of new clauses. To prove that a set S of conjunctive clauses is valid, we generate in advance a fixed number $k(C)$ of renamed versions $C\rho_1, \ldots, C\rho_{k(C)}$ of each clause C, so that no two clauses have any free variable in common, yielding a new set S' of conjunctive clauses. We then look for a substitution σ such that $S'\sigma$ is propositionally valid, by applying the propositional resolution rule, lifted to the quantifier-free case as follows. We associate a substitution with each clause, which is initially empty. We then apply the following rules, called the *V-resolution rules*:

- Factoring: given a clause $C \wedge A \wedge B$ with associated substitution σ, where A and B are atoms, such that θ is the mgu of A and B and σ' is the mgu of θ and σ, replace $C \wedge A \wedge B$ by $C\theta \wedge A\theta$; (and similarly on clauses $C \wedge \neg A \wedge \neg B$;)

- Binary resolution: given two clauses $C \wedge A$ (with substitution σ) and $\neg A' \wedge C'$ (with substitution σ'), such that θ is the mgu of A and A', θ' is the mgu of σ and σ', and σ'' is the mgu of θ and θ', generate the conjunctive clause $C\theta \wedge C'\theta$, with associated substitution σ''.

This procedure is sound and complete. Completeness, in particular, is a direct consequence of Herbrand's Theorem and completeness for propositional resolution. (For the definition of mgus of substitutions, as needed above, see Exercise 7.3, page 241.)

Factoring and binary resolution must be applied non-deterministically, just as it was the case with the tableaux expansion rules. To compare both methods, it is worthwhile to examine the effect of V-resolution on vertical paths.

Consider a set S_0 of conjunctive clauses. Choose an atom A_1 occurring in S_0, and compute the set S_1 of all resolvents of clauses inside S_0 on A_1 and $\neg A_1$. S_1 is then logically equivalent to $\forall A_1 \cdot S_0$. (Symmetrically, in the refutation case, we have seen that taking all resolvents on some atom amounted to computing $\exists A_1 \cdot S_0$, and that this was the Davis-Putnam splitting rule.) We now look at the way that fully expanded vertical paths in S_0 are transformed in S_1. See Figure 8.4 for an illustration of what happens, in the case when the vertical path goes through A but not $\neg A$. In general, given a vertical path P in S_0 that does not contain both A and $\neg A$, we shall find one or more paths in S_0 in which $P \setminus \{A, \neg A\}$ is included. On the other hand, if P contains both A and $\neg A$, then it will have no counterpart as a vertical path in S_1. (On the figure, such a path on the right-hand side should intersect the resolved clause $C' \wedge C$ at a place—a literal—where it intersected C or C' on the left-hand side, but such a place does not exists.) That is, computing the set of all resolvents on some atom A closes all vertical paths that can be closed on A and $\neg A$, while in general increasing the number of remaining paths to close.

(But don't forget that, as the paths remain implicit, they are automatically shared: we never have two identical copies of paths.)

V-resolution is not particularly interesting in practice. Indeed, it combines the worst aspects of tableaux (the non-deterministic character of the search, where different search paths may be redundant but are nevertheless explored totally independently) with the worst aspects of resolution (using propositional resolution as the basic propositional proof-search process). It is however interesting as an example of the connections that there might be between tableaux and resolution.

3.4 Model Elimination

A much more interesting interaction of resolution with tableaux is Donald Loveland's model-elimination procedure, also known as OL-resolution, a form of linear resolution with ordered clauses. (See Chapter 7, Section 4.5.) On the other hand, we can see it as a slight modification of the basic tableaux method, working on formulas in nnf.

Assume that we want to prove a set of conjunctive clauses S, containing a distinguished goal clause $L_1 \wedge \ldots \wedge L_k$. Let S' be S without the goal clause. Let T be a tableau, i.e. a set of paths. Initially, we let T consist of the unit paths $\{L_1\}$, $\ldots$, $\{L_k\}$. Let also σ be a substitution, initially empty. Model elimination works by transforming the tableau and the substitution by using clauses in S' to extend the paths of the tableau T, with the following two rules:

- Extension: select a path P from T, and a literal A (resp. $\neg A$) from P; choose non-deterministically a conjunctive clause $\neg A' \wedge L_1 \wedge \ldots L_n$ (resp. $A' \wedge L_1 \wedge \ldots \wedge L_n$) from S' (with its free variables renamed to fresh variables) such that $A\sigma$ and $A'\sigma$ are unifiable. Then, let σ' be the resulting mgu, replace σ by $\sigma\sigma'$, and replace P by the paths $P \cup \{L_i\}$, $1 \le i \le n$ in T. In short:

$$\frac{\overbrace{P_1,\ldots,P_m,P}^{T}/\sigma \quad \overbrace{L \wedge L_1 \wedge \ldots L_n}^{\text{in } S'}}{P_1,\ldots,P_m,P\cup\{L_1\},\ldots,P\cup\{L_n\}/\sigma\sigma'}$$

where $\sigma' = mgu(\neg L'\sigma, L\sigma)$, $L' \in P$.

- Reduction: select a path P from T, and a literal A (resp. $\neg A$) from P; choose non-deterministically a literal $\neg A'$ (resp. A') in P such that $A\sigma$ and $A'\sigma$ are unifiable. Then, let σ' be the resulting mgu, replace σ by $\sigma\sigma'$, and remove P from T. In short:

$$\frac{\overbrace{P_1,\ldots,P_m,P}^{T}/\sigma}{P_1,\ldots,P_m/\sigma\sigma'} \quad \sigma' = mgu(\neg A'\sigma, A\sigma), A \in P, \neg A' \in P$$

Note that, when we say "select" in the sentences above, we mean select an object by some arbitrary strategy. It does not matter in which order we extend or reduce paths, or in which order we choose the literals to extend on or to reduce by in paths. However, when we say "choose non-deterministically", the order may be important, but is in general unknown: this is where the procedure is truly non-deterministic, and may need to backtrack.

Model elimination is a sound and complete proof procedure, although we won't prove it. Instead, we shall show the relationship between model elimination, tableaux and linear resolution. The reader is invited to adapt the soundness and completeness proofs for tableaux or for linear resolution to model elimination after reading the following subsections.

Model Elimination as Tableaux

We can interpret these rules as tableaux rules. Extension consists of one γ-expansion step on some clause from S', followed by one α-expansion on the n-ary $\wedge$ conjunctive connective. β-expansion is automatically handled by selecting the clause from the set S': although the other clauses do not appear on the path P, they don't need to, as we are only interested in paths consisting of literals. More precisely, as we close paths by unifying atomic signed formulas only, the only part of the paths that interests us are the set of signed atomic formulas in them.

On the other hand, reduction steps are the analogue of the closing rule for tableaux. However, not all paths are closed by reduction steps, as extending a path with a unit clause ($m = 0$) also has the effect of closing it.

The constraints that model elimination imposes on the tableaux expansion process are the following. First, we have to guess a goal clause; this means starting off the whole tableau expansion by β-expanding the set S of clauses, and choosing one of the clauses to α-expand. Then, all other expansions are heavily constrained: we insist that, whenever we expand a clause, at least one of the resulting paths must close immediately. (This is the unification condition on the extension rule.) Finally, as in resolution, each α-expansion of a clause is preceded by systematically γ-expanding it, i.e. renaming all its free variables.

To illustrate this process, look at the following set of clauses:

$$A \wedge B$$
$$A \wedge \neg B$$
$$\neg A \wedge B$$
$$\neg A \wedge \neg B$$

Choose, say, $A \wedge B$ as goal clause. The initial tableau is then:

$$A \mid B$$

Extend the leftmost path by using, say, the clause $\neg A \wedge B$, we get:

$$
\begin{array}{cc|c}
 & A & B \\
\neg A & \big| \; B & \\
(*) & \big| & \\
\end{array}
$$

where extending on A has just closed the leftmost path. We represent this by the sign $*$. We continue the process, for example, by extending the leftmost unclosed path by using $\neg A \wedge \neg B$, and get:

$$
\begin{array}{ccc|c}
 & A & & B \\
\neg A & & B & \\
(*) & \neg A & \big| \; \neg B & \\
 & & (*) & \\
\end{array}
$$

But then, we can close the leftmost unclosed path $\{A, B, \neg A\}$ by the reduction rule. Continuing the process, we eventually get the following tableau, where all paths have been closed, either by extension (signaled by an asterisk $*$) or by reduction (signaled by a circle $\circ$):

$$
\begin{array}{cccc|cccc}
 & A & & & & B & & \\
\neg A & & B & & & A & & \neg B \\
(*) & \neg A & \big| & \neg B & \neg A & \big| & \neg B & (*) \\
 & (\circ) & & (*) & (*) & & (\circ) & \\
\end{array}
$$

Although the example is propositional, the first-order case is complicated only by the need for maintaining a substitution that applies to the whole tableau, and which is computed by unification.

Model Elimination as Linear Resolution

We can also interpret these rules as linear resolution rules. Indeed, a tableau T can be interpreted as a special representation for the current top clause: just read off the literals at the bottom of the unclosed paths in the tableau. For example, the first tableau above represents the clause $A \wedge B$ (the initial tableau, anyway, must represent the goal clause), the second one represents the clause $B \wedge B$ (before contraction), and the last one represents the empty clause.

Then, applying the extension rule means computing a binary resolvent between the current top clause as centre clause and an input clause (a clause from S'): extensions are input resolution steps.

In fact, the tableau structure is richer than just describing the current top clause. We can indeed extract tableaux generated before the current tableau, as sub-tableaux of the current one. (A tableau being a tree, we consider a sub-tableau to be any sub-tree of the current tableau.) Then, at least in the propositional case, the reduction rule can be seen as a binary resolution step between the centre clause and some ancestor, non-input clause, which remains encoded as a sub-tableau of the current tableau.

We demonstrate this on the example above. The first tableau corresponds to the goal clause $A \wedge B$. Then, the second tableau corresponds to the clause $B \wedge B$, but the leftmost B can be replaced by its parent literal A, so that we recover the original clause $A \wedge B$. The third tableau represents $\neg A \wedge B$, but also the previous clauses $B \wedge B$ and $A \wedge B$. Then, applying the reduction rule on A and $\neg A$ means resolving the top clause $\neg A \wedge B$ with the ancestor clause $A \wedge B$. This yields B, which is precisely the result that we read off the third tableau when we have closed the leftmost path by the reduction rule.

In the first-order case, the reduction rule does something actually a bit different than a binary resolution with some ancestor clause, because the set of free variables in the current tableau is left fixed. That is, the reduction rule computes a binary resolvent *without renaming* either the centre clause or the ancestor clause. Strangely enough, this does not harm the completeness of the procedure, and in fact it allows us to dispense completely with the factoring rule!

For example, consider the example of Chapter 7, Section 3 where we needed factoring:

$$P(x, a) \wedge P(a, x)$$
$$\neg P(y, a) \wedge \neg P(a, y)$$

Resolution was incomplete without factoring on this example, because the only binary resolvent we could then generate was $P(a, a) \wedge \neg P(a, a)$. In model elimination, choosing the first clause as goal clause for example, we would generate the following initial tableau:

$$P(x, a) \mid P(a, x)$$

Then, we would extend, say, the leftmost path by a new renamed version of the second clause, yielding:

$$
\begin{array}{c|c}
P(x, a) & \\
\neg P(y_1, a) \mid \neg P(a, y_1) & P(a, x) \\
(*) &
\end{array}
$$

with substitution $[a/x, a/y_1]$. Apply the reduction rule on the leftmost unclosed path, then extend the rightmost path by a renamed version of the second clause again, and finally reduce the rightmost path. We get:

$$
\begin{array}{c|c}
P(x, a) & P(a, x) \\
\neg P(y_1, a) \mid \neg P(a, y_1) & \neg P(y_2, a) \mid \neg P(a, y_2) \\
(\circ) \qquad (*) & (*) \qquad (\circ)
\end{array}
$$

with substitution $[a/x, a/y_1, a/y_2]$. We did not need to factor any clause, because the variable x we bind in the left half and the x we bind in the right half of the tableau are the same, thus instantiating the first clause to $P(a, a)$, i.e. the clause that we could get only by factoring the first clause in classical resolution.

Historically, model elimination was invented in the 1970s by Donald Loveland as a particular refinement of resolution, while various forms of linear resolution were developed by Robert Kowalski. It was named "model elimination" after the discovery that what the method basically did was to enumerate (sets of) interpretations as paths in the tableau, and then show that they are not models of the goal formula by closing the path. In fact, this is a general way of understanding tableaux as refutation methods. By duality, paths in refutation-style tableaux are now understood as conjunctions, so that fully expanded paths are conjunctions of literals. The instances of such paths where we don't have both an atom and its negation are partial Herbrand interpretations. To refute the goal formula, we have to find instances such that no path can be understood as a partial Herbrand interpretation, i.e. we close all paths.

CHAPTER 9

INCORPORATING KNOWLEDGE

1 MOTIVATIONS

The automated proof techniques that we have presented until now are a good basis. One of their main shortcomings, however, is that the only way we have to describe our knowledge of the world to the prover is by means of axioms. These proof methods indeed have no intrinsic knowledge, for example, that $\doteq$, $+$, $*$ or $\leq$ must be interpreted as equality, addition, multiplication, and an ordering respectively.

In fact, we don't usually want to prove that a given formula Φ is valid, but that it holds in every model in which $\doteq$ means equality, $+$ is addition, and so on. The only way we could do this until now was by axiomatising all symbols of interest as a hopefully finite, at least recursive set $\mathcal{A}$ of axioms, and then proving $\mathcal{A} \Rightarrow \Phi$. The problem is that to do this, a prover must look for a lot of possible ways of inferring $\mathcal{A} \Rightarrow \Phi$, including all the ways it can find of refuting $\mathcal{A}$. (Although the set-of-support strategy might help a bit, we have noticed that it was actually a very weak strategy in Chapter 7, Section 4.4.) Consider the following example:

$$coyote(x_1) \longrightarrow animal(x_1)$$
$$\longrightarrow coyote(Will)$$
$$bird(x_2) \longrightarrow animal(x_2)$$
$$\longrightarrow bird(Bip)$$
$$grain(x_3) \longrightarrow plant(x_3)$$
$$\longrightarrow grain(wheat)$$
$$bird(x_4), grain(y_4) \longrightarrow eats(x_4, y_4)$$
$$animal(x_5), plant(y_5), eats(x_5, y_5) \longrightarrow$$

where all but the last clause are axioms (a coyote is an animal, Will is a coyote, a bird is an animal, Bip is a bird, a grain is a plant, wheat is a grain, and every bird eats every grain), and the last clause is the negation of the question: is there an animal that eats some plant? Answering the question means attempting to refute the above set of clauses. But while answering the question is easy, the resolution process, even input resolution, gets lost quickly: it can unify $animal(x_5)$ of the last clause with, say, $animal(x_1)$ in the first clause, yielding $coyote(x_6), plant(y_6)$, $eats(x_6, y_6) \longrightarrow$ after renaming; then resolve with the second clause to get

$plant(y_7), eats(Will, y_7) \longrightarrow$; then resolve with the fifth to get $grain(y_8)$, $eats(Will, y_8) \longrightarrow$; then with the sixth to get $eats(Will, wheat) \longrightarrow$. Asking whether Will eats wheat is hopeless, but we can nonetheless go on, and resolve with the seventh clause and get $bird(Will), grain(wheat) \longrightarrow$, then with the sixth to get $bird(Will) \longrightarrow$, which finally fails.

This is an example of *taxonomical reasoning*, where the methods we have presented are poor, without some improvements. Another example where these techniques are even poorer are in handling well-known relationships like equality. Consider the proposition $a \doteq b \Rightarrow f^k(a) \doteq f^k(b)$, where $f^k(x)$ is shorthand for $\underbrace{f(\ldots f(x)\ldots)}_{k \text{ times}}$. With the axioms of reflexivity, symmetry, transitivity and congruence we gave in Chapter 6, Section 4, the shortest proof of this fact needs k applications of the rule of functional congruence, i.e. at least k resolution steps, or k applications of the γ-rule on the axiom of functional congruence followed by k non-trivial unifications to close paths in the tableaux method. We may have to generate much more resolvents, or in general to put some tremendous effort just to get to the desired conclusion. For example, for $k = 2$, which is still rather low, input resolution from the negated goals $\longrightarrow a \doteq b$ and $f(f(a)) \doteq f(f(b)) \longrightarrow$ will produce as level 1 clauses: $\longrightarrow b \doteq a$ and $f(f(b)) \doteq f(f(a)) \longrightarrow$ by resolving with the symmetry axiom, $b \doteq z \longrightarrow a \doteq z$ and $x \doteq a \longrightarrow x \doteq b$ and $f(f(a)) \doteq y, y \doteq f(f(b)) \longrightarrow$ by resolving with the transitivity axiom, and $\longrightarrow f(a) \doteq f(b)$ and $f(a) \doteq f(b) \longrightarrow$ by resolving with the functional congruence axioms. Notice that only the last two are relevant to the proof of the theorem at hand.

In fact, an automated theorem prover performs an almost blind search through the space of all possible refutations or proofs. A sensible solution is to incorporate some knowledge about the meaning of symbols in the very mechanism of the prover itself. À *tout seigneur, tout honneur*, we begin by looking at the equality relation in Section 2, which is probably the single most useful symbol whose meaning we want the prover to know. Incorporating equational properties about other symbols, like associativity, commutativity or other common algebraic properties, is dealt with in Section 3. Finally, we review other attempts to integrate some knowledge, about the sorts of objects (this will handle our taxonomical example), and about symbols described by non-equational theories notably, in Section 4.

This chapter is not in any way complete, and should be understood as an introduction. We provide pointers to the literature at the end of each major section.

2   EQUALITY AND REWRITING

2.1   Rewriting

Looking back at our small example on equality, we see that the main problem when k becomes large is that, to prove $a \doteq b \Rightarrow f^k(a) \doteq f^k(b)$, we derive it step by step, deducing $f(a) \doteq f(b)$ from $a \doteq b$, then $f^2(a) \doteq f^2(b)$ from the former, and so on. The problem is not so much that it may take $O(k)$ deduction steps to arrive at the conclusion, but that at each such step, we always have the choice between going on or trying some other proof step (i.e., resolving with some other axiom, or expanding some other axiom).

On the other hand, assuming $a \doteq b$, where $\doteq$ means equality, gives us the right to replace a by b anywhere we like, for instance in $f^k(a) \doteq f^k(b)$, yielding $f^k(b) \doteq f^k(b)$, which is then proved in one use of the reflexivity axiom only. Note that replacing a by b in $f^k(a)$ still takes $O(k)$ steps if we program in functional style: the gain is that there are now only two deduction steps, therefore reducing the number of non-deterministic choice points considerably.

Equational Logic

Replacing equals by equals is a deduction rule for the so-called *equational logic*:

Definition 9.1 (Equations, Rules) *An* equation *is a pair of terms* $\{s, t\}$, *written* $s \doteq t$ *or* $t \doteq s$. *A system of equations* E *is a finite set of equations.*

A rule *is a couple of terms* (s, t), *written* $s \rightarrow t$. *It is a* rewrite rule *if and only if* $\mathrm{fv}(t) \subseteq \mathrm{fv}(s)$.

The set of rules R *derived from a system* E *of equations is the set* R_E *of rules* $s \rightarrow t$ *and* $t \rightarrow s$ *such that* $(s \doteq t) \in E$.

The one-step reduction relation $\longrightarrow_R$ *is defined as follows:*

- *if* $s \rightarrow t$ *is in* R, *then* $s\sigma \longrightarrow_R t\sigma$ *for any substitution* σ;

- *for every n-ary function symbol* f, *if* $s \longrightarrow_R t$ *holds, then*

$$f(t_1, \ldots, t_{i-1}, s, t_{i+1}, \ldots, t_n) \longrightarrow_R f(t_1, \ldots, t_{i-1}, t, t_{i+1}, \ldots, t_n)$$

The rewrite relation $\longrightarrow_R^*$ *is the reflexive transitive closure of* $\longrightarrow_R$, *i.e.* $s \longrightarrow_R^* t$ *if and only if* $s = t_0$, $t_0 \longrightarrow_R t_1 \longrightarrow_R \ldots \longrightarrow_R t_n$ *and* $t_n = t$ *for some terms* t_0, $\ldots, t_n$, $n \in \mathbb{N}$.

Similarly, $\longrightarrow_R^+$ *is the transitive closure of* $\longrightarrow_R$ (*same, but with* $n \geq 1$), *and* $\longleftrightarrow_R^*$ *is the reflexive symmetric transitive closure of* $\longrightarrow_R$.

Observe that equations are now *unoriented* pairs, so as to account for symmetry in a natural way. That s' rewrites in one step to t', which we write $s' \longrightarrow_R t'$, means that there is an occurrence of s in s', where $s \rightarrow t$ is an instance of a rule in

R, such that if we replace s by t at that occurrence, we get t'. Taking instances of rules in R instead of just the rules in R means that, implicitly, we consider rules as representing *universally quantified* equations: for example, if we can rewrite $f(x)$ to $g(x, x)$, we can also rewrite $f(a)$ to $g(a, a)$, because we think of the rewrite rule $f(x) \longrightarrow g(x, x)$ as representing $\forall x \cdot f(x) \doteq g(x, x)$.

Finally, note that $\longleftrightarrow^*_R$ is reflexive, symmetric, transitive and verifies all congruence axioms. In fact, it is the smallest relation on terms having all these properties and containing all instances of rules in R.

Definition 9.2 (Equational Logic) *The formulas of* equational logic *are universally quantified equations* $\forall x_1 \cdot \ldots \forall x_n \cdot s \doteq t$, *where* $\{x_1, \ldots, x_n\} = \mathrm{fv}(s \doteq t)$. *We often leave the quantification implicit, and write* $s \doteq t$ *instead.*

An equational interpretation *is an interpretation I in which* $\doteq$ *is interpreted as the equality relation* $=$ *on D_I. An* equational model *of an equation $s \doteq t$ is an equational interpretation such that s and t have the same denotation for every possible assignment of their free variables to values. An equational model of a system E of equations is an equational model of all equations in E.*

Let E be a set of equations, and $s \doteq t$ be an equation. We say that $s \doteq t$ is an equational consequence *of E, and we write $E \models_= s \doteq t$ if and only if all equational models of E are equational models of $s \doteq t$.*

We say that $s \doteq t$ is equationally provable *from E, and we write $E \vdash_= s \doteq t$ if and only if $s \longleftrightarrow^*_{R_E} t$.*

Semantically, we have to restrict our classes of interpretations and models to those where $\doteq$ means equality, since in general interpretations, $\doteq$ may be any binary relation. Then, $E \models_= s \doteq t$ if and only if whenever all (universally quantified) equations in E hold, then s equals t. For example, the class of equational models of the following system of equations:

$$x + (y + z) \doteq (x + y) + z \quad 0 + x \doteq x \quad (-x) + x \doteq 0$$

is the class of all groups with $+$ as its law, 0 as neutral element, and $-$ as inverse function. An equation is then an equational consequence of the latter equations if and only if it holds in all groups.

Proof-theoretically, $E \vdash_= s \doteq t$ means that we can rewrite s into t by using finitely many instances of rewrite steps $s \rightarrow t$ or $t \rightarrow s$, where $s \doteq t$ is an instance of an equation in E. This is usually called *Leibniz's rule* of replacement of equals by equals. Garrett Birkhoff proved the following in 1935 (Birkhoff, 1935), as a consequence of a more general theorem on universal algebras (i.e., classes of equational models of systems of equations):

Theorem 9.3 (Soundness, Completeness) *Equational logic is sound and complete, i.e. $E \models_= s \doteq t$ if and only if $E \vdash_= s \doteq t$.*

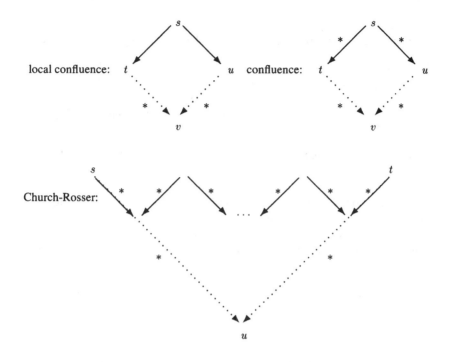

Figure 9.1. Confluence properties

▶ **EXERCISE 9.1**
Show that provability in equational logic is undecidable. (Look at the proof of Theorem 6.20.)

Rewrite Rules

In practice, we can hardly reason directly with equations for at least two reasons. First, we never know the direction in which we have to use an equation $s \doteq t$ in E for rewriting, i.e. whether we should replace s by t or t by s. Second, we never know which equation to choose and at which occurrence.

For the first reason at least, it is desirable to replace the system of equations E by a set of oriented rules R, such that we can prove $s \doteq t$ from E by just rewriting s and t by R until both are rewritten to the same term. That is, we want $\longleftrightarrow_R^*$ to be equal to $\longleftrightarrow_{R_E}^*$, plus the following:

Definition 9.4 (Church, Rosser) $\longrightarrow_R$ *is* Church-Rosser *if and only if, for any two terms s and t, $s \longleftrightarrow_R^* t$ if and only if $s \longrightarrow_R^* u$ and $t \longrightarrow_R^* u$ for some* *term u.*

The Church-Rosser property is in fact equivalent to the *confluence* property:

Definition 9.5 (Confluence) $\longrightarrow_R$ *is confluent if and only if whenever $s \longrightarrow^*_R t$ and $s \longrightarrow^*_R u$, then $t \longrightarrow^*_R v$ and $u \longrightarrow^*_R v$ for some v.*
    $\longrightarrow_R$ *is locally confluent if and only if whenever $s \longrightarrow_R t$ and $s \longrightarrow_R u$, then $t \longrightarrow^*_R v$ and $u \longrightarrow^*_R v$ for some v. (The difference with confluence is that s must rewrite to t and u in exactly one step.)*

The three notions are depicted on Figure 9.1. Being Church-Rosser and being confluent are two equivalent notions. A relation $\longrightarrow_R$ is confluent if we can turn any *peak* $s \longrightarrow^*_R t$ and $s \longrightarrow^*_R u$ proving that t equals u into a *valley* $t \longrightarrow^*_R v$ and $u \longrightarrow^*_R v$. A confluent, or Church-Rosser relation allows us to only consider valley proofs of equalities, instead of more general zig-zag ones.

    Local confluence is easier to check than confluence, but it is in general weaker than confluence. For example, the relation $a \to b, b \to a, a \to c, b \to d$ is locally confluent but not confluent:

    Finally, it is also desirable that the rewriting process terminates:

Definition 9.6 (Nœther) $\longrightarrow_R$ *is noetherian, or well-founded, or terminating, if and only if there is no infinite chain $t_1 \longrightarrow_R t_2 \longrightarrow_R \ldots t_k \longrightarrow_R \cdots$*
    $\longrightarrow_R$ *is convergent if it is Church-Rosser and terminating.*

A convergent set of rewrite rules is interesting because every term has a unique normal form:

Definition 9.7 (Normal Form) *Let R be a set of rewrite rules. A normal form is a term t that does not rewrite via $\longrightarrow_R$ to any term.*

Then indeed, if $\longrightarrow_R$ terminates, every term t rewrites to some normal form: any rewrite sequence $t \longrightarrow_R t_1 \longrightarrow_R \ldots t_k \longrightarrow_R \cdots$ terminates on some term t_n, which cannot be reduced further, i.e. t_n is a normal form. If, moreover, $\longrightarrow_R$ is Church-Rosser, then all possible normal forms must be the same; indeed, if t' and t'' are two normal forms for t, then $t' \longleftrightarrow^*_R t''$, so $t' \longrightarrow^*_R u$ and $t'' \longrightarrow^*_R u$ for some u, but because t' and t'' are both normal forms, the length of the reductions from t' to u and from t'' to u must be 0, that is, $t' = t''$. Therefore, as announced, with respect to a convergent set of rewrite rules, every term has a unique normal form; besides, this normal form is computable, by rewriting until the normal form is reached.

To conclude, if we manage to compute a convergent set R of rewrite rules equivalent to a given set E of equations (i.e., $\longrightarrow^{*}_{R}$ and $\longleftrightarrow^{*}_{R_E}$ coincide), we can check whether $E \vdash_= s \doteq t$ by rewriting s and t to their respective normal forms s' and t', using any strategy for selecting the rules and the occurrences of subterms to rewrite, and then comparing s' and t' syntactically: if $s' = t'$, then $E \vdash_= s \doteq t$, otherwise $E \not\vdash_= s \doteq t$.

The Knuth-Bendix Completion Process

We can then ask the following: is there always a convergent set R of rewrite rules equivalent to a given set E of equations? For example, take the case of groups, defined as the equational models of the following set E of equations:

$$x + (y + z) \doteq (x + y) + z \quad 0 + x \doteq x \quad (-x) + x \doteq 0$$

We can orient the equations to rules, say:

$$x + (y + z) \rightarrow (x + y) + z \quad 0 + x \rightarrow x \quad (-x) + x \rightarrow 0$$

But, although $x + 0 \doteq x$ is provable, there is no valley proof of it in the latter set of rewrite rules. We can still produce a zig-zag proof:

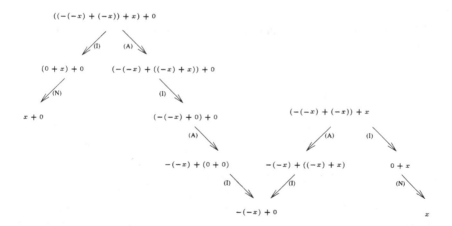

where we have labelled the rewrite steps by the names of the rules that we used: (A) for associativity, (N) for neutral, and (I) for inverse. But there is no way in which we can orient the equations in E so as to form a Church-Rosser set of rewrite rules.

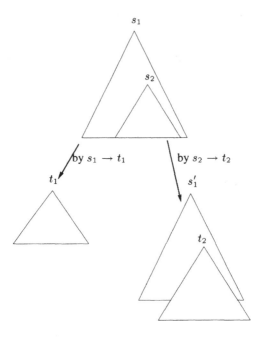

Figure 9.2. A critical pair

This can be solved by adding new rules to compensate for the lack of confluence. For example, the following set of rewrite rules:

$$0 + x \to x \qquad\qquad -0 \to 0$$
$$(-x) + x \to 0 \qquad\qquad -(-x) \to x$$
$$(x + y) + z \to x + (y + z) \qquad x + (-x) \to 0$$
$$(-x) + (x + y) \to y \qquad x + ((-x) + y) \to y$$
$$x + 0 \to x \qquad -(x + y) \to (-y) + (-x)$$

is a convergent set of rewrite rules for groups.

This process of adding new rules to force confluence is called *completion*. Can we always do this? Unfortunately, in general, no. Indeed, given a system E such that provability from E is undecidable (see Exercise 9.1), no convergent set R of rules can be equivalent to E, since the method above terminates and decides $\longleftrightarrow^*_R$. (By the way, this same argument, together with the fact that the rules above are a convergent set of rewrite rules for groups, proves that the theory of groups is decidable.)

The first and most influential method to find a convergent system of rewrite rules equivalent to a given system of equations by completion was proposed by Knuth and Bendix in 1970. To complete a set of equations E into a set of rewrite rules R, the Knuth-Bendix procedure does two things at once: first, it orients equations in E as rules so that the resulting relation $\longrightarrow_R^*$ terminates; and second, it identifies and repairs the faults in R with respect to local confluence. We start off by dealing with the latter:

Definition 9.8 (Critical Pair) *A* ground critical pair *between two ground rewrite rules* $s_1 \to t_1$ *and* $s_2 \to t_2$ *is an equation* $t_1 \doteq s_1'$, *where* s_1' *is the result of replacing one occurrence of* s_2 *by* t_2 *in* s_1.

A critical pair *between two rewrite rules* $s_1 \to t_1$ *and* $s_2 \to t_2$ *is an equation* $t_1\sigma \doteq s_1'\sigma$, *where* s_1' *is the result of replacing one occurrence of a non-variable subterm* s_1'' *by* t_2 *in* s_1, *and* $\sigma = mgu(s_1'', s_2)$.

The process of computing critical pairs of two rules is called superposition.

Ground critical pairs are illustrated on Figure 9.2. On the ground level, if we have $s_1 \longrightarrow_R t_1$ and $s_1 \longrightarrow_R s_1'$ as in the figure, then this is a possible source of lack of local confluence, unless both t_1 and s_1' already rewrite to the same term; if this is not the case, we should add new rules to R (i.e. *complete* the set R) so that $t_1 \longrightarrow_R^* u$ and $s_1' \longrightarrow_R^* u$ for some term u to ensure local confluence. Knuth and Bendix's solution is to add either $t_1 \to s_1'$ or $s_1' \to t_1$ to R.

The more complicated notion of critical pairs (not necessarily ground) is the same, lifted to full first-order terms. The constraint on s_1'' not being a variable is technical: we never need to produce critical pairs when s_1'' is a variable x, since this would just produce an equation $t_1[s_2/x] \doteq s_1[t_2/x]$, which can already be derived by a valley proof, and is therefore never an obstacle to local confluence.

The other point of the Knuth-Bendix completion procedure is to orient equations as rules, so that the resulting set of rewrite rules terminates. But if $\longrightarrow_R$ terminates, then $\longrightarrow_R^*$ is a well-founded ordering with some other properties like being stable and monotonic:

Definition 9.9 (Rewrite Orderings) *Let* $>$ *be a binary relation on terms.*

$>$ *is a* strict ordering *if and only if it is irreflexive and transitive.*

$>$ *is* monotonic *if and only if for every n-ary function symbol* f, *if* $s > t$, *then* $f(t_1, \ldots, t_{i-1}, s, t_{i+1}, \ldots, t_n) > f(t_1, \ldots, t_{i-1}, t, t_{i+1}, \ldots, t_n)$.

$>$ *is* stable *if and only if for every substitution* σ, $s > t$ *entails* $s\sigma > t\sigma$.

$>$ *has the* subterm property *if and only if whenever* s *is a proper subterm of* t, *then* $t > s$.

A simplification ordering *is a monotonic strict ordering having the subterm property.*

A rewrite ordering *is a monotonic and stable strict ordering.*

A reduction ordering *is a noetherian rewrite ordering.*

So, if R is a terminating rewrite system, then $\longrightarrow^*_R$ is a reduction ordering on the set of terms. In fact, Dallas Lankford noticed in 1977 that $\longrightarrow_R$ terminates if and only if there is a reduction ordering $>$ such that every rule $s \to t$ in R verifies $s > t$. (That is, if we see binary relations as sets of couples, $R \subseteq >$.)

To prove that R is a terminating rewrite system, the usual way is to guess a reduction ordering $>$ such that $R \subseteq >$. Then, a large class of reduction orderings from which we can choose are actually simplification orderings. This comes from a remark by Nachum Dershowitz that any monotonic ordering is noetherian whenever it has the subterm property (provided that the set of function symbols is finite).

For example, the *lexicographic path ordering* $>_{lpo}$, one of the simplest reduction orderings obtained this way, is parameterized by a strict ordering $>$ on function symbols (a set of *precedences*), and is defined as follows. Let s be $f(s_1, \ldots, s_m)$ and t be $g(t_1, \ldots, t_n)$ two application terms. Then $s >_{lpo} t$ if and only if:

- $s_i \geq_{lpo} t$ for some i, $1 \leq i \leq m$;

- or $f > g$ and $s >_{lpo} t_i$ for every i, $1 \leq i \leq n$;

- or $f = g$, $s_1 = t_1, \ldots, s_{i-1} = t_{i-1}$, and $s_i >_{lpo} t_i$, $s >_{lpo} t_{i+1}, \ldots$, $s >_{lpo} t_n$, for some i, $1 \leq i \leq n$.

where $\geq_{lpo}$ denotes the reflexive closure of $>_{lpo}$.

As the lexicographical path ordering demonstrates, we can even choose a reduction ordering that is total on ground terms: choose for $>$ any total ordering on the function symbols. (Compare with the lexicographic orderings of Chapter 7, Section 4.3.)

Many other reduction orderings exist, including the multiset path orderings, and the whole class of *recursive path orderings*, to which the latter two belong.

We now have all the tools we need to define Knuth-Bendix completion as a set of transformation rules on couples (E, R), where E is the current set of unoriented equations, and R is the current set of rewrite rules. The procedure consists in six rules, shown on Figure 9.3, which we can apply in any way we wish, that is, with any strategy for picking equations or rules. We assume that $>$ is any reduction ordering, used to orient the rules of R in rule **(Orient)**. We adopt the convention that $\cup$ denotes disjoint union on the left-hand sides of $\longmapsto$, so for instance, transforming $E \cup \{s \doteq s\}$ into E means erasing $s \doteq s$ from E.

To complete a set E of equations into a convergent set of rewrite rules, choose a reduction ordering $>$, and apply the rules of Figure 9.3, starting from the couple $(E, \varnothing)$ until no rule applies or we produce a couple of the form $(\varnothing, R)$. If we finally get $(\varnothing, R)$, then R is a convergent set of rewrite rules equivalent to E (it is convergent because $\longrightarrow_R \subseteq >$, and $>$ was chosen to be convergent). If we do not get such a couple, but no rule applies, we say that the procedure *fails*; we may then try another reduction ordering and start again. The third possibility is that the procedure loops forever: we cannot exclude it, as the problem of deciding provability is undecidable.

(Delete)	$(E \cup \{s \doteq s\}, R) \mapsto (E, R)$
(Compose)	$(E, R \cup \{s \to t\}) \mapsto (E, R \cup \{s \to u\})$
	if $t \longrightarrow_R u$
(Simplify)	$(E \cup \{s \doteq t\}, R) \mapsto (E \cup \{s \doteq u\}, R)$
	if $t \longrightarrow_R u$
(Orient)	$(E \cup \{s \doteq t\}, R) \mapsto (E, R \cup \{s \to t\})$
	if $s > t$
(Collapse)	$(E, R \cup \{s \to t\}) \mapsto (E \cup \{u \doteq t\}, R)$
	if $s \longrightarrow_R u$
(Deduce)	$(E, R) \mapsto (E \cup \{s \doteq t\}, R)$
	if $s \doteq t$ is a critical pair in R

Figure 9.3. Knuth-Bendix completion

Some theories can be completed rather nicely. In particular, the convergent system of rewrite rules that we have presented above for groups has been found by completing the initial set of three equations defining groups. This not only yields a convergent rewrite system for the theory of groups, but also proves automatically that the equational theory of groups is decidable.

In the case where completion does not terminate, we can still use the completion process, provided that we use a *fair* completion strategy, as shown by Gérard Huet in 1981:

Theorem 9.10 (Correctness) *Let* $(E_0, \varnothing) \mapsto (E_1, R_1) \mapsto \ldots (E_k, R_k) \mapsto \ldots be$ *a sequence of couples deduced by the rules of Figure 9.3.*

The set of persistent rules R is $\bigcup_{i \geq 0} \bigcap_{j \geq i} R_j$.

The sequence is fair if and only if all critical pairs of rules in R are in some E_i, $i \in \mathbb{N}$.

Then, if the sequence is fair, $E_0 \vdash_= s \doteq t$ if and only if $s \longrightarrow_R^* u$ and $t \longrightarrow_R^* u$ for some term u.

A persistent rule is one that has been fully reduced by (**Compose**) and (**Collapse**), whatever new rules we shall add later to the set of rewrite rules. A fair strategy is then a strategy where no critical pair of persistent rules is overlooked forever, that is, where every critical pair of persistent rules is eventually dealt with by the (**Deduce**) rule.

To prove $E_0 \vdash_= s \doteq t$, we therefore do two things in parallel: we complete E_0, and at each step i in a fair completion sequence, we use the terminating set of fully reduced rules in R_i to reduce s and t to their normal forms. If these normal forms are equal, then $E_0 \vdash_= s \doteq t$ is proved, otherwise, we increment i to continue the completion process.

This process is sound, but it is not complete, because the Knuth-Bendix completion procedure may fail. This is precisely the gist of the fairness condition: if the procedure fails, then there are critical pairs which will never be dealt with, so the sequence of couples (E_i, R_i) is not fair.

Unfailing Completion

The source of failure in Knuth-Bendix completion was identified by Jürgen Avenhaus in 1985:

Theorem 9.11 *Given a reduction ordering* $>$, *and a system of equations* E, *there is a set* R *of rewrite rules equivalent to* E *such that* $\longrightarrow_R \subseteq >$ *if and only if every congruence class of terms modulo* E *has a unique minimal element with respect to* $>$.

Since $>$ is noetherian, there are minimal elements for $>$ in any given congruence class. However, they might not be unique.

If $>$ was a total ordering, then they would be unique, and the Knuth-Bendix method would never fail. Unfortunately, we cannot ask for $>$ to be total on the set of all terms. For example, because $>$ is stable, any two distinct variables must be incomparable by $>$. (This is the same argument as in Chapter 7, Section 4.3.)

However, we can ask for $>$ to be a reduction ordering *total on ground terms*. We can then apply the Knuth-Bendix method on any set of ground terms, and it will never fail. Lifting this method to first-order terms yields the so-called *unfailing completion* method, due to Leo Bachmair, Nachum Dershowitz and David Plaisted.

The idea is the following. Take two equations $s_1 \doteq t_1$ and $s_2 \doteq t_2$. Take ground instances of these equations, say $s_1\sigma \doteq t_1\sigma$ and $s_2\sigma \doteq t_2\sigma$. Because we have chosen $>$ to be total on ground terms, there is a unique way of orienting these as rewriting rules, say $s_1\sigma \to t_1\sigma$ and $s_2\sigma \to t_2\sigma$. Then, a ground critical pair is an equation $t_1\sigma \doteq s_1'\sigma$, where s_1' is the result of replacing one occurrence of a non-variable subterm s_1'' by t_2 in s_1, with $s_1''\sigma = s_2\sigma$. (See Figure 9.2.)

To lift this to the non-ground case, we have to be sure that we generate at least one critical pair whose instance will be $t_1\sigma \doteq s_1'\sigma$. We therefore guess the way in which the equations $s_1\sigma \doteq t_1\sigma$ and $s_2\sigma \doteq t_2\sigma$ can be oriented, and compute all possible critical pairs, except those which cannot possibly correspond to any orientation of the corresponding ground terms. This motivates the following definition:

Definition 9.12 (Ordered Critical Pair) *Let* $>$ *be a reduction ordering, and* $>^\#$ *be the relation defined by* $s >^\# t$ *if and only if* $s > t$ *or* s *and* t *are incomparable by* $>$.

An ordered critical pair *of two equations* $s_1 \doteq t_1$ *and* $s_2 \doteq t_2$ *is an equation* $t_1\sigma \doteq s_1'\sigma$, *where* s_1' *is the result of replacing one occurrence of a non-variable subterm* s_1'' *by* t_2 *in* s_1, $\sigma = mgu(s_1'', s_2)$, *and where* $s_1\sigma >^\# t_1\sigma$ *and* $s_1\sigma >^\# s_1'\sigma$.

(Delete)	$E \cup \{s \doteq s\} \mapsto E$
(Simplify)	$E \cup \{s \doteq t\} \mapsto E \cup \{s \doteq u\}$
	if $t \longrightarrow_{E_>} u$ and $s > u$
(Collapse)	$E \cup \{s \doteq t\} \mapsto E \cup \{s \doteq u\}$
	if $t \longrightarrow_{E_>} u$
(Deduce)	$E \qquad\qquad \mapsto E \cup \{s \doteq t\}$
	if $s \doteq t$ is an ordered critical pair in E

Figure 9.4. Unfailing completion

The only added complexity with respect to ordinary critical pairs (Definition 9.8) is that we exclude the cases $s_1\sigma \leq t_1\sigma$ and $s_1\sigma \leq s_1'\sigma$. Indeed, taking ground instances of all terms by an arbitrary substitution θ, if we had one or the other, we would have $s_1\sigma\theta \leq t_1\sigma\theta$ or $s_1\sigma\theta \leq s_1'\sigma\theta$ by stability, but this is not compatible with the fact that we have essentially computed a critical pair of the ground rules $s_1\sigma\theta \to t_1\sigma\theta$ and $s_2\sigma\theta \to t_2\sigma\theta$.

Another important point to notice is that, because we are reasoning implicitly on ground terms, an equation $s \doteq t$ with incomparable s and t might be oriented as $s\sigma \to t\sigma$ for some ground substitutions σ and $t\sigma \to s\sigma$ for others. There is therefore no relevant separation between equations and rules. In fact, the completion procedure can be described as a transformation process on sets of equations only, as described in Figure 9.4. We have adopted the convention that $E_>$ was the set of rewrite rules $\{s \to t \mid (s \doteq t) \in E, s > t\}$, which takes the place of the R component in Knuth-Bendix completion. The **(Collapse)** rule can also be refined by requiring that t rewrite to u by some rule $t_1 \to u_1$ in $E_>$, where the subterm of t that we rewrite by the rule is a *strict* subterm of t. A similar but more complicated refinement can be done in the original Knuth-Bendix algorithm.

As we have announced, unfailing completion never fails, that is, it either returns a finite system of equations, or loops while generating an infinite system of equations. These equations $s \doteq t$ can be used as rewrite rules $s\sigma \to t\sigma$ whenever the chosen instance is such that $s\sigma > t\sigma$; otherwise we still have to use the equation both ways, i.e. we have to generate rewrites $s\sigma \to t\sigma$ and $t\sigma \to s\sigma$. In any case, we can use unfailing completion in much the same way as Knuth-Bendix completion to provide a semi-decision procedure for equality modulo a given set E of equations.

Further Reading

There is much more to say on rewriting. For a complete survey on the domain and a list of relevant publications, see (Dershowitz and Jouannaud, 1990).

2.2   Paramodulation, or Integrating Rewriting with Resolution

Rewriting and its associated techniques provide ways of reasoning in equational logic. Now, a set E of universally quantified equations is basically a set of positive unit clauses, where the predicate symbol is equality. Can we generalise these techniques to handle equality for more general sets of clauses?

It turns out that such a general technique for reasoning on sets of clauses with equality was invented in 1969 by George Robinson and Larry Wos, under the name of *paramodulation*.

Paramodulation

The paramodulation rule is the following, where we write clauses as disjunctions, and where C, C' denote disjunctions of literals, L, L' denote literals:

$$\frac{C \vee s \doteq t \qquad C' \vee L}{C\sigma \vee C'\sigma \vee L'\sigma}$$

where σ is the mgu of s with some non-variable subterm u of the literal L, and L' is the result of replacing one occurrence of u by t in L. (Remember that we have assumed equations to be pairs, that is, $s \doteq t$ and $t \doteq s$ are the same, so we can use any equation in any direction.)

Looking at the ground case, and ignoring any notions of orderings, paramodulation therefore rewrites L into L' by using the rule $s \rightarrow t$. This rewriting is done conditionally: indeed, we can see the premises of the rules as saying that if C is false, then $s \doteq t$ holds, and that if C' is false, then L holds; then the result of the inference rule (the *paramodulant*) says that if C and C' are both false, then the rewritten literal L' must hold.

Supplementing the resolution rule with the paramodulation rule and the following *negative reflection rule* (often encoded in the literature as one step of resolution with the extra clause $\longrightarrow x \doteq x$):

$$\frac{C \vee \neg s \doteq t}{C\sigma}$$

where $\sigma = mgu(s,t)$, yields a sound and complete refutation method for first-order logic with equality. That is, a set S of clauses has no *equational* model if and only if the empty clause can be derived from S by using resolution, paramodulation and the negative reflection rule.

Paramodulation alone is rather uneasy to work with because, just as basic resolution, it is highly non-deterministic and redundant. Crossing it with ordering strategies in the spirit of both ordered resolution and unfailing completion yields a much better rule for equality. Before we explain how this is done, it is beneficial to reinterpret unfailing completion as a case of paramodulation.

Unfailing Completion and Unit Paramodulation

The main tool in unfailing completion is the generation of oriented critical pairs. And computing the oriented critical pairs of $s_1 \doteq t_1$ and $s_2 \doteq t_2$ means finding a non-variable subterm s_1'' of s_1, computing the mgu σ of s_1'' with s_2, and replacing s_1'' by t_2 in s_1 before applying σ, to get a new term s_1'. This process is very similar to rewriting s_1 by the rule $s_2 \rightarrow t_2$, except that we compute σ so that $s_1''\sigma = s_2\sigma$ instead of $s_1'' = s_2\sigma$. This process is called *narrowing*, and we say that s_1 *narrows* to s_1' by the rule $s_2 \rightarrow t_2$ with mgu σ.

Notice that, to lift ground paramodulation to the non-ground case, we have not rewritten, but actually narrowed the literal L to the literal L'. That is, paramodulation is a generalisation of narrowing, except for the ordering constraints.

Notice also that rewriting is a special case of narrowing. Indeed, we can rewrite a term s by first replacing all its free variables by new constants, then narrowing it, and putting back the original variables in lieu of the constants in the result. This leads to the following idea to integrate unfailing completion as a way of refuting sets of equational unit clauses.

To prove $E \vdash_= s \doteq t$, we do as in resolution, and generate a positive unit clause $\longrightarrow u \doteq v$ for each equation $u \doteq v$ in E, and a negative unit clause $(s \doteq t)[c_1/x_1, \ldots, c_n/x_n] \longrightarrow$ where $x_1, \ldots, x_n$ are the free variables of s and t, and $c_1, \ldots, c_n$ are new constants. The latter negative unit clause is the skolemized version of the universally quantified equations $s \doteq t$.

To apply the unfailing completion procedure, we basically used the equations in E to generate new equations in E by narrowing, using Rule **(Deduce)**; meanwhile, we tried to rewrite $s \doteq t$ by the equations obtained thus far, i.e. to narrow $(s \doteq t)[c_1/x_1, \ldots, c_n/x_n] \longrightarrow$ by the equations in E. (Because of skolemization, narrowing it means rewriting it.) The **(Delete)**, **(Simplify)** and **(Collapse)** rules are not strictly needed for completeness, although they greatly help in simplifying the set of clauses. Finally, the way we conclude the proof that $E \vdash_= s \doteq t$ by rewriting s and t to the same term translates in the clausal framework into the fact that we deduce the empty clause from clauses of the form $u \doteq u \longrightarrow$, where u is closed.

The unfailing completion method therefore translates to the following method on equational unit clauses:

- (Oriented unit paramodulation)

$$\frac{\longrightarrow u \doteq v \qquad L}{L'\sigma}$$

  where L is an equational unit clause ($\longrightarrow s \doteq t$ or $s \doteq t \longrightarrow$), which narrows to L' by rule $u \rightarrow v$ (assuming $u >^{\#} v$) and mgu σ.

- (Ground unit negative reflection)

$$\frac{u \doteq u \longrightarrow}{\longrightarrow}$$

In the oriented unit paramodulation rule (or generalised narrowing rule), we have said that L narrowed to L', although we have only defined narrowing on terms. The extension to atoms and then clauses is however immediate. Notice that, as usual, we have to interpret $u \doteq v$ as $u \rightarrow v$ if $u > v$, but if u and v are incomparable, we may have to use $u \rightarrow v$ or $v \rightarrow u$.

Unit paramodulation between two equational unit clauses is then similar to computing critical pairs, and we shall simply equate unit paramodulation with superposition.

Ordered Paramodulation

Jieh Hsiang and Michaël Rusinowitch were the first, in 1989, to prove that resolution plus paramodulation and the negative reflection rule was complete for first-order logic with equality. Michaël Rusinowitch in fact managed to prove that an ordered refinement of resolution with paramodulation was still sound and complete (Rusinowitch, 1989), and generalised the ordered unit paramodulation technique (or the unfailing completion technique) to general sets of clauses.

The spirit is exactly the same as in unfailing completion for the equality part, and as in ordered resolution for the logical part. But we need a special kind of reduction ordering, as defined by Rusinowitch. Let $>$ be a strict ordering on terms and atoms. We say that $>$ is a *complete simplification ordering* if and only if it is a simplification ordering, and moreover:

- if $s > t$ or $s = t$, and s is a subterm of the non-equational atom A, then $(s \doteq t) < A$;

- if $s > t$ or $s = t$, and s is a proper subterm of s' or of t', then $(s \doteq t) < (s' \doteq t')$.

That is, $>$ is a complete simplification ordering if and only if it has a reinforced subterm property, which enforces every equality to be less than every atom that it can rewrite. (This is yet another sense of the word "complete".)

Let $>$ be a complete simplification ordering, total on ground atoms. The rules are:

- (Factoring)

$$\frac{\Gamma \longrightarrow \Delta, A, B}{\Gamma\sigma \longrightarrow \Delta\sigma, A\sigma} \ \sigma = mgu(A, B) \qquad \frac{A, B, \Gamma \longrightarrow \Delta}{A\sigma, \Gamma\sigma \longrightarrow \Delta\sigma} \ \sigma = mgu(A, B)$$

- (Ordered Resolution)

$$\frac{\Gamma \longrightarrow \Delta, A \qquad A', \Gamma' \longrightarrow \Delta'}{\Gamma\sigma, \Gamma'\sigma \longrightarrow \Delta\sigma, \Delta'\sigma} \ \sigma = mgu(A, A')$$

with the added restriction that $A\sigma$ and $A'\sigma$ must be maximal with respect to $>^{\#}$ in the respective clauses $\Gamma\sigma \longrightarrow \Delta\sigma, A\sigma$ and $A'\sigma, \Gamma'\sigma \longrightarrow \Delta'\sigma$. (Compare with Definition 7.25, page 252: this one is slightly more refined than the one we presented there, because it compares instances of A and A', not just A and A'.)

- (Oriented Paramodulation)

$$\frac{\Gamma \longrightarrow \Delta, s \doteq t \qquad \Gamma' \longrightarrow \Delta', A}{\Gamma\sigma, \Gamma'\sigma \longrightarrow \Delta\sigma, \Delta'\sigma, A'\sigma}$$

where σ is the mgu of s with some non-variable subterm u of A, A' is the result of replacing one occurrence of u by t in A, $s\sigma >^{\#} t\sigma$ and $A\sigma$ is maximal in $\Gamma'\sigma \longrightarrow \Delta'\sigma, A\sigma$ with respect to $>^{\#}$. We also constrain A to be a non-equational atom, as the case when A is a positive equality will be handled by the generalised superposition rule below.

We also have the symmetric rule:

$$\frac{\Gamma \longrightarrow \Delta, s \doteq t \qquad A, \Gamma' \longrightarrow \Delta'}{A'\sigma, \Gamma\sigma, \Gamma'\sigma \longrightarrow \Delta\sigma, \Delta'\sigma}$$

where again A is an atom (here, possibly equational) that narrows to A' with mgu σ, $s\sigma >^{\#} t\sigma$ and $A\sigma$ is a maximal atom in its clause.

(As Rusinowitch points out, this generalises narrowing to the clausal case.)

- (Generalised Superposition)

$$\frac{\Gamma \longrightarrow \Delta, s \doteq t \qquad \Gamma' \longrightarrow \Delta', s' \doteq t}{\Gamma\sigma, \Gamma'\sigma \longrightarrow \Delta\sigma, \Delta'\sigma, s''\sigma \doteq t'\sigma}$$

where σ is the mgu of s with some non-variable subterm u of s', s'' is the result of replacing one occurrence of u by t in s', $s\sigma >^{\#} t\sigma$, $s'\sigma >^{\#} t'\sigma$, $s\sigma \doteq t\sigma$ is maximal in $\Gamma\sigma \longrightarrow \Delta\sigma$, $s\sigma \doteq t\sigma$ for $>^{\#}$ and $s'\sigma \doteq t'\sigma$ is maximal in $\Gamma'\sigma \longrightarrow \Delta'\sigma, s'\sigma \doteq t'\sigma$ for $>^{\#}$.

(As Rusinowitch points out, this generalises the superposition of the unfailing completion case.)

- (Negative Reflection)

$$\frac{s \doteq t, \Gamma \longrightarrow \Delta}{\Gamma\sigma \longrightarrow \Delta\sigma} \quad \sigma = mgu(s, t)$$

As usual, we have assumed that distinct clauses were renamed so as to have no free variable in common.

This set of rules provides a sound and complete refutation method for first-order logic with equality. It can also be refined by deleting subsumed clauses, and by deleting tautologies (where not only clauses $\Gamma, A \longrightarrow A, \Delta$, but also clauses $\Gamma \longrightarrow \Delta, s \doteq s$ count as tautologies). The resulting method is significantly more efficient than the original paramodulation rule. The search space is still huge, but then the complexity of reasoning with equality seems to be able to reach appalling heights.

For a more detailed description of paramodulation, rewriting and ordered paramodulation, see Rusinowitch (1989).

2.3   Equality and Tableaux

Integrating equality reasoning with tableaux methods is tougher as with resolution. To integrate equality in the tableaux method, we have to take the special meaning of equality into account when closing paths. Now, a path is a sequent, or semantically, it is a disjunction. Intuitively, closing a path C then means finding a substitution σ such that $C\sigma$ holds in all equational models. This is however much harder to do than in the class of all possible (equational and non-equational) models. We now develop the theory from a semantic point of view, in much the same way as non-equational Herbrand theory.

Equational Herbrand Theory

First, we have to describe the propositional substrate of the theory:

Definition 9.13 *The* quantifier-free theory of equality *is defined as follows.*

The formulas are ordinary quantifier-free formulas, where $\doteq$ is understood as equality. In particular, we consider $s \doteq t$ and $t \doteq s$ as being the same atom.

The semantics is taken to be described by Herbrand interpretations, where $\doteq$ is constrained to denote a congruence on ground terms, that is, a relation verifying all the rules of equality below. Let us denote $\models_{0,=}$ the semantic consequence relation in this theory.

Let us denote $\vdash_{0,=}$ the provability relation in the quantifier-free theory of equality. Its deduction rules are those of propositional logic, plus the rules of equality:

- *(Reflexivity) $\vdash_{0,=} t \doteq t$ for every term t;*

- *(Symmetry) $s \doteq t \vdash_{0,=} t \doteq s$ for every terms s and t;*

- *(Transitivity) $s \doteq t, t \doteq u \vdash_{0,=} s \doteq u$ for every terms s, t, u;*

- *(Functional congruence) $s_1 \doteq t_1, \ldots, s_n \doteq t_n \vdash_{0,=} f(s_1, \ldots, s_n) \doteq f(t_1, \ldots, t_n)$ for every n-ary function symbol f, and every terms $s_1, \ldots, s_n$, $t_1, \ldots, t_n$;*

- *(Predicate congruence)* $s_1 \doteq t_1, \ldots, s_n \doteq t_n, P(s_1, \ldots, s_n) \vdash_{0,=} P(t_1, \ldots, t_n)$ *for every n-ary predicate symbol P, and every terms* $s_1, \ldots, s_n, t_1, \ldots, t_n$;

The quantifier-free theory of equality is basically the theory of equality on formulas where the free variables are considered as constants. (Not to say that we cannot substitute variables, but variables are not universally quantified, i.e. they do not denote the class of all ground terms, but exactly one ground term to be determined later.)

We can then transport the semantic construction of Herbrand's Theorem over to first-order logic with equality, as the following theorem, due to Jean Gallier, shows:

Theorem 9.14 (Herbrand-Gallier) *Let* Φ *be an existential formula* $\exists x_1 \cdot \ldots \exists x_n \cdot \Psi$, *where* Ψ *is quantifier-free. Assume moreover that there is at least one constant in the language.*

Then $\models_= \Phi$ *if and only if there is an integer k, and k closed instances* $\Psi\sigma_1, \ldots, \Psi\sigma_k$ *of* Ψ *such that* $\models_{0,=} \Psi\sigma_1 \vee \ldots \vee \Psi\sigma_k$.

Proof: Let (K) be the conjunction of the axioms of equality. Assume $\models_= \Phi$. Then $(K) \models \Phi$, i.e. $\models \neg(K) \vee \Phi$. Notice that $\neg(K)$ is equivalent to the disjunction of existential formulas, so the latter is equivalent to the existence of finitely many instances $K_1, \ldots, K_n$ of the axioms of equality and finitely many instances $\Psi\sigma_1, \ldots, \Psi\sigma_k$ of Ψ such that $\neg K_1 \vee \ldots \vee \neg K_n \vee \Psi\sigma_1 \vee \ldots \vee \Psi\sigma_k$ is propositionally valid. If the latter holds, then $\neg K \vee \Psi\sigma_1 \vee \ldots \vee \Psi\sigma_k$ is propositionally valid, where K is the conjunction of all ground instances of the axioms of equality. Therefore, $\models_{0,=} \Psi\sigma_1 \vee \ldots \vee \Psi\sigma_k$.

Conversely, if $\models_{0,=} \Psi\sigma_1 \vee \ldots \vee \Psi\sigma_k$, then trivially $\models_= \Phi$. □

We leave it as an exercise to the reader to verify that the same theorem can be proved with $\vdash_=$ instead of $\models_=$ and $\vdash_{0,=}$ instead of $\models_{0,=}$, that skolemization works, and that in fact **LK** plus the equality rules is a sound and complete deduction system for first-order logic with equality.

Equational Matings

Jean Gallier *et al.* have used this idea to devise an extension to the method of tableaux (Gallier *et al.*, 1992) to handle first-order logic with equality. We won't get into the details of this, but the idea is as in first-order logic without equality: given a tableau for a formula Φ, we can consider paths in the tableau as disjunctions of literals, whose conjunction is equivalent to Φ. Then Φ holds in all equational interpretations if and only if we can find a tableau for Φ, and a substitution σ that makes all paths true in all equational (Herbrand) interpretations. That is, the operation of closing a path becomes:

Given a path $\Gamma \longrightarrow \Delta$, let $\Gamma' \longrightarrow \Delta'$ be the subpath consisting of the atomic formulas in $\Gamma \longrightarrow \Delta$, and find a substitution σ such that $\models_{0,=} \Gamma'\sigma \longrightarrow \Delta'\sigma$,
$$\text{i.e. } \Gamma'\sigma \models_{0,=} \Delta'\sigma.$$

Because $\Gamma' \longrightarrow \Delta'$ contains only atomic formulas, Gallier *et al.* prove that $\Gamma'\sigma \models_{0,=} \Delta'\sigma$ if and only if there is an atom A in Δ' such that $\Gamma'\sigma \models_{0,=} A\sigma$.

However, contrarily to the case without equality, it is not the case that $\Gamma'\sigma \models_{0,=} A\sigma$ if and only if there is an atom B in Γ' such that $B\sigma \models_{0,=} A\sigma$. For instance, $a \doteq b, b \doteq c \models_{0,=} a \doteq c$, although neither $a \doteq b \models_{0,=} a \doteq c$ nor $b \doteq c \models_{0,=} a \doteq c$ holds. So closing a path cannot be done by choosing two complementary signed formulas in the path and unifying them: we have to find one positive atom A, and then, considering *all* negative atoms in the path as a set Γ, we must find a substitution σ such that $\Gamma\sigma \models_{0,=} A\sigma$.

2.4   Rigid E-Unification

The problem of finding σ such that $\Gamma\sigma \models_{0,=} A\sigma$ reduces to the subproblem where Γ consists of equations only, and A is an equation, by the following trick: create a new constant T, make every predicate symbol P a function symbol p and turn every non-equational atom $P(t_1, \ldots, t_n)$ into an equation $p(t_1, \ldots, t_n) \doteq T$.

To close a path, it therefore remains to solve the so-called *rigid E-unification* problem:

Given a finite set Γ of equations, and an equation $s \doteq t$, find all substitutions σ such that $\Gamma\sigma \models_{0,=} s\sigma \doteq t\sigma$.

Such a σ is called a *rigid E-unifier* of the sequent $\Gamma \longrightarrow s \doteq t$. Finding whether there is such a substitution is NP-complete, so knowing whether a given path can be closed is a NP-complete problem. This contrasts with the non-equational case, where closing a path could be done in polynomial time, even in linear time on conventional machines.

The problem is even more complicated. Indeed, there may be most general rigid E-unifiers, but in general they won't be unique. In fact, there may even be an infinite number of them. For example, the sequent $a \doteq f(a) \longrightarrow x \doteq a$ has infinitely many rigid E-unifiers, all incomparable, namely all substitutions $[f^k(a)/x]$, $k \in \mathbb{N}$.

This problem can be alleviated by relaxing the notion of being most general. Informally, the right notion of being more general seems to be the following. Given a substitution σ, let $E(\sigma)$ be the set of equations of the form $x \doteq x\sigma$, for every variable x in the domain of σ. Also, define two systems of equations to be *equivalent* if and only if their equational consequences (by $\models_{0,=}$ or $\vdash_{0,=}$) are the same. Then, instead of saying that σ is more general than σ' if $\sigma' = \sigma\sigma''$ for some substitution σ'', we say so if and only if $\Gamma, E(\sigma')$ and $\Gamma, E(\sigma), E(\sigma'')$ are equivalent systems of equations.

The notion of being more general is much more complicated and much less operational than in the non-equational case. Above all, it depends on the system of equations Γ. This complicates enormously the representation of most general unifiers, which cannot be mere substitutions.

The most annoying problem, in fact, is that we really want to close all paths, not just one. That is, we have to solve *simultaneous rigid E-unification*, or finding a common rigid E-unifier to several rigid E-unification problems at once. Because of the problem with the notion of more general above, we cannot compute simultaneous rigid E-unifiers by first computing a rigid E-unifier for the first path, then one for the second, instantiated path, and so on; or we should use the usual notion of being more general, and be prepared for enumerating infinite sets of rigid E-unifiers.

In fact, simultaneous rigid E-unification is undecidable (Degtyarev and Voronkov, 1996a). Compare this with the fact that trying to unify several couples simultaneously in the non-equality case can be done in polynomial time: equality pushes the problem outside the scope of decidability. As a consequence, there is no (terminating) algorithm that can find whether there is a substitution that closes all paths in a formula at once.

This remark should not be understood as a death sentence for the method of equational matings. As David Plaisted has noticed, there are particular cases that occur often in practice (namely, clausal forms where all clauses containing equations are unit or Horn clauses) where the simultaneous rigid E-unification problem remains in NP, and is therefore not much harder than, say, propositional satisfiability. Moreover, Degtyarev and Voronkov have devised a weaker version of simultaneous rigid E-unification that is decidable, and which is enough to ensure completeness of the resulting tableau system: see (Degtyarev and Voronkov, 1996b).

Congruence Closure

We end this section on equational matings by a remarkable tool, called *congruence closure* (Gallier, 1987). Whereas finding rigid E-unifiers is computationally hard, checking that a given substitution is a rigid E-unifier can be done very efficiently with congruence closure.

Given a substitution σ, checking whether it is a rigid E-unifier of $\Gamma \longrightarrow s \doteq t$ means checking whether $\Gamma\sigma \models_{0,=} s\sigma \doteq t\sigma$. Therefore, we can first apply σ, then check whether the equations $s\sigma \doteq t\sigma$ is an equational consequence of the finite set of equations $\Gamma\sigma$.

Our problem is therefore now: given a finite set of equations Γ, and an equation $s \doteq t$, does $\Gamma \models_{0,=} s \doteq t$ hold? Moreover, because variables behave as constants in the quantifier-free theory of equality, we can pretend that all equations are ground.

This problem is decidable, for example by rewriting. Indeed, take any reduction ordering $>$ total on ground terms. Apply the Knuth-Bendix (ground) completion procedure to Γ: this cannot fail, because all terms are linearly ordered. Moreover, the procedure terminates, because only finitely many terms and rules can be generated: the ordering is Noetherian, and we always replace terms by smaller ones. Therefore, the procedure produces, in finite time, a convergent set of ground rewrite

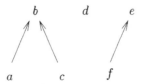

Figure 9.5. A Union-Find tree for $a = b = c, e = f$

rules R. We can then use R to reduce s and t to their normal forms and compare them. If they are equal, then $\Gamma \models_{0,=} s \doteq t$ holds, otherwise it does not hold.

This is a rather complicated decision algorithm. Congruence closure is more efficient, and also rests on simpler notions. Let Π be an equivalence relation on the set S of subterms of Γ, s and t. Initially, we let Π be the finest possible equivalence relation, i.e. $s \ \Pi \ t$ if and only if $s = t$. Assume that we have three operations $find$, $union$ and $list$ acting on Π:

- given a term t in S, $find(t)$ returns a *colour*, i.e. a canonical representative of the equivalence class of t in Π; that is, $s \ \Pi \ t$ if and only if $find(s) = find(t)$, i.e. if they have the same colour;

- given two colours u and v, $union(u, v)$ modifies Π so as to be the finest equivalence relation containing Π and relating u and v; informally, $union(u, v)$ "makes u and v equivalent" by merging their equivalence classes;

- finally, given a colour u, $list(u)$ returns a list of all application terms $f(t_1, \ldots, t_n)$ in S such that some t_i has colour u.

To implement $find$ and $union$, a data-structure known as a Union-Find structure is constructed. This is a collection of trees, where edges are links that go from the leaves to the roots; where nodes are terms in T; and where colours are those terms that are at the roots of trees. For example, Figure 9.5 shows a Union-Find structure for representing the equivalence relation with three classes $\{a, b, c\}$, $\{d\}$ and $\{e, f\}$.

The $find$ operation then takes its argument t, and follows the links until it comes to a root: this root is the colour of t. To implement $union$ on two colours u and v, either $u = v$ and we stop, or we draw a link from u to v, or from v to u. To keep the structure balanced, a common trick is to record at each node t the number $n(t)$ of nodes in the subtree rooted at t, and to always direct a link from the colour having the lowest such number to the colour having the highest such number (and then, we update the number of the new colour, by adding the number of the old one to it).

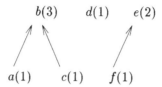

$$b(3) \qquad d(1) \qquad e(2)$$

$$a(1) \qquad c(1) \qquad f(1)$$

Figure 9.6. A Union-Find tree for $a = b = c$, $e = f$, with cardinals of classes

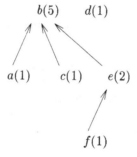

$$b(5) \qquad d(1)$$

$$a(1) \qquad c(1) \qquad e(2)$$

$$f(1)$$

Figure 9.7. Calling $union(b, e)$

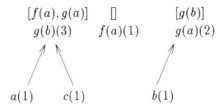

Figure 9.8. Union-Find structure with annotations for *list*

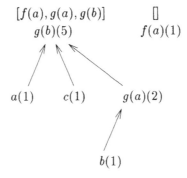

Figure 9.9. *union*, with annotations for *list*

For example, the Union-Find structure of Figure 9.5, when annotated with these numbers, is depicted in Figure 9.6. Now, calling $union(b, e)$ will add a link from e to b, since $n(e) = 2 < 3 = n(b)$, yielding the Union-Find structure of Figure 9.7. (Another common trick consists, then, in modifying $find$ so that computing $u = find(t)$ redirects the link from t directly to its colour u, whenever $t \neq u$; this ensures that paths from leaves to roots are lazily reduced to the shortest possible length.)

We also annotate colour nodes u by the list $list(u)$ of all application terms having some argument of colour u. For example, we have represented between square brackets above the root nodes u the annotations $list(u)$ in the Union-Find structure on Figure 9.8. When we do a $union$, we just concatenate the lists to compute the list of the new root node. Computing $union(g(b), g(a))$ for example yields the Union-Find structure of Figure 9.9.

Then, Union-Find structures implement $find$ in $O(\log n)$ average time, if there are n subterms, $union$ and $list$ in $O(1)$ time.

Union-Find structures are structures for representing equivalence classes. It remains to handle the congruence rules of equality. This is done by using the following procedure $merge$ instead of $union$, which propagates equalities upwards in the term structure, until the Union-Find structure reaches a fixed point. To compute $merge(s, t)$:

- let $u = find(s)$, $v = find(t)$;

- if $u = v$, then return;

- otherwise, let U be $list(u)$, V be $list(v)$;

- call $union(u, v)$;

- for all $s \in U, t \in V$, if $find(s) \neq find(t)$ and $congruent(s, t)$, then call $merge(s, t)$.

where the auxiliary expression $congruent(s, t)$ returns true if and only if $s = f(s_1, \ldots, s_m)$, $t = f(t_1, \ldots, t_m)$ and, for all i, $find(s_i) = find(t_i)$.

This code is rather rough, but it can be made quite efficient by using auxiliary data structures, and notably a *signature table*, i.e. a hash-table mapping $(m + 1)$-tuples $(f, c_1, \ldots, c_m)$, where f is an m-ary function symbol and the c_i are colours, to the common colour of all terms $f(t_1, \ldots, t_m)$ such that t_i has colour c_i, for every i. The signature table is used so as not to sweep the full Cartesian product $U \times V$ in the last step, but rather find all pairs of congruent pairs in $U \times V$ directly.

This yields a polynomial time algorithm that works extremely fast in practice, and decides all propositions $\Gamma \models_{0,=} s \doteq t$ in the quantifier-free theory of equality. See (Downey *et al.*, 1980) for details.

3   EQUATIONAL THEORIES

Going back to simpler theories than full equality, we may ask what we can do to help an automated prover reason with symbols like $+$, which are at least known to be associative, commutative, and to have a neutral element.

Notice that such properties are universally quantified equations: associativity is $\forall x \cdot \forall y \cdot \forall z \cdot (x + y) + z \doteq x + (y + z)$, commutativity is $\forall x \cdot \forall y \cdot x + y \doteq y + x$, and the fact that 0 denotes a neutral element is $\forall x \cdot x + 0 \doteq x, \forall x \cdot 0 + x \doteq x$. Such equational properties are quite frequent, and it is wise to take them into account when looking for a proof.

We could apply the full theory of equality (rewriting, rigid E-unification) to the problem of handling such equational theories, but it is in fact easier in principle to work modulo a system of equations E: the only thing we have to do to adapt the whole Herbrand theory is to replace the Herbrand universe $H(B_0)$ by its quotient modulo E.

3.1   Herbrand Theory Modulo an Equational Theory

Gordon Plotkin was the first to notice this fact. This led him to notice that the only thing we need to handle equational theories E is to change the unification process to *equational unification* modulo E (in short, E-unification), leaving the reasoning process (resolution, tableaux, whatever) unchanged:

Definition 9.15 (E-Unification) *Let E be a system of equations.*
We say that two terms s and t are E-equal if and only if $E \models_= s \doteq t$.
We say that σ is a E-unifier of s and t if and only if $E \models_= s\sigma \doteq t\sigma$.

The simplest equational theory is the empty theory $\varnothing$. By Theorem 9.3, two terms are $\varnothing$-equal if and only if they are syntactically equal. And σ is an $\varnothing$-unifier of s and t if and only if σ is a (syntactic) unifier of s and t.

We shall give a description of the E-unification problem, and how we can solve it in particular cases as well as in the general case, in Section 3.2. Before that, we first justify Plotkin's remark by extending Herbrand's Theorem to the case modulo an equational theory E:

Theorem 9.16 *Let E be an equational theory.*
Let Φ be an existential formula $\exists x_1 \cdot \ldots \exists x_n \cdot \Psi$, where Ψ is quantifier-free. Assume moreover that there is at least one constant in the language.
Then $E \models_= \Phi$ if and only if there is an integer k, and k closed instances $\Psi\sigma_1$, ..., $\Psi\sigma_k$ of Ψ such that $E \models_= \Psi\sigma_1 \vee \ldots \vee \Psi\sigma_k$.

Proof: This is an easy adaptation of Theorem 9.14. ☐

This theorem can be understood in the following way. If we look at the Herbrand universe $H(B_0)$, then what matters is the quotient $H(B_0)/\sim_E$, where $\sim_E$ is the equivalence relation of E-equality among ground terms.

Operationally speaking, this means that to reason modulo E in resolution, or the tableaux method, or in resolution with paramodulation, we just have to replace the unification procedure by some E-unification procedure, yielding corresponding E-resolution, E-tableaux or E-paramodulation rules.

In particular, if the equality symbol $\doteq$ does not appear at all in the formula Φ, but only in E, then $E \models_= \Psi\sigma_1 \vee \ldots \vee \Psi\sigma_k$ can be decided by using resolution or tableaux alone, with E-unification instead of unification, and we don't need any fancy rule like E-paramodulation to handle equality. Indeed, E-paramodulation in fact never applies, since there is no equality literal in Φ. In tableaux, the situation is similar: to close a path C, seen as a disjunction of literals, we have to find σ such that $E \models_= C\sigma$, but because of Theorem 9.3, this is so if and only if there are two literals A and $\neg A'$ in C such that $E \models_= A\sigma \doteq A'\sigma$.

The case where the equality symbol $\doteq$ does not appear in the formula Φ to prove is one of the most important particular cases of equational reasoning. Instead of using heavy and complicated rules like paramodulation, we simply use resolution or the tableaux method with E-unification.

Rewriting modulo an equational theory, and hence E-paramodulation, is also quite useful, but for other reasons. In particular, using E-unification instead of unification in rewriting and completion procedures may help produce convergent systems of rules modulo E for systems of equations E' that have no convergent system of rules modulo the empty theory. A typical example is the theory E' of *commutative groups*, which cannot be turned by Knuth-Bendix completion to a convergent rewrite system, but can be turned to a convergent rewrite system modulo the commutativity of $+$: we use Knuth-Bendix completion modulo the commutativity of $+$ to this end.

It remains to find whether E-unification procedures exist, and how to build them.

3.2   E-Unification

The first difficulty in E-unification is that mgus are not necessarily unique. For example, unifying $x + a$ and $y + b$ modulo associativity and commutativity of $+$ gives rise to mgus $[b + z/x, a + z/y]$ (where z is a new variable), and $[b/x, a/y]$, which are not instances of each other. In fact, b is not equal to any instance of $b + z$ modulo associativity and commutativity, so we cannot have unique mgus even if we take as equivalent any two E-equal substitutions.

The correct notion that replaces that of a unique mgu is the following:

Definition 9.17 (Complete Set of Unifiers) *Let E be an equational theory.*

An E-instance of s is any term E-equal to some instance of s.

A complete set of E-unifiers, or csu, of two terms s and t is a set Σ of idempotent E-unifiers σ of s and t, such that every E-unifier of s and t is an E-instance of some $\sigma \in \Sigma$.

That is, a complete set of E-unifiers is a (hopefully compact) way of describing all E-unifiers, by taking all instances of elements of the csu. For instance, the mgu in syntactic unification can be thought of as a csu consisting of exactly one unifier. (The word "Complete" assumes yet another, new meaning here; this is only indirectly related to the completeness of proof-search procedures, for example.)

Given a procedure that enumerates a complete set of unifiers, we therefore have to modify the resolution rule, for instance, by allowing the prover to generate resolvents for all possible mgus, and similarly for paramodulation and the tableau closing rule.

This poses a set of new problems. There are equational theories for which some terms have only infinite complete sets of unifiers. For example, assume that $+$ is only known to be associative. ($+$ then represents, say, string concatenation.) Then $x + a$ and $a + x$ have infinitely many mgus, none an instance of any other, namely

$[a/x], [a + a/x], \ldots, [\underbrace{a + \ldots + a}_{k \text{ times}}/x]$, and so on. The set of all these mgus is the

smallest possible csu of $x + a$ and $a + x$ modulo associativity of $+$, and it is infinite.

In the case where csus are infinite, E-unification can be used by *dovetailing* it with the proof-search process: for example in resolution, generate the resolvents of two clauses C_1 and C_2 corresponding to the first few elements of the csu of the selected literals, then go on resolving new clauses; now and then, we go back and generate the other elements of the csu to produce all other E-resolvents of C_1 and C_2. That is, we enumerate the elements of the csu and develop the set of clauses in parallel.

Complete sets of unifiers should be as small as possible, if only to control the number of resolvents—or the level of non-determinism—that we generate. A good way is to insist that complete sets of unifiers are minimal with respect to inclusion, or equivalently, that they contain only mgus. That is, if σ is more general than σ' in the csu Σ, we should only keep σ. Unfortunately, we cannot even do this, as some equational theories are nasty enough that the "more general than" relation is not well-founded. In this case, there are couples of terms for which all csus cannot consist of mgus only.

Equational theories, and corresponding E-unification problems, can then be classified as follows, in ascending difficulty order:

- *unitary*: there is always a csu having at most one mgu. For example, the empty theory $\varnothing$, or the theory of Boolean rings (see Chapter 2, Section 5.2);

- *finitary*: there is always a finite csu, and we can reduce it so that it consists only of mgus. This is a common case, which includes: the theory of one commutative symbol (nicknamed C), the theory of one associative and commutative symbol (AC), of one associative and commutative symbol with a neutral element (AC1), of one associative, commutative and idempotent symbol (ACI), of one associative, commutative and idempotent symbol with a neutral element (AC1I), the theory of groups (G), and the theory of Abelian groups (CG), among others.

- *infinitary*: we can always describe csus as sets of mgus, but the csus may not be finite any longer. This is the case for the theory of one associative symbol (A, a.k.a. semi-groups), or of one associative symbol with neutral element (A1, a.k.a. monoids). The latter is common enough, as this is for example the theory of concatenation on strings.

- *nullary* (a misleading name): for some E-unification problems, no csu is composed only of mgus; such csus are then necessarily infinite. An example is the theory of one associative and idempotent symbol (AI). Another example is higher-order unification, i.e. unification modulo the equational theory of the simply-typed λ-calculus. (This is not first-order, but can be made so

(Delete)	$S \cup \{s \doteq s\} \mapsto S$
(Decompose)	$S \cup \{s \doteq t\} \mapsto S \cup \{s_1 \doteq t_1, \ldots, s_n \doteq t_n\}$
	where $s = f(s_1, \ldots, s_n), t = f(t_1, \ldots, t_n)$
(Coalesce)	$S \cup \{x \doteq y\} \mapsto S[y/x] \cup \{x \doteq y\}$
	where x and y are distinct variables
(Eliminate)	$S \cup \{x \doteq s\} \mapsto S[s/x] \cup \{x \doteq s\}$
	where x is a variable outside $\mathrm{fv}(s)$,
	and s is not a variable
(LazyParamodulate)	$S \cup \{s \doteq t\} \mapsto S \cup \{s' \doteq u, s'' \doteq t\}$
	where s' is a non-variable subterm of s,
	$u \doteq v$ is an equation in E, and
	s'' is the result of replacing one occurrence of s' in s by v

Figure 9.10. Gallier-Snyder rules for E-unification

by translating to an adequate system of combinators, and using Curry's equations (Barendregt, 1984); or by using the $\lambda\sigma$-calculus as basic set of combinators (Dowek *et al.*, 1995).)

In general, E-unification is undecidable. Even the E-unifiability problem, where we ask whether two terms s and t have a E-unifier, but we don't require that we know them, can be undecidable. For example, unifiability modulo the theory DA of one associative symbol $+$ and one symbol $*$ distributing over $+$ is undecidable. Another example is higher-order unification, which is undecidable already at the second order —that is, when all λ-terms and subterms are restricted to have types of the form A or $A_1 \rightarrow \ldots \rightarrow A_n \rightarrow A$, with $A, A_1, \ldots, A_n$ base types. These unification problems are of infinitary type, but there are also theories of infinitary type whose unifiability problem is decidable, the most well-known ones being the theory A of one associative symbol and the theory AI of one associative idempotent symbol.

To solve E-unification problems, the simplest but one of the least usable method is to enumerate all possible $\longleftrightarrow^*_{R_E}$ derivations, and check that they prove $s\sigma \doteq t\sigma$ for some substitution σ, where s and t are the input terms to unify. This method, at least, proves that csus are always recursively enumerable.

To get a more efficient (possibly infinite) recursive enumeration of a csu, Jean Gallier and Wayne Snyder have proposed to use the set of rules of Figure 9.10. To unify s and t modulo E, apply the rules non-deterministically, starting from the set $\{s \doteq t\}$, and transforming it until we get a set of equations of the form $x \doteq u$. Such a set is called a *solved form*, and represents the substitution mapping each x to the corresponding u, provided that x does not occur free in u. Note that in the **(LazyParamodulate)** rule, $u \doteq v$ is assumed to be an equation in E, where

(Narrow)	$S \cup \{s \doteq t\} \mapsto \exists z_1 \cdots \ldots \exists z_n \cdot S \cup \{s' \doteq t\} \cup E(\sigma)$
	where s narrows to s' with mgu σ by rule $u \rightarrow v$ in R,
	and $\{z_1, \ldots, z_n\} = (\text{dom } \sigma \cup \text{yield } \sigma) \setminus \text{fv}(s)$

Figure 9.11. E-unification by narrowing from a set of rewrite rules R

all free variables have been renamed to fresh variables. This takes care of the fact that equations in E are thought of as implicitly universally quantified. The method can be refined by demanding that s' and u be two non-variable terms with the same topmost function symbol in the **(LazyParamodulate)** rules, and that **(Decompose)** be applied immediately afterwards.

(LazyParamodulate) is a version of paramodulation, where we do not yet know which substitution will be used to instantiate s' and u to the same term modulo E. It is a lazy version since, instead of first computing all possible E-unifiers of s' and u, it delays this computation by putting the equation $s' \doteq u$ in the set of equations to realize by the end of the computation.

Observe also that the **(LazyParamodulate)** rule introduces fresh variables, in the form of the free variables of the (renamed version of) the equation $u \doteq v$.

When E has a convergent equivalent system of rewrite rules R, we can improve on the previous algorithm by using narrowing, as introduced in Section 2.2. (The associated reduction ordering $>$ is simply $\longrightarrow_R^*$, here.) Take any algorithm for syntactic unification, for example the one of Figure 7.2, page 239, and add the rule of Figure 9.11. We have written $E(\sigma)$ for the set of equations defined by the substitution σ; that is, if $\sigma = [t_1/x_1, \ldots, t_m/x_m]$, then $E(\sigma) = \{x_1 \doteq t_1, \ldots, x_m \doteq t_m\}$.

Again, we assume that the rule $u \rightarrow v$ is renamed first in **(Narrow)**, so that all free variables in u and v are fresh. In particular, $z_1, \ldots, z_n$ are fresh variables. The purpose of the existential quantifier is to hide them: after some computation path in the algorithm terminates on a solved form for some substitution σ (i.e., on some set E' of equations between variables and terms such that $E' = E(\sigma)$), the E-unifier we were looking for is σ restricted to the set of non-fresh variables introduced during the computation. (Some freshly introduced variables may remain free in the range of the E-unifier; see the example at the beginning of the section.) This is represented formally by three new rules:

(Utilise1)	$\exists z \cdot S \quad\quad\quad\quad\quad \mapsto S$
	if z does not occur free in S
(Utilise2)	$\exists z \cdot S \cup \{z \doteq s\} \mapsto S$
	if z does not occur free in S
(ExistCongruence)	$\exists z \cdot Q \quad\quad\quad\quad\quad \mapsto \exists z \cdot Q'$
if the existential system Q transforms to Q' by some rule	

If E has no convergent rewrite system, then we can still use unfailing completion on E to produce more and more of the persistent rewrite rules that we need to apply the (**Narrow**) rule (this time, in the form where s narrows to s' by equation $u \doteq v$ in E, where $u >^{\#} v$, where $>$ is a reduction ordering total on ground terms), so narrowing can actually even be used in this case.

3.3   Particular Theories

Having a general procedure for enumerating all elements of a complete set of E-unifiers is not enough to deal with particular theories E that occur in practice. One of the most useful such theories is the theory AC of one associative and commutative symbol $+$. Extensions of this theory (AC1, ACI, AC1I, G, etc.) work mostly in the same fashion.

AC-unification is finitary, but the Gallier-Snyder method does not terminate in general, and the narrowing method does not apply since there is no convergent system of rules for AC. (The relaxation that we can apply unfailing completion while looking for a unifier does not terminate in general.) Mark Stickel proposed a clever method to AC-unify two terms in 1975 (Stickel, 1981), which is based on solving linear homogeneous Diophantine equations. Variants of the method have been used to solve the problem of AC1-unification, ACI-unification, CG-unification, and so on.

The idea is the following. First, we can assume without loss of generality that the terms s and t to unify are both applications of $+$. Then, we can also assume that s and t are built only with $+$, and no other function symbol: if $f(u_1, \ldots, u_m)$ is a maximal subterm of s or t with f another function symbol, we can replace this term by a new variable x, and add the new equation $x \doteq f(u_1, \ldots, u_m)$ to the set of unification problems to solve simultaneously. (This trick is called *variable abstraction*.) We have thus reduced the problem to that of unifying two terms in $T(X, \{+\})$.

But we can always see terms in $T(X, \{+\})$, where $+$ is AC, as multisets of variables, or formal linear combinations of variables with integer coefficients: for example, $(x_1 + x_2) + (x_3 + x_1)$ can be seen as the multiset containing x_1 twice, and x_2 and x_3 once each, or as the formal linear combination $2x_1 + x_2 + x_3$. Let $x_1, \ldots, x_n$ be the free variables in the terms s and t that we wish to unify. Any substitution σ with domain in $x_1, \ldots, x_n$ can then be represented as a formal linear combination of new variables $y_1, \ldots, y_k$, that is, we let $\sigma(x_i) = \sum_{j=1}^{k} a_{ij} y_j$, $1 \le i \le n$, where k and the a_{ij} are unknown natural integers.

We then write down the conditions for σ to be an AC-unifier of s and t. Assume that s is represented as $\sum_{i=1}^{n} s_i x_i$, and t as $\sum_{i=1}^{n} t_i x_i$:

- $\sum_{j=1}^{k} a_{ij} > 0$, $1 \le i \le n$; (σ maps x_i to a well-formed term)

- $\sum_{i=1}^{n} s_i \sum_{j=1}^{k} a_{ij} y_j = \sum_{i=1}^{n} t_i \sum_{j=1}^{k} a_{ij} y_j$; ($\sigma$ unifies s and t)

where the latter condition can also be rewritten by expressing that the coefficients of y_j must be the same on both sides of the equation as:

$$\sum_{i=1}^{n} s_i a_{ij} = \sum_{i=1}^{n} t_i a_{ij}$$

for each $1 \leq j \leq k$. That is, we have to find the solutions of the following equation in n natural integer variables X_i, $1 \leq i \leq n$:

$$(9.1) \quad \sum_{i=1}^{n} s_i X_i = \sum_{i=1}^{n} t_i X_i$$

If we find enough n-tuples $(a_{ij})_{1 \leq i \leq n}$, $1 \leq j \leq k$, that satisfy this equation, and such that $\sum_{j=1}^{k} a_{ij} > 0$, $1 \leq i \leq n$, then we are done. In fact, Stickel proved the following:

Theorem 9.18 (AC-Unification) *Let s and t be two terms in $T(X, \{+\})$.*

Let Γ be a set of n-tuples $a_j = (a_{ij})_{1 \leq i \leq n}$, $1 \leq j \leq k$, of natural integers satisfying Equation (3.3) (with a_{ij} in place of X_i), and assume that Γ contains all solutions of Equation (3.3) that are minimal with respect to the pointwise ordering on n-tuples of integers.

If Δ is a subset of Γ such that $\sum_{1 \leq j \leq k, a_j \in \Delta} a_{ij} > 0$ for each i, $1 \leq i \leq n$, then we let σ_Δ be the substitution mapping every x_i, $1 \leq i \leq n$, to any term represented by $\sum_{1 \leq j \leq k, a_j \in \Delta} a_{ij} y_j$ as a formal linear combination.

Then the set of all substitutions σ_Δ, where Δ is such a subset of Γ, is a complete set of AC-unifiers of s and t.

Then, solving a linear homogeneous Diophantine equation like Equation (3.3) can be done by first establishing an upper bound m on the values of the X_i's, and then exploring by brute force the whole space of all elements of $\{0, \ldots, m\}^n$.

The time for computing such a csu can be ghastly. In fact, although AC-unifiability is only NP-complete, computing a complete set of AC-unifiers needs a double exponential time in the sizes of s and t in the worst case. This requirement is a bit lower for ACI-unification, but we still need exponential time to compute a complete set of ACI-unifiers.

In automated theorem proving, moreover, we need to apply the unification algorithm at each step. A more efficient method is to collect all Diophantine equations as we go along instead of really unifying, and postponing their resolution until later. Solving systems of linear homogeneous Diophantine equations is indeed not more complicated, and even sometimes simpler than solving single equations, and does the work of several AC-unifications at once.

We end this section by mentioning that practical cases are much more difficult to implement than AC-unification alone. Indeed, in general we have several different symbols, all subject to some equational theories, and we have to combine

E-unification methods for each equational theory E. For example, we might have a CG-symbol $+$ and an AC1I-symbol $\cup$ in our language, so we would unify terms modulo both theories. More subtly, we may have two symbols, say $\cup$ and $\cap$, which are both ACI, and this also counts as the combination of two theories, the ACI theory of $\cup$ and the ACI theory of $\cap$.

We can combine several equational unification methods in many practical cases, provided that some restrictions are met. These restrictions enable us to clearly identify which equational theory applies to which terms. If E_0 and E_1 are two equational theories, then we require that the set of function symbols appearing in E_0 does not intersect the set of function symbols appearing in E_1. If we have terminating E_0-unification and E_1-unification algorithms, we can also find a terminating $E_0 \cup E_1$-unification algorithm, provided that E_0 and E_1 satisfy some additional restrictions, namely that they are *collapse-free* (no equation in E_0 or E_1 is of the form $x \doteq t$, with $x \in \mathrm{fv}(t)$) and *regular* (all equations $s \doteq t$ in E_0 or E_1 are such that $\mathrm{fv}(s) = \mathrm{fv}(t)$). The interested reader is referred to the bibliography.

3.4   Further Reading

The body of literature on E-unification, both in the case of general systems of equations E and in the particular cases of A-unification, C-unification, AC-unification, etc. is large. Furthermore, a lot of research has been done to explore questions related to the combination of several equational theories, to the introduction of sorts (see Section 4.2), to unification in the simply-typed λ-calculus and even more expressive frameworks. For a survey on these topics, see (Jouannaud and Kirchner, 1991).

4   OTHER THEORIES

There is a multitude of other kinds of theories that we might wish to incorporate into an automated theorem prover for first-order logic. Correspondingly, there are many other refinements of the basic automated proof-search techniques. We briefly describe three of them: sorts (Schmidt-Schauß, 1989; Walther, 1988), which help in taxonomical reasoning; Mark Stickel's theory resolution (Stickel, 1985), which provides the basis for integrating general theories, not only equational ones, in the code of the prover itself; and Gröbner bases (Davenport *et al.*, 1989), a variation on the theme of rewriting applied to algebras of polynomials over a field.

4.1   Sorts

Fundamentally, a *sort* is a piece of syntax that represents a set of values. In pure first-order logic, a sort is then any formula Φ with exactly one free variable x, representing the set of all x's such that Φ holds. A sort is what enables us to establish a *taxonomy* of values, i.e. a classification of values into different bins. Sorts are usually simple to reason about, but resolution or tableaux alone are inadequate for taxonomical reasoning, as the first example in Section 1 demonstrates. In fact, sorts as general formulas with one free variable form a class that is too big to really be usable, so we shall be interested in restrictions of the set of all possible sorts.

A common restriction is to consider only unary predicate symbols as being sorts. In the example:

$$coyote(x_1) \longrightarrow animal(x_1)$$
$$\longrightarrow coyote(Will)$$
$$bird(x_2) \longrightarrow animal(x_2)$$
$$\longrightarrow bird(Bip)$$
$$grain(x_3) \longrightarrow plant(x_3)$$
$$\longrightarrow grain(wheat)$$
$$bird(x_4), grain(y_4) \longrightarrow eats(x_4, y_4)$$
$$animal(x_5), plant(y_5), eats(x_5, y_5) \longrightarrow$$

the sorts would then be $coyote$, $animal$, $bird$, $grain$, and $plant$. This allows us to consider the sorts as being *type constants*. Then, because type-checking in simple enough type systems is efficiently decidable, incorporating a sort-checking algorithm in the unification process to take care of taxonomical reasoning should be quite a valuable extension.

For instance, we can read the above example as consisting of:

- sort inclusion judgements: $coyote \subseteq animal$, $bird \subseteq animal$, $grain \subseteq plant$;

- sorting judgements: $Will : coyote$, $Bip : bird$, $wheat : grain$;

- and finally, clauses where variables are quantified over all values of some sort:
$$(\forall x_4 : bird \cdot \forall y_4 : grain \cdot) \qquad \longrightarrow eats(x_4, y_4)$$
$$(\forall x_5 : animal \cdot \forall y_5 : plant \cdot) \ eats(x_5, y_5) \longrightarrow$$

Modifying unification so that bindings are restricted to obey the sort inclusion constraints and the sorting judgements, we then immediately derive the empty clauses from the above set of two clauses: unify $eats(x_4, y_4)$ with $eats(x_5, y_5)$ by the substitution $[x_4/x_5, y_4/y_5]$, which is legitimate as x_5 of sort $animal$ is mapped to x_4, which is of the smaller sort $bird$, and y_5 of sort $plant$ is mapped to y_4, of the smaller sort $grain$. This example is paradigmatic: many problems in logic contain a high proportion of taxonomic information that is best dealt with sorts.

This example illustrates the most common framework for handling sorts, namely the *order-sorted framework* with subsort declarations and function declarations:

Definition 9.19 (Order-sorted) *Let S be a non-empty set of so-called* sort symbols. *We assume that every variable x has a sort s, and occasionally write it $x : s$
to resolve ambiguities concerning its sort.*

*Let $\subseteq$ be a partial order on sorts, which we assume to be generated from a set
of* subsort declarations *of the form $s_1 \subseteq s_2$, $s_1 \in \mathcal{S}$, $s_2 \in \mathcal{S}$.*

Let $\mathcal{D}$ be a mapping from function symbols f to non-empty sets of function declarations *of the form $f : s_1 \times \ldots \times s_n \to s$, where n is the arity of f, and
from predicate symbols P to non-empty sets of* predicate declarations *of the form
$P : s_1 \times \ldots \times s_n \to \mathbb{B}$, where n is the arity of P and $\mathbb{B}$ is a distinguished constant
not in S.*

An order-sorted signature Σ *is a triple $(\mathcal{S}, \subseteq, \mathcal{D})$ of a set of sorts, a partial order
on sorts and a set of function and predicate declarations.*

We say that a term t is a well-formed term *of sort s, and we write $t : s$, if and
only if either:*

- *t is a variable $x : s'$, with $s' \subseteq s$,*

- *or t is a well-formed application $f(t_1, \ldots, t_n)$, where $t_1 : s_1$, ..., $t_n : s_n$,
  $s_1 \subseteq s_1'$, ..., $s_n \subseteq s_n'$, there is a declaration $f : s_1' \times \ldots \times s_n' \to s'$, and
  $s' \subseteq s$.*

Similarly, we say that an atom $P(t_1, \ldots, t_n)$ is a well-formed atom *if and only if
$t_1 : s_1$, ..., $t_n : s_n$, $s_1 \subseteq s_1'$, ..., $s_n \subseteq s_n'$, and there is a declaration $P :
s_1' \times \ldots \times s_n' \to \mathbb{B}$. Well-formed formulas are formulas where every atom is well-
formed.*

We say that a signature Σ is regular *if and only if every term t has a least sort
$S(t)$.*

The semantics of order-sorted terms and formulas is then directly adapted from
the usual first-order semantics, with a caveat: not only do we require that the domain D_I of the interpretation is not empty, but in fact the denotation of any sort, as
a set of values, must also be non-empty. This is to ensure that the deduction rules
remain valid: the formula $(\forall x : s \cdot \Phi) \Rightarrow (\exists x : s \cdot \Phi)$, which will remain provable,
has to remain valid.

Definition 9.20 (Semantics) *Let $\Sigma = (\mathcal{S}, \subseteq, \mathcal{D})$ be an order-sorted signature.
An interpretation I consists in:*

- *a map from sort symbols s to non-empty domains $I(s)$, such that if $s \subseteq s'$ is
  a subsort declaration in Σ, then $I(s) \subseteq I(s')$; (let $D_I = \bigcup_{s \in \mathcal{S}} I(s)$ be the
  carrier of the interpretation;)*

- *a map from function symbols f to functions $I(f)$ from D_I^m to D_I, where m
  is the arity of f, such that for every sort declaration $f : s_1 \times \ldots \times s_m \to
  s$ in $\mathcal{D}$, for every $v_1 \in I(s_1)$, ..., $v_m \in I(s_m)$, we have $I(f)(v_1, \ldots,
  v_m) \in I(s)$;*

- *a map from predicate symbols P to functions $I(P)$ from D_I^m to $\mathbb{B}$, where m is the arity of P.*

A well-formed assignment ρ is a function from V to D_I, such that $\rho(x : s) \in I(s)$.

In an interpretation I, and modulo a well-formed assignment ρ, the semantics of terms and formulas is defined by:

- $\llbracket x \rrbracket I \rho = \rho(x);$

- $\llbracket f(t_1, \ldots, t_m) \rrbracket I \rho = I(f)(\llbracket t_1 \rrbracket I \rho, \ldots, \llbracket t_m \rrbracket I \rho);$

- $\llbracket P(t_1, \ldots, t_m) \rrbracket I \rho = I(P)(\llbracket t_1 \rrbracket I \rho, \ldots, \llbracket t_m \rrbracket I \rho);$

- $\llbracket \neg \Phi \rrbracket I \rho = \bar{\neg} \llbracket \Phi \rrbracket I \rho;$

- $\llbracket \Phi \vee \Phi' \rrbracket I \rho = \llbracket \Phi \rrbracket I \rho \bar{\vee} \llbracket \Phi' \rrbracket I \rho;$

- $\llbracket \Phi \wedge \Phi' \rrbracket I \rho = \llbracket \Phi \rrbracket I \rho \bar{\wedge} \llbracket \Phi' \rrbracket I \rho;$

- $\llbracket \Phi \Rightarrow \Phi' \rrbracket I \rho = \llbracket \Phi \rrbracket I \rho \bar{\Rightarrow} \llbracket \Phi' \rrbracket I \rho;$

- $\llbracket \forall x : s \cdot \Phi \rrbracket I \rho = \bigwedge_{v \in I(s)} \llbracket \Phi \rrbracket I(\rho[v/x]);$

- $\llbracket \exists x : s \cdot \Phi \rrbracket I \rho = \bigvee_{v \in I(s)} \llbracket \Phi \rrbracket I(\rho[v/x]);$

where $\bigwedge$ denotes distributed conjunction and $\bigvee$ denotes distributed disjunction.

The definition of $\llbracket \_ \rrbracket$ is almost the same as in the unsorted case; the only difference is in the domain of variation of v in the denotations of the quantifiers, which is now $I(s)$ instead of D_I. The notions of validity, satisfiability, models, semantic consequence all carry through modulo the sorts.

Beside the semantic notions, we also have proof theories. All deduction systems carry through, with only one difference: substitutions are now restricted to be well-formed substitutions:

Definition 9.21 (Substitution) *A well-formed substitution σ is a substitution mapping variables $x : s$ to well-formed terms $t : s'$, such that $s' \subseteq s$.*

Substitutions are the syntactical counterpart of assignments. A well-formed substitution is then the syntactic representation of a well-formed assignment.

The whole of Chapter 6 translates directly to the order-sorted framework. First-order order-sorted logic has sound and complete deduction systems, is compact, verifies the Löwenheim-Skolem Theorem. We can herbrandize and skolemize: to convert a variable $x : s$ into a Skolem term $f(x_1 : s_1, \ldots, x_n : s_n)$, create a Skolem function f with the unique function declaration $f : s_1 \times \ldots \times s_n \to s$.

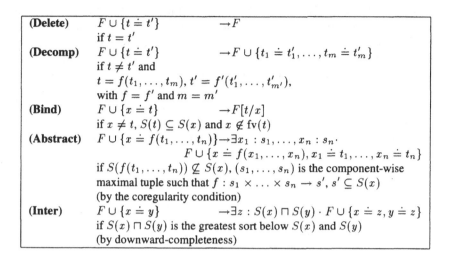

(Delete)	$F \cup \{t \doteq t'\}$	$\rightarrow F$
	if $t = t'$	
(Decomp)	$F \cup \{t \doteq t'\}$	$\rightarrow F \cup \{t_1 \doteq t'_1, \dots, t_m \doteq t'_m\}$
	if $t \neq t'$ and	
	$t = f(t_1, \dots, t_m), t' = f'(t'_1, \dots, t'_{m'})$,	
	with $f = f'$ and $m = m'$	
(Bind)	$F \cup \{x \doteq t\}$	$\rightarrow F[t/x]$
	if $x \neq t$, $S(t) \subseteq S(x)$ and $x \notin \mathrm{fv}(t)$	
(Abstract)	$F \cup \{x \doteq f(t_1, \dots, t_n)\} \rightarrow \exists x_1 : s_1, \dots, x_n : s_n.$	
	$F \cup \{x \doteq f(x_1, \dots, x_n), x_1 \doteq t_1, \dots, x_n \doteq t_n\}$	
	if $S(f(t_1, \dots, t_n)) \not\subseteq S(x)$, $(s_1, \dots, s_n)$ is the component-wise	
	maximal tuple such that $f : s_1 \times \dots \times s_n \rightarrow s'$, $s' \subseteq S(x)$	
	(by the coregularity condition)	
(Inter)	$F \cup \{x \doteq y\}$	$\rightarrow \exists z : S(x) \sqcap S(y) \cdot F \cup \{x \doteq z, y \doteq z\}$
	if $S(x) \sqcap S(y)$ is the greatest sort below $S(x)$ and $S(y)$	
	(by downward-completeness)	

Figure 9.12. Rules for order-sorted unification (regular, coregular, downward complete case)

To translate Chapter 7 and Chapter 8 to the order-sorted case, we only have to change the unification procedure into an order-sorted one, i.e. a unification procedure that only returns well-formed substitutions with respect to the signature. If the signature is regular —a common case— then unification is a variant of the usual unsorted unification, described in (Jouannaud and Kirchner, 1991) for example. This unification procedure has added rules, which must be applied non-deterministically, and indeed order-sorted unification may not be unitary any longer. The reason is that function symbols with several function declarations may force us to guess the sorts that the arguments will have when instantiated by the final unifier.

In fact, the following example, due to Manfred Schmidt-Schauß, shows that order-sorted unification may even be infinitary: take three sorts s_1, s_2 and s_3 with subsort declarations $s_1 \subseteq s_3$, $s_2 \subseteq s_3$, one constant a with declarations $\rightarrow s_1$ and $\rightarrow s_2$, and one function symbol f with declarations $s_1 \rightarrow s_1$ and $s_2 \rightarrow s_2$; then a minimal complete set of unifiers for $x : s_1$ and $y : s_2$ is the set of all $[\underbrace{f(\dots f(}_{n \text{ times}} a)\dots)/x, \underbrace{f(\dots f(}_{n \text{ times}} a)\dots)/y], n \in \mathbb{N}$.

In the simpler but quite common subcase where the signature is not only regular but *coregular* (i.e., the set of all $(s_1, \dots, s_n)$ such that $f : s_1 \times \dots \times s_n \rightarrow s'$ and $s' \subseteq s$ has at most one component-wise maximal element) and *downward complete* (i.e., the set of all sorts s such that $s \subseteq s_1$ and $s \subseteq s_2$ has at most one maximal element), then unification becomes unitary again. Figure 9.12 shows a deterministic algorithm for computing unifiers in this case. (**Bind**) has to be split into the three rules (**Bind**) (binding variables to terms of lower sorts), (**Abstract**) (forcing instances of application terms to lower sorts so that unification becomes possible) and (**Inter**) (forcing instances of variables to a common lower sort to make unification possible). Unification fails when trying to unify two terms with distinct top function symbols, when the occurs-check ordering fails, or when there is no n-tuple $(s_1, \ldots, s_n)$ of sorts such that $f : s_1 \times \ldots \times s_n \to s'$, $s' \subseteq S(x)$ in the (**Abstract**) rule, or when $S(x)$ and $S(y)$ have no common lower sort in Rule (**Inter**).

To translate the techniques of Section 2 requires more care. Indeed, it is not enough to use paramodulation or rewriting with order-sorted unification. Assume the order-sorted signature is regular, to simplify matters, and that we are dealing with paramodulation.

It can happen that we can prove $t \doteq t'$ after a few rewriting and narrowing steps, where t and t' are of different sorts. A first case that did not occur in the unsorted case is when there is no sort s such that $s \subseteq S(t)$ and $s \subseteq S(t')$, in which case $t \doteq t'$ must be false in all interpretations respecting the sorts. This means that we need the following new rules:

- (Deletion of Tautologies) Erase all clauses of the form $t \doteq t', \Gamma \longrightarrow \Delta$, where $S(t)$ and $S(t')$ have no common lower sort;

- (Positive Reflection)

$$\frac{\Gamma \longrightarrow \Delta, t \doteq t'}{\Gamma \longrightarrow \Delta}$$

when $S(t)$ and $S(t')$ have no common lower sort.

But this is not the only thing that can happen. Consider the following example: we choose two sorts $\mathbb{Z}$ and $\mathbb{R}$, meant to represent the relative integers and the real numbers respectively, so that we produce the subsort declaration $\mathbb{Z} \subseteq \mathbb{R}$; let $sqrt$ be a unary function symbol denoting the square root function, which we declare by $sqrt : \mathbb{R} \to \mathbb{R}$; let $* : \mathbb{R} \times \mathbb{R} \to \mathbb{R}$ denote multiplication, and $\geq : \mathbb{R} \times \mathbb{R} \to \mathbb{B}$ denote the usual ordering on the real line; produce the following axiom:

$$\forall x : \mathbb{R} \cdot x \geq 0 \Rightarrow sqrt(x * x) \doteq x$$

If a is of sort $\mathbb{Z}$, then $sqrt(a * a)$ is of sort $\mathbb{R}$, and cannot be given the sort $\mathbb{Z}$ by the sorting rules. In particular, unifying $z : \mathbb{Z}$ with $sqrt(a * a)$ always fails. However, applying the axiom above proves that $sqrt(a * a) \doteq a$, where a is of sort $\mathbb{Z}$, meaning that binding z to $sqrt(a * a)$ was actually meaningful in every model. The problem lies in the fact that the sorting process is capable of very weak inferences only, which are not enough to deduce all valid sort judgements: sorting is incomplete in the presence of equality.

In fact, what is lacking in order-sorted unification in the presence of equality is a rule like:

If $\neg C_1 \sigma_1' \vee \ldots \vee \neg C_m \sigma_m' \vee \neg L_1 \sigma_1 \vee \ldots \vee \neg L_n \sigma_n$ is a theorem in theory T:

$$
\overbrace{\begin{array}{ccc} \text{clauses } C_i \vee C_i', \text{ with } C_i \text{ non empty} \end{array}}^{} \qquad \overbrace{\text{unit clauses}}^{}
$$

$$
\frac{C_1 \vee C_1' \quad \ldots \quad C_m \vee C_m' \qquad L_1 \quad \ldots \quad L_n}{C_1' \sigma_1' \vee \ldots \vee C_m' \sigma_m' \vee \neg L_1 \sigma_1 \vee \ldots \vee \neg L_n \sigma_n}
$$

Figure 9.13. T-resolution (partial, large)

$$
\frac{\longrightarrow t \doteq t' \qquad t : A}{t' : A} (=)
$$

where the sequent comes from the deduction process and the sorting claims are output by the sorting process. Manfred Schmidt-Schauß has shown, however, that we could dispense with such rules by adding a new sort TOP containing all the others, and adding the new clause $x \doteq x \longrightarrow$, with x : TOP. The major disadvantage of this technique is that this clause can be paramodulated into every other clause, leading to an explosion in the size of the search space of proofs. Another disadvantage of the method is that the introduction of TOP usually breaks the downward-completeness property, making order-sorted unification harder.

Another solution is to restrict the use of sorts so that any provable equality $t \doteq t'$ must be so that t and t' have the same sorts. This is already what is done in typed programming languages, where both sides of an equality are usually constrained to have the same types, even when the equality does not hold. This is also the preferred case in order-sorted rewriting, and in particular in the important field of algebraic specifications, where specifications of programs are given as sort declarations together with sets of equational axioms on terms in the signature, and programs are terms that execute by rewriting them modulo the equational theory: it is then required that the equational theory be sort-preserving, i.e. that all provable equations are between terms of the same sort, an undecidable property in general. The square root example above, however, shows that we cannot always choose to be in this particular nice case.

Rules like (=) are integral parts of very powerful deduction systems like Per Martin-Löf's Intuitionistic Theory of Types, which aimed at giving a constructivist foundation for all mathematics, or for Thierry Coquand and Gérard Huet's Calculus of Constructions, which is a conservative extension of (intuitionistic) higher-order logic. Such rules therefore play an important rôle in other parts of logic as well.

4.2 Theory Resolution

One method that generalises both E-unification and sorts is Mark Stickel's *theory resolution* (Stickel, 1985). The idea is to augment resolution with an oracle that de-

If $\neg L_1 \sigma_1 \vee \ldots \vee \neg L_m \sigma_m$ is a theorem in theory T:

$$\frac{L_1 \vee C_1 \qquad \ldots \qquad L_m \vee C_m}{C_1 \sigma_1 \vee \ldots \vee C_m \sigma_m}$$

Figure 9.14. T-resolution (total, narrow)

cides whether some quantifier-free formula has a valid instance in the given theory T. This theory may be equational, in which case we shall find back E-unification, or it may be a taxonomic theory, in which case we shall find back a mechanism doing the same as sorting constraints. The reader is referred to (Stickel, 1985) for details, we shall only present the main ideas.

Look at the basic ground resolution rule: the reason why resolution is sound is that we can deduce the clause $C \vee C'$ from the two clauses $C \vee A$ and $C' \vee \neg A$, which comes from the fact that A is either true or false, i.e. that $A \vee \neg A$ is valid (if A holds, then C' holds, otherwise C holds). The idea of theory resolution is, given a theory T, to use more expressive well-known theorems of T to combine clauses together.

For general T, the basic rule of T-resolution is described in Figure 9.13. The main problems of this general case are that we may need to combine more than two clauses together (the $m + n$ clauses on top of the rule), that we cannot combine only literals of each clause but in general full subclauses C_i, and that we may need to attach some unit clauses that participated in the resolution process to the resulting resolvent.

This rule is called *partial* T-resolution, because of the last point: resolvents are generated under the condition that some unit clauses $L_1, \ldots, L_n$ hold. *Total* T-resolution is the restriction of T-resolution where only unconditional resolvents are generated, i.e. $n = 0$. Total and partial T-resolution are both sound and complete, in the sense that they can prove all and only the sentences that are valid in all models of T. However, total T-resolution may need longer proofs than partial T-resolution.

The rule of Figure 9.13 is also an instance of *wide* T-resolution, because we may need to take not only literals, but subclauses $C_1, \ldots, C_m$. The restriction where we combine only literals from each clause, as in resolution or paramodulation, is called *narrow* T-resolution. The total, narrow T-resolution rule is shown in Figure 9.14. (Compare with semantic resolution as presented in Chapter 7, Section 4.4.)

Although it is a much simpler rule, it is not always complete. Indeed, combining literals from different clauses morally means choosing a path to close modulo the theory T, as in tableaux methods. If the theory T treats paths as being independent, this is complete: this is the case for the empty theory (where we only need the resolution rule), for the theory of equality (where we only need resolution and paramodulation), or for the theory of a strict order $<$. The latter is a good example where total narrow T-resolution is complete, but where we cannot combine clauses pairwise in general: the only theorems we need to close paths in the theory of a strict order $<$ are of the form $\neg(t_1 < t_2 \wedge \ldots \wedge t_{n-1} < t_n \wedge t_n < t_1)$; they are disjunctions of literals, so narrow T-resolution is enough, but we cannot by any means reduce the number of literals in the disjunction, hence the number of clauses to T-resolve to some number fixed in advance.

All that we need in theory resolution is an auxiliary procedure that selects subclauses of some of the clauses at hand, instantiates them and checks that the disjunction of their negations holds in all models of T. One of the most promising uses of this technique is to simplify quickly a set of clauses modulo a theory, so as to get more quickly to a refutation. It is probably of little help to show that the current set of clauses has no refutation, as the T-resolution rule is highly non-deterministic in general. What we need are good heuristics to apply T-resolution on valuable cases.

4.3   Gröbner Bases

We finish this small catalogue of valuable specialisations of the basic automated theorem proving techniques by mentioning Bruno Buchberger's notion of *Gröbner bases* for reasoning in rings of multivariate polynomials over a given field. This tool is one of the fundamentals of computer algebra, another domain where we ask the computer to find proofs of theorems.

Computer algebra is specialised to proving theorems in particular algebraic theories, like theories of polynomials over the rationals or the complex numbers, as well as other theorems about integrals, derivatives and solutions of differential equations over complex-valued functions. (See (Davenport *et al.*, 1989) for a good introduction.) Computer algebra specialists have developed a set of remarkably ingenious algorithms to deal with the most common algebraic problems, in quite special cases. Computer algebra systems often embed these algorithms so as to deal with more general cases, for which these techniques may be neither sound nor complete.

In short, although the techniques developed in automated theorem proving are sound and complete, they are usually slow; and although the techniques developed in computer algebra are usually much faster, they may happen to be unsound or incomplete. A very important open problem is to bridge the gap between both domains, either by integrating computer algebra techniques in automated proof tools, or by constraining computer algebra systems by more logical rigour.

Among all computer algebra tools, Gröbner bases almost allow us to reason on the equational theory of polynomial functions over a given field $\mathbb{K}$. We say "almost" because it is our intention to do so, but we cannot, and have to actually work on ideals in the ring of polynomials over $\mathbb{K}$. Let us first recapitulate a few basic concepts of algebra:

Definition 9.22 (Polynomials) *A* ring *is a set A with two binary operations $+$ (addition) and $.$ (multiplication), such that $(A, +)$ is an Abelian group with neutral element 0, $.$ is associative, distributive with respect to $+$ and has a unit element 1 distinct from 0.*

A division ring *is a ring where every non-0 element has a multiplicative inverse. A* field *is a division ring where multiplication is commutative. The* characteristic *of a field $\mathbb{K}$ is the least non-zero integer p such that $px = 0$, where $px = \underbrace{x + x + \ldots + x}_{p \text{ times}}$, for all $x \in \mathbb{K}$ if it exists, or 0 otherwise.*

The set $\mathbb{K}[X_1, \ldots, X_n]$ of n-variate polynomials *over the field $\mathbb{K}$ is the set of all almost zero functions from $\mathbb{N}^n$ to $\mathbb{K}$ (mapping any n-tuple of exponents of variables to a coefficient). A function P from $\mathbb{N}^n$ to $\mathbb{K}$ is said to be* almost zero *if and only if the set of tuples c such that $P(c) \neq 0$ is finite.*

The zero polynomial 0 *is the function mapping every n-tuple to 0. A* monomial *is a polynomial P such that there is a unique tuple $(i_1, \ldots, i_n)$ with $P(i_1, \ldots, i_n) \neq 0$; if $a = P(i_1, \ldots, i_n)$, we also write $aX_1^{i_1} \ldots X_n^{i_n}$, or $a\vec{X}^{(i_1, \ldots, i_n)}$, for the monomial P. If $a = 1$, then we say that P is* monic.

The sum $P_1 + P_2$ *of two polynomials is defined as the function mapping every tuple c to $P_1(c) + P_2(c)$. Every polynomial can be expressed as the sum $\sum_c a_c \vec{X}^c$ of finitely many monomials.*

The product *of a polynomial $P = \sum_c a_c \vec{X}^c$ by a scalar $a \in \mathbb{K}$ is defined as $a.P = \sum_c (a.a_c) \vec{X}^c$. The product of two polynomials $P_1 = \sum_c a_{1c} \vec{X}^c$ and $P_2 = \sum_c a_{2c} \vec{X}^c$ is defined as $P_1.P_2 = \sum_{c_1, c_2} a_{1c_1} a_{2c_2} \vec{X}^{c_1 + c_2}$, where $c_1 + c_2$ denotes the component-wise sum of the tuples c_1 and c_2.*

$(\mathbb{K}[X_1, \ldots, X_n], +, .)$ is a commutative ring containing $\mathbb{K}$ (as the set of all monomials $aX_1^0 \ldots X_n^0$). With its scalar multiplication, it is also a vector space over $\mathbb{K}$, of infinite dimension if $n \neq 0$.

We say that P divides *Q, or that P is a* divisor *of Q, if and only if there is a polynomial P' such that $PP' = Q$. A polynomial P is said to be* irreducible *if and only if its only divisors are of the form a or aP, where $a \in \mathbb{K} \setminus \{0\}$.*

If $n = 1$, $\mathbb{K}[X_1]$ is a Euclidean ring, *i.e. every two polynomials P and Q have a* greatest common divisor, *or gcd $P \wedge Q$, where greatest is with respect to the "divides" pre-ordering.*

If $n \geq 2$, $\mathbb{K}[X_1, \ldots, X_n]$ is not Euclidean, but is a unique factorisation domain, *i.e. every non-zero polynomial can be written as a product of irreducible polynomials in a unique fashion (modulo multiplication by non-zero elements of $\mathbb{K}$, and modulo commutativity and associativity of polynomial multiplication).*

A polynomial function *is a function f from $\mathbb{K}^n$ to $\mathbb{K}$ such that there is a polynomial $\sum_c a_c \vec{X}^c$ with $f(x_1, \ldots, x_n) = \sum_{i_1, \ldots, i_n} a_{i_1, \ldots, i_n} x_1^{i_1} \ldots x_n^{i_n}$ for every $(x_1, \ldots, x_n) \in \mathbb{K}^n$.*

If $\mathbb{K}$ has characteristic 0, then the algebra of polynomials and the algebra of polynomial functions are isomorphic.

Polynomials can be seen as terms over the alphabet $\{+, .\}$ plus ground terms a to represent the elements of $\mathbb{K}$. These terms are normalised as sums of monomials by the usual rules of algebra. Semantically speaking, if we want to interpret the ground terms a as elements of $\mathbb{K}$, and $+$ and $.$ as addition and multiplication in the field, then terms with free variables $x_1, \ldots, x_n$ represent polynomial functions on the n variables $x_1, \ldots, x_n$.

Fields $\mathbb{K}$ of characteristic 0 are most comfortable to work with, like the field $\mathbb{Q}$ of rational numbers, the field $\mathbb{R}$ of real numbers or the field $\mathbb{C}$ of complex numbers, since we can safely interchange the notions of polynomial functions and of polynomials. But this is not the case in other fields, like $\mathbb{Z}/p\mathbb{Z}$, or the field of p-adics, for p prime. Assume anyway that $\mathbb{K}$ has characteristic 0 for the rest of the section.

Some of the questions that are of value to us are:

1. can we decide equality of two polynomial functions?

2. if so, is unification of two polynomial functions decidable?

3. given a set of polynomial equations $P_i \doteq Q_i$, can we decide whether they imply $P \doteq Q$?

4. given a set of polynomial equations $P_i \doteq Q_i$, can we decide whether there is a substitution σ such that the equations $P_i\sigma \doteq Q_i\sigma$ imply $P\sigma \doteq Q\sigma$?

Question 2 is E-unification modulo the theory of polynomial functions. If this problem is decidable, then we can use it to extend resolution and paramodulation so as to deal with polynomial equations. Question 4 is the analogue of rigid E-unification, and can most probably be used as an extension to determine whether we can close paths in tableaux modulo deductions on polynomial functions.

Equality of Polynomials

The answer to question 1 is yes. Because we are working in a field $\mathbb{K}$ of characteristic 0, we can work with polynomials instead of polynomial functions. Then, we can represent polynomials P, say, by maps of finite domain from the n-tuples of integers $(i_1, \ldots, i_n)$ to the non-zero coefficient $a_{i_1,\ldots,i_n}$ of $X_1^{i_1} \ldots X_n^{i_n}$ in P. This map can then be represented as a list of couples $((i_1, \ldots, i_n), a_{i_1,\ldots,i_n})$, sorted by any given total order on n-tuples $(i_1, \ldots, i_n)$ for instance. Then, equality of polynomials corresponds to equality of their representations, and is therefore decidable.

Unification of Polynomials

Question 2 is more difficult. Because deciding whether $P \doteq Q$ is equivalent to deciding whether $P - Q \doteq 0$, and because ground substitutions must map variables to ground terms representing values in $\mathbb{K}$, the question is equivalent to the decidability of the existence of roots in $\mathbb{K}$ of a polynomial over $\mathbb{K}$. When $\mathbb{K}$ is the field $\mathbb{Q}$ of rational numbers, this is known as Hilbert's tenth problem. More precisely, Hilbert's tenth problem deals with integer roots of polynomials with integer coefficients, but this is in fact the same problem.

A set of tuples of integers is called *Diophantine* if and only if it is the set of integer roots of some polynomial with integer coefficients. Yuri Matyasevitch showed in 1970 that all recursively enumerable sets are Diophantine, which implies that Hilbert's tenth problem is in fact undecidable. (See (Davis, 1973) for a gentle introduction, an easy proof and a discussion.)

If $\mathbb{K}$ is the field $\mathbb{R}$ of real numbers, or the field $\mathbb{C}$ of complex numbers, the difficulty is that we cannot represent the uncountably many real or complex numbers as ground terms; if we restrict ourselves to rationals, then the problem is the same as before, and is undecidable.

On the other hand, we can enrich the set of all ground terms in $\mathbb{K}$ so as to contain not only representations for rational numbers, but in general for all algebraic numbers. By definition, an algebraic number is a root of a polynomial with rational coefficients. If we do so, we actually deal with polynomials over the subfield $A_\mathbb{R}$ (resp. $A_\mathbb{C}$) of real (resp. complex) algebraic numbers. Computer algebra people usually consider the $A_\mathbb{C}$ case, because it is simpler and generally useful.

Now, if P is a polynomial with rational (resp. complex rational, that is, complex with rational real and imaginary parts) coefficients only, the question of unifiability of P with 0 is trivially decidable: if P is the polynomial 0, then $[]$ is a mgu; if P is any other constant polynomial a, then there is no unifier; and if P contains free variables, then by D'Alembert's Theorem it has roots in $\mathbb{C}$, hence in $A_\mathbb{C}$, so P is unifiable with 0.

In the last case, it is also quite simple to give a description of all algebraic numbers that are roots of P: just take the polynomial P itself. In fact, to represent all polynomials over $A_\mathbb{C}$, it is enough to know how to represent all algebraic numbers. And we have just seen that algebraic numbers could be represented as their defining polynomials. There is a twist, though. An example will show what the representation is, and where the problem lies: to represent the polynomial $\sqrt{2}XY - \sqrt{3}X$, we produce the polynomial $\alpha_1 XY - \alpha_2 X$, where α_1 and α_2 are two new variables, and generate two polynomial equations that express constraints over the new variables, namely $\alpha_1^2 - 2 \doteq 0$ and $\alpha_2^2 - 3 \doteq 0$. Therefore, $\sqrt{2}XY - \sqrt{3}X$ is $\alpha_1 XY - \alpha_2 X$ *provided that* $\alpha_1^2 - 2 \doteq 0$ and $\alpha_2^2 - 3 \doteq 0$ both hold.

The twist is that, under these constraints, $\alpha_1 XY - \alpha_2 X$ also denotes the three other polynomials $\sqrt{2}XY + \sqrt{3}X$, $-\sqrt{2}XY - \sqrt{3}X$ and $-\sqrt{2}XY + \sqrt{3}X$. But, if we start from polynomials with rational coefficients (or complex rational coefficients) only, and we add new polynomials over $A_\mathbb{C}$ only as the result of unification, then we actually do not care about which of the algebraic roots we wish to represent. In a sense, this is a feature, since it allows us to have exactly one representation for all possible unifiers, thus mimicking the main property of mgus in the syntactic case.

On the other hand, the problem of representing exactly $\sqrt{2}XY - \sqrt{3}X$ and not the other polynomials above is more complicated. Either we enrich the language of coefficients to include square roots, and possible other algebraic numbers, or we enrich the constraints with bounds —say, $\alpha_1 \geq 0$, $\alpha_2 \geq 0$— to constraint the solutions of $\alpha_1^2 - 2 \doteq 0$ and $\alpha_2^2 - 3 \doteq 0$ to be unique. The latter is an instance of Collins' cylindrical decomposition algorithm (Davenport *et al.*, 1989).

Equality of Polynomials modulo a System of Polynomial Equations

Working in $A_{\mathbb{C}}$ means that questions 1 and 2 are now to be understood modulo a set of polynomial equations defining algebraic numbers. In particular, this turns question 1 over the algebraics into question 3 over the rationals.

Question 3, i.e. the question of deciding whether two polynomials P and Q over the rationals are equal whenever $P_i \doteq Q_i$ holds, $1 \leq i \leq m$, for $2m$ given rational polynomials P_i and Q_i, reduces by using subtraction to the following question:

Given $m + 1$ polynomials P and P_i, $1 \leq i \leq m$, over $\mathbf{Q}$, does
$P_1 \doteq 0 \wedge \ldots \wedge P_m \doteq 0 \Rightarrow P \doteq 0$ always hold?

or alternatively:

Does the set of zeros of P contain the intersection of the sets of zeros of the P_i's,
$1 \leq i \leq m$?

This problem is, unfortunately, undecidable, as can be shown by a reduction from Hilbert's tenth problem. However, we can decompose the problem in much the same way as Herbrand's Theorem decomposes the undecidable problem of first-order validity. This is due to David Hilbert, long before Herbrand invented his own theorem. We need a few extra definitions first:

Definition 9.23 *An* ideal *I in a ring R is an additive subgroup of R such that ax and xa are in I whenever $x \in I$ and $a \in R$.*

The ideal generated *by a set S is the smallest ideal containing S in R. The sum $I_1 + I_2$ of two ideals I_1 and I_2 is the ideal generated by the sums $x_1 + x_2$, $x_1 \in I_1$ and $x_2 \in I_2$. We write $\overline{\{x_1, \ldots, x_n\}}$ the ideal generated by n elements $x_1, \ldots, x_n$; this is also the set of all elements $a_1 x_1 + \ldots + a_n x_n$, $a_1 \in R, \ldots, a_n \in R$.*

The radical *$\sqrt{I}$ of a set I is the set of all elements $x \in R$ such that $x^k \in I$ for some k. If I is an ideal, then so is $\sqrt{I}$.*

The property that David Hilbert proved was the following:

Theorem 9.24 (Nullstellensatz) *Given $m+1$ polynomials P and P_i, $1 \leq i \leq m$, over $\mathbf{Q}$, $P_1 \doteq 0 \wedge \ldots \wedge P_m \doteq 0 \Rightarrow P \doteq 0$ holds if and only if $P \in \sqrt{\overline{\{P_1, \ldots, P_m\}}}$.*

This property is quite similar to Herbrand's Theorem: it says that all common zeros of the P_i's are zeros of P if and only if P^k, for some integer k, can be written as a linear combination of the P_i's with polynomial coefficients. Note that, similarly to the case of Herbrand's Theorem, the condition is not that P itself can be written as a linear combination of the P_i's, but that one of its power, with an unknown exponent, can.

The analogy with Herbrand's Theorem goes even farther, in that given k, checking whether $P^k \in \overline{\{P_1, \ldots, P_m\}}$ is decidable, by a method similar to Knuth-Bendix completion. The method is due to Bruno Buchberger, and is called the computation of a Gröbner base. To show the analogy, we first have to get a notion of rewrite rules:

Definition 9.25 (Polynomials as Rules) *Let $<$ be a linear (i.e., total) ordering on monic monomials. We say that $<$ is* admissible *if and only if $\vec{X}^c < \vec{X}^{c'}$ implies $\vec{X}^{c+c''} < \vec{X}^{c'+c''}$ for any n-tuple c'', and $\vec{X}^{(0,\ldots,0)} < \vec{X}^c$ for every $c \neq (0, \ldots, 0)$.*

If P is a non-zero polynomial, it can be written in a unique way as $a\vec{X}^c + P'$, where $\vec{X}^c$ is the greatest monomial in P with respect to $<$, and a is its coefficient in P. The monomial $a\vec{X}^c$ is called the leading monomial *of P.*

Then, any equation $P \doteq 0$ can be understood as a rewrite rule $\vec{X}^c \rightarrow -a^{-1}P'$, where $a\vec{X}^c$ is the leading monomial of P and $P = a\vec{X}^c + P'$.

An example of an admissible ordering is the lexicographic ordering on n-tuples. Rewriting itself should be mostly self-explanatory. Consider a rule $\vec{X}^{(k_1,\ldots,k_n)} \rightarrow Q$, where Q is a polynomial containing monomials less than $\vec{X}^{(k_1,\ldots,k_n)}$. We can use this rule to rewrite any monomial $a\vec{X}^{(i_1,\ldots,i_n)}$ by first computing the greatest integer q such that $(i_1, \ldots, i_n) = q(k_1, \ldots, k_n) + (r_1, \ldots, r_n)$ for some non-negative integers $r_1, \ldots, r_n$ (this is similar to Euclidean division), then producing $a\vec{X}^{(r_1,\ldots,r_n)}Q^q$. To rewrite any polynomial P by Q seen as a rewrite rule, rewrite each of its monomials, and sum the results.

As in ordinary rewriting, given a finite set E of polynomial equations, seen as rewrite rules, we can define the rewriting relations $\longrightarrow_E$, $\longrightarrow_E^*$. But any admissible ordering $<$ on monomials is well-founded, and so $\longrightarrow_E^*$ always terminates. (To prove that $<$ is well-founded, notice that if $(i_1, \ldots, i_n) < (j_1, \ldots, j_n)$, then $i_k < j_k$ for some k; then, use Dickson's Lemma: for any sequence $u_p, p \geq 0$, of n-tuples of non-negative integers, if for every p, q, with $p < q$, the kth component of u_q is less than the kth component of u_p for some k, then there are only finitely many u_p's, i.e. the sequence terminates.)

So, in contrast with general rewriting, $\longrightarrow_E^*$ always terminates, whatever the system E is. As in rewriting, however, $\longrightarrow_E^*$ may not be confluent. (Equivalently, locally confluent since we already know that it terminates.) It therefore remains to find suitable notions of critical pairs and of a completion process:

Definition 9.26 (S-Polynomials) *Let E be a finite set of polynomial equations, seen as rules.*

Given two rules $\vec{X}^{c_1} \rightarrow Q_1$ and $\vec{X}^{c_2} \rightarrow Q_2$ in E, we let their S-polynomials be the normal forms of $\vec{X}^{c_1'} Q_1 - \vec{X}^{c_2'} Q_2$ by $\longrightarrow_E^*$, where c_1' and c_2' are the least (component-wise) n-tuples such that $c_1' + c_1 = c_2' + c_2$.

A critical pair *is any non-zero S-polynomial of two rules in E.*

A set E with no critical pairs *represents a convergent set of rewrite rules, and is called a* Gröbner base.

The algorithm then proceeds mostly like a simpler version of the Knuth-Bendix completion algorithm: while we can generate non-zero S-polynomials from pairs of polynomials in the set E, do so and add them to E. Since $<$ is well-founded, this process eventually terminates, in contrast with the Knuth-Bendix algorithm: this is because any non-zero S-polynomial is a normal form for the current relation $\longrightarrow_E^*$, and therefore has a leading term less in $<$ than all other leading terms of polynomials in E. The output of the procedure is then a convergent rewrite system equivalent to E, modulo the rules of algebra.

Then, we have:

Theorem 9.27 *Let $P_1, \ldots, P_m$ be m polynomials, and E a Gröbner basis for the P_is. Then $P \in \overline{\{P_1, \ldots, P_m\}}$ if and only if $P \longrightarrow_E^* 0$.*

And to check whether $P \in \sqrt{\{P_1, \ldots, P_m\}}$, we can use the following procedure:

- Initialise Q to P;

- Reduce Q to a normal form Q' by $\longrightarrow_E^*$;

- If $Q' = 0$, then stop on success;

- Otherwise, set Q to Q'^2, and go back to step 2.

This terminates on success if and only if P indeed is in the radical of the ideal, although it is still rough.

Although the construction may seem similar to, but simpler than the general construction due to Knuth and Bendix, there is however a difference. In usual rewrite systems, the rewrite relation $\longrightarrow_R^*$ is not only stable but also monotonic (or compatible, as some say): if $t \longrightarrow_R^* t'$, then $f(t_1, \ldots, t_{i-1}, t, t_{i+1}, \ldots, t_m) \longrightarrow_R^* f(t_1, \ldots, t_{i-1}, t', t_{i+1}, \ldots, t_m)$. Here, the only thing we can ask for is that $\longrightarrow_E^*$ is *semi-compatible*, i.e. that if $t \longrightarrow_E^* t'$, then $f(t_1, \ldots, t_{i-1}, t, t_{i+1}, \ldots, t_m)$ and $f(t_1, \ldots, t_{i-1}, t', t_{i+1}, \ldots, t_m)$ reduce to the same polynomial by $\longrightarrow_E^*$, where f is any polynomial operation.

A number of improvements on the algorithm are possible. In particular, we can inter-reduce all polynomials (rewrite all polynomials in E by all new polynomials), and eliminate the zero polynomials from E (both strategies provide an analogue of subsumption in resolution). Several heuristics have been designed so as to minimise the number of S-polynomials that we have to generate during the completion process. Even then, however, there are examples that need time doubly-exponential in the size of the original set E of polynomials to be completed.

An interesting point that has been made by Buchberger on this completion algorithm is that it generalises both the Gaussian elimination algorithm for inverting matrices and Euclid's algorithm for computing greatest common divisors of univariate polynomials. In the case of Euclid, the restriction is that we only take polynomials on one variable X; the only admissible ordering is then the natural ordering on exponents, and the S-polynomial of two polynomials is then the remainder of one by the other in Euclidean division. In the case of Gauss, the restriction is that all polynomials must be linear combinations of the variables (i.e., the only exponents we allow are all zeros except for the ith component, which is 1); the admissible orderings are simply all possible total orderings on variables, the polynomials in E are linear equations (or rows in a matrix), and the S-polynomial of two polynomials is the linear combination of the two corresponding rows that cancels the entry for the first variable with a non-zero coefficient (the pivot, in Gaussian elimination jargon).

Unifying Polynomials modulo a System of Polynomial Equations

We are now in a position where we can answer question 4: given a set of polynomial equations $P_i \doteq 0$, can we decide whether there is a substitution σ such that the equations $P_i\sigma \doteq 0$ imply $P\sigma \doteq 0$?

This problem, as we have seen, is useful for closing paths in tableaux augmented with reasoning on polynomials over $\mathbf{Q}$. It is also useful for unifying polynomials over $A_{\mathbf{C}}$: if P and P_i, $1 \leq i \leq m$, are polynomials over $\mathbf{Q}[\alpha_1, \ldots, \alpha_k, X_1, \ldots, X_n]$ representing some polynomial P over $A_{\mathbf{C}}[X_1, \ldots, X_n]$, where the P_i's are constraints on the algebraic numbers $\alpha_1, \ldots, \alpha_k$, then asking whether there exists σ that instantiates P as a polynomial over $A_{\mathbf{C}}[X_1, \ldots, X_n]$ to 0 is the same as asking whether there exists σ such that dom $\sigma \subseteq \{X_1, \ldots, X_n\}$ (i.e., σ does not bind the algebraic constants α_j) and such that the equations $P_i \doteq 0$ imply $P\sigma \doteq 0$.

We don't know whether this problem is decidable or not. In the case where it is not, we can always use the Nullstellensatz and try to solve the following problem:

is there a substitution σ such that $P\sigma$ is expressible as a linear combination of the polynomials $P_i\sigma$, $1 \leq i \leq m$?

As far as we know, the solution to this problem is not known. More research is necessary, not only to see if we can solve it, but also to see how we can apply it to incorporate algebraic reasoning into existing automated proof methods. Combining ideas from the Gallier-Snyder procedure, or better, from narrowing, with Buchberger's algorithm seems to be a good start. Methods other than Buchberger's completion may also bear some fruit in this area; in particular, Ritt-Wu's decomposition method is particularly promising (Chou and Gao, 1990).

CHAPTER 10

LOGIC PROGRAMMING LANGUAGES

1 INTRODUCTION

In this final chapter we will give an overview of some of the contributions of automated deduction techniques to the development of programming languages, specifically *logic programming languages.*

There are a lot of connections between automated deduction and logic programming. Many of the theorem proving techniques that we have studied in the previous chapters can be realised as concrete programming languages, for instance the language Prolog. On the other side, many of the techniques developed for logic programming languages can be applied to automated deduction, for example specific strategies, the use of parallelism to improve efficiency, etc. This is an example where an application of a theory gives feedback to the theory itself.

To be precise, logic programming is not just a derivate from automated theorem proving, but is the point of convergence of several disciplines:

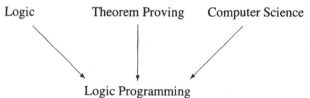

We can summarise the main contributions of each of these disciplines as follows.

Logic has contributed by giving an "economy of syntax" in the sense that it provides a formal language that allows us to express, in a brief and concise way, the problem that we are studying. It has also provided the notions of *derivation* and *interpretation* of sentences, which give an *operational* and *denotational* understanding of logic programming languages. Most logic programming languages are based on *first order classical logic*, more precisely the fragment called *clausal form*, but there are others based on higher-order logics for example.

Logic programming has benefited from theorem proving by the work on resolution. In fact one could say that this is one of the most important contributions to logic programming, since it forms the basis of most implementations of logic programming languages.

341

Finally, to obtain an actual programming paradigm, these results had to be adapted to the needs of computer programming. This work was undertaken by Robert Kowalski, Alain Colmerauer and Carl Hewitt and led to the development of the first logic programming language, Prolog.

Some of the key connections between logic programming and automated deduction are:

- Logic programs *are* in fact just logical sentences, and hence, with a different perspective, we have been looking at these all along!

- Implementation of programming languages based on this paradigm are essentially concerned with finding proofs in some particular logic (Horn clause fragment of classical logic in the case of Prolog). Thus evaluation of logic programs *is* automated theorem proving.

- Logic programming languages have benefited from a great wealth of research with respect to efficiency; perhaps much more than other theorem provers. Hence one would hope to glean some insight from this work which could be applied to other classes of theorem provers. Indeed this is already the case. Many modern automated theorem provers apply optimisation techniques that were originally developed for logic programming languages.

- Finally, logic programming is a nice realisation and a practical application of the material presented in the previous chapters, in terms of concrete programming languages.

All of these issues provide evidence of the importance of logic programming for automated deduction. Moreover, an implementation of a logic programming language, for example Prolog, can also be used as an automated theorem prover itself. In addition, it provides a good setting for the implementation of theorem provers. See for example (Fitting, 1996), where the reader can find logic programs for many of the algorithms used for automated deduction.

Logic programming started by using Horn clauses as a programming language. Recently other fragments of classical logic have been used as a basis for the development of new logic programming languages. For instance *hereditary Harrop formulas* in λProlog (Nadathur and Miller, 1988), which allows unrestricted use of quantifiers and implication in the clause (not just at the outer level). Also recently other logics have served as a basis for the definition of logic programming languages, for example linear logic is the basis for the languages Lolli (Hodas and Miller, 1994), Linear Objects (LO) (Andreoli and Pareschi, 1991) and Forum (Miller, 1994). The language Lolli is a linear refinement of the language λProlog, Linear Objects are an extension of Horn clauses in which literals are generalised to be *multisets* of literals connected by the $\otimes$ connective of linear logic. Finally, the language Forum can be seen as a generalisation of all of these linear logic programming languages, which contains all of linear logic. Here, however, we will restrict our study to the Horn clause fragment of classical logic, and we will consider two particular extensions that are directly applicable to automated theorem provers:

- Constraint logic programming (CLP) — based on a generalisation of the unification-based notion of computation that is characteristic of logic languages.

- Parallelism: logic programming languages are well suited to parallel evaluation. Proof search can be implemented in a more efficient way by splitting the task between many processors.

We will now give a brief overview of Prolog and then present these extensions.

2 PROLOG

The programming language Prolog is based on a fragment of first order classical logic, specifically *Horn clauses*. This fragment contains logical sentences that are built from the following usual connectives:

$$\forall, \exists, \land, \lor, \Rightarrow, \neg$$

We have already seen that there are many different kinds of normal form for classical logic (dnf, nnf, etc.). Here we will consider yet another which is called *clausal form*. For Prolog we will be interested in a restricted class of clausal forms, which are called *Horn clauses*. In Chapter 2, Section 4.2 we introduced a notion of clause for propositional logic which we generalised to predicate logic in Chapter 7. We begin by recalling the notion of clausal form, for which we adopt an alternative notation in that quantifiers are written explicitly. We then give the restriction to obtain Horn clauses.

Definition 10.1 (Clausal forms) *A sentence is said to be in clausal form iff it is of the form*

(universal prefix)(matrix in conjunctive normal form)

where:

- universal prefix *is a string of universal quantifiers* $\forall$, *each followed by a variable, and*

- matrix in conjunctive normal form *is a quantifier-free conjunction of formulas where each is a disjunction of literals.*

We can write a clausal form as a conjunction of universally quantified disjunctions, each of which is called a clause.

For example, $(\forall X \forall Y)((p(X,Y) \lor q(Y,X)) \land \ldots)$ is a clausal form, but $(\forall X)$ $((p(X) \lor (\forall Y)q(Y)) \land \ldots)$ is not because of the $\forall Y$.

We now give the algorithm for finding the clausal form of a sentence in classical logic in the form of a set of transformation rules, denoted by $\longrightarrow$ (not to be confused with sequent calculus entailment relation), that we assume are applied exhaustively to the sentence from top-to-bottom.

1. *Elimination of* $\Longleftrightarrow$.

$$A \Longleftrightarrow B \longrightarrow (A \Rightarrow B) \land (B \Rightarrow A)$$

2. *Elimination of* $\Rightarrow$.

$$A \Rightarrow B \longrightarrow \neg A \lor B$$

3. *Push* $\neg$ *into literals.*

$$
\begin{aligned}
\neg\neg A &\longrightarrow A \\
\neg(A \land B) &\longrightarrow \neg A \lor \neg B \\
\neg(A \lor B) &\longrightarrow \neg A \land \neg B \\
\neg(\exists x . A) &\longrightarrow \forall x . \neg A \\
\neg(\forall x . A) &\longrightarrow \exists x . \neg A
\end{aligned}
$$

4. *Skolemization.*

$$\exists x . A \longrightarrow A[f(t)/x]$$

where f is a new function symbol, and t is a tuple of those distinct variables other than x occurring in A.

5. *Distribute.* This rule distributes $\lor$ maximally over $\land$. For example,

$$A \lor (B \land C) \longrightarrow (A \lor B) \land (A \lor C)$$

Note that since we have the logical equivalence $\forall X.(A \land B) \Longleftrightarrow \forall X.A \land \forall X.B$ by distributivity of $\forall$ over conjunction, it is possible to think of clausal forms as a set of conjoined clauses.

It is important to note that of the above, only rules 1,2,3 and 5 preserve equivalence — they respect the standard logical equivalences. However, rule 4 does not preserve logical equivalence, but it does satisfy the (sufficient) weaker property that: if C rewrites to C' using the previous set of transformation rules, then C has a valid model iff C' does too, and $C \models C'$.

Definition 10.2 (Horn clause) *A* Horn clause *is a clause with either zero or one positive literal. In the first case the clause is called a* denial *or a* goal, *and in the second case it is called a* definite clause.

Any clause can be written in *conditional format*, which is of the form

$$(disjunction) \leftarrow (conjunction)$$

where *disjunction* contains the positive literals of the clause and *conjunction* the atomic parts of the negative literals, and the universal quantifiers are omitted. For example, the conditional format of $A \vee \neg C \vee \neg D \vee B \vee \neg E$ is $A \vee B \leftarrow C \wedge D \wedge E$. In the case of Horn clauses *disjunction* contains zero or one literal.

Horn clauses, specifically definite clauses written in conditional format, are claimed to be the most intuitive kind of logical sentence. This format organises logical sentences in such a way that they are easy to read, and they also lend themselves well to mechanical evaluation. It is this kind of clause that the programming language Prolog is based on.

We will assume some familiarity with basic concepts from Prolog, and just remind the reader of the notations that we will use. In Prolog we usually adopt a different syntax for the logical symbols. In the following table we show the language that we use.

Classical logic	Prolog
$\vee$	;
$\wedge$	,
$\Rightarrow$	$\rightarrow$
$\neg$	not

In addition, it is more common to write $B \leftarrow A$ (read B if A) rather than $A \rightarrow B$ (read A then B), as in the conditional format. It is standard to use capital letters for variables in Prolog, and lower case for predicates (which incidentally is directly opposite to the standard notation used in logic). In our examples, we also use capitals A, B, etc. to denote literals in the meta-language.

A Prolog program consists of two entities, a *program* and a *query*.

- A *program* is a set of definite clauses, written in conditional format. Hence programs will invariably look like:

$$P \leftarrow P_1, \ldots, P_n$$
$$Q \leftarrow Q_1, \ldots, Q_n$$
$$\vdots$$

- A *query* is a denial (a clause with zero positive literals), for example $?P_1, \ldots, P_n$, denoting $\leftarrow P_1, \ldots, P_n$.

Think of the program as the set of assumptions, and the query as the *goal* — the formula that we want to prove from the assumptions. Computation is then the process of proof search. We give a simple example:

$$
\begin{array}{rcl}
A & \leftarrow & B, C \\
B & \leftarrow & D, E \\
C & & \\
D & & \\
E & &
\end{array}
$$

$$?A$$

If we apply resolution to the goal and the first clause, we are left with a sub-problem of the "same form", i.e. $?B, C$. It is a straightforward exercise to show that the resolvent of a Horn clause is still a Horn clause. Horn clauses are therefore very convenient, since the "execution strategy" is a very simple one.

A *derivation* is a sequence of goals $G_1, \ldots, G_n$ such that G_i is obtained from G_{i-1} by an application of resolution (in fact *linear* resolution, see Chapter 7, Section 4.5). Resolution will be directed by a *selection function*, which will choose an atom in the goal to build a resolvent. If the last goal in the derivation is the empty query then we have a *successful derivation*. A derivation that finishes with a non-empty goal, where the selected atom does not unify with any of the heads of the clauses in P is a *failure*. For the goal $?A$ in the previous program there is a successful derivation:

$$
\begin{array}{l}
?A \\
?B, C \\
?D, E, C \\
?E, C \\
?C
\end{array}
$$

which then terminates with the empty query.

Resolution, implemented in terms of unification, turns out to give a very powerful programming paradigm. One can express algorithms in a very natural way. Consider a program which represents the connectivity of a graph in terms of directed edges. We write $edge(a, b)$ if there is an edge from a to b. We can compute all the directed paths in the graph with a very simple Prolog program:

$$
\begin{array}{rcl}
path(X, Y) & \leftarrow & edge(X, Y) \\
path(X, Y) & \leftarrow & edge(X, Z), path(Z, Y)
\end{array}
$$

Binary resolution alone is sufficient for all Horn clause problem solving (we need factoring for non-Horn clauses). However, as we have seen in Chapter 7, it is not the most efficient method of proof search. There have been many studies in making resolution more efficient by adding a strategy. For Horn clause programming, SLD resolution is predominant. This is the one that we now outline.

SLD resolution is a particular case of the resolution principle: S means Selection, indicating the use of a special function (the selection function) which chooses the literal to resolve; L means Linear, and refers to the fact that the two clauses used in the resolution step are the most recent resolvent and one of the other input or ancestor clauses (however, we do not need to consider ancestor clauses in Horn clause programming); D means Definite, as all clauses are definite, except the query.

Definition 10.3 *Let P be a Prolog program. The SLD resolution rule with selection function S takes a query $?A_1, \ldots, A_n$, and a clause $B \leftarrow B_1, \ldots, B_m$ in P such that*

$$S(?A_1, \ldots, A_n) = A_i \ (1 \leq i \leq n)$$

and there is a substitution $\sigma = mgu(A_i, B)$, and gives the new query

$$?\sigma(A_1, \ldots, A_{i-1}, B_1, \ldots, B_m, A_{i+1}, \ldots, A_n)$$

An *SLD derivation* is a sequence of goals $G_1, \ldots, G_n$ such that G_i is obtained from G_{i-1} by an application of the SLD resolution rule. If the last goal in the derivation is the empty query then we have a *successful SLD derivation*. An SLD derivation that finishes with a non-empty goal, where the selected atom does not unify with any of the heads of the clauses in P is a *failure*. In general, the selection function used is the one that chooses the leftmost literal in the query. It is easy to implement SLD with this selection function, since it treats the query as a stack. Each time we apply an SLD resolution step, we take the top element of the stack, and place the new queries at the top of the stack.

The SLD resolution rule is both sound and complete for proof search in the Horn case, however, in Prolog implementations the search of a successful SLD derivation for a goal does not explore all the possibilities. In fact Prolog uses a depth first search strategy (with backtracking) and therefore is not complete. For example, in the following program

$$nat(s(X)) \quad \leftarrow \quad nat(X)$$
$$nat(0)$$

$$?nat(Y)$$

Prolog would fail to find a solution since clauses are treated in the order they appear in the program and depth first evaluation causes an infinite number of applications of resolution with the first clause, as shown below.

$$?nat(Y)$$
$$?nat(X_1)$$
$$?nat(X_2)$$
$$\vdots$$

with Y bound to $s(X_1)$, X_1 bound to $s(X_2)$, etc.

However, there are successful SLD resolution derivations for this query, and a
breadth first search strategy (or any other fair strategy) would find them. Although
breadth first search is complete, depth first search is preferred in most of the implementations because if it terminates it is faster in general in finding the first solution.

What we have outlined here is in fact what is called *pure* Prolog. To make it
into a useful programming paradigm a number of non-logical features have been
added to allow the programmer some control over the evaluation and to make some
computations more efficient. We end this section with several examples of these.

First the *cut* (usually written as !) which is a mechanism for controlling the backtracking in proof search. For example, writing a clause as: $P \leftarrow A_1, \ldots,$
$A_n, !, B_1, \ldots, B_m$ has the effect of abandoning all the choices $A_1, \ldots, A_n$ that were
possible up to the cut. The ! has no logical meaning, and as a literal it always succeeds. However, it has the side effect of pruning the search tree. This feature allows
programs to be written in a more control-directed way, as the following coding of
a conditional shows:

$$\begin{aligned} \text{if}(X, Y, Z) &\leftarrow X, !, Y \\ \text{if}(X, Y, Z) &\leftarrow Z \end{aligned}$$

Here, if X succeeds, then the cut causes all other possibilities to be abandoned (here
the second clause) and now the result just depends on Y. On the other-hand, if X
fails then we move to the second clause and all that we have to do is try to show Z,
since we know that X already failed.

Remark that the notion of cut in Prolog has nothing to do with the notion of a
cut rule or *cut elimination* in the sequent calculus. Moreover, there is also no
connection with the 'of course' modality ! of linear logic.

Another way of controlling Prolog programs is with *assert*. With this feature
we can dynamically change the program, analogous to assignments in imperative
programming languages. The effect of assert(X) is to add the clause X to the program. A simple example shows that it can be used to avoid recomputation. Assume
that we have a program together with a query $?Q$. Writing the query as

$$?Q, \text{assert}(Q)$$

has the effect of memoising the query Q, so that future queries $?Q$ are not computed, but are facts in the program. Similarly, there is a mechanism for removing
facts in a program, called *retract*. This has the effect of removing a clause from
a program, for example removing a redundant clause. Intuitively, this has the effect of garbage collection. In summary, assert and retract together provide a way
of programming in Prolog with *side effects*.

Another non-logical feature of Prolog is arithmetic. Of course it is possible to
perform arithmetic in a logical way. For instance we can define natural numbers in

Prolog as follows:

$$nat(0)$$
$$nat(s(X)) \quad \leftarrow \quad nat(X)$$

which gives the usual inductive definition of the naturals. We can also define operations on numbers, for example addition:

$$add(0, X, X) \qquad \leftarrow \quad nat(X)$$
$$add(s(X), Y, s(Z)) \quad \leftarrow \quad add(X, Y, Z)$$

We leave it to the reader to work through an example, such as $add(s(s(0)),$ $s(s(0)), Z)$, to verify that this does indeed compute addition. It is worth remarking at this point a very powerful programming feature of Prolog. We can write subtraction using the same program as addition, but using it in a different way. For example we have the intended meaning of $add(X, Y, Z)$ to be that Z is the sum of X and Y. Now, given a goal $add(s(0), X, s(s(0)))$ we can compute the value of X (usually thought of as the input) which will be the difference between $s(0)$ and $s(s(0))$. More easily, we can write subtraction as a predicate $sub(X, Y, Z) \leftarrow add(Z, Y, X)$.

However, we would hope to be able to perform arithmetic calculations in a more efficient way using the underlying machine operations. Indeed Prolog has built-in natural numbers and *functions* over them, which takes us out of the pure logical framework. The main consequence is that the built-in arithmetic operations are not as general as their logical counterparts. In particular, we cannot define subtraction in terms of addition as in the example above. For example we can define addition as:

$$add(X, Y, Z) \leftarrow Z = X + Y$$

where $+$ is the built-in addition operation, and $=$ is a system predicate which is interpreted as a query to the evaluator. The effect of this predicate is to evaluate the expression on the right-hand side (here $X + Y$) and unify its value with the left-hand side (here Z), succeeding or failing accordingly. Now it is important to note that the evaluation of the expression $X + Y$ requires that the values X and Y are both ground. This is the reason why we cannot use the predicate *add* to compute subtraction: the goal $add(s(0), X, s(s(0)))$ will fail since X is not ground. Additional built-in system predicates are $\leq$, $\geq$, etc.

There are many other non-logical features that have been added to Prolog to provide a convenient programming environment: input/output to provide interactive programs, file handling and interfacing to a windows environment, etc. However, we will not cover any of these issues here. For general background reading on logic programming, we suggest (Hogger, 1984; Hogger, 1990; Lloyd, 1984). Specifically for Prolog, there are many texts, for example (Sterling and Shapiro, 1986) where the reader can find more details, and examples of use, of the non-logical features briefly sketched above.

▶ **EXERCISE 10.1**
Which of the following are Horn clauses:

$$(i) \quad \leftarrow A \qquad\qquad (ii) \quad A \leftarrow A$$
$$(iii) \quad A \vee B \leftarrow C \qquad (iv) \quad A \leftarrow$$

▶ **EXERCISE 10.2**
Given two *Horn clauses* $P \leftarrow A_1, \ldots, A_m$ and $Q \leftarrow B_1, \ldots, B_n$, show that the resolvent is also a Horn clause.

▶ **EXERCISE 10.3**
Given the following Prolog program:

$$edge(a, b)$$
$$edge(a, c)$$
$$edge(b, d)$$
$$edge(c, d)$$

$$path(X, Y) \quad \leftarrow \quad edge(X, Y)$$
$$path(X, Y) \quad \leftarrow \quad edge(X, Z), path(Z, Y)$$

compute the following goals:

$$(i) \qquad ?path(A, B)$$
$$(ii) \qquad ?path(A, B), path(B, A)$$
$$(iii) \qquad ?path(d, A)$$

3   CONSTRAINTS

In recent years a new class of programming languages have gained immense popularity: *constraint logic programming languages*. These languages consist of a *constraint language* combined with a Horn clause language. This gives rise to a more general notion of logic program, that is governed by the constraint language. It is possible to see conventional logic programming languages (for example Prolog) as constraint logic programming languages where constraints are equations which are solved in the Herbrand Universe by means of term unification. Extensions to Prolog, for example Prolog II, which is based on unification of *rational* terms, can also be seen as a constraint programming language where constraints are again equations, but which are solved in the algebra of rational trees. Hence constraint languages provide a more general notion of logic programming language, which includes as a particular case the notion of logic programming that we presented in the previous section. We will now explain this idea in more detail.

A general Prolog clause is of the following form:

$$p(t_1, \ldots, t_n) \leftarrow Q_1, \ldots, Q_m$$

where $t_1, \ldots, t_n$ are terms. The process of computation is resolution (more precisely SLD resolution), which includes unification: to apply resolution to a query $?p(t'_1, \ldots, t'_n)$ we have to find a unifier σ such that $\sigma t_i = \sigma t'_i$ (for all $1 \leq i \leq n$), then continue with the goal $\sigma Q_1, \ldots, \sigma Q_m$.

However, this process of computation can be described from a different perspective which we present using the factorial function as an example:

$$\text{fact}(0, 1)$$
$$\text{fact}(N, F) \quad \leftarrow \quad N > 0, \text{fact}(N - 1, M), F = N * M$$

$$?\text{fact}(3, F)$$

The process of solving the goal $?\text{fact}(3, F)$ using resolution can be described as the process of generating the following sequence of constraints, which in this case are arithmetic equations and inequations.

$$
\begin{aligned}
c_1 &= \{3 > 0, F = 3 * M_1\} \\
c_2 &= c_1 \wedge \{2 > 0, M_1 = 2 * M_2\} \\
c_3 &= c_2 \wedge \{1 > 0, M_2 = 1 * M_3\} \\
c_4 &= c_3 \wedge \{0 = 0, M_3 = 1\}
\end{aligned}
$$

Solving these constrains gives the required result: $F = 6$. This example shows that Prolog computation can be expressed in terms of constraint solving combined with resolution.

Prolog programs can be transformed into constraint logic programs by replacing each program clause

$$p(t_1, \ldots, t_n) \leftarrow Q_1, \ldots, Q_m$$

by the more explicit

$$p(X_1, \ldots, X_n) \leftarrow X_1 = t_1, \ldots, X_n = t_n, Q_1, \ldots, Q_m$$

where the unification steps are represented as equations. Unification is now straightforward since the head of the clause only contains variables. However, the work has been pushed inside the body of the clause, and unification will be required to solve the equations $X_i = t_i$ $(i \leq 1 \leq n)$, which we see as constraints. In fact constraint languages go one step further in that the constraints are not evaluated until the end of the computation (if at all). The normal output of an evaluation is just a constraint, which can of course, be the representation of an infinite object (as can be done in the language Prolog II (Colmerauer, 1984)).

This view of computation provides a solution for another drawback in Prolog, which is the evaluation of goals containing arithmetic expressions. If we consider

the previous program with the query ?fact($N, 6$), then Prolog will fail to find a solution. This is because fact($N - 1, M$) will only succeed in the original program
when N has been instantiated. The origin of this problem is in the way Prolog han-
dles evaluation of arithmetic expressions (using built-in operators). As we already
mentioned, the arithmetic in Prolog is based on primitive functions that are not defined in the same logical framework. In particular, in this case, the distinction between input and output is important; only closed expressions can be evaluated. In a
constraint system, this however causes no problems what-so-ever, as we shall now
explain. First, let us be more precise about the language of constraints.

Definition 10.4 (Constraint language) *A constraint language consists of: a set of
variables, a set of function symbols (or constants), and a set of predicates, usually
containing* true *and equality* ($=$).
   An atomic constraint *is simply an atom of the constraint language. We define a*
constraint *as a conjunction (or a set) of atomic constraints.*

Definition 10.5 *A constraint logic program consists of two entities, a* program *and
a* query.

- *A program is a set of clauses of the form*

$$P \leftarrow c_1, \ldots, c_n \mid A_1, \ldots, A_n$$

   where $c_1, \ldots, c_n$ are atomic constraints.

- *A query is denoted by*

$$?c_1, \ldots, c_n \mid A_1, \ldots, A_n$$

   where again $c_1, \ldots, c_n$ are atomic constraints.

Pure Prolog programs are interpreted in the Herbrand universe (cf. Chapter 6).
In contrast, the semantics of a constraint logic programming language is parame-
terised by a mathematical structure which gives the interpretation of the constraint
language. For example, if we write the factorial program in a constraint logic programming language, the domain of interpretation of the constants will be the natural
numbers.
   A *constraint solver* takes a set of constraints and gives a set of results which are
the solutions of the constraints with respect to the chosen interpretation (we see now
why open arithmetic expressions cause no problems here). A constraint is *satisfiable* if it has solutions. It is assumed that the constraint language is decidable.
   Rather than developing the formal theory associated to the evaluation of constraint logic programs, we will give a simple example to show the basic principle.
Consider the following constraint program, where the bodies of the clauses contain
only constraints.

$$p(S,T) \quad \leftarrow \quad S + T = 8 \mid$$
$$q(U,V) \quad \leftarrow \quad U - V = 3 \mid$$

The query $?p(X,Y), q(X,Y)$ will produce the following constraint:

$$\{X = S, Y = T, S + T = 8, X = U, Y = V, U - V = 3\}$$

Since this constraint is satisfiable, we could justify having just this as an answer to the query. However, there are ways to simplify this constraint, for example the following is clearly equivalent and simpler:

$$\{X + Y = 8, X - Y = 3\}$$

We leave it as an exercise to the reader to verify that the solution to this constraint, in the domain of real numbers, is $\{X = 5.5, Y = 2.5\}$.

Implementations of constraint logic programming systems normally only check satisfiability of the constraint, and have these last two phases of the computation (simplification and solving) optional; they are left to the user interface of the language.

The previous was an example of constraint solving in a specific domain: constraints are arithmetic expressions, and solutions are real numbers. Some standard domains for constraint solving are: linear arithmetic, real numbers, or any finite domain.

We can also consider the Herbrand universe as the domain of interpretation of the constraint language (this is in fact the case if we want to see Prolog as a constraint programming language, as already mentioned at the beginning of this section). Moreover, we can think of an extended language where the user can define the domain where the constraints have to be interpreted. This can be done, for example, by combining Prolog with the λ-calculus, or with rewrite systems. If we have this facility in the language, then we need a more general (parametric) constraint solver. These are generally based on extensions of the unification procedure, for instance E-unification, or higher-order unification (which are not decidable in general). Since constraints are not restricted to equations, other techniques might be of use, for instance disunification and negation elimination.

Constraints are another means of incorporating knowledge about the domain in automated theorem provers. In this respect they are a recent contribution from logic programming to the field of automated theorem proving.

We refer the reader to (Jaffar and Lassez, 1987) for a survey and further references on Constraint Logic Programming.

▶ **EXERCISE 10.4**

Given the following constraint logic program:

$$\text{fact}(0, 1)$$
$$\text{fact}(N, F) \quad \leftarrow \quad N > 0, F = N * M, N' = N - 1 \mid \text{fact}(N', M)$$

show a computation trace for the goal ?fact($V, 2$)

4   PARALLELISM

We have seen that Prolog is essentially a theorem prover for a specific fragment (Horn clauses) of classical logic. Much work has been done in terms of implementation techniques to make this into an efficient language, and hence new (specific) theorem proving techniques have been developed. Of all the methods investigated, one of the greatest potentials for speeding up proof search is the shift to parallel implementations, breaking the problem up and distributing the work over many processors. However, the evaluation strategy for Prolog, SLD resolution, is totally sequential (stack based). Hence, a new approach is needed.

Here we will just identify the parallelism in logic programming, and refer the reader to the appropriate literature for a more complete exposition of parallel logic programming. There are basically two main different types of parallelism that have been studied in the scope of logic programming languages which are basically taking profit of the *order* of both the clauses in a program and the conjuncts in a clause.

4.1   OR-parallelism

A logic program consists of a *set* of clauses, for which the order is therefore unimportant. When we consider a specific sequential implementation, for example Prolog, we actually see this as an ordered sequence, and evaluate in some order, for example top to bottom.

However, if more than one clause matches the query, then this clearly has potential for parallelism — we can try to do a proof search using all the clauses that match. Take for example the following program:

$$P \leftarrow Q$$
$$P \leftarrow R$$

$$?P$$

Clearly, both clauses match the query, and there is no reason why we cannot try to continue with two parallel queries $?Q$ and $?R$. Now the question is how do we collect the solutions, since we have no way of knowing which clause will succeed (or fail) first. There are essentially two different types of *or-parallelism*.

- Don't know parallelism.

- Don't care parallelism (same solution).

We will first look at the case of don't care parallelism which causes the least problems. Consider the program:

$$min(A, B, A) \leftarrow A \leq B$$
$$min(A, B, B) \leftarrow B \leq A$$

which, given a goal $?min(n_1, n_2, V)$ will compute the minimum of the two numbers n_1 and n_2. However, given a goal $?min(n, n, V)$ both the above clauses match, since they overlap, but the result will be the same for each clause. If we were to evaluate this goal in parallel, then whichever clause succeeded first can be chosen correctly as the result of the program, and all other (potential) solutions should be ignored.

This suggests that for or-parallelism, the first clause to succeed should be the result of the program. Only if all clauses fail will the whole query fail.

However, for don't know parallelism this might not be what we require since once a single clause has succeeded, this result will preclude any further solutions which might be required later. Take for example the following program:

$$p(A) \leftarrow q(A)$$
$$p(A) \leftarrow r(A)$$
$$q(a)$$
$$r(b)$$

$$?p(X), r(X)$$

Here, the query $?p(X)$ will match on both the first two clauses, and will in fact succeed in either, but giving different bindings for X ($X = a$ in the first, $X = b$ for the second). Assuming that we execute the two in parallel and the first clause returned success first with the binding $X = a$, then the whole clause will fail since the query $r(a)$ will fail. However, if the second clause returned success first with the binding $X = b$, then the query $?r(b)$ will indeed succeed and thus the whole query will succeed.

The above example demonstrates that taking only the first answer from a parallel execution can cause a query to fail in cases where we would expect success. There are two ways around this problem. The first is to program only applications that use don't care parallelism which are quite common for data-base retrieval applications where many searches can be run in parallel. A second solution is that we can introduce a notion of *mode declaration* into our programs so that we specify that certain clauses should only be used for input bindings, for example. Using this approach it is possible to suspend unification in the above example until the second part of the goal $?r(X)$ binds X to b and then $p(b)$ can be evaluated correctly.

An example of such a mode declaration is $p(+)$ which specifies that p only takes bound arguments (it will not give output bindings). Similarly, we could designate

r as an output binding predicate $r(-)$ which indicates that r is to be used to generate bindings. Mode declarations are, in some sense, similar to types of functional languages that we studied in Chapter 4.

4.2 AND-parallelism

The second potential for parallelism is the evaluation within a clause. Recall that the body of a Horn clause is a sequence of conjuncts.

$$P \leftarrow A_1, \ldots, A_n$$

Since the order of conjuncts is not specified by the semantics of Horn clauses, we are free to evaluate a conjunct in any order, and in particular, in parallel (however, the order may affect the efficiency and even worse, in the presence of non-logical features like assert, cut, etc. it can also affect the semantics; we are assuming pure logic programming now).

If we evaluate the conjuncts in parallel, there is a potential problem on how to handle the output when calls share variables. Consider the following program:

$$p(a)$$
$$p(b)$$
$$q(b)$$
$$q(c)$$

$$?p(X), q(X)$$

$p(X)$ has solutions $X = \{a, b\}$ and $q(X)$ has solutions $X = \{b, c\}$, and we would expect that the query would get the (unique) solution $X = b$. A parallel execution of the query $?p(X), q(X)$ must agree on any binding of the *shared variable* X. There is a need to place some control here which could be achieved by making either p or q the *producer* and the other the *consumer*. Note how this example is similar to the example used for or-parallelism above. In the same way as mode declarations were used to provide control information in that case, we can use them here to specify one of p or q as the producer and the other as consumer.

Parallelism in logic programming is mentioned in many texts on general logic programming. There are also several implementations, for example Parlog (Conlon, 1989) and Concurrent Prolog (CP) (see for example (Shapiro, 1988)).

CHAPTER A

ANSWERS TO EXERCISES

▶ **1.1.**

1. P = "go to Asterix Park", W = "might get wet", then we can write $P \Rightarrow W$.

2. $(L \wedge P) \Rightarrow D$.

3. $(R \Rightarrow T) \vee (\neg R \Rightarrow E)$.

4. "A only if B" is generally understood to mean $\neg B \Rightarrow \neg A$. Hence, $\neg A \Rightarrow \neg P$.

5. Write $N(n)$ for n is a number, then $N(0) \wedge \forall n, (N(n) \Rightarrow N(n+1))$.

6. $(T \vee C) \wedge ((S \vee E) \wedge (F \vee P)) \wedge (M \vee (P \vee S))$.

   $\vee$ is used to mean a choice, but for the entrée and plat, it is us who decide (internal choice), for the sorbet option, it is the season that chooses (external choice).

▶ **1.2.**

1. $\wedge$ and $\vee$ are associative, $\Rightarrow$ is not. The same is true also for symmetry.

2. They are equivalent.

3. Not equivalent: take $A = B$ = false, the left hand side is true, and the right hand side is false.

4. They are equivalent.

▶ **2.1.**

Observe that for every x and y in $\mathbb{B}$, $x \overline{\Rightarrow} y = \overline{\neg} x \overline{\vee} y$ (the truth-tables are identical). By definition of $[\![\_]\!]_{\neg}$, the claim is proved.

▶ **2.2.**

The truth-table for logical equivalence is:

$\Leftrightarrow$	$\perp$	$\top$
$\perp$	$\top$	$\perp$
$\top$	$\perp$	$\top$

whence the result.

▶ **2.3.**

(i), (ii), (iii), $(viii)$, (ix) and (x) are valid. (v) is unsatisfiable. (iv) is invalid and satisfiable (taking B to be $\top$ makes it true, but taking A true, B false and C false makes it false). (vii) is also both invalid and satisfiable (taking A false makes it true, but taking A true and B false makes it false). (vi) is again invalid and satisfiable (taking B true makes it true, but taking B false, and taking for A and C anything but A true and C false, makes it false).

> Notice that $(viii)$ is paradoxical in nature, as it entails that for any two formulas Φ and Φ', either Φ implies Φ', or Φ' implies Φ. In classical logic, the direction of the implication that holds depends on the interpretation of propositional variables that we choose. In particular, notice that saying that $(A \Rightarrow B) \vee (B \Rightarrow A)$ is valid does *not* imply that $A \Rightarrow B$ or $B \Rightarrow A$ is valid. A typical wrong argument would be to say that whenever $(A \Rightarrow B) \vee (B \Rightarrow A)$ is true, $A \Rightarrow B$ or $B \Rightarrow A$ is true, too, by definition, and to erroneously replace "true" by "valid". The moral of this story is: never use the terms "true" and "false" in an absolute sense, this is a sure sign of a hidden error.

> (i), (ii) and (iii), with the modus ponens rule are enough to define a calculus which is sound and complete for classical propositional logic, i.e. that is able to prove exactly all true formulas; (i) and (ii) alone define what is called *minimal logic*, which is basically intuitionistic logic restricted to the $\Rightarrow$ connective. (The difference is that we cannot define false in minimal logic.)

▶ **2.4.**

The truth-table is:

$\mid$	$\perp$	$\top$
$\perp$	$\top$	$\top$
$\top$	$\top$	$\perp$

By comparing the truth-tables, we see that we could have defined $\neg\Phi$ as an abbreviation for $\Phi \mid \Phi$; $\Phi \wedge \Phi'$ as one for $\neg(\Phi \mid \Phi')$, i.e. for $(\Phi \mid \Phi') \mid (\Phi \mid \Phi')$; $\Phi \vee \Phi'$ as one for $(\neg\Phi) \mid (\neg\Phi')$, i.e. for $(\Phi \mid \Phi) \mid (\Phi' \mid \Phi')$; and $\Phi \Rightarrow \Phi'$ as one for $\Phi \mid (\neg\Phi')$, i.e. for $\Phi \mid (\Phi' \mid \Phi')$. (This is one possible set of definitions amongst many others.)

▶ **2.5.**

The truth-table is:

$\downarrow$	$\perp$	$\top$
$\perp$	$\top$	$\perp$
$\top$	$\perp$	$\perp$

By comparing the truth-tables, we see that we could have defined $\neg\Phi$ as an abbreviation for $\Phi \downarrow \Phi$; $\Phi \vee \Phi'$ as one for $\neg(\Phi \downarrow \Phi')$, i.e. for $(\Phi \downarrow \Phi') \downarrow (\Phi \downarrow \Phi')$; $\Phi \wedge \Phi'$ as one for $(\neg\Phi) \downarrow (\neg\Phi')$, i.e. for $(\Phi \downarrow \Phi) \downarrow (\Phi' \downarrow \Phi')$; and $\Phi \Rightarrow \Phi'$ as one for $(\neg\Phi) \vee \Phi'$, i.e. for $((\Phi \downarrow \Phi) \downarrow \Phi') \downarrow ((\Phi \downarrow \Phi) \downarrow \Phi')$. (This is one possible set of definitions amongst many others.)

▶ **2.6.**

Whatever the truth-value for A, $\top$ or $\perp$, the truth-value of **F** is $\perp$, and that of **T** is $\top$. Then, the truth-table of the if/then/else connective is given by the value of if Φ then Φ' else Φ'' as follows:

Φ'	$\bot$	$\bot$	$\top$	$\top$
Φ''	$\bot$	$\top$	$\bot$	$\top$
Φ				
$\bot$	$\bot$	$\top$	$\bot$	$\top$
$\top$	$\bot$	$\bot$	$\top$	$\top$

so we can define $\Phi \wedge \Phi'$ as if Φ then Φ' else $\mathbf{F}$, $\Phi \vee \Phi'$ as if Φ then $\mathbf{T}$ else Φ', $\neg \Phi$ as if Φ then $\mathbf{F}$ else $\mathbf{T}$, $\Phi \Rightarrow \Phi'$ as if Φ then Φ' else $\mathbf{T}$.

▶ **2.7.**

We present two proofs.

A first proof is the following: by completeness, $\models \Phi$, then by Corollary 2.11, $\models \Phi \sigma$, hence by soundness $\vdash^P \Phi \sigma$.

There is also a syntactic proof, by induction on the length of a proof of Φ in $\mathcal{P}$: notice that if Φ was an axiom, then $\Phi \sigma$ is again an instance of the same axiom; and if Φ was obtained by modus ponens from $\Phi' \Rightarrow \Phi$ and Φ', where the latter have strictly shorter proofs, then by induction hypothesis $(\Phi' \Rightarrow \Phi)\sigma$ (i.e., $\Phi'\sigma \Rightarrow \Phi\sigma$) and $\Phi'\sigma$ are provable, so by (MP) again, we get a proof of $\Phi\sigma$.

▶ **2.8.**

Append the proofs of $\Phi \Rightarrow \Phi'$ and of $\Phi'' \vee \Phi$, and add the following new facts: $(\Phi \Rightarrow \Phi') \Rightarrow (\Phi'' \vee \Phi \Rightarrow \Phi' \vee \Phi'')$, which is an instance of (3); by modus ponens (MP) with $\Phi \Rightarrow \Phi'$, generate $\Phi'' \vee \Phi \Rightarrow \Phi' \vee \Phi''$. By modus ponens with $\Phi'' \vee \Phi$, produce $\Phi' \vee \Phi''$.

▶ **2.9.**

$(\Phi \vee \Phi \Rightarrow \Phi) \Rightarrow (\neg \Phi \vee (\Phi \vee \Phi) \Rightarrow \Phi \vee \neg \Phi)$ is an instance of Axiom (3) (replace Φ' by Φ, Φ'' by $\neg \Phi$ and Φ by $\Phi \vee \Phi$). By modus ponens with $\Phi \vee \Phi \Rightarrow \Phi$, which is an instance of (1), generate $\neg \Phi \vee (\Phi \vee \Phi) \Rightarrow \Phi \vee \neg \Phi$, which is also $(\Phi \Rightarrow \Phi \vee \Phi) \Rightarrow \Phi \vee \neg \Phi$ by definition of $\Rightarrow$ as an abbreviation; by modus ponens with $\Phi \Rightarrow \Phi \vee \Phi$, which is an instance of (2), generate $\Phi \vee \neg \Phi$.

As regards $\Phi \Rightarrow \neg \neg \Phi$, this is an instance of the latter where Φ is replaced by $\neg \Phi$. (Recall that $\Phi \Rightarrow \neg \neg \Phi$ abbreviates $\neg \Phi \vee \neg \neg \Phi$.)

▶ **2.10.**

Notice that $(\neg \Phi \Rightarrow \neg \neg \neg \Phi) \Rightarrow (\Phi \vee \neg \Phi) \Rightarrow (\neg \neg \neg \Phi \vee \Phi)$ is an instance of Axiom (3). By modus ponens with $\neg \Phi \Rightarrow \neg \neg \neg \Phi$, which is an instance of Exercise 2.9, produce $(\Phi \vee \neg \Phi) \Rightarrow (\neg \neg \neg \Phi \vee \Phi)$. Then, by modus ponens with $\Phi \vee \neg \Phi$ (from Exercise 2.9 again), produce $\neg \neg \neg \Phi \vee \Phi$, that is, $\neg \neg \Phi \Rightarrow \Phi$.

▶ **2.11.**

If $\Gamma \vdash^{\mathcal{SKC}} \Phi \Rightarrow \Phi'$, as by definition $\Gamma, \Phi \vdash^{\mathcal{SKC}} \Phi$, we apply (MP) to get a proof of Φ' from Γ.

Conversely, assume that $\Gamma, \Phi \vdash^{\mathcal{SKC}} \Phi'$, and prove $\Gamma \vdash^{\mathcal{SKC}} \Phi \Rightarrow \Phi'$ by induction on the length of a proof of the latter in $\mathcal{SKC}$. Consider the last proposition Φ' in the proof. If it was an axiom or a member of Γ, then generate $\Phi' \Rightarrow \Phi \Rightarrow \Phi'$ as an instance of (K), and by modus ponens with Φ', generate $\Phi \Rightarrow \Phi'$. If Φ' was Φ itself, generate $(\Phi \Rightarrow (\Phi \Rightarrow \Phi) \Rightarrow \Phi) \Rightarrow (\Phi \Rightarrow (\Phi \Rightarrow \Phi)) \Rightarrow (\Phi \Rightarrow \Phi)$ (from (S)), then by modus ponens with $\Phi \Rightarrow (\Phi \Rightarrow \Phi) \Rightarrow \Phi$ (instance of (K)), and again modus ponens with $\Phi \Rightarrow \Phi \Rightarrow \Phi$ (instance of (K), again), generate $\Phi \Rightarrow \Phi$. Finally, if Φ' was deduced by (MP) from $\Phi'' \Rightarrow \Phi'$ and Φ'' respectively for some Φ'' (with shorter proofs), then by induction hypothesis we have $\Gamma \vdash^{\mathcal{SKC}} \Phi \Rightarrow \Phi'' \Rightarrow \Phi'$ and $\Gamma \vdash^{\mathcal{SKC}} \Phi \Rightarrow \Phi''$, so generate $(\Phi \Rightarrow \Phi'' \Rightarrow \Phi') \Rightarrow (\Phi \Rightarrow \Phi'') \Rightarrow (\Phi \Rightarrow \Phi')$ (from (S)), and then by two applications of (MP), produce $\Phi \Rightarrow \Phi'$.

Observe that (C) was not needed to prove this. ((S) and (K) alone actually define what is known as *minimal logic*, which is more akin to intuitionistic logic than to classical logic.)

▶ **2.12.**

From the set $\Phi, \neg\Phi$, we can deduce both Φ and $\neg\Phi$, i.e. $\Phi \Rightarrow \mathbf{F}$, so by (MP) we infer $\mathbf{F}$. This means that $\Phi, \neg\Phi \vdash^{\mathcal{SKC}} \mathbf{F}$. Now, by the Deduction Theorem, this entails that $\Phi \vdash^{\mathcal{SKC}} \neg\Phi \Rightarrow \mathbf{F}$, i.e. $\Phi \vdash^{\mathcal{SKC}} \neg\neg\Phi$, and by the Deduction Theorem again, $\vdash^{\mathcal{SKC}} \Phi \Rightarrow \neg\neg\Phi$.

▶ **2.13.**

We can recast the problem as showing that $\Phi \Rightarrow \Phi', \Phi' \Rightarrow \Phi'' \vdash^{\mathcal{SKC}} \Phi \Rightarrow \Phi''$. By the Deduction Theorem, this means showing that $\Phi \Rightarrow \Phi', \Phi' \Rightarrow \Phi'', \Phi \vdash^{\mathcal{SKC}} \Phi''$. But by (MP) between $\Phi \Rightarrow \Phi'$ and Φ, we generate Φ', and by (MP) between $\Phi' \Rightarrow \Phi''$ and the latter, we generate Φ''.

▶ **2.14.**

$\neg\Phi \lor \Phi'$ is $\neg\neg\Phi \Rightarrow \Phi'$ by definition. From $\Phi \Rightarrow \Phi'$, we deduce $\neg\neg\Phi \Rightarrow \Phi'$ by using Exercise 2.13 with $\neg\neg\Phi \Rightarrow \Phi$, i.e. Axiom (C). Conversely, from $\neg\neg\Phi \Rightarrow \Phi'$, we deduce $\Phi \Rightarrow \Phi'$ by using Exercise 2.13 with $\Phi \Rightarrow \neg\neg\Phi$, i.e. Exercise 2.12.

Then, let Ψ and Ψ' be two formulas as in the text. It is enough to prove that any formula is logically equivalent to itself (trivial), and that if Φ_1 and Φ_2 are logically equivalent, and Φ_1' and Φ_2' are logically equivalent, then $\Phi_1 \Rightarrow \Phi_1'$ and $\Phi_2 \Rightarrow \Phi_2'$ are logically equivalent. Assume $\Phi_1 \Rightarrow \Phi_1'$ and Φ_2 as hypotheses; because Φ_2 is equivalent to Φ_1, we infer Φ_1; by modus ponens with $\Phi_1 \Rightarrow \Phi_1'$, we infer Φ_1', hence Φ_2' by logical equivalence. Therefore $\Phi_1 \Rightarrow \Phi_1', \Phi_2 \vdash^{\mathcal{SKC}} \Phi_2'$, hence by the Deduction Theorem for $\mathcal{SKC}$, $\Phi_1 \Rightarrow \Phi_1' \vdash^{\mathcal{SKC}} \Phi_2 \Rightarrow \Phi_2'$. The converse is similar.

▶ **2.15.**

Soundness is clear from the truth-tables.

For completeness, it is enough to show that we can prove the axioms of $\mathcal{P}$ in $\mathcal{SKC}$. But, in $\mathcal{P}$ we made the assumption that $\Phi \Rightarrow \Phi'$ was an abbreviation for $\neg\Phi \lor \Phi'$, so we have to prove that we can always replace $\Phi \Rightarrow \Phi'$ by $\neg\Phi \lor \Phi'$ in formulas of $\mathcal{SKC}$ without changing their deducibility status. This is precisely what we proved in Exercise 2.14.

Then, we prove the axioms of $\mathcal{P}$ themselves. (1) is $(\neg\Phi \Rightarrow \Phi) \Rightarrow \Phi$: first, generate $(\neg\Phi \Rightarrow \Phi \Rightarrow \mathbf{F}) \Rightarrow (\neg\Phi \Rightarrow \Phi) \Rightarrow (\neg\Phi \Rightarrow \mathbf{F})$, which is an instance of (S). Then, generate $\neg\Phi \Rightarrow \Phi \Rightarrow \mathbf{F}$ (which is $\neg\Phi \Rightarrow \neg\Phi$; this can be proved, because of the Deduction

Theorem and the fact that $\neg\Phi \vdash^{\mathcal{SKC}} \neg\Phi$), and apply (MP) to get $(\neg\Phi \Rightarrow \Phi) \Rightarrow (\neg\Phi \Rightarrow F)$, i.e. $(\neg\Phi \Rightarrow \Phi) \Rightarrow \neg\neg\Phi$. Then, by using Exercise 2.13 with Axiom (C), generate $(\neg\Phi \Rightarrow \Phi) \Rightarrow \Phi$.

(2) is $\Phi \Rightarrow \Phi' \vee \Phi$, i.e. $\Phi \Rightarrow \neg\Phi' \Rightarrow \Phi$, and this is an instance of (K).

(3) is $(\Phi\Rightarrow\Phi')\Rightarrow(\Phi''\vee\Phi)\Rightarrow(\Phi'\vee\Phi'')$, i.e. $(\Phi\Rightarrow\Phi')\Rightarrow(\neg\Phi''\Rightarrow\Phi)\Rightarrow(\neg\Phi'\Rightarrow\Phi'')$. By the Deduction Theorem for $\mathcal{SKC}$, to prove this in $\mathcal{SKC}$ is equivalent to proving $\Phi \Rightarrow \Phi', \neg\Phi'' \Rightarrow \Phi, \neg\Phi' \vdash^{\mathcal{SKC}} \Phi''$. From $\Phi \Rightarrow \Phi'$ and $\neg\Phi'$, i.e. $\Phi' \Rightarrow F$, use Exercise 2.13 to generate $\Phi \Rightarrow F$. From the latter and $\neg\Phi'' \Rightarrow \Phi$, use Exercise 2.13 again to generate $\neg\Phi'' \Rightarrow F$, i.e. $\neg\neg\Phi''$. Then, by modus ponens with Axiom (C), generate Φ''.

▶ **2.16.**

By Lemma 2.20, let (π) be a derivation of $\Gamma, \neg\Phi \longrightarrow \Phi$, then produce:

$$
\cfrac{\cfrac{\begin{array}{c}(\pi)\\ \vdots\end{array}}{\Gamma, \neg\Phi \longrightarrow \Phi} \qquad \cfrac{}{\Gamma, \neg\Phi \longrightarrow \neg\Phi}\,(Ax)}{\cfrac{\Gamma, \neg\Phi \longrightarrow F}{\Gamma \longrightarrow \neg\neg\Phi}\,(\neg I)}\,(\neg E)
$$

▶ **2.17.**

Assume $\vdash^{\mathcal{ND}} \Gamma \longrightarrow \Phi$. We build a proof of $\Gamma \longrightarrow \Phi$ in $\mathbf{LK_0}$ by structural induction on the natural deduction proof of $\Gamma \longrightarrow \Phi$ (this directly yields an algorithm):

- if $\Gamma \longrightarrow \Phi$ was proved by (Ax), then it is proved by Ax in $\mathbf{LK_0}$.

- similarly, rules $(\wedge I)$, $(\Rightarrow I)$ translate to $\wedge R$, $\Rightarrow R$ respectively.

- Rules $(\wedge E_1)$ and $(\wedge E_2)$ translate to:

$$
\cfrac{\begin{array}{c}\vdots\\ \Gamma \longrightarrow \Phi \wedge \Phi'\end{array} \qquad \cfrac{\cfrac{}{\Phi \wedge \Phi' \longrightarrow \Phi}\,Ax}{\Phi \wedge \Phi' \longrightarrow \Phi}\,\wedge L}{\Gamma \longrightarrow \Phi}\,Cut
$$

and

$$
\cfrac{\begin{array}{c}\vdots\\ \Gamma \longrightarrow \Phi \wedge \Phi'\end{array} \qquad \cfrac{\cfrac{}{\Phi, \Phi' \longrightarrow \Phi'}\,Ax}{\Phi \wedge \Phi' \longrightarrow \Phi'}\,\wedge L}{\Gamma \longrightarrow \Phi'}\,Cut
$$

respectively.

- Rule $(\Rightarrow E)$ translates to:

$$
\cfrac{
  \cfrac{
    \cfrac{\vdots}{\Gamma \longrightarrow \Phi}
    \qquad
    \cfrac{}{\Phi' \longrightarrow \Phi'}\;\mathrm{Ax}
  }{\Gamma, \Phi \Rightarrow \Phi' \longrightarrow \Phi'}\;\Rightarrow\!\mathrm{L}
  \qquad
  \cfrac{\vdots}{\Gamma \longrightarrow \Phi \Rightarrow \Phi'}
}{\Gamma \longrightarrow \Phi'}\;\mathrm{Cut}
$$

- Rules $(\vee I_1)$ and $(\vee I_2)$ translate to:

$$
\cfrac{
  \cfrac{\vdots}{\Gamma \longrightarrow \Phi}
  \qquad
  \cfrac{
    \cfrac{}{\Phi \longrightarrow \Phi, \Phi'}\;\mathrm{Ax}
  }{\Phi \longrightarrow \Phi \vee \Phi'}\;\vee\mathrm{R}
}{\Gamma \longrightarrow \Phi \vee \Phi'}\;\mathrm{Cut}
$$

and

$$
\cfrac{
  \cfrac{\vdots}{\Gamma \longrightarrow \Phi}
  \qquad
  \cfrac{
    \cfrac{}{\Phi \longrightarrow \Phi', \Phi}\;\mathrm{Ax}
  }{\Phi \longrightarrow \Phi' \vee \Phi}\;\vee\mathrm{R}
}{\Gamma \longrightarrow \Phi' \vee \Phi}\;\mathrm{Cut}
$$

respectively.

- Rule $(\vee E)$ translates to:

$$
\cfrac{
  \cfrac{
    \cfrac{\vdots}{\Gamma \longrightarrow \Phi \vee \Phi'}
    \quad
    \cfrac{
      \cfrac{}{\Phi \longrightarrow \Phi, \Phi'}\;\mathrm{Ax} \quad \cfrac{}{\Phi' \longrightarrow \Phi, \Phi'}\;\mathrm{Ax}
    }{\Phi \vee \Phi' \longrightarrow \Phi, \Phi'}\;\vee\mathrm{L}
  }{\Gamma \longrightarrow \Phi, \Phi'}\;\mathrm{Cut}
  \quad
  \cfrac{\vdots}{\Gamma, \Phi \longrightarrow \Phi''}
}{
  \cfrac{\Gamma \longrightarrow \Phi'', \Phi' \qquad \cfrac{\vdots}{\Gamma, \Phi' \longrightarrow \Phi''}}{\Gamma \longrightarrow \Phi''}\;\mathrm{Cut}
}\;\mathrm{Cut}
$$

- $(\neg I)$ translates to (recall that $\bot$ abbreviates $\Phi' \wedge \neg\Phi'$):

$$
\cfrac{
  \cfrac{
    \cfrac{\vdots}{\Gamma, \Phi \longrightarrow \Phi' \wedge \neg\Phi'}
  }{\Gamma \longrightarrow \neg\Phi, \Phi' \wedge \neg\Phi'}\;\neg\mathrm{R}
  \qquad
  \cfrac{
    \cfrac{
      \cfrac{}{\Phi' \longrightarrow \Phi'}\;\mathrm{Ax}
    }{\Phi', \neg\Phi' \longrightarrow}\;\neg\mathrm{L}
  }{\Phi' \wedge \neg\Phi' \longrightarrow}\;\wedge\mathrm{L}
}{\Gamma \longrightarrow \neg\Phi}\;\mathrm{Cut}
$$

- $(\neg E)$ translates to:

$$\dfrac{\dfrac{\vdots \qquad \vdots}{\dfrac{\Gamma \longrightarrow \Phi \qquad \Gamma \longrightarrow \neg\Phi}{\Gamma \longrightarrow \Phi \wedge \neg\Phi}\ \wedge\text{R}} \qquad \dfrac{\dfrac{\dfrac{\overline{\Phi \longrightarrow \Phi, \Phi'}}{\Phi, \neg\Phi \longrightarrow \Phi'}\ \text{Ax}}{\Phi \wedge \neg\Phi \longrightarrow \Phi'}\ \neg\text{L}}{\Phi \wedge \neg\Phi \longrightarrow \Phi'}\ \wedge\text{L}}{\Gamma \longrightarrow \Phi'}\ \text{Cut}$$

- $(\neg\neg I)$ translates to:

$$\dfrac{\dfrac{\vdots}{\Gamma \longrightarrow \Phi} \qquad \dfrac{\dfrac{\dfrac{\overline{\Phi \longrightarrow \Phi}}{\Phi, \neg\Phi \longrightarrow}\ \text{Ax}}{\Phi, \neg\Phi \longrightarrow}\ \neg\text{L}}{\Phi \longrightarrow \neg\neg\Phi}\ \neg\text{R}}{\Gamma \longrightarrow \neg\neg\Phi}\ \text{Cut}$$

- $(\neg\neg E)$ translates to:

$$\dfrac{\dfrac{\vdots}{\Gamma \longrightarrow \neg\neg\Phi} \qquad \dfrac{\dfrac{\dfrac{\overline{\Phi \longrightarrow \Phi}}{\longrightarrow \Phi, \neg\Phi}\ \text{Ax}}{\neg\neg\Phi \longrightarrow \Phi}\ \neg\text{R}}{\neg\neg\Phi \longrightarrow \Phi}\ \neg\text{L}}{\Gamma \longrightarrow \Phi}\ \text{Cut}$$

▶ **2.18.**

Assume $\models \Gamma \longrightarrow \Delta$. If Δ consists in the sole formula Φ, then by the Completeness Theorem for $\mathcal{ND}$, $\vdash^{\mathcal{ND}} \Gamma \longrightarrow \Phi$. By Exercise 2.17, $\vdash^{\text{LK}_0} \Gamma \longrightarrow \Phi$.

In the general case, $\models \Gamma \longrightarrow \Delta$ entails $\models \Gamma, \neg\Delta \longrightarrow \mathbf{F}$, where $\neg\Delta$ is the set of negated formulas of Δ. By the Compactness Theorem and by Lemma 2.31, we may without loss of generality assume Γ and Δ finite. And by the Completeness Theorem for Natural Deduction, $\vdash^{\mathcal{ND}} \Gamma, \neg\Delta \longrightarrow \mathbf{F}$, so by the previous case $\vdash^{\text{LK}_0} \Gamma, \neg\Delta \longrightarrow \mathbf{F}$. Apply Cut between the latter and:

$$\dfrac{\dfrac{\overline{\Phi' \longrightarrow \Phi'}}{\ }\ \text{Ax}}{\Phi', \neg\Phi' \longrightarrow}\ \neg\text{L}$$

to get $\vdash^{\text{LK}_0} \Gamma, \neg\Delta \longrightarrow$. Then, write Δ as $\Phi_1, \ldots, \Phi_n$, and write down the following proofs π_i:

$$\dfrac{\dfrac{\overline{\Phi_i \longrightarrow \Phi_i}}{\ }\ \text{Ax}}{\longrightarrow \Phi_i, \neg\Phi_i}\ \neg\text{R}$$

for each $1 \leq i \leq n$. Apply Cut between Δ and π_1, then with $\pi_2, \ldots$, with π_n, to get $\Gamma \longrightarrow \Delta$.

▶ **2.19.**

We assume the reasonable fact that the expansion strategy works in time independent of the size of terms in the path l, and that if f chose Φ from l, then f chooses $\Phi\sigma$ from $l\sigma$ (i.e., the list of substituted formulas). For example, we might always choose the first non-variable formula of the list l. This is usually what is being done in actual implementations, where the current path l is actually represented as two lists, one list containing only atomic formulas, and the other containing possibly non-atomic formulas.

We then show that `prove` works exactly in the same way for $\Phi\sigma$ as for Φ by structural induction on the tableaux for Φ: whenever we used any α or β rule for Φ, we use the same on $\Phi\sigma$, and transform a path l into $l\sigma$; and as `prove` closes a path as soon as possible, whenever we closed a path for Φ, we also close the corresponding path for $\Phi\sigma$.

On the other hand, if Φ is not a tautology, then some path could not be closed (some sequent was not recognised as an instance of Ax), and this stopped the procedure immediately. On $\Phi\sigma$, further expansion may still be possible, as σ may be arbitrarily big. Therefore, the expansion of the tableau may take considerably more time. Moreover, $\Phi\sigma$ may either be unprovable (say, Φ is $A \Rightarrow B$, and we replace A by $C \Rightarrow C$ and B by $D \vee E$), or provable (replace A and B by $C \Rightarrow C$ in $A \Rightarrow B$, for example.)

▶ **2.20.**

Let C be an unclosed path consisting of signed variables only. As in the proof of Theorem 2.32 (with a little rewording), we can view C as corresponding to the set of assignments ρ mapping A to $\top$ whenever $-A$ occurs in C, and to $\bot$ whenever $+A$ occurs in C (all the variables that do not occur in C are mapped to arbitrary truth-values). Alternatively, if we view C as a sequent, this is the set of all countermodels to C.

By definition, all paths in S are unclosed and consist of signed variables only. Moreover, by the same argument as in the proof of Theorem 2.32, the set of countermodels to Φ is exactly the union of the set of counter-models of sequents in S; we can then read these counter-models directly from the shape of the sequents, as indicated above.

▶ **2.21.**

The size of Ψ_1 is 1. Then $\Psi_n = (A_n \Rightarrow \Psi_{n-1}) \wedge (\Psi_{n-1} \Rightarrow A_n)$, so its size is 4 plus the size of Ψ_{n-1}. By induction, the size of Ψ_n is $4n - 3$. By a similar argument, the size of Φ_n is then $8n - 3$. (If we represented formulas as trees, i.e. if we counted the number of distinct *occurrences* of subformulas in Φ_n, we would get an exponential number of nodes.)

Consider Ψ_n: we claim that $\rho \models \Psi_n$ if and only if ρ makes an even number of the free variables in Ψ_n false. Indeed, this is true for $n = 1$. Assume this true for

$n - 1$, and prove it for n. $\rho \models \Psi_n$ if and only if either $\rho(A_n) = \top$ and $\rho \models \Psi_{n-1}$ or $\rho(A_n) = \bot$ and $\rho \not\models \Psi_{n-1}$.

In particular, if $\rho \models \Psi_n$, then either $\rho(A_n) = \top$ and by induction hypothesis ρ maps an even number of variables among $A_1, \ldots, A_{n-1}$ to $\bot$, hence also an even number of variables among $A_1, \ldots, A_n$ to $\bot$; or $\rho(A_n) = \bot$, and the number of variables among $A_1, \ldots, A_{n-1}$ that ρ maps to $\bot$ is odd, so the number of variables among $A_1, \ldots, A_n$ that ρ maps to $\bot$ is even.

Conversely, if the number of variables among $A_1, \ldots, A_n$ that ρ maps to $\bot$ is even, then either $\rho(A_n) = \top$ and there remains an even number of variables mapped to $\bot$, so $\rho \models \Psi_{n-1}$, or $\rho(A_n) = \bot$ and there remains an odd number of variables mapped to $\bot$, so $\rho \not\models \Psi_{n-1}$. In either case, $\rho \models \Psi_n$.

Now, Φ_n is just Ψ_{n+1} where we have replaced (in parallel) A_{n+1} by A_n, A_n by $A_{n-1}, \ldots, A_2$ by A_1 and A_1 by Ψ_n. Therefore $\rho \models \Phi_n$ if and only if ρ makes an even number of formulas among $A_1, \ldots, A_n$ and Ψ_n false. But for all assignments ρ, either ρ makes an even number of $A_1, \ldots, A_n$ false, then it makes Ψ_n true, hence Φ_n true; or ρ makes an odd number of $A_1, \ldots, A_n$ false, so it makes Ψ_n false, and Φ_n true again. Therefore Φ_n is valid.

Let $\Phi_{n,m}$ be $A_m \Leftrightarrow (A_{m-1} \Leftrightarrow \ldots \Leftrightarrow (A_1 \Leftrightarrow \Psi_n) \ldots)$. Then Φ_n is $\Phi_{n,n}$, and $\Phi_{n,m} = (A_m \Rightarrow \Phi_{n,m-1}) \wedge (\Phi_{n,m-1} \Rightarrow A_m)$ for all $m \geq 1$. The only way we can expand $+\Phi_{n,m}$ in a tableau is the following:

$$+(A_m \Rightarrow \Phi_{n,m-1}) \wedge (\Phi_{n,m-1} \Rightarrow A_m) \ (*)$$

$+(A_m \Rightarrow \Phi_{n,m-1})$ $(*)$	$+(\Phi_{n,m-1} \Rightarrow A_m)$ $(*)$
$-A_m$	$-\Phi_{m-1}$
$+\Phi_{m-1}$	$+A_m$

where the only next thing we can do is re-expand $+\Phi_{m-1}$ on the left and $-\Phi_{m-1}$ on the right (we cannot close the paths as we have only distinct signed variables on the rest of the paths), and the only way we can expand $-\Phi_{n,m}$ is to get a new path containing two formulas $-(A_m \Rightarrow \Phi_{n,m-1})$ and $-(\Phi_{n,m-1} \Rightarrow A_m)$. There are several ways in which we can expand this, but in essence none can be essentially shorter than:

$$-(A_m \Rightarrow \Phi_{n,m-1}) \wedge (\Phi_{n,m-1} \Rightarrow A_m) \ (*)$$
$$-(A_m \Rightarrow \Phi_{n,m-1}) \ (*)$$
$$-(\Phi_{n,m-1} \Rightarrow A_m) \ (*)$$

$+A_m$		$-\Phi_{n,m-1}$	
$+\Phi_{n,m-1}$	$-A_m$	$+\Phi_{n,m-1}$	$-A_m$
	(closed)	(closed)	

where the only thing we can do is re-expand $+\Phi_{m-1}$ on the left and $-\Phi_{m-1}$ on the right (only the two paths in the middle are closed). By induction on m, we see that we shall generate a tableau that has at least 2^n paths (it has already that many paths when we come to $m = 0$, i.e. before we expand Ψ_n), generated in at least 3.2^n operations, before we can close any path. This is not a proof, since we are not forced to continue to expand on the new formulas created on each branch: for instance, it might be the case that by avoiding to expand $-(\Phi_{n,m-1} \Rightarrow A_m)$ right away on the rightmost two branches, we could get a shorter proof. Proving that this is in fact not the case is much more difficult.

▶ **2.22.**

By Exercise 2.20, the set of counter-models of $\neg\Phi$ is exactly the union of the set of counter-models of sequents in S. Therefore, if we see sequents as disjunctions of literals, and consider them to be in conjunction, we get a formula Φ' that is logically equivalent to Φ. Moreover, all sequents in S are atomic, so Φ' is in nnf. Finally, by construction no conjunction occurs as an argument to a disjunction in Φ', so Φ' is a cnf.

Notice that with the tableaux method, we essentially convert Φ on demand to a cnf, which we compare to the empty conjunction (true). In resolution, we do something rather weird: we convert $\neg\Phi$ to a cnf (as though we wished to prove $\neg\Phi$ instead of Φ), then we apply cuts in the hope of deriving the empty disjunction (false).

This exercise implies that taking the cnf of a formula this way is at least as complex as looking for a proof of it by tableaux, and may take exponential time in the size of the original formula, by Exercise 2.21. It is paradoxical that we would want to use a preprocessing step (conversion to cnf) that can in general be as costly as doing the proof directly. The reason why it is interesting is that most mathematical theorems have the form $C_1 \wedge \ldots \wedge C_n \Rightarrow \Phi$, where the C_i are axioms, and Φ is the formula to prove, and that the C_i and $\neg\Phi$ are usually already clauses or expand to very few clauses. Another solution to the paradox is given by Exercise 2.23.

▶ **2.23.**

The idea of the construction is that we create a new variable $\xi(\Phi')$ to represent the truth status of Φ', and we express the semantics of the logical connectives by constraining the ξ variables by the given clauses, which basically express the truth-tables for each connective. If Φ is satisfiable, then let ρ be an assignment satisfying Φ, and let ρ' be the assignment mapping each $\xi(\Phi')$ to $[\![\Phi']\!]\rho$. By inspection, ρ' satisfies S. Conversely, if S is satisfiable, then the restriction ρ of any assignment ρ' satisfying S to $\mathrm{fv}(\Phi)$ must satisfy Φ: an easy structural induction on Φ' shows that $\rho'(\xi(\Phi')) = [\![\Phi']\!]\rho$ for every non-variable subformula of Φ, so $\top = \rho'(\xi(\Phi)) = [\![\Phi]\!]\rho$.

Recall that the size n of Φ is the number of its subformulas. Assuming that the size of S is the sum of the sizes of its clauses, we see that the cardinal of S is less than or equal to $3n + 1$, and that the sizes of its clauses does not exceed 7 (for $\neg\xi(\Psi) \vee \neg\xi(\Psi') \vee \xi(\Phi')$), so that the size of S does not exceed $21n + 7$.

Finally, let Φ be $\neg A$, for instance. Then S consists of $B \vee A$, $\neg B \vee \neg A$, B, where B is $\xi(\Phi)$. But under the assignment mapping A and B to $\bot$, Φ is true, while $B \vee A$, hence the conjunction of the clauses, is false. So S is not logically equivalent to Φ.

▶ **2.24.**

We number the resolvents we get on the left. We annotate these numbers by an equality sign to signal a duplicate clause (for example, $(9) = (7)$ means that the ninth clause is identical to the seventh, so we delete it), and we put on the right the two clauses that we cut: see Figure A.1. So, the proof is reached at level 2 only, but we used 30 cut rules to get to S_2. The $\square$ result could have been found at any step between 13 and 34, depending on the order in which we combine pairs of clauses: in particular, although there is a short derivation of $\square$ (from $A \vee B$, $A \vee \neg B$, deduce A; from $\neg A \vee B$, $\neg A \vee \neg B$, deduce $\neg A$; from A, $\neg A$, deduce $\square$; this takes 3 cuts only), the shortest derivation of $\square$ by level saturation needs 13 steps at least. Redundancy is high: all clauses with an equality annotation above are redundant (only 7 distinct clauses were generated from S_0, for 30 cuts). Notice that cutting with tautologies like (7) and (8) *always* produces redundant clauses.

▶ **2.25.**

We would be tempted to say that it is better to implement a function BDDand for conjunction in a direct recursive style, mimicking BDDor, instead of first negating its two BDDs in arguments, then computing the disjunction, and finally taking the negation of the result. This is because we save three negation operations.

But, on the other hand, BDDor and BDDand would be memo-functions with separate hash-tables, so neither could share its results with the other. If we implement BDDand in terms of BDDor and BDDneg, we don't have to manage a separate hash-table for BDDand, and reuse all the results we have computed previously. The gains can be much bigger than the loss. Moreover, if we use the variant of Exercise 2.28, negation will be a constant time operation, so the latter solution is definitely the better in this case.

▶ **2.26.**

$\rho \models A$ if and only if either $\rho \models A$ and $\rho \models \Phi[\mathbf{T}/A]$, or $\rho \not\models A$ and $\rho \models \Phi[\mathbf{F}/A]$. That is, $\rho \models A$ if and only if $\rho \models (A \wedge \Phi[\mathbf{T}/A]) \vee (\neg A \wedge \Phi[\mathbf{F}/A])$. As this holds for all ρ, the formulas are equivalent.

▶ **2.27.**

The BDD is shown in Figure A.2 (notice that its shape does not actually depend on the ordering). It has $2n+1$ nodes (including $\mathbf{F}$ and $\mathbf{T}$), and 2^n paths (half of them lead to $\mathbf{T}$, the other half to $\mathbf{F}$). With the given ordering, it takes time proportional to n to build it, while with the converse ordering, it takes time proportional to n^2 (sum of times of the form $1 + 2 + \ldots + n$).

If $n = 20$, for example, it has 41 nodes, for around one million paths; if $n = 30$, this means 61 nodes, for around one billion paths, and so on.

▶ **2.28.**

To convert a TDG to a BDD, translate $+\mathbf{T}$ to $\mathbf{T}$, $-\mathbf{T}$ to $\mathbf{F}$; and to translate $(s, A \longrightarrow \Phi_+; \Phi_-)$, first convert Φ_+ and Φ_- to BDDs Φ'_+ and Φ'_- respectively, then output $A \longrightarrow \Phi'_+; \Phi'_-$ if s is $+$, otherwise $A \longrightarrow \text{BDDneg}(\Phi'_+); \text{BDDneg}(\Phi'_-)$.

Define negation on TDGs by the function TDGneg such that $\text{TDGneg}(+v) = -v$, $\text{TDGneg}(-v) = +v$.

To convert a BDD to a TDG, translate $\mathbf{T}$ to $+\mathbf{T}$, $\mathbf{F}$ to $-\mathbf{T}$; and to translate $A \longrightarrow$

S_0 :		
(1)	$A \lor B$	
(2)	$\neg A \lor B$	
(3)	$A \lor \neg B$	
(4)	$\neg A \lor \neg B$	
S_1 :		
(5)	B	$\leftarrow (1),(2)$
(6)	A	$\leftarrow (1),(3)$
(7)	$B \lor \neg B$	$\leftarrow (1),(4)$
(8)	$A \lor \neg A$	$\leftarrow (1),(4)$
$(9) = (7)$	$B \lor \neg B$	$\leftarrow (2),(3)$
$(10) = (8)$	$A \lor \neg A$	$\leftarrow (2),(3)$
(11)	$\neg A$	$\leftarrow (2),(4)$
(12)	$\neg B$	$\leftarrow (3),(4)$
S_2 :		
$(13) = (1)$	$A \lor B$	$\leftarrow (1),(7)$
$(14) = (1)$	$A \lor B$	$\leftarrow (1),(8)$
$(15) = (5)$	B	$\leftarrow (1),(11)$
$(16) = (6)$	A	$\leftarrow (1),(12)$
$(17) = (5)$	B	$\leftarrow (2),(6)$
$(18) = (2)$	$\neg A \lor B$	$\leftarrow (2),(7)$
$(19) = (2)$	$\neg A \lor B$	$\leftarrow (2),(8)$
$(20) = (11)$	$\neg A$	$\leftarrow (2),(12)$
$(21) = (6)$	A	$\leftarrow (3),(5)$
$(22) = (3)$	$A \lor \neg B$	$\leftarrow (3),(7)$
$(23) = (3)$	$A \lor \neg B$	$\leftarrow (3),(8)$
$(24) = (12)$	$\neg B$	$\leftarrow (3),(11)$
$(25) = (11)$	$\neg A$	$\leftarrow (4),(5)$
$(26) = (12)$	$\neg B$	$\leftarrow (4),(6)$
$(27) = (4)$	$\neg A \lor \neg B$	$\leftarrow (4),(7)$
$(28) = (4)$	$\neg A \lor \neg B$	$\leftarrow (4),(8)$
$(29) = (5)$	B	$\leftarrow (5),(7)$
(30)	$\square$	$\leftarrow (5),(12)$
$(31) = (6)$	A	$\leftarrow (6),(8)$
$(32) = (30)$	$\square$	$\leftarrow (6),(11)$
$(33) = (12)$	$\neg B$	$\leftarrow (7),(12)$
$(34) = (11)$	$\neg A$	$\leftarrow (8),(11)$

Figure A.1. Level saturation search in propositional resolution

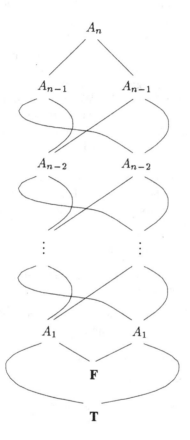

Figure A.2. A BDD for Ψ_n

Φ'_+; Φ'_-, first convert Φ'_+ and Φ'_- to TDGs Φ_+ and Φ_- respectively, then output $+A \longrightarrow \Phi_+$; Φ_- if the sign of Φ_+ is $+$, or $-A \longrightarrow$ TDGneg(Φ_+); TDGneg(Φ_-) if the sign of Φ_- is $-$.

Notice that the conversion functions we defined are mutually inverse: if we convert a BDD to a TDG, then reconvert the TDG into a BDD, we get the original BDD; and if we convert a TDG to a BDD, then again to a TDG, we get the original TDG. So, if Φ and Φ' are two logically equivalent TDGs, their conversions to BDDs are equivalent, hence equal to some BDD Φ''. Then, if we reconvert Φ'' to a TDG, by the remark above, we must have $\Phi = \Phi'' = \Phi'$. Therefore, TDGs are canonical representations of formulas modulo logical equivalence.

Negation on TDGs is defined by the TDGneg function. We still have to allocate and share couples (s, v) of a sign and a vertex to guarantee canonicity, and this cannot be done in constant time easily. We can instead use reserve a sign bit in addresses (for example, the lowest significant bit is usually 0 on modern computers, because of alignment considerations), so that if a is the (even) address of a vertex v, we can take a to represent $+v$ as a TDG, and $a + 1$ to represent $-v$ as a TDG. Then negation works in constant time, as we only need to flip one bit in a number.

Binary connectives on TDGs work in the same way as on BDDs, except we have to massage the sign bit somehow. Here is a definition for disjunction, described as a binary function TDGor. We first define the analogue of BDDmake, which must also ensure sign normalisation:

- if $\Phi' = \Phi''$, then TDGmake(A, Φ', Φ'') returns Φ';

- otherwise, if Φ'' has positive sign, then TDGmake(A, Φ', Φ'') allocates, shares and returns $A \longrightarrow \Phi'$; Φ'';

- or if Φ'' has negative sign, then TDGmake(A, Φ', Φ'') allocates, shares and returns $-(A \longrightarrow$ TDGneg(Φ'); TDGneg(Φ'')).

The following function TDGsign will be useful:

- if s is $+$, then TDGsign(s, Φ) is Φ;

- if s is $-$, then TDGsign(s, Φ) is TDGneg(Φ).

Then:

- TDGor($+\mathbf{T}, \Phi$) = $\mathbf{T}$, TDGor($-\mathbf{T}, \Phi$) = Φ;

- TDGor($\Phi, +\mathbf{T}$) = $\mathbf{T}$, TDGor($\Phi, -\mathbf{T}$) = Φ;

- if $\Phi = (s, A \longrightarrow \Phi_+; \Phi_-)$, and $\Phi' = (s', A' \longrightarrow \Phi'_+; \Phi'_-)$, then:

  - if $A < A'$, then TDGor(Φ, Φ') =

    TDGmake(A,TDGor(TDGsign(s, Φ_+), TDGsign(s, Φ')),
               TDGor(TDGsign(s, Φ_-), TDGsign(s, Φ')));

– if $A > A'$, then $\text{TDGor}(\Phi, \Phi') =$

$\text{TDGmake}(A', \text{TDGor}(\text{TDGsign}(s', \Phi), \text{TDGsign}(s', \Phi'_+)),$
$\qquad \text{TDGor}(\text{TDGsign}(s', \Phi), \text{TDGsign}(s', \Phi'_-)));$

– if $A = A'$, then $\text{TDGor}(\Phi, \Phi') =$

$\text{TDGmake}(A, \text{TDGor}(\text{TDGsign}(s, \Phi_+), \text{TDGsign}(s', \Phi'_+)),$
$\qquad \text{TDGor}(\text{TDGsign}(s, \Phi_-), \text{TDGsign}(s', \Phi'_-))).$

And it is much more efficient to code conjunction and the other connectives in terms of TDGneg and TDGor, as this will make the memo-functions share a unique hash-table, that of TDGor.

▸ **3.1.**

There are six structural rules for classical logic: exchange, weakening and contraction on both the left and right of the $\longrightarrow$. The missing rules from Section 1.4 are the following right rules:

$$\frac{\Gamma \longrightarrow \Delta, \Phi, \Phi', \Delta'}{\Gamma \longrightarrow \Delta, \Phi', \Phi, \Delta'} \, (RX) \qquad \frac{\Gamma \longrightarrow \Delta}{\Gamma \longrightarrow \Phi, \Delta} \, (RW) \qquad \frac{\Gamma \longrightarrow \Phi, \Phi, \Delta}{\Gamma \longrightarrow \Phi, \Delta} \, (RC)$$

One can easily recover the usual axiom from classical logic using the new axiom together with weakening on both left and right:

$$\frac{\dfrac{\dfrac{}{\Phi \longrightarrow \Phi} \, (Ax)}{\Gamma, \Phi \longrightarrow \Phi} \, (LW)}{\Gamma, \Phi \longrightarrow \Phi, \Delta} \, (RW)$$

▸ **3.2.**

All these follow in a quite straightforward manner from the logical rules. We give the solutions to parts (i) and (vi), without being so formal about the exchange rule.

(i)

$$\frac{\dfrac{\dfrac{\dfrac{}{A \longrightarrow A} \, (Ax)}{A, B \longrightarrow A} \, (W)}{A \longrightarrow B \Rightarrow A} \, (\Rightarrow I)}{\longrightarrow A \Rightarrow B \Rightarrow A} \, (\Rightarrow I)$$

(vi) Rather than giving one, quite large, derivation, we split it up into several parts.

First we give a proof π_1 of $A \Rightarrow B \wedge C \longrightarrow A \Rightarrow B$:

$$\dfrac{\dfrac{}{(A \Rightarrow B \wedge C) \longrightarrow (A \Rightarrow B \wedge C)}\,(Ax) \qquad \dfrac{}{A \longrightarrow A}\,(Ax)}{\dfrac{A \Rightarrow B \wedge C, A \longrightarrow B \wedge C}{\dfrac{A \Rightarrow B \wedge C, A \longrightarrow B}{A \Rightarrow B \wedge C \longrightarrow A \Rightarrow B}\,(\Rightarrow I)}\,(\wedge E)}\,(\Rightarrow E)$$

There is a similar proof π_2 of the formula $A \Rightarrow B \wedge C \longrightarrow A \Rightarrow C$. Putting these together, we can now give a proof of the required formula in the following way:

$$\dfrac{\dfrac{\pi_1}{A \Rightarrow B \wedge C \longrightarrow A \Rightarrow B} \qquad \dfrac{\pi_2}{A \Rightarrow B \wedge C \longrightarrow A \Rightarrow C}}{\dfrac{A \Rightarrow B \wedge C, A \Rightarrow B \wedge C \longrightarrow (A \Rightarrow B) \wedge (A \Rightarrow C)}{\dfrac{A \Rightarrow B \wedge C \longrightarrow (A \Rightarrow B) \wedge (A \Rightarrow C)}{\longrightarrow (A \Rightarrow B \wedge C) \Rightarrow (A \Rightarrow B) \wedge (A \Rightarrow C)}\,(\Rightarrow I)}\,(C)}\,(\wedge I)$$

▶ **3.3.**

All of these hold in only one direction (from right to left), with the exception of the last one, $\Phi \wedge \neg\Phi \Leftrightarrow \bot$, which is indeed an equivalence. Only the details of the last two are given here.

To show $\neg\Phi \vee \Phi' \longrightarrow \Phi \Rightarrow \Phi'$ we need a derivation ending in the following:

$$\dfrac{\dfrac{}{\neg\Phi \vee \Phi' \longrightarrow \neg\Phi \vee \Phi'}\,(Ax) \qquad \dfrac{\pi_1}{\neg\Phi \longrightarrow \Phi \Rightarrow \Phi'}\,(\Rightarrow I) \qquad \dfrac{\pi_2}{\Phi' \longrightarrow \Phi \Rightarrow \Phi'}}{\neg\Phi \vee \Phi' \longrightarrow \Phi \Rightarrow \Phi'}\,(\vee E)$$

But a proof π_1 of $\neg\Phi \longrightarrow \Phi \Rightarrow \Phi'$ and a proof π_2 of $\Phi' \longrightarrow \Phi \Rightarrow \Phi'$ can easily be given by the following two derivations:

$$\dfrac{\dfrac{\dfrac{}{\neg\Phi \longrightarrow \neg\Phi}\,(Ax) \qquad \dfrac{}{\Phi \longrightarrow \Phi}\,(Ax)}{\dfrac{\neg\Phi, \Phi \longrightarrow \bot}{\dfrac{\neg\Phi, \Phi \longrightarrow \Phi'}{\neg\Phi \longrightarrow \Phi \Rightarrow \Phi'}\,(\Rightarrow I)}\,(\bot)}\,(\Rightarrow E)}{} \qquad \dfrac{\dfrac{\dfrac{}{\Phi' \longrightarrow \Phi'}\,(Ax)}{\dfrac{\Phi', \Phi \longrightarrow \Phi'}{\Phi' \longrightarrow \Phi \Rightarrow \Phi'}\,(\Rightarrow I)}\,(W)}{}$$

The other direction does not hold. An attempt to build a derivation ending in $\Phi \Rightarrow \Phi' \longrightarrow \neg\Phi \vee \Phi'$ requires the use of double negation elimination.

The last part of the question is indeed an equivalence. $\bot \longrightarrow \Phi \wedge \neg\Phi$ is proved as follows:

$$\cfrac{\cfrac{\cfrac{\overline{\bot \longrightarrow \bot}\,(Ax)}{\bot \longrightarrow \Phi}\,(\bot) \qquad \cfrac{\overline{\bot \longrightarrow \bot}\,(Ax)}{\bot \longrightarrow \neg\Phi}\,(\bot)}{\bot, \bot \longrightarrow \Phi \wedge \neg\Phi}\,(\wedge I)}{\bot \longrightarrow \Phi \wedge \neg\Phi}\,(C)$$

and $\Phi \wedge \neg\Phi \longrightarrow \bot$ can be proved as:

$$\cfrac{\cfrac{\cfrac{\overline{\Phi \wedge \neg\Phi \longrightarrow \Phi \wedge \neg\Phi}\,(Ax)}{\Phi \wedge \neg\Phi \longrightarrow \Phi}\,(\wedge E) \qquad \cfrac{\overline{\Phi \wedge \neg\Phi \longrightarrow \Phi \wedge \neg\Phi}\,(Ax)}{\Phi \wedge \neg\Phi \longrightarrow \neg\Phi}\,(\wedge E)}{\Phi \wedge \neg\Phi, \Phi \wedge \neg\Phi \longrightarrow \bot}\,(\Rightarrow E)}{\Phi \wedge \neg\Phi \longrightarrow \bot}\,(C)$$

▸ **3.4.**

Using the $(\wedge I)$ rule with an axiom (Ax) allows us to add an additional element to the context. We can then use the $(\wedge E_1)$ rule to recover the same formula that we are required to prove, leaving the additional assumption in the context; hence simulating weakening.

$$\cfrac{\cfrac{\Gamma \longrightarrow \Phi \qquad \overline{\Phi' \longrightarrow \Phi'}\,(Ax)}{\Gamma, \Phi' \longrightarrow \Phi \wedge \Phi'}\,(\wedge I)}{\Gamma, \Phi' \longrightarrow \Phi}\,(\wedge E_1)$$

It is worth noting that this only works because we are concatenating contexts (using them in a *multiplicative* way). In different presentations of logics that use the *additive* style, where contexts are merged, we are not able to do this.

▸ **3.5.**

(i) and (iv) are not derivable, (ii), (iii) and (v) are. We just show part (ii):

$$\cfrac{\cfrac{\cfrac{\overline{A \longrightarrow A}\,(Ax) \qquad \cfrac{\overline{A \longrightarrow A}\,(Ax)}{\neg A, A \longrightarrow \bot}\,(\neg L)}{A \Rightarrow \neg A, A, A \longrightarrow \bot}\,(\Rightarrow L)}{A \Rightarrow \neg A, A \longrightarrow \bot}\,(C)}{A \Rightarrow \neg A \longrightarrow \neg A}\,(\neg R)$$

▸ **3.6.**

We just show one direction:

$$\cfrac{\cfrac{}{\Phi \longrightarrow \Phi}\,(Ax)}{\cfrac{\neg\neg\Phi, \Phi \longrightarrow \Phi}{\cfrac{\Phi \longrightarrow \neg\neg\Phi \Rightarrow \Phi}{}\,(\Rightarrow R)}\,(W)} \qquad \cfrac{\cfrac{\cfrac{\cfrac{}{\neg\Phi \longrightarrow \neg\Phi}\,(Ax)}{\neg\Phi, \neg\neg\Phi \longrightarrow \bot}\,(\neg L)}{\cfrac{\neg\Phi, \neg\neg\Phi \longrightarrow \Phi}{\cfrac{\neg\Phi \longrightarrow \neg\neg\Phi \Rightarrow \Phi}{}\,(\Rightarrow R)}\,(\bot)}}{\Phi \vee \neg\Phi \longrightarrow \neg\neg\Phi \Rightarrow \Phi}\,(\vee L)$$

$$\cfrac{\Phi \vee \neg\Phi \longrightarrow \neg\neg\Phi \Rightarrow \Phi}{\longrightarrow (\Phi \vee \neg\Phi) \Rightarrow (\neg\neg\Phi \Rightarrow \Phi)}\,(\Rightarrow R)$$

▶ **3.7.**

All these are quite straightforward. We just give a selection of cases. See also Exercise 3.2 for the first two axioms!

$$\cfrac{\cfrac{\cfrac{\cfrac{}{\Phi \longrightarrow \Phi}\,(Ax)}{\Phi, \Phi' \longrightarrow \Phi}\,(W)}{\cfrac{\Phi \longrightarrow \Phi' \Rightarrow \Phi}{\longrightarrow \Phi \Rightarrow (\Phi' \Rightarrow \Phi)}\,(\Rightarrow I)}\,(\Rightarrow I)}{} \qquad \cfrac{\cfrac{\cfrac{}{\Phi \wedge \Phi' \longrightarrow \Phi \wedge \Phi'}\,(Ax)}{\Phi \wedge \Phi' \longrightarrow \Phi}\,(\wedge R_1)}{\longrightarrow \Phi \wedge \Phi' \Rightarrow \Phi}\,(\Rightarrow R) \qquad \cfrac{\cfrac{}{\bot \longrightarrow \bot}\,(Ax)}{\bot \longrightarrow \Phi}\,(\bot)$$

▶ **3.8.**

$$
\begin{aligned}
x \models \Phi \Rightarrow \bot \quad &\text{iff} \quad \text{for all } y, x \leq y, y \models \Phi \text{ implies } y \models \bot \\
&\text{iff} \quad \text{for all } y, x \leq y, y \not\models \Phi \\
&\text{iff} \quad x \models \neg\Phi
\end{aligned}
$$

▶ **3.9.**

We are required to prove: If $x \models \Phi$ and $x \leq y$, then $y \models \Phi$, which can be demonstrated by an induction over Φ. We just show a selection of cases.

- Case: $\Phi = \Psi \wedge \Psi'$.

$$
\begin{aligned}
x \models \Psi \wedge \Psi' \quad &\text{iff} \quad x \models \Psi \text{ and } x \models \Psi' \\
&\text{iff} \quad y \models \Psi \text{ and } y \models \Psi' \text{ (by hypothesis twice)} \\
&\text{iff} \quad y \models \Psi \wedge \Psi'
\end{aligned}
$$

- Case: $\Phi = \Psi \Rightarrow \Psi'$. Given $x \models \Psi \Rightarrow \Psi'$ and $x \leq y$ we have to show that $y \models \Psi \Rightarrow \Psi'$. But $y \models \Psi \Rightarrow \Psi'$ iff for all $z, y \leq z, z \models \Psi$ implies $z \models \Psi'$. Now the result follows by transitivity of $\leq$, since $x \leq y \leq z$.

- Case: $\Phi = \neg\Psi$. Given $x \models \neg\Psi$ and $x \leq y$ we have to show that $y \models \neg\Psi$. But $y \models \neg\Psi$ iff for all $z, y \leq z, z \not\models \Psi$. Again the result follows by transitivity of $\leq$.

The remaining cases follow in a similar style.

▶ **3.10.**

The tableaux for $A \vee \neg A$ is given by:

$$-0 \models A \vee \neg A$$
$$-0 \models A$$
$$-0 \models \neg A$$
$$+00 \models A$$

where $0 \leq 00$ for some new 00. Now there is no way to expand the tableaux further, and we cannot close the path. Hence $A \vee \neg A$ is not intuitionistically valid.

The tableaux for $A \Rightarrow \neg\neg A$ is given by following, where we have numbered the nodes of the tree for easy reference.

(1)	$-0 \models A \Rightarrow \neg\neg A$	
(2)	$+00 \models A$	
(3)	$-00 \models \neg\neg A$	(new 00, $0 \leq 00$ (1))
(4)	$+000 \models \neg A$	(new 000, $00 \leq 000$ (3))
(5)	$-000 \models A$	(any 000, $000 \leq 000$ (4))
(6)	$+000 \models A$	(using monotonicity (2))

and the (only) path is closed showing that $A \Rightarrow \neg\neg A$ is intuitionistically valid. Note in particular the way that monotonicity can be used to close the path.

The case for $\neg\neg A \Rightarrow A$ is similar to the above, with the exception that we cannot apply the monotonicity rule to close the path.

▶ **3.11.**

All are intuitionistically valid, with the exception of (iv) and (vi). We consider parts (i),(ii) and (v).

(i) A proof of $A \Rightarrow A$ is a function π such that if a is a proof of A, then $\pi(a) = a$. Hence, π is the identity function.

(ii) The empty function with image A.

(v) The interpretation of $A \Rightarrow A \vee \neg A$ is a function f such that $f(a) = (0, a)$.

▶ **3.12.**

To apply the translation function we first build the corresponding proofs in $\mathbf{LK_0}$, which we leave for the reader, and translate into $\mathbf{LJ_0}$. The first, $\Phi \vee \neg\Phi$ gives a derivation ending in $\neg(\neg\neg\Phi \vee \neg\neg\neg\Phi) \longrightarrow \bot$, as shown below:

$$\cfrac{\cfrac{\cfrac{\cfrac{\cfrac{\cfrac{\cfrac{\cfrac{}{\neg\Phi \longrightarrow \neg\Phi}\,(Ax)}{\neg\Phi, \neg\neg\Phi \longrightarrow \bot}\,(\neg L)}{\neg\Phi \longrightarrow \neg\neg\neg\Phi}\,(\neg R)}{\neg\Phi \longrightarrow \neg\neg\Phi \vee \neg\neg\neg\Phi}\,(\vee R_1)}{\neg(\neg\neg\Phi \vee \neg\neg\neg\Phi), \neg\Phi \longrightarrow \bot}\,(\neg L)}{\neg(\neg\neg\Phi \vee \neg\neg\neg\Phi) \longrightarrow \neg\neg A}\,(\neg R)}{\cfrac{\cfrac{\neg(\neg\neg\Phi \vee \neg\neg\neg\Phi) \longrightarrow \neg\neg\Phi \vee \neg\neg\neg\Phi}{\neg(\neg\neg\Phi \vee \neg\neg\neg\Phi), \neg(\neg\neg\Phi \vee \neg\neg\neg\Phi) \longrightarrow \bot}\,(\neg L)}{\neg(\neg\neg\Phi \vee \neg\neg\neg\Phi) \longrightarrow \bot}\,(C)}}\,(\vee R_2)$$

The second, $\neg\neg\Phi \Rightarrow \Phi$, gives a derivation ending in $\neg(\neg\Phi \Rightarrow \neg\neg\neg\Phi) \longrightarrow \bot$, which can be done as follows:

$$\cfrac{\cfrac{\cfrac{\cfrac{\cfrac{}{\neg\Phi \longrightarrow \neg\Phi}\,(Ax)}{\neg\Phi, \neg\neg\Phi \longrightarrow \bot}\,(\neg L)}{\neg\Phi \longrightarrow \neg\neg\neg\Phi}\,(\neg R)}{\longrightarrow \neg\Phi \Rightarrow \neg\neg\neg\Phi}\,(\Rightarrow R)}{\neg(\neg\Phi \Rightarrow \neg\neg\neg\Phi) \longrightarrow \bot}\,(\neg L)$$

▶ **3.13.**

We simply apply the translation to the two formulas:

$$[\Phi \vee (\Phi' \vee \Phi'')] \;=\; \neg\neg[\Phi] \vee \neg\neg(\neg\neg[\Phi'] \vee \neg\neg[\Phi''])$$
$$[(\Phi \vee \Phi') \vee \Phi''] \;=\; \neg\neg(\neg\neg[\Phi] \vee \neg\neg[\Phi']) \vee \neg\neg[\Phi'']$$

Since we know that there is no way to remove the outer $\neg\neg$, there is no way we can prove these two formulas equivalent.

▶ **3.14.**

Here are the multiplicative and additive versions respectively. It is clear that the additive case is rather more complicated.

$$\cfrac{\cfrac{\cfrac{}{\Phi \longrightarrow \Phi}\,(Ax)}{\longrightarrow \neg\Phi, \Phi}\,(\neg R)}{\longrightarrow \neg\Phi \vee \Phi}\,(\vee R) \qquad \cfrac{\cfrac{\cfrac{\cfrac{\cfrac{}{\Phi \longrightarrow \Phi}\,(Ax)}{\longrightarrow \neg\Phi, \Phi}\,(\neg R)}{\longrightarrow \neg\Phi \vee \Phi, \Phi}\,(\vee R_1)}{\longrightarrow \neg\Phi \vee \Phi, \neg\Phi \vee \Phi}\,(\vee R_2)}{\longrightarrow \neg\Phi \vee \Phi}\,(C)$$

▶ **3.15.**

We can only do this in the additive case, for the other we need to make a contraction. The additive case is:

$$\frac{\dfrac{}{\Phi \longrightarrow \Phi}\,(Ax) \qquad \dfrac{}{\Phi \longrightarrow \Phi}\,(Ax)}{\Phi \longrightarrow \Phi \wedge \Phi}\,(\wedge I)$$

The multiplicative case would be:

$$\frac{\dfrac{}{\Phi \longrightarrow \Phi}\,(Ax) \qquad \dfrac{}{\Phi \longrightarrow \Phi}\,(Ax)}{\Phi, \Phi \longrightarrow \Phi \wedge \Phi}\,(\wedge I)$$

But, we have no way of removing the additional Φ.

▶ **3.16.**

The first can be simulated in the second:

$$\frac{\dfrac{\Gamma, \Phi \longrightarrow \Phi'' \qquad \Gamma, \Phi' \longrightarrow \Phi''}{\Gamma, \Gamma, \Phi \vee \Phi' \longrightarrow \Phi''}\,(\vee L)}{\Gamma, \Phi \vee \Phi' \longrightarrow \Phi''}\,(C)$$

The second can be simulated in the first:

$$\frac{\dfrac{\Gamma, \Phi \longrightarrow \Phi''}{\Gamma, \Delta, \Phi \longrightarrow \Phi''}\,(W) \qquad \dfrac{\Delta, \Phi' \longrightarrow \Phi''}{\Gamma, \Delta, \Phi' \longrightarrow \Phi''}\,(W)}{\Gamma, \Delta, \Phi \vee \Phi' \longrightarrow \Phi''}\,(\vee L)$$

▶ **3.17.**

There are in fact ten rules, since we need two versions of the $(\&L)$ and $(\oplus R)$. For the $\wedge$ connective there are the multiplicative $(\otimes)$ and additive $(\&)$ cases:

$$\frac{\Gamma \longrightarrow \Delta \qquad \Gamma' \longrightarrow \Delta', \Phi'}{\Gamma, \Gamma' \longrightarrow \Delta, \Delta', \Phi \otimes \Phi'}\,(\otimes L) \qquad \frac{\Gamma, \Phi, \Phi' \longrightarrow \Delta}{\Gamma, \Phi \otimes \Phi' \longrightarrow \Delta}\,(\otimes R)$$

$$\frac{\Gamma \longrightarrow \Delta \qquad \Gamma \longrightarrow \Delta, \Phi'}{\Gamma \longrightarrow \Delta, \Phi \& \Phi'}\,(\& R)$$

$$\frac{\Gamma, \Phi \longrightarrow \Delta}{\Gamma, \Phi \& \Phi' \longrightarrow \Delta}\,(\& L_1) \qquad \frac{\Gamma, \Phi' \longrightarrow \Delta}{\Gamma, \Phi \& \Phi' \longrightarrow \Delta}\,(\& L_2)$$

Similarly, for the $\vee$ connective, there are the multiplicative $(\wp)$ and additive $(\oplus)$ cases:

$$\frac{\Gamma, \Phi \longrightarrow \Delta \qquad \Gamma', \Phi' \longrightarrow \Delta'}{\Gamma, \Gamma', \Phi \wp \Phi' \longrightarrow \Delta, \Delta'}\,(\wp L) \qquad \frac{\Gamma \longrightarrow \Delta, \Phi, \Phi'}{\Gamma \longrightarrow \Delta, \Phi \wp \Phi'}\,(\wp R)$$

$$\frac{\Gamma,\Phi \longrightarrow \Delta \quad \Gamma,\Phi' \longrightarrow \Delta}{\Gamma,\Phi \oplus \Phi' \longrightarrow \Delta}\;(\oplus L)$$

$$\frac{\Gamma \longrightarrow \Delta,\Phi}{\Gamma \longrightarrow \Delta,\Phi \oplus \Phi'}\;(\oplus R_1) \qquad \frac{\Gamma \longrightarrow \Delta,\Phi'}{\Gamma \longrightarrow \Delta,\Phi \oplus \Phi'}\;(\oplus R_2)$$

▶ **3.18.**

We show a selection of cases, most are very straightforward and are included just to get used to the logical system.

Implicational Fragment. Only (i) and (iii) are provable. If we try to prove (ii) we are forced into a situation where we have to use a weakening on B to complete the proof, and similarly for (iv) we are forced into using a contraction on A. Since neither A nor B were marked with an exponential, there is no way to complete the proof.

> In this fragment one can in fact guess if formulas are provable or not simply by counting the number of positive and negative occurrences of every formula. If a formula occurs an equal number of times both positive and negative, then there is a good chance that it is a provable formula.
>
> It is also worth remarking that every formula in this fragment can be written in a form where each formula appears exactly twice: once positive and once negative. Take the formula $(A \multimap A) \multimap (A \multimap A)$ as an example. This formula is indeed provable, and the proof is the *same* as the formula $(A \multimap B) \multimap (A \multimap B)$.

Multiplicative/Additive. Only (ii), (iv) and (v) are provable.

Exponentials. All are provable with the exception of (v).

(ii) We prove $!(A^{\perp} \wp A) = !(A \multimap A)$.

$$\frac{\dfrac{\quad}{A^{\perp}A}\;(Ax)}{\dfrac{A^{\perp}\wp A}{!(A^{\perp}\wp A)}\;(\wp)}\;(!)$$

(iii) Note that $!(A\&B)\multimap !A \otimes !B = ?(A^{\perp}\oplus B^{\perp})\wp !A \otimes !B$, which was shown as an example derivation on page 125.

(vi) We prove $!A \multimap !!A = ?A^{\perp}\wp !!A$.

$$\frac{\dfrac{\quad}{?A^{\perp},!A}\;(Ax)}{\dfrac{?A^{\perp},!!A}{?A^{\perp}\wp !!A}\;(\wp)}\;(!)$$

► **3.19.**

There are *three* main cases that we need to consider.

1. An axiom cut:

$$\cfrac{\cfrac{}{\Phi^\perp, \Phi}\,(Ax) \qquad \cfrac{}{\Gamma, \Phi}}{\Gamma, \Phi}\,(Cut)$$

is reduced to simply:

$$\overline{\Gamma, \Phi}$$

2. A cut between two proofs, one ending with $\otimes$ and the other ending with $\invamp$:

$$\cfrac{\cfrac{\Gamma, \Phi \qquad \Delta, \Phi'}{\Gamma, \Delta, \Phi \otimes \Phi'}\,(\otimes) \qquad \cfrac{\Theta, \Phi^\perp, \Phi'^\perp}{\Theta, \Phi^\perp \invamp \Phi'^\perp}\,(\invamp)}{\Gamma, \Delta, \Theta}\,(Cut)$$

becomes the following:

$$\cfrac{\cfrac{\Gamma, \Phi \qquad \Theta, \Phi^\perp, \Phi'^\perp}{\Gamma, \Theta, \Phi'^\perp}\,(Cut) \qquad \Delta, \Phi'}{\Gamma, \Delta, \Theta}\,(Cut)$$

3. A cut between a proof ending in $\oplus$ and one ending in $\&$.

$$\cfrac{\cfrac{\Gamma, \Phi \qquad \Gamma, \Phi'}{\Gamma, \Phi \& \Phi'}\,(\&) \qquad \cfrac{\Delta, \Phi^\perp}{\Delta, \Phi^\perp \oplus \Phi'^\perp}\,(\oplus_1)}{\Gamma, \Delta}\,(Cut)$$

becomes the following

$$\cfrac{\Gamma, \Phi \qquad \Delta, \Phi^\perp}{\Gamma, \Delta}\,(Cut)$$

► **3.20.**

One other translation is based on the following:

$$
\begin{aligned}
A^\circ &= \ !A \\
(\Phi \Rightarrow \Psi)^\circ &= \ !(\Phi^\circ \multimap \Psi^\circ) \\
(\Phi \wedge \Psi)^\circ &= \ \Phi^\circ \otimes \Psi^\circ \\
(\Phi \vee \Psi)^\circ &= \ \Phi^\circ \oplus \Psi^\circ
\end{aligned}
$$

This translation extends to sequents in the following way:

$$
\Gamma \longrightarrow \Phi \ = \ (\Gamma^\perp)^\circ, \Phi^\circ
$$

It is possible to prove formally that this translation works. Here we will just hint at the proof by showing how to translate proofs in intuitionistic logic into linear logic sequents under this translation. We omit the translation function $(\cdot)^\circ$ for clarity.

- The axiom is translated as

$$
\cfrac{\cfrac{\cfrac{}{\Phi^\perp, \Phi}\ (Ax)}{?\Phi^\perp, \Phi}\ (D)}{?\Phi^\perp, !\Phi}\ (!)
$$

- The $(\Rightarrow I)$ rule:

$$
\cfrac{\cfrac{?\Gamma^\perp, ?\Phi^\perp, !\Phi'}{?\Gamma^\perp, ?\Phi^\perp \otimes\!\!\!\!\! \mathtt{?} \,!\Phi'}\ (\otimes\!\!\!\!\!\mathtt{?})}{?\Gamma^\perp, !(?\Phi^\perp \otimes\!\!\!\!\!\mathtt{?}\,!\Phi')}\ (!)
$$

- The $(\Rightarrow E)$ rule:

$$
\cfrac{?\Gamma^\perp, !(?\Phi^\perp \otimes\!\!\!\!\!\mathtt{?}\,!\Phi') \qquad \cfrac{\cfrac{?\Delta^\perp, !\Phi \qquad \cfrac{}{?\Phi'^\perp, !\Phi'}\ (Ax)}{?\Delta^\perp, !\Phi \otimes ?\Phi'^\perp, !\Phi'}\ (\otimes)}{?\Delta^\perp, ?(!\Phi \otimes ?\Phi'^\perp), !\Phi'}\ (D)}{?\Gamma^\perp, ?\Delta, !\Phi'}\ (Cut)
$$

These translations are *safe* in that they make all formulae *non-linear*. A "better" translation would try to analyse which formulae are really non-linear, and only use exponentials there. A simple example would be $\Phi \Rightarrow \Phi' \Rightarrow \Phi$. This would get translated to $!\Phi \multimap !\Phi' \multimap \Phi$, but a "better" translation would code this is $\Phi \multimap !\Phi' \multimap \Phi$. However, this is *much* harder!

▶ **3.21.**

We show just the first and last to demonstrate the principle.

The translation for the first one gives $[\Phi \Rightarrow \Phi' \Rightarrow \Phi] = ![\Phi] \multimap ![\Phi'] \multimap [\Phi]$. Unpacking the abbreviation for $\multimap$ requires that we have to prove $?\Phi^{\perp} \invamp (?\Phi'^{\perp} \invamp \Phi)$. The proof of this is given by:

$$
\dfrac{\dfrac{\dfrac{\dfrac{\dfrac{\rule{2cm}{0.4pt}}{\Phi^{\perp},\Phi}\,(Ax)}{?\Phi^{\perp},\Phi}\,(D)}{?\Phi^{\perp},?\Phi'^{\perp},\Phi}\,(W)}{?\Phi^{\perp},?\Phi'^{\perp}\invamp\Phi}\,(\invamp)}{?\Phi^{\perp}\invamp(?\Phi'^{\perp}\invamp\Phi)}\,(\invamp)
$$

For the last case,

$$[(\Phi \Rightarrow (\Phi' \wedge \Phi'')) \Rightarrow ((\Phi \Rightarrow \Phi') \wedge (\Phi \Rightarrow \Phi''))]$$

becomes the following linear formula:

$$![\Phi] \otimes ([\Phi']^{\perp} \oplus [\Phi'']^{\perp}), (?[A]^{\perp}\invamp[\Phi'])\&(?[\Phi]^{\perp}\invamp[\Phi''])$$

which we can prove as follows using the translation given.

$$
\dfrac{
\dfrac{
\dfrac{
\dfrac{\rule{1.5cm}{0.4pt}}{!\Phi,?\Phi^{\perp}}(Ax)\quad
\dfrac{\dfrac{\rule{1.5cm}{0.4pt}}{\Phi'^{\perp},\Phi'}(Ax)}{\Phi'^{\perp}\oplus\Phi''^{\perp},\Phi'}(\oplus_1)
}{!\Phi\otimes(\Phi'^{\perp}\oplus\Phi''^{\perp}),?\Phi^{\perp},\Phi'}(\otimes)
}{!\Phi\otimes(\Phi'^{\perp}\oplus\Phi''^{\perp}),?\Phi^{\perp}\invamp\Phi'}(\invamp)
\qquad
\dfrac{
\dfrac{
\dfrac{\rule{1.5cm}{0.4pt}}{!\Phi,?\Phi^{\perp}}(Ax)\quad
\dfrac{\dfrac{\rule{1.5cm}{0.4pt}}{\Phi''^{\perp},\Phi''}(Ax)}{\Phi'^{\perp}\oplus\Phi''^{\perp},\Phi''}(\oplus_2)
}{!\Phi\otimes(\Phi'^{\perp}\oplus\Phi''^{\perp}),?\Phi^{\perp},\Phi''}(\otimes)
}{!\Phi\otimes(\Phi'^{\perp}\oplus\Phi''^{\perp}),?\Phi^{\perp}\invamp\Phi''}(\invamp)
}{!\Phi\otimes(\Phi'^{\perp}\oplus\Phi''^{\perp}),(?\Phi^{\perp}\invamp\Phi')\&(?\Phi^{\perp}\invamp\Phi'')}(\&)
$$

The first point to note is that proofs get bigger. We were required to use the structural rules to manipulate the exponential formulas. However, this additional information is propagated down the proof, and the resulting formula has more information about how the proof was built. In particular, we know which formulas had structural rules applied to them (the ones with an exponential). More precisely, a linear proof ends in a formula that contains more information about the *history* of the deduction.

▶ **4.1.**

We prove the claim: If x and y are distinct variables and $x \notin FV(v)$, then

$$t[u/x][v/y] = t[v/y][u[v/y]/x]$$

by induction over the structure of the term t.

1. If t is a variable, then there are three cases

   (a) $t = x$ then both sides are $u[v/y]$.

   (b) $t = y$ then both sides are v, since $x \notin FV(v)$.

   (c) $t = z$, where z is distinct from both x and y, then both sides are z.

2. If t is an abstraction $\lambda z.t'$, where z is different from x and y, and $x \notin FV(uv)$, then

$$
\begin{aligned}
(\lambda z.t')[u/x][v/y] &= \lambda z.t'[u/x][v/y] \\
&= \lambda z.t'[v/y][u[v/y]/x] \text{ (by hypothesis)} \\
&= (\lambda z.t')[v/y][u[v/y]/x]
\end{aligned}
$$

3. If t is an application, then the same reasoning as for abstraction gives the required result.

4. If t is the projection $\mathsf{fst}(t')$ or $\mathsf{snd}(t')$ then the result follows almost immediately.

5. Similarly for the case when t is a pair $\langle t_1, t_2 \rangle$.

Remark that this proof takes no account of the types, hence holds equally well for the untyped λ-calculus.

▸ **4.2.**

A very simple example would be the term $(\lambda f x. f(f(f(x))))N$, where N is a very large term. After one reduction we have $\lambda x.N(N(Nx))$ which looks considerably larger. The sense that this is simpler than the original term is in the same way at cut elimination works for logics. What we have done is expanded the proof by putting lemmas "in-line".

▸ **4.3.**

We first show the derivation of $(\vee I_1)$ followed by $(\vee E)$, and assigning the terms:

$$
\dfrac{\dfrac{\Gamma \longrightarrow t : \sigma}{\Gamma \longrightarrow \mathsf{inl}(t) : \sigma \vee \tau}(INL) \quad \Delta, x : \sigma \longrightarrow u : \gamma \quad \Delta, y : \tau \longrightarrow v : \gamma}{\Gamma, \Delta \longrightarrow \mathsf{case}\ \mathsf{inl}(t)\ \mathsf{of}\ \mathsf{inl}(x) \Longrightarrow u \mid \mathsf{inr}(y) \Longrightarrow v : \gamma}(CASE)
$$

Recalling the normalisation step for the logic, we have the following derivation:

$$
\overline{\Gamma, \Delta \longrightarrow u[t/x] : \gamma}
$$

This gives one of the reductions. The other case, inr is just as easy to read off the derivation, as is the η-rule. Altogether, we have two β-reductions:

$$\text{case inl}(t) \text{ of inl}(x) \Longrightarrow u \mid \text{inr}(y) \Longrightarrow v \quad \longrightarrow \quad u[t/x]$$
$$\text{case inr}(t) \text{ of inl}(x) \Longrightarrow u \mid \text{inr}(y) \Longrightarrow v \quad \longrightarrow \quad v[t/y]$$

and one η-reduction:

$$\text{case } t \text{ of inl}(x) \Longrightarrow \text{inl}(x) \mid \text{inr}(y) \Longrightarrow \text{inr}(y) \quad \longrightarrow \quad t$$

▶ **4.4.**

We give the terms, together with the types. The reader will be able to reconstruct the derivations from this information.

$(i) \quad \lambda xy.x : \alpha \to \beta \to \alpha$

$(ii) \quad \lambda xyz.xz(yz) : (\alpha \to \beta \to \gamma) \to (\alpha \to \beta) \to (\alpha \to \gamma)$

$(iii) \quad \lambda xyz.x(y,z) : (\alpha \times \beta \to \gamma) \to (\alpha \to \beta \to \gamma)$

$(iv) \quad \lambda xy.x(\text{fst}(y))(\text{snd}(y)) : (\alpha \to \beta \to \gamma) \to (\alpha \times \beta \to \gamma)$

▶ **4.5.**

$$(i) \quad \mathbf{I} \qquad\qquad (ii) \quad \mathbf{S(KK)I}$$
$$(iii) \quad \mathbf{KI} \qquad\qquad (iv) \quad \mathbf{S(KK)II}$$

Note that in part (iv), the term can be reduced to $\mathbf{KI}$.

$(v) \quad (\lambda xy.x)(\lambda z.z) \to \lambda yz.z \qquad (vi) \quad xy$

$(vii) \quad x(\lambda z.z) \qquad\qquad\qquad\qquad (viii) \quad \lambda xyz.x(yz)$

▶ **4.6.**

We define $\text{fst} : \sigma \times \tau \to \sigma$ and $\text{snd} : \sigma \times \tau \to \tau$ as the following:

$$\text{fst}(t) \quad \text{as} \quad t\sigma(\lambda x^\sigma.\lambda y^\tau.x)$$
$$\text{snd}(t) \quad \text{as} \quad t\tau(\lambda x^\sigma.\lambda y^\tau.y)$$

To verify that these are indeed the correct definitions, we show that $\text{fst}(\langle t, u \rangle) \to^* t$

$$\text{fst}(\langle t, u \rangle) = (\Lambda X.\lambda x^{(\sigma \to \tau \to \sigma)}.xtu)\sigma(\lambda x^\sigma.\lambda y^\tau.x)$$
$$\longrightarrow \quad (\lambda x^{(\sigma \to \tau \to \sigma)}.xtu)(\lambda x^\sigma.\lambda y^\tau.x)$$
$$\longrightarrow \quad (\lambda x^\sigma.\lambda y^\tau.x)tu$$
$$\longrightarrow^* \quad t$$

▶ **4.7.**

This calculus corresponds to the *intuitionistic* fragment of linear logic (without products, sums or exponentials). Thus we have the following type assignment system with only three rules:

$$\frac{}{x:\sigma \longrightarrow x:\sigma} \quad \frac{\Gamma, x:\sigma \longrightarrow t:\tau}{\Gamma \longrightarrow \lambda x.t:\sigma \multimap \tau} \quad \frac{\Gamma \longrightarrow t:\sigma \multimap \tau \quad \Delta \longrightarrow u:\sigma}{\Gamma,\Delta \longrightarrow tu:\tau}$$

Note in particular that there are no structural rules here.

We can show that the calculus is normalising by assigning a system of weights as follows:

$$
\begin{aligned}
|x| &= 1 \\
|\lambda x.t| &= |t| + 1 \\
|tu| &= |t| + |u| + 1
\end{aligned}
$$

We now have to show that for the β-rule we have

$$|(\lambda x.t)u| < |t[u/x]|$$

which is trivial since x occurs exactly once in t. It is obvious that usual ordering on $|\cdot|$ is well-founded. Remark that this proof of normalisation doesn't depend on the types of the terms — the syntactic linearity constraint is sufficient to guarantee termination.

▶ **5.1.**

if $\Box\Phi \Rightarrow \Phi'$ is provable, then from the only hypothesis Φ, we deduce $\Box\Phi$ by necessitation, then we apply modus ponens with $\Box\Phi \Rightarrow \Phi'$ to get a proof of Φ'.

Conversely, let π be a proof of Φ' from Φ; we prove that $\Box\Phi \Rightarrow \Phi'$ is provable from the empty set of hypotheses, by induction on the length of π. If Φ' is an instance of an axiom of L, $\Box\Phi \Rightarrow \Phi'$ is trivially provable. If Φ' is Φ, then $\Box\Phi \Rightarrow \Phi'$ is an instance of (T). If Φ' was deduced by (MP), then there are formulas Φ'' and $\Phi'' \Rightarrow \Phi'$ with shorter proofs, which translate by induction hypothesis to proofs of $\Box\Phi \Rightarrow \Phi''$ and $\Box\Phi \Rightarrow \Phi'' \Rightarrow \Phi'$ respectively. Using modus ponens twice with the (non-modal) tautology $(A \Rightarrow B \Rightarrow C) \Rightarrow (A \Rightarrow B) \Rightarrow (A \Rightarrow C)$, we get $\Box\Phi \Rightarrow \Phi'$.

▶ **5.2.**

if $(\Box\Phi \wedge \Phi) \Rightarrow \Phi'$ is provable, then from the only hypothesis Φ, we deduce $\Box\Phi$ by necessitation, hence $\Box\Phi \wedge \Phi$ from the latter two by propositional reasoning, then we apply modus ponens with $(\Box\Phi \wedge \Phi) \Rightarrow \Phi'$ to get a proof of Φ'.

Conversely, let π be a proof of Φ' from Φ; we prove that $(\Box\Phi \wedge \Phi) \Rightarrow \Phi'$ is provable from the empty set of hypotheses, by induction on the length of π. If Φ' is an instance of an axiom of L, $\Box\Phi \Rightarrow \Phi'$ is trivial. If Φ' is Φ, then $\Box\Phi \Rightarrow \Phi'$ is an instance of (T), and $\Phi \Rightarrow \Phi'$ is provable by propositional reasoning, so we infer $(\Box\Phi \wedge \Phi) \Rightarrow \Phi'$ again by propositional reasoning. If Φ' was deduced by (MP), then there are formulas Φ'' and $\Phi'' \Rightarrow \Phi'$ with shorter proofs, which translate by induction hypothesis to proofs of $(\Box\Phi \wedge \Phi) \Rightarrow \Phi''$ and $(\Box\Phi \wedge \Phi) \Rightarrow \Phi'' \Rightarrow \Phi'$ respectively. Using modus ponens twice with the (non-modal) tautology $(A \Rightarrow B \Rightarrow C) \Rightarrow (A \Rightarrow B) \Rightarrow (A \Rightarrow C)$, we get $(\Box\Phi \wedge \Phi) \Rightarrow \Phi'$.

▶ **5.3.**

Let $(W, \leq)$ be an arbitrary reflexive transitive Kripke frame, and ρ be an assignment. We say that ρ is *upward-closed* if and only if $w \in \rho(A) \wedge w' \geq w$ implies $w' \in \rho(A)$. We let $\models$ be the satisfaction relation for S4 over Kripke frames, and $\models_i$ be the satisfaction relation for intuitionistic logic over the same Kripke frames.

We first show that if ρ is upward-closed, then $w, \rho \models_i \Phi$ if and only if $w, \rho \models \overline{\Phi}$. Assuming $w, \rho \models_i \Phi$, we prove $w, \rho \models \overline{\Phi}$ by structural induction on Φ: if Φ is a variable A, then $w, \rho \models_i A$ if and only if $w, \rho \models A$, that is if and only if, for every $w' \geq w$, $w', \rho \models A$, since ρ is upward-closed; but the latter is equivalent to $w, \rho \models \Box A$. The other cases are straightforward. Conversely, $w, \rho \models_i \Phi$ follows from $w, \rho \models \overline{\Phi}$ by a trivial structural induction on Φ.

So, if $w, \rho \models \overline{\Phi}$ for every ρ, in particular this holds for every upward-closed ρ, and then $w, \rho \models_i \Phi$. In particular, if $\overline{\Phi}$ is valid in S4, then Φ is intuitionistically valid.

Conversely, if ρ is an assignment, let $\overline{\rho}$ be the assignment mapping each variable A to $\{w \mid \forall w' \geq w \cdot w' \in \rho(A)\}$. We show that $w, \overline{\rho} \models \overline{\Phi}$ if and only if $w, \rho \models \overline{\Phi}$ by structural induction on Φ. If Φ is a variable A, then $w, \overline{\rho} \models \overline{A}$ if and only if for every $w' \geq w$, $w' \in \overline{\rho}(A)$, that is if and only if for every $w' \geq w$, for every $w'' \geq w'$, $w'' \in \rho(A)$; but, as $\leq$ is a preorder, this is equivalent to saying that for every $w' \geq w$, $w' \in \rho(A)$, i.e., $w, \rho \models \overline{A}$. The other cases are straightforward.

In particular, if Φ is intuitionistically valid, then for every assignment ρ, $\overline{\rho}$ is upward-closed, so necessarily $w, \overline{\rho} \models_i \Phi$ for every world w in any Kripke frame, so $w, \overline{\rho} \models \overline{\Phi}$, hence $w, \rho \models \overline{\Phi}$ for every w and every Kripke frame, that is, $\overline{\Phi}$ is valid in S4.

▶ **5.4.**

A occurs in the scope of two more negation signs in Φ', but this does not change the parity of the number of negation signs in which A is wrapped up.

Now, $\mu A \cdot \Phi$ is equivalent to $\Phi[(\mu A \cdot \Phi)/A]$, hence to $\Phi[\neg A/A][\neg(\mu A \cdot \Phi)/A]$, so $\neg \mu A \cdot \Phi$ is a fixed point of Φ'. Moreover, $\mu A \cdot \Phi$ is the least fixed point of Φ, i.e. from $\Psi \Leftrightarrow \Phi[\Psi/A]$ we can deduce $(\mu A \cdot \Phi) \Rightarrow \Psi$ for every formula Ψ. Therefore, replacing Ψ by $\neg\Psi$, from $\neg\Psi \Leftrightarrow \Phi[\neg\Psi/A]$ we can deduce $(\mu A \cdot \Phi) \Rightarrow \neg\Psi$. But $\neg\Psi \Leftrightarrow \Phi[\neg\Psi/A]$ is equivalent to $\Psi \Leftrightarrow \neg\Phi[\neg\Psi/A]$, which is $\Psi \Leftrightarrow \Phi'[\Psi/A]$. On the other hand, $(\mu A \cdot \Phi) \Rightarrow \neg\Psi$ is equivalent to $\neg(\mu A \cdot \Phi) \vee \neg\Psi$, hence to $\Psi \Rightarrow \neg \mu A \cdot \Phi$. So, from $\Psi \Leftrightarrow \Phi'[\Psi/A]$ we can deduce $\Psi \Rightarrow \neg \mu A \cdot \Phi$. (I.e., $\neg \mu A \cdot \Phi$ is a greatest fixed point of Φ' with respect to A.) But we could also deduce $\Psi \Rightarrow \nu A \cdot \Phi'$ by definition. Taking Ψ to be $\neg \mu A \cdot \Phi$, then $\nu A \cdot \Phi'$, shows that $(\neg \mu A \cdot \Phi)$ is logically equivalent to $\nu A \cdot \Phi'$.

By negating both sides of the equivalence, we conclude that $\mu A \cdot \Phi$ is logically equivalent to $\neg \nu A \cdot \Phi'$. This is the first desired equivalence. This also means that $\neg \mu A \cdot \neg\Phi[\neg A/A]$ is logically equivalent to $\nu A \cdot \neg\neg\Phi[\neg A/A][\neg A/A]$, hence to $\nu A \cdot \Phi$; this is the second equivalence that was asked for.

Moreover, this translation preserves the positivity of occurrences of variables.

Indeed, if B is a propositional variable occurring free in $\neg \nu A \cdot \Phi' = \neg \nu A \cdot \neg \Phi[\neg A/A]$, it is in the scope of two more negations than in $\mu A \cdot \Phi$, hence the parity is conserved. Similarly for $\neg \mu A \cdot \Phi'$ and $\nu A \cdot \Phi$.

Therefore, we could have defined only μ, and let $\nu A \cdot \Phi$ be an abbreviation for $\neg \mu A \cdot \neg \Phi[\neg A/A]$, or defined only ν, and let $\mu A \cdot \Phi$ be an abbreviation for $\neg \nu A \cdot \neg \Phi[\neg A/A]$.

▶ **5.5.**

Any HML formula can be seen as a PDL formula, where Hennessy-Milner transitions are PDL action constants. Then we show that any HML formula Φ is satisfiable in HML if and only if it is satisfiable in PDL. (From which, it follows that Φ is valid in HML if and only if it is valid in PDL.)

If Φ is satisfiable in HML, there is a labelled state-transition graph and an assignment of sets of states to propositional variables such that Φ holds at some state w_0 of the graph. But this graph defines a PDL model, where the relation R_a for each transition a is defined by $w \, R_a \, w'$ if and only if there is a transition labelled a from state w to state w'. By a straightforward structural induction on Φ, Φ also holds at the world w_0 of the latter PDL frame.

Conversely, if Φ is satisfiable in PDL, there is a PDL Kripke frame (W, R), where R is a set of binary relations R_a for each transition a, and an assignment under which Φ holds at some world w_0 in W. Define the labelled state-transition graph G as follows: the states are the worlds in W, and there is a transition from w to w' labelled a if and only if $w \, R_a \, w'$. By a straightforward structural induction on Φ, Φ holds at state w_0 in the HML frame G.

▶ **5.6.**

Choose an action constant a, and translate an S4 formula Φ to a PDL formula $\overline{\Phi}$ by structural recursion in the following way:

- if Φ is a propositional variable, then $\overline{\Phi} = \Phi$;

- if $\Phi = \Phi' \vee \Phi''$, then $\overline{\Phi} = \overline{\Phi'} \vee \overline{\Phi''}$ (and similarly for $\wedge$, $\neg$, $\Rightarrow$);

- if $\Phi = \Box \Phi'$, then $\overline{\Phi} = [a^*]\overline{\Phi'}$.

We show that the translation preserves validity (hence satisfiability).

If Φ is satisfiable in S4, there is a Kripke frame $(W, \leq)$, with $\leq$ a preorder, and an assignment under which Φ holds at some world w_0. Define a PDL frame (W, R) by letting R_a be $\leq$ itself. Notice that R_a is a preorder, so in particular $[\![a^*]\!]$ coincides with $[\![a]\!]$ on this frame. A straightforward structural induction on Φ then shows that Φ is satisfied at w_0 in the PDL frame (W, R).

Conversely, if Φ is satisfiable in PDL, there is a PDL frame (W, R) and an assignment such that Φ is satisfied at some world w_0. Define a preorder $\leq$ on W as the reflexive transitive closure of R_a. A straightforward induction on Φ then shows that Φ is satisfied at w_0 in the S4 frame $(W, \leq)$.

▶ **5.7.**

The trick is the following: as the Hilbert-style axioms show, a formula of the form $[P]\Phi$ can almost be expressed as a formula where the only modalities are $[a]$, where a is an action constant. The defect lies in programs of the form P^*, which cannot be expressed this way in PDL itself. However, we can do this by using fixed point operators. We define the translation from PDL formulas Φ to strong modal μ-calculus formulas $\overline{\Phi}$ in the following way:

- $\overline{A} = A$ for every propositional variable A;

- if $\Phi = \Phi' \vee \Phi''$, then $\overline{\Phi} = \overline{\Phi'} \vee \overline{\Phi''}$, and similarly for $\neg, \wedge, \Rightarrow$;

- if $\Phi = [a]\Phi'$, where a is an action constant, we let $\overline{\Phi}$ be $[a]\overline{\Phi'}$;

- if $\Phi = [P \cup P']\Phi'$, then $\overline{\Phi} = \overline{[P]\Phi'} \wedge \overline{[P']\Phi'}$;

- if $\Phi = [P; P']\Phi'$, then $\overline{\Phi} = \overline{[P][P']\Phi'}$;

- if $\Phi = [\Psi?]\Phi'$, then $\overline{\Phi} = \overline{\Psi} \Rightarrow \overline{\Phi'}$;

- if $\Phi = [P^*]\Phi'$, then $\overline{\Phi} = \nu A \cdot \overline{\Phi'} \wedge \overline{[P]A}$, where A is a variable not free in Φ';

This process terminates. Indeed, define the degree $d(\Phi)$ (resp. $d(P)$) of a PDL formula Φ (resp. program P) by structural recursion in the following way: $d(A) = 0$ for all variables A, $d(\Phi' \vee \Phi'') = \max(d(\Phi'), d(\Phi'')) + 1$ (and similarly for $\wedge, \neg, \Rightarrow$), $d([P]\Phi') = 1 + d(P) + d(\Phi')$, $d(a) = 0$ for all action constants a, $d(P \cup P') = 1 + \max(d(P), d(P'))$, $d(P; P') = 2 + d(P) + d(P')$, $d(\Psi?) = d(\Psi)$, $d(P^*) = 1 + d(P)$. Then the procedure above defines $\overline{\Phi}$ in terms of a finite number of expressions of the form $\overline{\Phi'}$, where $d(\Phi') < d(\Phi)$. Therefore, the recursion stops at some finite level, so the translation terminates.

Finally, $\overline{\Phi}$ has only finitely many transition letters (the action constants). We now show that Φ is satisfiable in PDL if and only if $\overline{\Phi}$ is satisfiable in the strong modal μ-calculus.

We first need to make the following remarks:

1. The translation commutes with substitution, i.e., $\overline{\Psi[\Psi'/B]} = \overline{\Psi}[\overline{\Psi'}/B]$. (A straightforward induction on $d(\Phi)$.)

2. In every μ-calculus frame, $w, \rho \models \nu A \cdot \overline{\Phi'} \wedge \overline{[P]A}$ (where $A \notin \mathrm{fv}(\Phi')$) if and only if for every integer n, $w, \rho \models \overline{[P]^n \Phi'}$ (where $[P]^n$ abbreviates $\underbrace{[P] \ldots [P]}_{n \text{ times}}$).

Indeed, assume $w, \rho \models \nu A \cdot \overline{\Phi'} \wedge \overline{[P]A}$. Notice that $\nu A \cdot \overline{\Phi'} \wedge \overline{[P]A}$ is equivalent to $\overline{\Phi'} \wedge [P]\nu A \cdot \overline{\Phi'} \wedge \overline{[P]A}$ since ν is a fixed point operator. In particular, it implies $\overline{\Phi'}$; also, it implies $[P]\nu A \cdot \overline{\Phi'} \wedge \overline{[P]A}$, so that whenever

$\nu A \cdot \overline{\Phi'} \wedge \overline{[P]A}$ implies Ψ, it also implies $\overline{[P]\Psi}$. By induction, it must imply every $\overline{[P]^n \Phi'}$, so that $w, \rho \models \overline{[P]^n \Phi'}$ for every $n \in \mathbb{N}$.

Conversely, assume that $w, \rho \models \overline{[P]^n \Phi'}$ for every integer n. We let ρ' be the assignment that is identical with ρ on all variables but A, and which maps A to $\bigcap_{n \in \mathbb{N}} \{w' \mid w', \rho \models \overline{[P]^n \Phi'}\}$. Because A is not free in $\overline{[P]^n \Phi'}$, we have $w, \rho' \models \overline{[P]^n \Phi'}$ for every integer n. Now, $w, \rho' \models \overline{[P]^{n+1} \Phi'}$ holds for every $n \in \mathbb{N}$, so $w, \rho' \models \overline{[P][P]^n \Phi'}$ holds, i.e. $w, \rho' \models \overline{[P]A}$. Then, $w, \rho' \models A$ because $w \in \rho'(A)$ by construction, and $w, \rho' \models \overline{\Phi'}$ by assumption, so $w, \rho' \models A \Leftrightarrow \overline{\Phi'} \wedge \overline{[P]A}$. We can therefore apply the rule defining ν as a greatest fixed point, and infer $w, \rho' \models A \Rightarrow \nu A \cdot \overline{\Phi'} \wedge \overline{[P]A}$. But, as $w, \rho' \models A$, this means that $w, \rho' \models \nu A \cdot \overline{\Phi'} \wedge \overline{[P]A}$, hence $w, \rho \models \nu A \cdot \overline{\Phi'} \wedge \overline{[P]A}$ since A is not free in the latter formula.

If Φ is satisfiable in PDL, then there is a PDL frame (W, R) and an assignment under which Φ holds at some world w_0. Recall that R is a set of binary relations R_a for each action constant a. Define the labelled state-transition graph G as follows: the states are the worlds in W, and there is a transition from w to w' labelled a if and only if $w R_a w'$. By induction on $d(\Phi)$, we show that $w, \rho \models \Phi$ if and only if $w, \rho \models \overline{\Phi}$, so that Φ holds at state w_0 in the strong modal μ-calculus frame G. The only non-obvious step is when $\Phi = [P^*]\Phi'$, but then $w, \rho \models [P^*]\Phi'$ if and only if, for every $n \in \mathbb{N}$, $w, \rho \models [P]^n \Phi'$ by definition, which is equivalent to saying $w, \rho \models \nu A \cdot \overline{\Phi'} \wedge \overline{[P]A}$ by induction hypothesis and Remark 2 above.

Conversely, if $\overline{\Phi}$ is satisfiable in the strong modal μ-calculus, then there is a state-transition graph and an assignment under which $\overline{\Phi}$ holds at some world w_0. We let (W, R) be the PDL frame defined by taking W equal to the set of states and letting R_a be such that $w R_a w'$ if and only if there is a transition labelled a from w to w'. The proof that Φ holds at world w_0 is again a straightforward induction on $d(\Phi)$, using Remark 2 for the $\Phi = [P^*]\Phi'$ case.

▶ **5.8.**

$\forall A \cdot \Phi$ is defined in the same way as substitution:

- $\forall A \cdot \mathbf{T} = \mathbf{T}, \forall A \cdot \mathbf{F} = \mathbf{F}$;

- if $\Phi = A \longrightarrow \Phi_+; \Phi_-$, then $\forall A \cdot \Phi = \Phi_+ \wedge \Phi_-$;

- if $\Phi = A' \longrightarrow \Phi_+ : \Phi_-$, with $A' \neq A$, then $\forall A \cdot \Phi = \texttt{BDDmake}(A', \forall A \cdot \Phi_+, \forall A \cdot \Phi_-)$.

Symmetrically, we could define $\exists A \cdot \Phi$ as the above, replacing $\wedge$ by $\vee$, but it is more economical to define it as $\neg \forall A \cdot \neg \Phi$, because then the computation will share its cache with that of $\forall$.

▶ **5.9.**

A first way is to define $\Phi[\Phi'/A]$ as $\forall A \cdot (\Phi' \Leftrightarrow A) \Rightarrow \Phi$.

Indeed, by Shannon's decomposition theorem, we can write Φ as $(A \Rightarrow \Phi[\mathbf{T}/A]) \wedge (\neg A \Rightarrow \Phi[\mathbf{F}/A])$, then $\Phi[\Phi'/A]$ is $(\Phi' \Rightarrow \Phi[\mathbf{T}/A]) \wedge (\neg \Phi' \Rightarrow \Phi[\mathbf{F}/A])$, which is equivalent to $((\Phi' \Leftrightarrow \mathbf{T}) \Rightarrow \Phi[\mathbf{T}/A]) \wedge ((\Phi' \Leftrightarrow \mathbf{F}) \Rightarrow \Phi[\mathbf{F}/A])$, i.e. $\forall A \cdot (\Phi' \Leftrightarrow A) \Rightarrow \Phi$.

Another way is to define it as $\exists A \cdot (\Phi' \Leftrightarrow A) \wedge \Phi$, using Shannon's dual decomposition to justify it (Exercise 2.26).

We can then use the results of Exercise 5.8 to compute substitutions on BDDs in another way than given in the text.

▸ **6.1.**

We must first axiomatise the accessibility relation, which is a preorder $\leq$ on worlds. To express that $\leq$ is reflexive, we may think that we need an additional equality symbol, axiomatised by the equality axioms. This is not the case, and we can axiomatise $\leq$ by: $\forall w \cdot w \leq w$ (reflexivity) and $\forall x \cdot \forall w' \cdot \forall w'' \cdot w \leq w' \wedge w' \leq w'' \Rightarrow w \leq w''$ (transitivity).

Then, we translate modal formulas Φ to first-order formulas $\Phi(w)$ having exactly one free world variable in the following way:

- propositional variables A translate to $\hat{A}(w)$, where $\hat{A}$ is a unary predicate symbol;

- conjunctions $\Phi' \wedge \Phi''$ translate to $\Phi'(w) \wedge \Phi''(w)$, and similarly for disjunctions, implications and negations;

- finally, $\Box\Phi$ translates to $\forall w' \cdot w \leq w' \Rightarrow \Phi(w')$.

Finally, Φ is valid in S4 if and only if $\forall w \cdot \Phi(w)$ is valid in first-order logic. Similarly, Φ is satisfiable in S4 if and only if $\exists w \cdot \Phi(w)$ is satisfiable in first-order logic.

▸ **6.2.**

That D is a filter is a routine check. We then apply the ultrafilter existence theorem to conclude that $\mathcal{F}$ exists.

I is called an *ultrapower* construction, because it is essentially a product of $\aleph_0$ copies of the model $\mathbb{N}$, indexed by an ultrafilter. Let I_0 denote the standard interpretation of $\mathbf{PA}_1$ over $\mathbb{N}$.

Let ρ denote an arbitrary assignment on N, and for each $n \in \mathbb{N}$, let ρ_n denote the assignment on $\mathbb{N}$ mapping each variable x to $\rho(x)(n)$. We show by structural induction on any first-order formula Φ that $[\![\Phi]\!]I\rho$ holds if and only if $\{n \mid [\![\Phi]\!]I_0\rho_n\}$ is in $\mathcal{F}$.

If Φ is atomic, i.e. $\Phi = P(t_1, \ldots, t_m)$, then $[\![\Phi]\!]I\rho$ holds whenever $\{n \mid [\![\Phi]\!]I_0\rho_n\}$ is in $\mathcal{F}$, by construction.

If $\Phi = \Phi' \wedge \Phi''$, then if $[\![\Phi]\!]I\rho$ holds, so do $[\![\Phi']\!]I\rho$ and $[\![\Phi'']\!]I\rho$, so by induction hypothesis $\{n \mid [\![\Phi']\!]I_0\rho_n\}$ and $\{n \mid [\![\Phi'']\!]I_0\rho_n\}$ are both in $\mathcal{F}$. Because $\mathcal{F}$ is a filter (and is thus stable by intersection), their intersection is also in $\mathcal{F}$, i.e. $\{n \mid [\![\Phi]\!]I_0\rho_n\}$ is in $\mathcal{F}$. Conversely, if $[\![\Phi]\!]I\rho$ does not hold then either $[\![\Phi']\!]I\rho$ or $[\![\Phi'']\!]I\rho$ does not hold, say the former. By induction hypothesis $\{n \mid [\![\Phi']\!]I_0\rho_n\}$ is not in $\mathcal{F}$. Because

$\mathcal{F}$ is a filter (and is thus stable by supersets), its subset $\{n \mid [\![\Phi]\!]I_0\rho_n\}$ cannot be in $\mathcal{F}$ either.

If $\Phi = \neg\Phi'$, then if $[\![\Phi]\!]I\rho$ holds, then by induction hypothesis $\{n \mid [\![\Phi']\!]I_0 \rho_n\}$ is not in $\mathcal{F}$. Because $\mathcal{F}$ is an ultrafilter, its complement $\{n \mid \text{not } [\![\Phi']\!]I_0\rho_n\}$ is in $\mathcal{F}$, i.e. $\{n \mid [\![\Phi]\!]I_0\rho_n\}$ is in $\mathcal{F}$. Conversely, if $[\![\Phi]\!]I\rho$ does not hold, then by induction hypothesis $\{n \mid [\![\Phi']\!]I_0\rho_n\}$ is in $\mathcal{F}$; then its complement cannot be in $\mathcal{F}$ (because otherwise $\varnothing$ would be in $\mathcal{F}$), so $\{n \mid [\![\Phi]\!]I_0\rho_n\}$ is not in $\mathcal{F}$.

The cases of disjunctions and implications are similar.

If $\Phi = \forall x \cdot \Phi'$, then if $[\![\Phi]\!]I\rho$ holds, so does $[\![\Phi']\!]I(\rho[v/x])$ for every $v \in N$. Now, $\{n \mid [\![\Phi]\!]I_0\rho_n\} = \{n \mid \forall v_n \in \mathbb{N} \cdot [\![\Phi']\!]I_0(\rho_n[v_n/x])\} \supseteq \{n \mid \forall v \in N \cdot [\![\Phi']\!]I_0(\rho_n[v(n)/x])\}$, which is therefore in $\mathcal{F}$. Conversely, if $[\![\Phi]\!]I\rho$ does not hold, then there is a function v in N such that $[\![\Phi']\!]I(\rho[v/x])$ does not hold. By induction hypothesis, $\{n \mid [\![\Phi']\!]I_0(\rho_n[v(n)/x])\}$ is not in $\mathcal{F}$, so $\{n \mid \text{not } [\![\Phi']\!]I_0(\rho_n[v(n)/x])\}$ is in $\mathcal{F}$. The intersection of the latter with $\{n \mid [\![\Phi]\!]I_0\rho_n\}$ is empty by definition of Φ, so $\{n \mid [\![\Phi]\!]I_0\rho_n\}$ cannot be in $\mathcal{F}$.

The case of existential quantifications is similar.

Now let Φ be any axiom of $\mathbf{PA}_1$: because $\mathbb{N}$ is a model of $\mathbf{PA}_1$, $\{n \mid [\![\Phi]\!]I_0 \rho_n\} = \mathbb{N}$ for every ρ. By the above considerations, and since $\mathbb{N} \in \mathcal{F}$, it follows that $[\![\Phi]\!]I\rho$ holds. So N is a model of $\mathbf{PA}_1$.

We embed $\mathbb{N}$ in N by mapping every integer m to $\lambda n \cdot m$ (a constant function). It is readily seen that this mapping is a morphism for all arithmetical operations (0, s, $+$, $*$, $\leq$, $\doteq$), and that this function is injective. So this is really an embedding. Let $\overline{m}$ denote the integer m as viewed through this embedding, namely $\overline{m} = \lambda n \cdot m$, and let $\underline{m}$ denote the term $\underbrace{s(\ldots s(0)\ldots)}_{m \text{ times}}$.

Let c be the element $\lambda n \cdot n$ in N. For each $m \in \mathbb{N}$, the set $\{n \mid \overline{m}n \leq n\}$ is the complement of the finite set $\{0, 1, \ldots, m-1\}$, so it is in D, hence in $\mathcal{F}$. That is, letting ρ be any assignment, $\{n \mid [\![\hat{m} \leq x]\!]I_0(\rho[c/x])_n\}$ in in $\mathcal{F}$. In particular, $\hat{m} \leq x$ holds under any interpretation mapping x to c, for any integer m. Intuitively, c is an element of N that is greater than any (standard) integer. c is in fact called a *non-standard* integer. It is easily checked that no such element exists in $\mathbb{N}$.

One objection to this construction is that $\doteq$ is not interpreted as equality. The remedy is simple: take all equivalence classes of elements of N modulo $I(\doteq)$, which is an equivalence relation. The element c above cannot be made equal to any (standard) integer, so the quotient set is not isomorphic to $\mathbb{N}$ either.

This construction is not particular to $\mathbb{N}$. In general, take any first-order theory T with a model M. Let I be an infinite index set, and define a (non-standard) model as follows: the domain if the set M^I of all functions from I to M, and $[\![\Phi]\!]I\rho$ if and only if $\{n \in I \mid [\![\Phi]\!]I\rho_n\}$ is in some ultrafilter $\mathcal{F}$ containing all complements of all finite subsets of I. Then the latter interpretation is also a model of T, but it is strictly larger.

▶ **6.3.**

-
$$\cfrac{\cfrac{\cfrac{\cfrac{\cfrac{\cfrac{}{\Psi \longrightarrow \Psi}\;\text{Ax}}{\longrightarrow \Psi, \neg\Psi}\;\neg\text{R}}{\longrightarrow \Psi, \exists x \cdot \neg\Psi}\;\exists\text{R}}{\longrightarrow \forall x \cdot \Psi, \exists x \cdot \neg\Psi}\;\forall\text{R}}{\neg\forall x \cdot \Psi \longrightarrow \exists x \cdot \neg\Psi}\;\neg\text{L}}$$

$$\cfrac{\cfrac{\cfrac{\cfrac{\cfrac{\cfrac{}{\Psi \longrightarrow \Psi}\;\text{Ax}}{\neg\Psi, \Psi \longrightarrow}\;\neg\text{L}}{\neg\Psi, \forall x \cdot \Psi \longrightarrow}\;\forall\text{L}}{\exists x \cdot \neg\Psi, \forall x \cdot \Psi \longrightarrow}\;\exists\text{L}}{\exists x \cdot \neg\Psi \longrightarrow \neg\forall x \cdot \Psi}\;\neg\text{R}}$$

-
$$\cfrac{\cfrac{\cfrac{\cfrac{\cfrac{\cfrac{}{\Psi \longrightarrow \Psi}\;\text{Ax}}{\longrightarrow \Psi, \neg\Psi}\;\neg\text{R}}{\longrightarrow \exists x \cdot \Psi, \neg\Psi}\;\exists\text{R}}{\longrightarrow \exists x \cdot \Psi, \forall x \cdot \neg\Psi}\;\forall\text{R}}{\neg\exists x \cdot \Psi \longrightarrow \forall x \cdot \neg\Psi}\;\neg\text{L}}$$

$$\cfrac{\cfrac{\cfrac{\cfrac{\cfrac{\cfrac{}{\Psi \longrightarrow \Psi}\;\text{Ax}}{\neg\Psi, \Psi \longrightarrow}\;\neg\text{L}}{\forall x \cdot \neg\Psi, \Psi \longrightarrow}\;\forall\text{L}}{\forall x \cdot \neg\Psi, \exists x \cdot \Psi \longrightarrow}\;\exists\text{L}}{\forall x \cdot \neg\Psi \longrightarrow \neg\exists x \cdot \Psi}\;\neg\text{R}}$$

-
$$\cfrac{\cfrac{\cfrac{\cfrac{\cfrac{}{\Psi[x'/x], \Psi' \longrightarrow \Psi[x'/x]}\;\text{Ax} \qquad \cfrac{}{\Psi[x'/x], \Psi' \longrightarrow \Psi'}\;\text{Ax}}{\Psi[x'/x], \Psi' \longrightarrow \Psi[x'/x] \wedge \Psi'}\;\wedge\text{R}}{\forall x \cdot \Psi, \Psi' \longrightarrow \Psi[x'/x] \wedge \Psi'}\;\forall\text{L}}{(\forall x \cdot \Psi) \wedge \Psi' \longrightarrow \Psi[x'/x] \wedge \Psi'}\;\wedge\text{L}}{(\forall x \cdot \Psi) \wedge \Psi' \longrightarrow \forall x' \cdot \Psi[x'/x] \wedge \Psi'}\;\forall\text{R}}$$

and:

$$\cfrac{\cfrac{\cfrac{\cfrac{\cfrac{}{\Psi, \Psi' \longrightarrow \Psi}\;\text{Ax}}{\Psi \wedge \Psi' \longrightarrow \Psi}\;\wedge\text{L}}{\forall x' \cdot \Psi[x'/x] \wedge \Psi' \longrightarrow \Psi}\;\forall\text{L}}{\forall x' \cdot \Psi[x'/x] \wedge \Psi' \longrightarrow \forall x \cdot \Psi}\;\forall\text{R} \qquad \cfrac{\cfrac{\cfrac{}{\Psi, \Psi' \longrightarrow \Psi'}\;\text{Ax}}{\Psi \wedge \Psi' \longrightarrow \Psi'}\;\wedge\text{L}}{\forall x' \cdot \Psi[x'/x] \wedge \Psi' \longrightarrow \Psi'}\;\forall\text{L}}{\forall x' \cdot \Psi[x'/x] \wedge \Psi' \longrightarrow (\forall x \cdot \Psi) \wedge \Psi'}\;\wedge\text{R}$$

where we play on the fact that $x' \notin \text{fv}(\Psi')$ in all uses of ∀R (and similarly to prove $\forall x' \cdot \Psi \wedge \Psi'[x'/x]$ equivalent to $\Psi \wedge (\forall x \cdot \Psi')$).

•

$$\dfrac{\dfrac{}{\Psi,\Psi'\longrightarrow\Psi}\text{ Ax}\qquad\dfrac{}{\Psi,\Psi'\longrightarrow\Psi'}\text{ Ax}}{\dfrac{\dfrac{\dfrac{\dfrac{\Psi,\Psi'\longrightarrow\Psi\wedge\Psi'}{\Psi,\Psi'\longrightarrow\exists x'\cdot\Psi[x'/x]\wedge\Psi'}\text{ }\exists R}{\exists x\cdot\Psi,\Psi'\longrightarrow\exists x'\cdot\Psi[x'/x]\wedge\Psi'}\text{ }\exists L}{\exists x\cdot\Psi)\wedge\Psi'\longrightarrow\exists x'\cdot\Psi[x'/x]\wedge\Psi'}\text{ }\exists L}{}}$$

and:

$$\dfrac{\dfrac{\dfrac{\dfrac{\dfrac{}{\Psi[x'/x],\Psi'\longrightarrow\Psi[x'/x]}\text{ Ax}}{\Psi[x'/x]\wedge\Psi'\longrightarrow\Psi[x'/x]}\text{ }\wedge L}{\Psi[x'/x]\wedge\Psi'\longrightarrow\exists x\cdot\Psi}\text{ }\exists R}{\exists x'\cdot\Psi[x'/x]\wedge\Psi'\longrightarrow\exists x\cdot\Psi}\text{ }\exists L\qquad\dfrac{\dfrac{\dfrac{}{\Psi[x'/x],\Psi'\longrightarrow\Psi'}\text{ Ax}}{\Psi[x'/x]\wedge\Psi'\longrightarrow\Psi'}\text{ }\wedge L}{\exists x'\cdot\Psi[x'/x]\wedge\Psi'\longrightarrow\Psi'}\text{ }\exists L}{\exists x'\cdot\Psi[x'/x]\wedge\Psi'\longrightarrow(\exists x\cdot\Psi)\wedge\Psi'}\text{ }\wedge R$$

where we play on the fact that $x'\notin\text{fv}(\Psi')$ in all uses of $\exists L$ (and similarly to prove $\Psi\wedge(\exists x\cdot\Psi')$ equivalent to $\exists x'\cdot\Psi\wedge\Psi'[x'/x]$).

•

$$\dfrac{\dfrac{\dfrac{\dfrac{\dfrac{}{\Psi[x'/x]\longrightarrow\Psi[x'/x],\Psi'}\text{ Ax}}{\Psi[x'/x]\longrightarrow\Psi[x'/x]\vee\Psi'}\text{ }\vee R}{\forall x\cdot\Psi\longrightarrow\Psi[x'/x]\vee\Psi'}\text{ }\forall L}{\forall x\cdot\Psi\longrightarrow\forall x'\cdot\Psi[x'/x]\vee\Psi'}\text{ }\forall R\qquad\dfrac{\dfrac{\dfrac{}{\Psi'\longrightarrow\Psi[x'/x],\Psi'}\text{ Ax}}{\Psi'\longrightarrow\Psi[x'/x]\vee\Psi'}\text{ }\vee R}{\Psi'\longrightarrow\forall x'\cdot\Psi[x'/x]\vee\Psi'}\text{ }\forall R}{(\forall x\cdot\Psi)\vee\Psi'\longrightarrow\forall x'\cdot\Psi[x'/x]\vee\Psi'}\text{ }\vee L$$

and:

$$\dfrac{\dfrac{}{\Psi\longrightarrow\Psi,\Psi'}\text{ Ax}\qquad\dfrac{}{\Psi'\longrightarrow\Psi,\Psi'}\text{ Ax}}{\dfrac{\dfrac{\dfrac{\dfrac{\Psi\vee\Psi'\longrightarrow\Psi,\Psi'}{\forall x'\cdot\Psi[x'/x]\vee\Psi'\longrightarrow\Psi,\Psi'}\text{ }\forall L}{\forall x'\cdot\Psi[x'/x]\vee\Psi'\longrightarrow\forall x\cdot\Psi,\Psi'}\text{ }\forall R}{\forall x'\cdot\Psi[x'/x]\vee\Psi'\longrightarrow(\forall x\cdot\Psi)\vee\Psi'}\text{ }\vee R}{}}\text{ }\vee L$$

where we play on the fact that $x'\notin\text{fv}(\Psi')$ in all uses of $\forall R$ (and similarly to prove $\Psi\vee(\forall x\cdot\Psi')$ equivalent to $\forall x'\cdot\Psi\vee\Psi'[x'/x]$).

●

$$
\dfrac{\dfrac{\dfrac{\dfrac{\overline{\Psi \longrightarrow \Psi, \Psi'}\ \text{Ax}}{\Psi \longrightarrow \Psi \vee \Psi'}\ \vee\text{R}}{\Psi \longrightarrow \exists x' \cdot \Psi[x'/x] \vee \Psi'}\ \exists\text{R}}{\exists x \cdot \Psi \longrightarrow \exists x' \cdot \Psi[x'/x] \vee \Psi'}\ \exists\text{L} \qquad \dfrac{\dfrac{\overline{\Psi' \longrightarrow \Psi, \Psi'}\ \text{Ax}}{\Psi' \longrightarrow \Psi \vee \Psi'}\ \vee\text{R}}{\Psi' \longrightarrow \exists x' \cdot \Psi[x'/x] \vee \Psi'}\ \exists\text{R}}{(\exists x \cdot \Psi) \vee \Psi' \longrightarrow \exists x' \cdot \Psi[x'/x] \vee \Psi'}\ \vee\text{L}
$$

and:

$$
\dfrac{\dfrac{\dfrac{\dfrac{\dfrac{\overline{\Psi[x'/x] \longrightarrow \Psi[x'/x], \Psi'}\ \text{Ax} \qquad \overline{\Psi' \longrightarrow \Psi[x'/x], \Psi'}\ \text{Ax}}{\Psi[x'/x] \vee \Psi' \longrightarrow \Psi[x'/x], \Psi'}\ \vee\text{L}}{\Psi[x'/x] \vee \Psi' \longrightarrow \exists x \cdot \Psi, \Psi'}\ \exists\text{R}}{\exists x' \cdot \Psi[x'/x] \vee \Psi' \longrightarrow \exists x \cdot \Psi, \Psi'}\ \exists\text{L}}{\exists x' \cdot \Psi[x'/x] \vee \Psi' \longrightarrow (\exists x \cdot \Psi) \vee \Psi'}\ \vee\text{R}}
$$

where we play on the fact that $x' \notin \mathrm{fv}(\Psi')$ in all uses of $\exists$L (and similarly to prove $\Psi \vee (\exists x \cdot \Psi')$ equivalent to $\exists x' \cdot \Psi \vee \Psi'[x'/x]$).

The additional rules for conversion to prenex form are justified as follows:

●

$$
\dfrac{\dfrac{\dfrac{\dfrac{\dfrac{\overline{\Psi, \Psi'[x/x'] \longrightarrow \Psi}\ \text{Ax} \qquad \overline{\Psi, \Psi'[x/x'] \longrightarrow \Psi'[x/x']}\ \text{Ax}}{\Psi, \Psi'[x/x'] \longrightarrow \Psi \wedge \Psi'[x/x']}\ \wedge\text{R}}{\Psi, \forall x' \cdot \Psi' \longrightarrow \Psi \wedge \Psi'[x/x']}\ \forall\text{L}}{\forall x \cdot \Psi, \forall x' \cdot \Psi' \longrightarrow \Psi \wedge \Psi'[x/x']}\ \forall\text{L}}{(\forall x \cdot \Psi) \wedge (\forall x' \cdot \Psi') \longrightarrow \Psi \wedge \Psi'[x/x']}\ \wedge\text{L}}{\forall x \cdot \Psi) \wedge (\forall x' \cdot \Psi') \longrightarrow \forall x \cdot \Psi \wedge \Psi'[x/x']}\ \forall\text{R}
$$

and:

$$
\dfrac{\dfrac{\dfrac{\dfrac{\overline{\Psi, \Psi'[x/x'] \longrightarrow \Psi}\ \text{Ax}}{\Psi \wedge \Psi'[x/x'] \longrightarrow \Psi}\ \wedge\text{L}}{\forall x \cdot \Psi \wedge \Psi'[x/x'] \longrightarrow \Psi}\ \forall\text{L}}{\forall x \cdot \Psi \wedge \Psi'[x/x'] \longrightarrow \forall x \cdot \Psi}\ \forall\text{R} \qquad \dfrac{\dfrac{\dfrac{\overline{\Psi[x'/x], \Psi' \longrightarrow \Psi'}\ \text{Ax}}{\Psi[x'/x] \wedge \Psi' \longrightarrow \Psi'}\ \wedge\text{L}}{\forall x \cdot \Psi \wedge \Psi'[x, x'] \longrightarrow \Psi'}\ \forall\text{L}}{\forall x \cdot \Psi \wedge \Psi'[x, x'] \longrightarrow \forall x' \cdot \Psi'}\ \forall\text{R}}{\forall x \cdot \Psi \wedge \Psi'[x/x'] \longrightarrow (\forall x \cdot \Psi) \wedge (\forall x' \cdot \Psi')}\ \wedge\text{R}
$$

where the fact that $x' \notin \mathrm{fv}(\Phi')$ has been used in the $\forall$L rule on the right-hand side.

$$\dfrac{\dfrac{}{\Psi \longrightarrow \Psi, \Psi'[x/x']}\ \mathrm{Ax} \qquad \dfrac{}{\Psi'[x/x'] \longrightarrow \Psi, \Psi'[x/x']}\ \mathrm{Ax}}{\dfrac{\dfrac{\Psi \vee \Psi'[x/x'] \longrightarrow \Psi, \Psi'[x/x']}{\dfrac{\Psi \vee \Psi'[x/x'] \longrightarrow \Psi, \exists x' \cdot \Psi'}{\dfrac{\Psi \vee \Psi'[x/x'] \longrightarrow \exists x \cdot \Psi, \exists x' \cdot \Psi'}{\dfrac{\Psi \vee \Psi'[x/x'] \longrightarrow (\exists x \cdot \Psi) \vee (\exists x' \cdot \Psi')}{\exists x \cdot \Psi \vee \Psi'[x/x'] \longrightarrow (\exists x \cdot \Psi) \vee (\exists x' \cdot \Psi')}\ \exists \mathrm{L}}\ \vee \mathrm{R}}\ \exists \mathrm{R}}\ \exists \mathrm{R}}\ \vee \mathrm{L}}$$

and:

$$\dfrac{\dfrac{\dfrac{\dfrac{}{\Psi \longrightarrow \Psi, \Psi'[x/x']}\ \mathrm{Ax}}{\dfrac{\Psi \longrightarrow \Psi \vee \Psi'[x/x']}{\Psi \longrightarrow \exists x \cdot \Psi \vee \Psi'[x/x']}\ \exists \mathrm{R}}\ \vee \mathrm{R}}{\exists x \cdot \Psi \longrightarrow \exists x \cdot \Psi \vee \Psi'[x/x']}\ \exists \mathrm{L} \qquad \dfrac{\dfrac{\dfrac{\dfrac{}{\Psi' \longrightarrow \Psi[x'/x], \Psi'}\ \mathrm{Ax}}{\dfrac{\Psi' \longrightarrow \Psi[x'/x] \vee \Psi'}{\Psi' \longrightarrow \exists x \cdot \Psi \vee \Psi'[x, x']}\ \exists \mathrm{R}}\ \vee \mathrm{R}}{\longrightarrow \exists x \cdot \Psi \vee \Psi'[x, x']}\ \exists \mathrm{L}}{(\exists x \cdot \Psi) \vee (\exists x' \cdot \Psi') \longrightarrow \exists x \cdot \Psi \vee \Psi'[x/x']}\ \vee \mathrm{L}$$

where the fact that $x' \notin \mathrm{fv}(\Phi')$ has been used in the $\exists \mathrm{R}$ rule on the right-hand side.

▶ **6.4.**

We show by structural induction on the formula Φ that:

1. $\Phi \Rightarrow h(\Phi)$ holds in every interpretation, for some assignment;

2. $s(\Phi) \Rightarrow \Phi$ holds in every interpretation, for every assignment;

3. if $h(\Phi)$ holds in every interpretation under some assignment, then Φ holds for every interpretation under some assignment;

4. if Φ holds in some interpretation for all assignments, then $s(\Phi)$ holds in some interpretation for all assignments.

The only non-trivial steps are when Φ is $\forall x \cdot \Phi'$ or $\exists x \cdot \Phi'$. We deal with the latter, as the former is dual by negation. We have:

1. for every interpretation, if under some assignment ρ, $\Phi' \Rightarrow h(\Phi')$ holds, we infer $\Phi' \Rightarrow h(\Phi)$ since $h(\Phi) = h(\Phi')$; but since $\Phi = \exists x \cdot \Phi'$, $\Phi \Rightarrow \Phi'$ holds under the assignment $\rho[v/x]$ for some value v in the domain of the interpretation, hence $\Phi \Rightarrow h(\Phi)$ holds under the assignment $\rho[v/x]$.

2. from $s(\Phi') \Rightarrow \Phi'$, we infer $s(\Phi')[f(x_1, \ldots, x_n)/x] \Rightarrow \Phi'[f(x_1, \ldots, x_n)/x]$; moreover $\Phi'[f(x_1, \ldots, x_n)/x] \Rightarrow \exists x \cdot \Phi'$ holds, so $s(\Phi')[f(x_1, \ldots, x_n)/x] \Rightarrow \exists x \cdot \Phi'$ holds, i.e. $s(\Phi) \Rightarrow \Phi$ holds.

3. if $h(\Phi)$ holds in every interpretation for some assignment, then $h(\Phi')$ does, too. By induction hypothesis, Φ' holds in every interpretation for some assignment of $x_1, \ldots, x_n, x$, so in particular Φ holds for some assignment of $x_1, \ldots, x_n$.

4. if Φ holds in some interpretation I for all assignments, then for all assignments mapping $x_1, \ldots, x_n$ to respective values $v_1, \ldots, v_n$, there is a value v for x such that Φ' holds in this interpretation. Let G be the function mapping n-tuples $(v_1, \ldots, v_n)$ to the set of values v. By the Axiom of Choice, there is a function g such that for every n-tuple $(v_1, \ldots, v_n)$, $g(v_1, \ldots, v_n) \in G(v_1, \ldots, v_n)$. Build an interpretation I' extending I by interpreting the function symbol f as the above function g. By construction, $\Phi'[f(x_1, \ldots, x_n)/x]$ holds for all assignments in I', hence $s(\Phi')[f(x_1, \ldots, x_n)/x]$, i.e., $s(\Phi)$, holds for all assignments in I' by induction hypothesis.

Notice that, if Φ' and Φ'' are two equivalent quantified subformulas for which we have to create Skolem symbols, the function F above will be the same for both in every interpretation, so we can choose their Skolem symbols to denote the same function f. In particular, we can safely choose the same Skolem symbol for both. A simple application of this remark is, instead of creating fresh function symbols, to create the symbol f as the textual representation of Φ' (resp. Φ'') normalized so that variables have canonical names. (We can push this further by first simplifying Φ' and Φ'', say by using BDDs and canonical representations for variables.)

▶ **6.5.**

Use the Compactness Theorem. Let S be any finite subset of T. S contains only finitely many of the axioms $\hat{m} \overset{.}{\leq} c$, so S is satisfied in the standard interpretation of arithmetic extended so that c's value if any integer greater than or equal to the maximum of all $\hat{m}$ occurring in S. So S is consistent. Because S is arbitrary, every finite subset of T is consistent, so by compacity T is consistent. Any model of T is of course a model of $\mathbf{PA}_1$, and is also non-standard because it contains an element c greater than any (standard) integer. Moreover, this model induces a countable Herbrand model, whereas the model of Exercise 6.2 is uncountable.

▶ **6.6.**

Skolemize the formula to prove: this means replacing the x_is by new constants $c_i, 1 \leq i \leq n$. Now, the Herbrand universe $H(B_0)$ is finite: its cardinal is $\max(1, n)$. (To simplify, assume $n \geq 1$, so that the cardinal is n.) There are therefore at most n^m closed substitutions of domain $\{y_1, \ldots, y_m\}$, hence at most n^m closed instances of Ψ. By Herbrand's Theorem, the formula is unsatisfiable if and only if some conjunction of closed instances of Ψ is propositionally unsatisfiable, or equivalently, if the conjunction of all of its finitely many closed instances is propositionally unsatisfiable. The latter is then decidable, by enumerating all Herbrand interpretations on the finite number of closed atoms that we get.

▶ **6.7.**

If $\exists x_1 \cdots \exists x_n \cdot \Psi$ is provable, it has a cut-free proof, and the last n rules must therefore be $\exists$R. This directly shows the required result.

▶ **6.8.**

There are eleven impermutabilities in **LJ**, eight of which stem from the fact that we cannot put more than one formula on the right-hand side of sequents (see (Shankar, 1992)):

1. $\forall$L/$\forall$R,

2. $\forall$L/$\exists$L,

3. $\exists$R/$\forall$R, as in the classical case;

4. $\Rightarrow$L/$\Rightarrow$R (to prove $A \Rightarrow \neg A \longrightarrow A \Rightarrow B$, we have to use $\Rightarrow$R last, otherwise we would have to produce the sequent $\longrightarrow A$, which is not provable);

5. $\neg$L/$\Rightarrow$R (for example, $\neg A \longrightarrow A \Rightarrow B$);

6. $\Rightarrow$L/$\neg$R (for example, $A \Rightarrow \neg A \longrightarrow \neg A$);

7. $\neg$L/$\neg$R (for example, $\neg A \Rightarrow \neg (A \wedge A)$);

8. $\Rightarrow$L/$\vee$L (to prove $A \vee B, A \Rightarrow B \longrightarrow B$, we can use $\vee$L last, but we cannot use $\Rightarrow$L last, or else we would be stuck with the unprovable sequent $A \vee B \longrightarrow A$);

9. $\vee$R/$\vee$L (to prove $A \vee B \longrightarrow B \vee A$, we cannot use $\vee$R last, as we would be stuck with the unprovable sequents $A \vee B \longrightarrow B$ and $A \vee B \longrightarrow A$);

10. $\neg$L/$\vee$L (to prove $A \vee B, \neg A \longrightarrow B$, we cannot use $\neg$L because $A \vee B \longrightarrow A$ is unprovable);

11. $\exists$R/$\vee$L (to prove $\Phi[a/x] \vee \Phi[b/x] \longrightarrow \exists x \cdot \Phi$, we can use $\vee$L from proofs of $\Phi[a/x] \longrightarrow \exists x \cdot \Phi$ and $\Phi[b/x] \longrightarrow \exists x \cdot \Phi$, but using $\exists$R last forces us to find a term t such that $\Phi[a/x] \vee \Phi[b/x] \longrightarrow \Phi[t/x]$, which is impossible if a and b are distinct constants).

Because of impermutabilities 4–10, we cannot in general put a (propositional) intuitionistic formula in clausal form. Because of impermutability 11, we cannot put formulas in prenex form either. (The implication $\Phi[a/x] \vee \Phi[b/x] \Rightarrow \exists x \cdot \Phi$ is provable, but $\exists x \cdot \Phi[a/x] \vee \Phi[b/x] \Rightarrow \Phi$ is not; indeed, otherwise there would be a term t such that $\Phi[a/x] \vee \Phi[b/x] \Rightarrow \Phi[t/x]$ would be provable, which is impossible by Exercise 6.7.) In particular, we cannot herbrandize or skolemize in general.

▶ **7.1.**

$\prec^+$ is transitive by construction. We have to show that it is irreflexive. First, there is a permutation p of $\{1, \ldots, n\}$ such that $\sigma = [t_{p(1)}/x_{p(1)}] \cdots [t_{p(n)}/x_{p(n)}]$. Without loss of generality, assume that p is the identity, i.e. $\sigma = [t_1/x_1]$

$\ldots [t_n/x_n]$. Because σ is idempotent, for every i, $x_i \notin \mathrm{fv}(t_i[t_{i+1}/x_{i+1}]\ldots$ $[t_n/x_n])$, and because x_i is different from all x_j, $j > i$, $x_i \notin \mathrm{fv}(t_i) \cup \mathrm{fv}(t_{i+1}) \cup \ldots \cup \mathrm{fv}(t_n)$. Therefore, the relation $<$ defined by $x_i < x$ if and only if $x \in \mathrm{fv}(t_i) \cup \mathrm{fv}(t_{i+1}) \cup \ldots \cup \mathrm{fv}(t_n)$ is a strict ordering.

We claim that $\prec^+$ is included in $<$, i.e. $x \prec^+ y$ implies $x < y$. First, if $x \prec y$, then clearly $x < y$. Because $<$ is transitive, we can conclude that $x \prec^+ y$ implies $x < y$.

But then, if there was a variable x such that $x \prec^+ x$, we would have $x < x$, which is impossible.

▶ **7.2.**

Let σ' be $[t_1/x_1]\ldots[t_n/x_n]$. We have $\sigma = [t_{p(1)}/x_{p(1)}]\ldots[t_{p(n)}/x_{p(n)}]$ for some permutation p of $\{1, \ldots, n\}$. Now, the latter is also sorted with respect to $\prec^+$: indeed, if $x_{p(i)} \prec^+ x_{p(j)}$ and $i \geq j$, then σ maps x_i to some term where $x_{p(j)}$ is free, contradicting the idempotence of σ. Then, we can sort the list $[p(1), \ldots, p(n)]$ with respect to the usual ordering on integers by, say, bubble sort (which proceeds by only exchanging pairs of adjacent elements in the list), getting $[1, \ldots, n]$ in the end. Notice that the sort procedure exchanges elements i and j (where i was initially just before j in the list) only when $i > j$, and in particular only when $x_j \prec^+ x_i$ or when x_j and x_i are incomparable via $\prec^+$. But $x_j \prec^+ x_i$ can never happen, because at each step, the list is sorted with respect to $\prec^+$. So the bubble sort procedure only exchanges indices of variables that are incomparable with respect to $\prec^+$. Summing up, we can get σ' from σ by changing the order of application of the $[t_i/x_i]$'s, in finitely many steps of the form $[t_i/x_i][t_j/x_j] \longrightarrow [t_j/x_j][t_i/x_i]$, where x_i and x_j are incomparable with respect to $\prec^+$. In particular, they are incomparable with respect to $\prec$, so the two compound substitutions $[t_i/x_i][t_j/x_j]$ and $[t_j/x_j][t_i/x_i]$ are in fact equal. By induction on the number of exchange steps the bubble sort needs, we see that $\sigma = \sigma'$.

Therefore we can get back an idempotent substitution from one of its triangular forms by first sorting it with respect to the occurs-check ordering, getting $[t_1/x_1;$ $\ldots; t_n/x_n]$, then computing $[t_1/x_1]\ldots[t_n/x_n]$. In fact, this is only theoretical: in most applications where unification is needed, we never need to compute the substitution represented by the triangular form, since we shall always be able to compute on the triangular forms directly.

▶ **7.3.**

Notice that σ'' is a unifier σ and σ' if and only if it simultaneously unifies $x\sigma$ with $x\sigma'$, for every variable x.

The answer of the theoretician would then be the following trick: let $x_1, \ldots, x_n$ be the variables in $\mathrm{dom}\,\sigma \cup \mathrm{dom}\,\sigma'$; create a new n-ary function symbol f, then the unifiers of σ and σ' are those of $f(x_1\sigma, \ldots, x_n\sigma)$ with $f(x_1\sigma', \ldots, x_n\sigma')$. This reduces the result to Theorem 7.5.

The answer of the computer scientist would be the following: launch Martelli and Montanari's algorithm on the couple $(\{x_1\sigma \doteq x_1\sigma', \ldots, x_n\sigma \doteq x_n\sigma'\}, [])$.

This computes a most general simultaneous unifier or fails, hence proving the result constructively.

▸ **7.4.**

Let S be any finite set of Horn clauses having non-empty left-hand sides. Any factor of such a clause again has a non-empty left-hand side, and similarly any binary resolvent of such clauses has a non-empty left-hand side. But the empty clause has an empty left-hand side, so resolution cannot derive it. By the completeness of resolution, S must be satisfiable. This generalises to any set of clauses with non-empty left-hand sides (not only finite sets of clauses) by the compactness of first-order logic.

Now, let S be a finite set of Horn clauses. If it contains the empty clause, then unit resolution immediately concludes that S is unsatisfiable. Otherwise, S consists of clauses with non-empty left-hand sides and positive unit clauses $\longrightarrow A$ only. As we showed above, it is pointless to resolve between clauses with non-empty left-hand sides, so we can concentrate on resolving between positive unit clauses and clauses with non-empty left-hand sides. In particular, unit resolution is complete for sets of Horn clauses. In fact, positive unit resolution is itself complete for this class of problems, and even more, we don't need factoring, since here the unification step in binary resolution is enough to create all needed factors.

▸ **7.5.**

Build the Herbrand tree (decision tree) on the enumeration A_0, A_1,$\ldots$. Consider an inference node N on A_i, and its two sons N_1 and N_2, which are by definition failure nodes. There are two clauses $C \vee A_i$ and $C' \vee \neg A_i$ such that the first is falsified at N_1 and the second is falsified at N_2, but which are not falsified at the inference node. Since $C \vee A_i$ is falsified at N_1, all its literals are made false under the associated interpretation; in particular, all the literals in C were made false already at node N. It follows that all the atoms in C must have been met above in the tree, or alternatively that A_i is maximal in C. Similarly, A_i is maximal in C', so we are really doing an ordered resolution step here. Completeness follows by standard lifting arguments.

The first thing that is weaker in this result is the fact that we needed $>$ to be total on ground terms. If $>$ is not total on ground terms, however, we can always extend it to an ordering $>^*$ that is total on ground terms: first extend its restriction to ground terms to a total ordering (by Zorn's Lemma), then define $>^*$ on all terms by $A >^* B$ if and only if $A\sigma >^* B\sigma$ for every ground σ. $>^*$ may not be computable any longer, but we may then apply the above argument; then any $>^*$-ordered resolution is a $>$-ordered resolution, so the result still holds.

The second thing that is weaker is the fact that $>$ needed to correspond to some enumeration of the Herbrand universe. In particular, the order type of $>$ must be ω. This excludes orderings like the lexicographic ordering as soon as there are at least two predicate symbols and infinitely many terms. For instance, letting $P > Q$ and $f > a$, the lexicographic ordering produces $Q(a) < Q(f(a)) < Q(f(f(a))) <$

$\dots P(a) < P(f(a)) < P(f(f(a))) < \dots$, which cannot correspond to an enumeration of the ground atoms, since there are infinitely many atoms before $P(a)$. This can be corrected by using transfinite semantic trees (Rusinowitch, 1989).

Finally, by using this technique, we did not prove that we could actually choose the maximal atom among all non-pure atoms in the entire clause set. In particular, we did not prove that eliminating pure clauses was compatible with the ordering strategy.

▸ **7.6.**

The proof of completeness is exactly the same as for Theorem 7.26.

If $A_1\sigma$ is maximal among $\Gamma\sigma, \Delta\sigma, \Gamma'\sigma, \Delta'\sigma$, then it is clear that $A_1, \dots, A_m$ are maximal among Γ, Δ. So this strategy is at least as restrictive as the ordering strategy. It is in fact more restrictive. For instance, consider the lexicographic ordering built on $a > b$, and the clauses:

$$P(x, y) \longrightarrow P(y, x)$$
$$P(b, a) \longrightarrow$$

then $P(b, a)$ is trivially maximal in the second clause, and $P(y, x)$ is maximal in the first (as well as $P(x, y)$, but this does not matter here). The mgu is $[a/x, b/y]$, so the ordering strategy would produce the resolvent $P(a, b) \longrightarrow$. Since $P(x, y)[a/x, b/y] > P(y, x)[a/x, b/y]$, the literal resolved upon, once instantiated, is not maximal among the remaining instantiated literals, hence the refinement would not produce any clause.

▸ **7.7.**

Apply positive hyperresolution. Since there is only one positive clause, $\longrightarrow A, B$, we must resolve it with some other clause. In ordered positive hyperresolution, we must resolve on B, so we must resolve this clause with $B \longrightarrow C$. But B is not maximal in the latter, so there is no ordered positive hyperresolvent of this set of clauses. In particular, $\longrightarrow$ is not derivable.

However, this set is unsatisfiable: derive $\longrightarrow A, C$ from the above two clauses, then $\longrightarrow A$ by resolving with $C \longrightarrow$, then $\longrightarrow$ by resolving with $A \longrightarrow$. (And this refutation is a positive hyperresolution refutation, in fact, the only semi-ordered positive hyperresolution derivation.)

So ordered semantic resolution, and also ordered hyperresolution, is incomplete.

▸ **7.8.**

The interpretation mapping every atom to true clearly satisfies every definite clause. In fact, we don't need them to be Horn.

Now, let S be a finite set of Horn clauses. If it contains the empty clause, then input resolution immediately concludes that S is unsatisfiable. Otherwise, S consists of definite clauses and of clauses of the form $\Gamma \longrightarrow$ only. We can therefore apply linear resolution by choosing C_0 to be one of the clauses $\Gamma \longrightarrow$. Then, any resolvent between such a clause and a definite clause is again of the form $\Gamma \longrightarrow$, since a definite clause has exactly one formula on the right, which must be the formula we resolve upon. By induction on the length of the linear derivation, we therefore conclude that all centre clauses are of this form. In particular, no centre clause can ever be resolved with another centre clause, hence the linear resolution derivation is in fact an input resolution derivation.

▶ **8.1.**

Permuting the goal sequent translates to a permutation of expansion steps in the tableau expansion, and to a permutation of variables in the substitution that the procedure finds. As there are $k!$ permutations of $\{1, \ldots, k\}$, and the procedure may test them all, this means that we already have an overhead of order $k!$ for not recognising these symmetries. To limit this waste, it is therefore useful to restrict expansions and closings in the following way: first, expand only subformulas of Ψ_1, then expand Ψ_2 and its subformulas only if no subformulas of Ψ_1 remain to expand, and so on. As for closing paths, while we are in the process of expanding Ψ_i, $i \geq 2$, and its subformulas, we can safely restrict unification to occur between subformulas of $\Psi_1, \ldots, \Psi_{i-1}$ on one hand, and subformulas of $\Psi_1, \ldots, \Psi_i$ on the other hand.

▶ **8.2.**

First-order tableaux are not scalable, because we can only close paths by unifying *atomic* formulas, whereas in the propositional case we could recognise identical *non-atomic* formulas as well. This forces us to delay, sometimes considerably, the time when we close the paths, and also the number of paths to close.

A solution, in view of the problem, is to extend unification to a process acting on whole formulas, by regarding the propositional connectives as new function symbols, regardless of their logical properties. (I.e., we approximate the algebra of quantifier-free formulas by the free algebra generated by the logical symbols, the predicate symbols and the function symbols over the variables.) In particular, to unify $\Phi_1 \wedge \Phi_1'$ with $\Phi_2 \wedge \Phi_2'$, we would simultaneously unify Φ_1 with Φ_2 and Φ_1' with Φ_2', but not Φ_1 with Φ_2' and Φ_1' with Φ_2 for example.

This clearly corrects the problem in terms of the number of expansions needed to arrive at a tableaux that we can close. Moreover, the method remains sound and complete. However, this increases the level of non-determinism caused by increasing the number of different ways that we can close paths now, because of the added non-atomic complementary pairs. In practice, there does not seem to be any advantage in doing so.

▶ **9.1.**

We encode Post's correspondence problem in equational logic. Using the notations of Theorem 6.20, we produce the equations:

$$P(u_i(x), v_i(y)) \doteq P(x, y)$$

for every Post pair (u_i, v_i), $1 \leq i \leq n$, and:

$$P(a(x), a(x)) \doteq T$$
$$P(b(x), b(x)) \doteq T$$

Let E be this set of equations. We ask whether $E \models_= P(\epsilon, \epsilon) = T$.

On the one hand, let i_1, i_2, ..., i_k be a solution to Post's problem, that is, $u_{i_1} u_{i_2} \ldots u_{i_k} = v_{i_1} v_{i_2} \ldots v_{i_k}$. Let w be the latter word. If the last letter of w is a (recall that w is not the empty word), then $w = w'a$ for some word w', or alternatively $w(x) = a(w'(x))$. But $E \models_= P(a(w'(\epsilon)), a(w'(\epsilon)))$, so $E \models_= P(w(\epsilon), w(\epsilon))$. By induction on k, it follows easily that $E \models_= P(\epsilon, \epsilon) \doteq T$.

On the other hand, if $E \models_= P(\epsilon, \epsilon) = T$, then let R be the smallest set of couple of words (w_1, w_2) such that:

- for every non-empty word w, $(w, w) \in R$;

- and for every i, $1 \leq i \leq n$, if $(w_1 u_i, w_2 v_i) \in R$, then $(w_1, w_2) \in R$.

Observe that: $(*)$ $(w_1, w_2) \in R$ if and only if for some sequence i_1, ..., i_k, $w_1 u_{i_1} \ldots u_{i_k}$ and $w_2 v_{i_1} \ldots v_{i_k}$ are the same non-empty word. Indeed, the latter trivially implies the former. And conversely, if the latter does not hold but the former does, then R minus (w_1, w_2) would be a strictly smaller set obeying the same conditions.

Then, we choose a particular equational model. Let the domain be the set of all words over $\{a, b\}$, ϵ denote the empty word, $a(\_)$ denote the function $w \mapsto wa$, $b(\_)$ denote the function $w \mapsto wb$, T denote any arbitrary element, and P denote the function mapping any couple (w_1, w_2) to T if $(w_1, w_2) \in R$, and to some other arbitrary, fixed element otherwise. By definition of R, all the equations in E are clearly satisfied. By assumption, $P(\epsilon, \epsilon) = T$ holds in the model, therefore $(\epsilon, \epsilon) \in R$, hence by $(*)$ there is a Post pair.

As Post's correspondence problem is undecidable, so is provability in equational logic.

▸ **10.1.**

(i) is a *denial*, (ii) and (iv) are a *definite clauses*, and (iii) is *not* a Horn clause.

▸ **10.2.**

By definition, P and Q each consist of either zero (denial) or one (definite clause) positive literal. Resolving eliminates one of the positive literals, say Q, leaving $P \leftarrow \Gamma$, where Γ contains all the right-hand sides except the element A_i which was eliminated by resolution. Since P has zero or one positive literals by assumption, the resolvent is also a Horn clause.

▸ **10.3.**

The goal $?path(A, B)$ will compute all paths from the data given. Hence the solutions are the following pairs (A, B): $(a, b), (a, c), (b, d), (c, d), (a, d), (a, d)$.

The goal $?path(A, B), path(B, A)$ will compute all symmetric paths, for which there are none so resolution will fail.

The goal $?path(d, A)$ will compute all paths starting from d, for which there are none, so the program will again fail.

▸ **10.4.**

Resolving the goal with the first clause will fail immediately. Hence, resolving with the second clause yields: $V > 0, 2 = V * M, N' = V - 1 \mid \text{fact}(N', M)$ where the constraint is satisfiable. Resolving this with the first clause yields an unsatisfiable constraint. However, resolving with the second clause again gives: $V > 0, 2 = V * M, N' = V - 1, N' > 0, M = N' * M_2, N_2' = N' - 1 \mid \text{fact}(N_2', M_2)$. Now, resolving with the first clause gives $N_2' = 0$, $M_2 = 1$ and the constraints are solvable giving the result $V = 2$, as required.

CHAPTER B

BASICS OF TOPOLOGY

For more information about point-set topology, we refer the interested reader to (Bourbaki, 1990). We only recapitulate the basic notions that we need in this book.

Definition B.1 *A* topological space E *is a non-empty set with a* topology, *i.e. a non-empty set $\mathcal{O}$ of subsets, called the* open sets, *such that:*

- *every finite intersection of open sets is open,*

- *every union of open sets is open.*

Notice that arbitrary unions are permitted, while only finite intersections are guaranteed to yield open sets from open sets. By finite intersection, we also mean intersection of an empty family of open sets, so E itself is an open set. Also, arbitrary union includes the case of the union of no set, so $\varnothing$ is also open.

Definition B.2 *Let E be a topological space. The* closed sets *are the complements of the open sets, i.e. F is closed if and only if $E \setminus F$ is open.*

We can always equip any set with at least two (trivial) topologies:

Definition B.3 *Let E be a set.*
The discrete topology *on E is that in which all subsets of E are open.*
The coarse topology *on E is that in which exactly $\varnothing$ and E only are open.*

Notice that in both these topologies, the open sets are exactly the closed sets.

Definition B.4 *Let E be a topological space. We say that E is* Hausdorff-separated *if and only if for every x, y in T, with $x \neq y$, there are two open sets X and Y such that $x \in X$, $y \in Y$ and $X \cap Y = \varnothing$.*

We say that X and Y are *open neighbourhoods* of x and y respectively. A space is Hausdorff if and only if we can separate any two distinct points by disjoint neighbourhoods.

Notice that spaces with the discrete topology are always Hausdorff-separated, while spaces with the coarse topology are Hausdorff-separated only when they only contain one element.

Topological spaces give rise to the notion of continuity:

403

Definition B.5 *Let f be a function from the topological space E to the topological space F. We say that f is* continuous *if and only if for every open set O of F, $f^{-1}(O)$ is an open set of E.*

In short, if f maps x to y, for any neighborhood Y of y, the set of x' such that $f(x') \in Y$ is a neighborhood of x. Or in informal terms, the set of points "close to" y is the image of a set of points "close to" x.

We can define the product of a family of topological spaces:

Definition B.6 *Let $(E_i)_{i \in I}$ be a family of topological spaces indexed by some set I.*

Then $\Pi_{i \in I} E_i$, the cartesian product of the E_is, is a topological space whose open sets are by definition the unions of sets of the form $\Pi_{i \in I} O_i$, where O_i is an open set of E_i for each $i \in I$, and all but finitely many O_i are equal to E_i.

The fact that this defines a topology is straightforward. The curiosity in the definition is that we would expect to define the open sets as (unions of) products of open sets. But for infinite products, this would not have nice properties, hence we insist that only finitely many open sets in the product $\Pi_{i \in I} O_i$ are proper open sets (different from E_i).

The fundamental notion of compacity is defined as follows:

Definition B.7 *An* open cover *of E is a family $\mathcal{C}$ of open sets such that $\bigcup_{O \in \mathcal{C}} O = E$.*

A topological space E is said to be compact *if and only if it is Hausdorff separated and if from every open cover of E, we can extract a finite open cover of E.*

Compact spaces are interesting first because of the property stated in the definition, and second because of the following properties:

Theorem B.8 *Let f be a continuous function from a compact space E to a topological space F. Then the image of E by f is compact in F.*

(By a set $G \subseteq F$ being compact in F, we mean that G with open sets the $G \cap O$, for O open in F, is a compact space.) Therefore, continuous functions map compact spaces to compact spaces.

Theorem B.9 (Tychonoff) *Any product of compact spaces is compact.*

It is understood that the product is taken with the product topology. This theorem is a bit hard to prove. It is equivalent to the ultrafilter theorem, or to the Axiom of Choice.

A *filter* $\mathcal{F}$ on a non-empty set I is any set of subsets of I such that:

(i) $\varnothing \notin \mathcal{F}$;

(ii) for every A, B in $\mathcal{F}$, $A \cap B$ is in $\mathcal{F}$;

(iii) for every A in $\mathcal{F}$, every superset of A in I is in $\mathcal{F}$;

Filters are ordered by inclusion, and maximal filters are called *ultrafilters*. A direct consequence of Zorn's Lemma (every inductively ordered set has a maximal element) is:

Theorem B.10 *Every filter is included in some ultrafilter.*

Another characterisation of an ultrafilter is as a filter satisfying the extra condition:

(iv) for every A in I, either A or its complement in I is in $\mathcal{F}$ (but not both).

BIBLIOGRAPHY

(Abramsky, 1993) Samson Abramsky. Computational Interpretations of Linear Logic. *Journal of Theoretical Computer Science*, 111(1):3–57, 1993.

(Andreoli and Pareschi, 1991) Jean-Marc Andreoli and Remo Pareschi. Linear objects: Logical processes with built-in inheritance. *New Generation Computing*, 9(3–4):445–473, 1991.

(Andrews, 1986) Peter B. Andrews. *An Introduction to Mathematical Logic and Type Theory: To Truth through Proof*. Computer Science and Applied Mathematics. Academic Press, 1986.

(Barendregt, 1984) Henk P. Barendregt. *The Lambda Calculus, Its Syntax and Semantics*, volume 103 of *Studies in Logic and the Foundations of Mathematics*. North-Holland Publishing Company, Amsterdam, revised edition, 1984.

(Barendregt, 1992) Henk P. Barendregt. Lambda calculi with types. In Samson Abramsky, Dov M. Gabbay, and Thomas S. E. Maibaum, editors, *Handbook of Logic in Computer Science*, volume 2, pages 117–309. Oxford University Press, 1992.

(Beckert and Posegga, 1994) Bernhard Beckert and Joachim Posegga. lean$T^A P$: Lean tableau-based theorem proving. In Alan Bundy, editor, *12th International Conference on Automated Deduction*, volume 814 of *Lecture Notes in Artificial Intelligence*, pages 793–795, Nancy, France, June 1994. Springer Verlag.

(Beckert et al., 1993) Bernhard Beckert, Reiner Hähnle, and Peter H. Schmidt. The even more liberalized δ-rule in free variable semantic tableaux. In Alexander Leitsch, editor, *Third Kurt Gödel Colloquium on Computational Logic and Proof Theory*, Lecture Notes in Computer Science, page 108. Springer Verlag, 1993.

(Bibel, 1987) Wolfgang Bibel. *Automated Theorem Proving*. Vieweg, second, revised edition, 1987.

(Billon, 1996) Jean-Paul Billon. The disconnection method — a confluent integration of unification in the analytic framework. In Pierangelo Miglioli, Ugo Moscato, Daniel Mundici, and Mario Ornaghi, editors, *5th International Workshop on Theorem Proving with Analytic Tableaux and Related Methods*, number 1071 in Lecture Notes in Artifical Intelligence, pages 110–126, Terrasini, Palermo, Italy, May 1996. Springer Verlag.

(Birkhoff, 1935) Garrett Birkhoff. On the structure of abstract algebras. *Proceedings of the Cambridge Philosophy Society*, 31:433–454, 1935.

(Bourbaki, 1990) Nicolas Bourbaki. *Eléments de Mathématiques — Topologie 1–4*. Hermès, 1990.

(Chang and Lee, 1973) Chin-Liang Chang and Richard Char-Tung Lee. *Symbolic Logic and Mechanical Theorem Proving*. Computer Science Classics. Academic Press, 1973.

(Chellas, 1980) Brian F. Chellas. *Modal Logics: an Introduction*. Cambridge University Press, 1980.

(Chou and Gao, 1990) Shan-Ching Chou and Xiao-Shan Gao. Ritt-Wu's decomposition algorithm and geometry theorem proving. In Mark E. Stickel, editor, *10th International Conference on Automated Deduction*, volume 449 of *Lecture Notes in Artifical Intelligence*, pages 207–220, Kaiserslautern, RFA, July 1990. Springer Verlag.

(Colmerauer, 1984) Alain Colmerauer. Equations and inequations on finite and infinite trees. In *Proceedings of the International Conference on Fifth Generation Computer Systems (FGCS-84)*, pages 85–99, Tokyo, Japan, November 1984.

(Conlon, 1989) Tom Conlon. *Programming in PARLOG*. Addison-Wesley, 1989.

(Coudert and Madre, 1995) Olivier Coudert and Jean-Christophe Madre. The implicit set paradigm: A new approach to finite state system verification. *Formal Methods in System Design*, 6(2):133–145, 1995.

(Dalen, 1986) Dirk van Dalen. Intuitionistic logic. In Dov M. Gabbay and F. Guenthner, editors, *Handbook of Philosophical Logic: Alternatives to Classical Logic*, volume III, chapter 4, pages 225–339. Reidel Publishing Company, Dordrecht, 1986.

(Davenport et al., 1989) James H. Davenport, Yvon Siret, and Evelyne Tournier. *Computer Algebra, Systems and Algorithms for Algebraic Computation*. Academic Press, 1989.

(Davis and Weyuker, 1985) Martin D. Davis and Elaine J. Weyuker. *Computability, Complexity and Languages*. Academic Press, New York, 1985.

(Davis, 1973) Martin D. Davis. Hilbert's tenth problem is unsolvable. *American Mathematic Monthly*, 80:233–269, March 1973.

(Degtyarev and Voronkov, 1996a) Anatoli Degtyarev and Andrei Voronkov. Simultaneous rigid E-unification is undecidable. In H. Kleine Büning, editor, *Computer Science Logic (CSL'95), 9th International Workshop, Paderborn, Germany*, number 1092 in Lecture Notes in Computer Science, pages 178–190. Springer Verlag, September 1996.

(Degtyarev and Voronkov, 1996b) Anatoli Degtyarev and Andrei Voronkov. What you always wanted to know about rigid E-unification. In J.J. Alferes, L. Moniz Pereira, and E. Orlowska, editors, *Logics in Artificial Intelligence (JELIA'96)*, number 1126 in Lecture Notes in Artificial Intelligence. Springer Verlag, 1996.

(Dershowitz and Jouannaud, 1990) Nachum Dershowitz and Jean-Pierre Jouannaud. Rewrite systems. In Leeuwen (1990), chapter 6, pages 243–320.

(Dowek *et al.*, 1995) Gilles Dowek, Thérèse Hardin, and Claude Kirchner. Higher-order unification via explicit substitutions. In *Proceedings of the 10th IEEE Symposium on Logics in Computer Science*, 1995.

(Downey *et al.*, 1980) Peter K. Downey, Ravi Sethi, and Robert Endre Tarjan. Variations on the common subexpression problem. *Journal of the ACM*, 27(4):758–771, 1980.

(Emerson, 1990) E. Allen Emerson. Temporal and modal logics. In Leeuwen (1990), chapter 16, pages 995–1072.

(Felleisen, 1988) Matthias Felleisen. The theory and practice of first-class prompts. In *Conference Record of the Fifteenth Annual ACM Symposium on Principles of Programming Languages*, pages 180–190, San Diego, California, January 1988.

(Fitting, 1983) Melvin C. Fitting. *Proof Methods for Modal and Intuitionistic Logics*. Reidel Publishing Company, Dordrecht, 1983.

(Fitting, 1996) Melvin C. Fitting. *First-Order Logic and Automated Theorem Proving*. Graduate Texts in Computer Science. Springer Verlag, second edition, 1996.

(Gabbay and Guenthner, 1984) Dov M. Gabbay and F. Guenthner. *Handbook of Philosophical Logic, Extensions of Classical Logic*, volume II. Reidel Publishing Company, Dordrecht, 1984.

(Gallier *et al.*, 1992) Jean H. Gallier, Paliath Narendran, Stan Raatz, and Wayne Snyder. Theorem proving using equational matings and rigid E-unification. *Journal of the Association for Computing Machinery*, 39(2):377–429, April 1992.

(Gallier, 1987) Jean H. Gallier. *Logic for Computer Science — Foundations of Automatic Theorem Proving*. John Wiley and Sons, 1987.

(Gallier, 1990) Jean H. Gallier. On Girard's "Candidats de Reductibilité". In Piergiorgio Odifreddi, editor, *Logic and Computer Science*, pages 123–203. Academic Press, 1990.

(Garey and Johnson, 1979) Michael R. Garey and David S. Johnson. *Computers and Intractability — A Guide to the Theory of NP-Completeness*. W.H. Freeman and Co., San Francisco, 1979.

(Girard *et al.*, 1989) Jean-Yves Girard, Yves Lafont, and Paul Taylor. *Proofs and Types*, volume 7 of *Cambridge Tracts in Theoretical Computer Science*. Cambridge University Press, 1989.

(Girard, 1972) Jean-Yves Girard. *Interprétation fonctionelle et élimination des coupures dans l'arithmétique d'ordre supérieur*. PhD thesis, University of Paris VII, 1972.

(Girard, 1987a) Jean-Yves Girard. Linear Logic. *Theoretical Computer Science*, 50(1):1–102, 1987.

(Girard, 1987b) Jean-Yves Girard. *Proof Theory and Logical Complexity*, volume 1. Bibliopolis, 1987.

(Gödel, 1931) Kurt Gödel. Über formale unentscheidbare Sätze der Principia Mathematica und verwandter System I. *Monatshefte für Mathematik und Physik*, 38:173–198, 1931. Translated into English under the title 'On formally undecidable propositions of the *Principia Mathematica*' in (Heijenoort, 1967), pages 596–616.

(Goré, 1991) Rajeev Prabhakar Goré. Semi-analytic tableaux for propositional normal modal logics with applications to nonmonotonicity. *Logique et Analyse*, 133–134:73–104, 1991.

(Griffin, 1990) Timothy G. Griffin. A formulae-as-types notion of control. In *Conference Record of the Seventeenth Annual ACM Symposium on Principles of Programming Languages*, pages 47–58, San Francisco, California, January 1990.

(Hankin, 1994) Chris L. Hankin. *Lambda Calculi, A Guide for Computer Scientists*. Number 3 in Graduate texts in Computer Science. Oxford University Press, September 1994.

(Heijenoort, 1967) Jean van Heijenoort. *From Frege to Gödel*. Harvard University Press, Cambridge, Massachussetts, 1967.

(Hodas and Miller, 1994) Joshua Hodas and Dale A. Miller. Logic programming in a fragment of intuitionistic linear logic. *Journal of Information and Computation*, 110(2):327–365, 1994.

(Hodges, 1976) Wilfrid Hodges. *Logic: An introduction to elementary logic*. Penguin books Ltd, 1976.

(Hogger, 1984) Christopher John Hogger. *Introduction to Logic Programming*. APIC Studies in Data Processing series. Academic Press, 1984.

(Hogger, 1990) Christopher John Hogger. *Essentials of Logic Programming*. Number 1 in Graduate Texts in Computer Science. Oxford University Press, 1990.

(Hooker and Vinay, 1995) J. N. Hooker and V. Vinay. Branching rules for satisfiability. *Journal of Automated Reasoning*, 15(3):359–383, 1995.

(Howard, 1980) William A. Howard. The formulae-as-types notion of construction. In Jonathan P. Hindley and J. Roger Seldin, editors, *To H. B. Curry: Essays on Combinatory Logic, Lambda Calculus and Formalism*, pages 479–490. Academic Press, 1980.

(Hsiang, 1985) Jieh Hsiang. Refutational theorem proving using term-rewriting systems. *Artificial Intelligence*, 25(1):255–300, 1985.

(Huet, 1973) Gérard P. Huet. A mechanization of type theory. In *Proceedings of the 3rd International Joint Conference on Artificial Intelligence*, pages 139–146, Stanford University, Stanford, California, August 1973.

(Huet, 1975) Gérard P. Huet. A unification algorithm for typed λ-calculus. *Journal of Theoretical Computer Science*, 1(1):27–57, 1975.

(Hughes and Creswell, 1968) G.E. Hughes and M.J. Creswell. *An Introduction to Modal Logics*. Methuen and Co., 1968.

(Hughes and Creswell, 1984) G.E. Hughes and M.J. Creswell. *A Companion to Modal Logics*. Methuen and Co., 1984.

(Jaffar and Lassez, 1987) Joxan Jaffar and Jean-Louis Lassez. Constraint logic programming. In *Proceedings of the 14th ACM Symposium on Principles of Programming Languages, Munich*, 1987.

(Johnstone, 1992) Peter T. Johnstone. *Notes on Logic and Set Theory*. Cambridge University Press, 1992.

(Jouannaud and Kirchner, 1991) Jean-Pierre Jouannaud and Claude Kirchner. Solving equations in abstract algebras: a rule-based survey of unification. In Jean-Louis Lassez and Gordon D. Plotkin, editors, *Computational Logic. Essays in Honor of Alan Robinson*, pages 257–321. MIT Press, 1991.

(Kleene, 1967) Stephen Cole Kleene. *Mathematical Logic*. John Wiley and Sons, 1967.

(Knuth, 1973) Donald Ervin Knuth. *Sorting and Searching*, volume 3 of *The Art of Computer Programming*. Addison-Wesley, 1973.

(Kohlhase, 1995) Michael Kohlhase. Higher-order tableaux. In *Workshop on Theorem Proving with Analytic Tableaux and Related Methods*, 1995.

(Kozen and Tiuryn, 1990) Dexter Kozen and Jerzy Tiuryn. Logics of programs. In Leeuwen (1990), chapter 14, pages 789–840.

(Lee and Plaisted, 1992) Shie-Jue Lee and David A. Plaisted. Eliminating duplication with the hyper-linking strategy. *Journal of Automated Reasoning*, 9(1):25–42, August 1992.

(Leeuwen, 1990) Jan van Leeuwen, editor. *Handbook of Theoretical Computer Science*, volume B. Elsevier Science Publishers b.v., 1990.

(Lemmon, 1965) E. J. Lemmon. *Beginning Logic*. Van Nostrand Reinhold (UK), 1965.

(Lewis, 1918) Clarence I. Lewis. *A Survey of Symbolic Logic*. University of California, 1918.

(Lloyd, 1984) John W. Lloyd. *Foundations of Logic Programming*. Springer Verlag, 1984.

(Manin, 1977) Youri I. Manin. *A Course in Mathematical Logic*. Graduate Texts in Mathematics. Springer Verlag, 1977.

(Martelli and Montanari, 1982) Alberto Martelli and Ugo Montanari. An efficient unification algorithm. *ACM Transactions on Programming Languages and Systems*, 4(2):258–282, April 1982.

(McMillan, 1993) Kenneth L. McMillan. *Symbolic Model Checking*. Kluwer Academic Publishers, 1993.

(Miller, 1987) Dale A. Miller. A compact representation of proofs. *Studia Logica*, 46(4), 1987.

(Miller, 1994) Dale A. Miller. A multiple-conclusion meta-logic. In *Proceedings, Ninth Annual IEEE Symposium on Logic in Computer Science*, pages 272–281. IEEE Computer Society Press, 1994.

412   BIBLIOGRAPHY

(Milner, 1989) Robin Milner. *Communication and Concurrency*. International Series in Computer Science. Prentice Hall, 1989.

(Minato, 1993) Shin-Ichi Minato. Zero-suppressed BDDs for set manipulation in combinatorial problems. In *Proceedings of the 30th ACM/IEEE Design Automation Conference*, pages 272–277, Dallas, TX, June 1993. ACM Press.

(Nadathur and Miller, 1988) Gopalan Nadathur and Dale A. Miller. An overview of λProlog. In Robert A. Kowalski and Kenneth A. Bowen, editors, *Logic Programming: Proceedings of the Fifth International Conference and Symposium, Volume 1*, pages 810–827. MIT Press, August 1988.

(Nerode, 1990) Anil Nerode. Some lectures on intuitionistic logic. In Piergiorgio Odifreddi, editor, *Logic and Computer Science*, number 1429 in Lecture Notes in Mathematics. Springer Verlag, 1990.

(Ohlbach, 1993) Hans-Jürgen Ohlbach. Translation methods for non-classical logics: An overview. *Bulletin of the Interest Group in Pure and Applied Logics*, 1(1):69–89, 1993. Available by ftp on the IGPL server, at theory.doc.ic.ac.uk:/theory/forum/igpl/Bulletin.

(Parigot, 1992) Michel Parigot. λμ-calculus: An algorithmic interpretation of classical natural deduction. In *Proceedings of the International Conference on Logic Programming and Automated Deduction, St. Petersburg (Russia)*, number 624 in Lecture Notes in Computer Science, pages 190–201. Springer Verlag, 1992.

(Peyton Jones, 1987) Simon L. Peyton Jones. *The Implementation of Functional Programming Languages*. Prentice Hall International, 1987.

(Prawitz, 1965) Dag Prawitz. *Natural Deduction, A Proof-Theoretical Study*. Almqvist and Wiskell, Stockholm, 1965.

(Rusinowitch, 1989) Michaël Rusinowitch. *Démonstration automatique — Techniques de réécriture*. Interéditions, Paris, 1989.

(Ryan and Sadler, 1992) Mark D. Ryan and Martin R. Sadler. Valuation systems and consequence relations. In Samson Abramsky, Dov M. Gabbay, and Thomas S. E. Maibaum, editors, *Handbook of Logic in Computer Science*, volume 1, pages 1–78. Oxford University Press, 1992.

(Schmidt-Schauß, 1989) Manfred Schmidt-Schauß. *Computational Aspects of an Order-Sorted Logic with Term Declarations*, volume 395 of *Lecture Notes in Artificial Intelligence*. Springer Verlag, 1989.

(Schwichtenberg, 1977) Helmut Schwichtenberg. Proof theory: Some applications of cut-elimination. In Jon Barwise, editor, *Handbook of Mathematical Logic*, chapter D.2, pages 867–895. North-Holland Publishing Company, 1977.

(Shankar, 1992) Natarajan Shankar. Proof search in the intuitionistic sequent calculus. In Deepak Kapur, editor, *11th International Conference on Automated Deduction*, volume 607 of *Lecture Notes in Artificial Intelligence*, pages 522–536, Saratoga Springs, New York, USA, June 1992. Springer Verlag.

(Shapiro, 1988) Ehud Shapiro, editor. *Concurrent Prolog: Collected Papers*. MIT Press, 1988.

(Shoenfield, 1967) Joseph R. Shoenfield. *Mathematical Logic*. Addison Wesley, 1967.

(Sterling and Shapiro, 1986) Leon Sterling and Ehud Shapiro. *The Art of Prolog*. MIT Press, 1986.

(Stickel, 1981) Mark E. Stickel. A unification algorithm for associative-commutative functions. *Journal of the ACM*, 28:423–434, 1981.

(Stickel, 1985) Mark E. Stickel. Automated deduction by theory resolution. *Journal of Automated Reasoning*, 1(4):333–355, 1985.

(Stirling, 1992) Colin Stirling. Modal and temporal logics. In Samson Abramsky, Dov M. Gabbay, and Thomas S. E. Maibaum, editors, *Handbook of Logics in Computer Science*, volume 2, pages 477–563. Clarendon Press, Oxford, 1992.

(Szabo, 1969) M. E. Szabo. *The Collected Papers of Gerhard Gentzen*. North-Holland Publishing Company, Amsterdam, 1969.

(Troelstra and Dalen, 1988a) Anne Sjerp Troelstra and Dirk van Dalen. *Constructivism in Mathematics: An introduction (volume 1)*, volume 121 of *Studies in Logic and the Foundations of Mathematics*. North-Holland Publishing Company, 1988.

(Troelstra and Dalen, 1988b) Anne Sjerp Troelstra and Dirk van Dalen. *Constructivism in Mathematics: An introduction (volume 2)*, volume 123 of *Studies in Logic and the Foundations of Mathematics*. North-Holland Publishing Company, 1988.

(Walther, 1988) Christoph Walther. Many-sorted unification. *Journal of the ACM*, 35(1):1–17, 1988.

Index

APPLIED LOGIC SERIES

KLUWER ACADEMIC PUBLISHERS – DORDRECHT / BOSTON / LONDON